Toward a New Philosophy of Biology

TOWARD A NEW PHILOSOPHY OF BIOLOGY ❧ ❧ ❧

Observations of an Evolutionist

ERNST MAYR ❧

The Belknap Press of Harvard University Press

Cambridge, Massachusetts, and London, England 1988

Library of Congress Cataloging-in-Publication Data

Mayr, Ernst, 1904–
Toward a new philosophy of biology.

Bibliography: p.
Includes index.
1. Biology—Philosophy.   2. Evolution—Philosophy.
I. Title.
QH331.M374 1988   574'.01   87-31892
ISBN 0-674-89665-3 (alk. paper)

Designed by Gwen Frankfeldt

Preface

Since the Scientific Revolution, the philosophy of science has been characterized by an almost exclusive reliance on logic, mathematics, and the laws of physics. But in recent years we have witnessed a laudable state of ferment in the field. This unrest stems mainly from the growing realization that any sound philosophy of science must do justice to the living world as well as to the physical one.

Yet, is the conceptual framework of biology sufficiently established to be made part of a philosophy of science? How far have we come in creating a respectable philosophy of biology? As I will attempt to show in the following essays, a great deal of conceptual confusion or at least vagueness persists in biology, and we will not arrive at a satisfying philosophy of this field until more clarity has been achieved. To make a contribution to this clarification is my objective in this volume.

The building of an autonomous philosophy of biology started with Darwin, but this fact was quite universally ignored. Twenty or thirty years ago, almost everything about evolutionary biology, but particularly the concept of selection, seemed incompatible with the axioms of logical positivism, the dominant philosophy of the time. Even as recently as 1974 Karl Popper said that "Darwinism is not a testable scientific theory, but a metaphysical research program." One philosopher and historian after the other would say with conviction: "Darwin was no philosopher."

Actually, Darwin scholars have now clearly established that Darwin carefully read the writings of William Whewell, John Herschel, and other philosophers and conscientiously attempted to conduct his research along the lines prescribed by them. However, he could not help it that biology—especially evolutionary biology—has numerous specific properties not met with in the sciences of inanimate objects. Most important, the very essence of that which characterizes living organisms was left out in the analyses of the logicians and positivists—namely, a historical component in the form of an inherited genotype. The role which this historical program plays in development and behavior was completely ignored. As Max Delbrück observed in 1949, every organism carries with it the information acquired by its ancestors during the past three billion years. A philosophy of science that would include the study of the living world must emphasize this fact, along with many other phenomena and processes that are either nonexistent or unimportant in inanimate matter.

Eventually a much needed change took place. Beckner, Hull, and Ruse included evolution in their philosophies of biology, and a new generation of philosophers arose who specifically concentrated on evolution: Beatty, Brandon, Burian, Munson, Rosenberg, Sober, Wimsatt, and M. Williams, to mention some names. At the same time a handful of biologists— Simpson, Dobzhansky, Rensch, Ghiselin, and myself—began to write on philosophical aspects of biology. Finally, in 1985, the time was right for the founding of a new journal, *Biology and Philosophy*.

These recent philosophical developments are distinguished not only by the incorporation of uniquely biological phenomena (such as life, selection, coded information, and teleology) but also by a pronounced shift in methodology and *Fragestellung*. One important change is the deemphasis of laws. In most classical philosophies of science, explanation consists in connecting phenomena with laws. Although laws are also encountered in biology, particularly in physiological and developmental processes, most regularities encountered in the living world lack the universality of the laws of physics. Consequently, biologists nowadays make use of the word *law* only rarely. But a reliance on laws is unnecessary, according to the so-called semantic view of philosophy (as J. Beatty has pointed out).

The philosophy of biology is characterized by its emphasis on *concepts* and their clarification. The importance of concepts has been articulated by M. J. West-Eberhard: "The spectacular recent progress in this field [evolution of social behavior] has been primarily due to conceptual advances in biology (especially "adaptationist" thinking), not improvements in the precision of measurements." The demonstration of the significance of concept clarification was one of the main themes in my *Growth of Biological Thought*. Alas, the older generation of philosophers of science still ignore this insight. Too many of them persist in dissecting with the sharpest tools of logic the grueness or not of emeralds or the blackness or not of ravens. Fortunately, the younger philosophers of biology entirely agree with the scientists that a careful analysis of the underlying concepts has primacy in philosophy over exercises in logic.

Because of these recent shifts in the interests of philosophers, the analysis of concepts undertaken in Parts 2 and 8 of this volume—dealing with such specific biological subjects as speciation, adaptation, and macro-evolution—will, I hope, be of as much interest to the philosopher as to the biologist. In fact, philosophers of biology have written on all of these topics in recent years. I am convinced that the joint efforts of both classes of scholars are required if we are to achieve a full understanding of these difficult aspects of living organisms. Too often in the past the biologists have ignored the analyses of the philosophers, and the philosophers have ignored the discoveries of the biologists. My hope is that this book will help to strengthen the bridge between biology and philosophy, and point to the direction in which a new philosophy of biology will move.

E.M.

Acknowledgments

Words cannot express my deep gratitude to the many friends and associates whose suggestions and criticisms have substantially improved this volume. Some of the previously unpublished essays were scrutinized by F. J. Ayala, W. J. Bock, R. Brandon, G. W. Cottrell, Jr., D. Futuyma, D. Hull, and M. Ruse, who spotted errors, omissions, and inconsistencies and helped me to clarify my arguments. My coverage of the literature was facilitated by many colleagues who kindly sent me reprints of their publications.

Walter Borawski cheerfully typed the numerous drafts and final version of all the essays and greatly assisted in the compilation of the bibliographies and index. The staff of Harvard University Press, who have done their utmost to achieve a product of quality, have my sincere appreciation for their efforts. Most of all I wish to thank Susan Wallace, whose imaginative editorial suggestions, particularly in the new essays, have materially improved the internal cohesion and readability of this book.

As emeritus professor, I would have been quite unable to undertake this project had I not received a grant (83-10-11) through the generosity of the Alfred P. Sloan Foundation.

E. M.

Contents

part one ॐ

PHILOSOPHY

Introduction

In order to qualify for my Ph.D. at the University of Berlin I had to take an examination in philosophy. I continued the study of philosophy throughout my biological career, and during the past twenty-five years have attempted some contributions of my own (Mayr 1969; 1976). One of my special concerns has been the neglect of biology in works claiming to be philosophies of science. From the 1920s to the 1960s the logical positivists and physicalists who dominated the philosophy of science had little interest in and even less understanding of biology, because it simply did not fit their methodology. Their endeavors to solve all scientific problems by pure logic and refined measurements were unproductive, if not totally irrelevant, when applied to biological phenomena.

The assumption that it should be possible to "reduce" the theories and concepts of all other sciences, including biology, to those of the physical sciences has clearly dominated not only philosophy but science itself, from the days of Galileo and Descartes. But the further the study of biological systems advanced during the past 200 years, the more evident it became how different living systems are from inanimate systems, no matter how complex the inanimate system or how simple the organism. Attempts to "reduce" biological systems to the level of simple physico-chemical processes have failed because during the reduction the systems lost their specifically biological properties. Living systems, as we shall see in Essay 1, have numerous properties that are simply not found in the inanimate world.

These pecularities of organisms must be duly considered in a balanced philosophy of science. The new generation of philosophers of science is fully aware of this. By contrast, some members of the older generation seem to assume that any reference to the autonomy of biology is an attempt to reintroduce the philosophy of vitalism by the back door. It must therefore be emphasized that the modern biologist rejects in any form whatsoever the notion that a "vital force" exists in living organisms which does not obey the laws of physics and chemistry. All processes in organisms, from the interaction of molecules to the complex functions of the brain and other whole organs, strictly obey these physical laws. Where organisms differ from inanimate matter is in the organization of their systems and especially in the possession of coded information.

The greatest advance in our thinking on this subject has been the recognition that organisms have a dualistic nature. They consist of (1) a genetic program represented by the nucleic acids, in eukaryotes largely in the cell nucleus (genotype), and (2) an extended body or soma (phenotype) produced under the instructions of the genetic program. The entire ontogeny as well as the physiological processes and the behavior of organisms are directly or indirectly controlled by the information encoded in the genes (Essay 2). There is nothing in any nonliving (except man-made) system that corresponds to the genotype, a system that has selectively stored vital information during the billions of years that life has existed on the earth.

The development of completely new disciplines—evolutionary biology and genetics—was necessary before the centuries-old battle between mechanists and their opponents could be resolved. To the distress of both camps, the conclusion reached was that both were, to some extent, correct. The finding that all processes in living organisms strictly obey the laws of physics and chemistry—that there is no residue of "vital forces" outside the realm of the physical sciences—meant that the mechanists were right. But the finding that the coded information system of living organisms has no equivalent in inanimate nature meant that the antimechanists were also right. This genotype-phenotype duality of the living organism is the reason why it is not sufficient in biology to search for a single cause in the study of a phenomenon, as is often sufficient in the physical sciences.

In Essay 2 I emphasize the need to distinguish two causations underlying all phenomena or processes in organisms. These have been referred to by earlier authors as *proximate* and *ultimate* causations. The proximate causes consist of answers to "how?" questions; they are responsible for all physiological and developmental processes in the living organism, and their domain is the phenotype. The ultimate or evolutionary causes consist of answers to "why?" questions, and provide the historical explanation for the occurrence of these phenomena. Their domain is the genotype. For instance, when T. H. Morgan (1932) asserted that sexual dimorphism could be fully explained by a study of the physiological, including hormonal, factors causing the dimorphism, he completely ignored the ultimate causes. Only evolutionary explanations can account for the fact that different species display pronounced, slight, or no sexual dimorphism. Many famous controversies in various fields of biology have been due to a failure of opponents to realize that one of them was interested in proximate, the other in evolutionary, causes.

The clear recognition of two types of causation in organisms has helped to solve an important problem in biology, the problem of teleology. What is teleology, and to what extent is it a valid concept? These have been burning questions since the time of Aristotle. Kant based his explanation of biological phenomena, particularly of the perfection of adaptations, on teleology (Löw 1980)—the notion that organisms were designed for some purpose. Teleology was the principal argument used by some of Darwin's major opponents (Sedgwick and von Baer). And the numerous autogenetic theories of evolution, such as orthogenesis (Eimer), nomogenesis (Berg), aristogenesis (Osborn), and omega principle (Teilhard de Chardin), were all based on a teleological world view. Indeed, as Jacques Monod (1971) rightly stressed, almost all of the most important ideologies of the past and the present are built on a belief in teleology.

It is my belief that the pervasive confusion in this subject has been due to a failure to discriminate among very different processes and phenomena, all labeled "teleological." In Essay 3 I show that the word *teleological* has been indiscriminately applied to four entirely different phenomena or processes. By partitioning so-called teleological phenomena into these four categories, and by introducing an appropriate terminology for each, it is possible to study each of them separately and show that three of them can be explained scientifically. On the other hand, no evidence whatsoever has been found for the existence of the fourth one, cosmic teleology.

The most important conclusion of the recent research on teleology is that it is illegitimate to extrapolate from the existence of *teleonomic* processes (that is, those directed or controlled by the organism's own DNA) and *teleomatic* processes (those resulting from physical laws) to an existence of cosmic teleology. There is neither a program nor a law that can explain and predict biological evolution in any teleological manner. Nor is there, since 1859, any need for a teleological explanation: The Darwinian mechanism of natural selection with its chance aspects and constraints is fully sufficient.

If we had to name a single person as responsible for the refutation of cosmic teleology, it would be Charles Darwin. Natural selection, as he showed in the *Origin,* can explain all the phenomena for which, up to that time, a principle of finality had been invoked (Essay 14). But this conclusion was completely unacceptable to some of Darwin's contemporaries, such as Adam Sedgwick, K. E. von Baer, and Louis Agassiz. And even today, more than 125 years later, there are some philosophers who not only uphold the existence of a teleological principle in the world but seem to be quite unable to develop a philosophy of life without teleology.

The study of genetics has shown that seemingly goal-directed processes in a living organism (teleonomic processes) have a strictly material basis, being controlled by a coded genetic program. Curiously, the coded program is a concept philosophers with a background in logic, physics, or mathematics seem to have great difficulty in understanding and accepting. Since the term *program* was taken over from the field of informatics, it is sometimes rejected as an anthropomorphism. Yet, the use of the term in biology is fully justified (Beniger 1986). Even though the mechanism by which the DNA stores and codifies information is of course different from that of a computer, the basic principle is remarkably similar, as demonstrated by the researches of molecular biology.

Returning for a moment to the rift between the physicalists and biologists, we must note that advances during the last 150 years not just in biology but in the physical sciences as well have greatly helped to narrow the gap that existed between the two camps. Many of the concepts of classical mechanics and the traditional philosophy of science that were questioned by biologists, such as strict determinism (vs. high frequency of probability), the predictiveness of all processes, or the universality of laws, have now also been either given up entirely by modern physics or at least restricted in applicability.

Classical physics was strictly deterministic. Laplace's boast that he would be able to predict the future course of events on earth *ad infinitum* if he had a complete catalogue of the existing situation was symptomatic of this attitude. Not surprisingly, natural selection with its emphasis on the chance nature of variation was not palatable to the physicists. This is why John Herschel referred to it as the "law of the higgledy-piggledy." Modern physics has theoretically abandoned such determinism, and yet physicists still are far more deterministic in their thinking than biologists.

This difference is amusingly reflected in the divergent positions taken by most physicists and biologists vis-à-vis the probability of the existence of life—particularly intelligent life—on other planets. Physicists tend to reason that if life originated anywhere else in the universe (and most of them think that this might have happened millions of times), then it is a virtual certainty that in many or most cases intelligent beings would have evolved. Since space travel to other solar systems and other galaxies is impossible, in 1959 various radio astronomers initiated a project to communicate with possibly existing extraterrestrial intelligent beings with the help of signals. Thousands of hours and hundreds of thousands, if not millions, of dollars have been devoted to the search for extraterrestrial intelligence (SETI). Although the costs are minimal compared with other NASA projects, various scientists began to question the rationality of the whole enterprise.

Looking at the SETI project from a biologist's point of view in Essay 4, I demonstrate that each step leading to the evolution of intelligent life on earth was highly improbable and that the evolution of the human species was the result of a sequence of thousands of these highly improbable steps. It is a miracle that man ever happened, and it would be an even greater miracle if such a sequence of improbabilities had been repeated anywhere else. The real message of this essay is to call attention to the difference in the psychology of physical scientists and evolutionists. This difference is not, of course, universal. A few biologists, particularly molecular biologists, believe that the probability of extraterrestrial intelligence is sufficiently great to justify the search for it, and a few physicists have forcefully called attention to the unbelievably large improbability of success. The arguments of these physicists, however, are rather different from those of a biologist such as myself. Rather than questioning the probability of the existence of extraterrestrial intelligence, they question the feasibility of sending out enough directional beams for long enough a time to make a call-response probable, or else they call attention to the extremely short duration of civilizations as compared with astronomical time. The difference in the approach of physical scientists and biologists to this problem ought to be of considerable interest to the philosopher.

In Essay 5 I enter a field where angels fear to tread—the origin of human ethics. An enormous amount of recent literature exists on this highly controversial subject, and I have studied only a very small part of it. Being anything but an authority in the field of ethics, I have not been concerned with finding definitive answers in this essay. Rather, I have attempted to develop some previously neglected aspects and to ask open questions.

No simple answer can be given to the most frequently posed question: What portion of human ethics is part of mankind's primate heritage? If the individual is the target of selection, that is, if natural selection rewards only that which is of benefit to the individual, then it is a puzzle how any altruism beyond parental care could ever have evolved. The origin of human ethics posed a formidable problem for Darwin. Human morality, for the natural theologians, was part of the creation. To replace God's design by the strictly material process of natural selection, said Sedgwick and others, deprived morality of its very foundation. Ever since Darwin, efforts have been made either to derive human ethics from evolution or at least to demonstrate that no conflict exists between Darwinism and the origin of human ethics.

Any acceptable solution would seem to require the recognition that the human species is indeed unique in having a culture and in thus possessing the capacity to transmit ethical norms from generation to generation without their being encoded in genes. Genetics is not entirely uninvolved, however, because there is the evolutionarily selfish altruism provided by inclusive fitness (Essay 5), and because, as Waddington (1960) has suggested, there must have been a premium on the evolution of an open behavior program capable of accepting culturally transmitted ethical norms.

The controversies surrounding sociobiology have renewed the old argument. It is evident from the extent of the recent literature on this subject, and the seeming irreconcilability of opposing opinions, that we are still far from a resolution of questions surrounding the role of genetics in human ethics. I hope I have been able to articulate some of these issues a little more concisely than has been done by the more passionate writers. Here is one of the many areas in philosophy where it is of the utmost importance to ask well-defined questions. This much is certain: The problem of evolution and ethics can be solved only by the most careful analysis of the underlying biological processes.

Additional problems in the philosophy of biology will be treated in other parts of this volume. Nearly all of them deal with topics that had been poorly dealt with or entirely ignored in the traditional volumes of the philosophy of science. A broader, more adequate philosophy of science requires, however, the development of a critical philosophy of biology. Endeavors to fill this gap have been made by a number of philosophers, including Ruse (1973), Hull (1974), Rosenberg (1985), Sattler (1986), and Smith (1976). As praiseworthy as the writings of these philosophers are, they seem to lack the balance and perspective one would hope for. Perhaps this type of synthesis cannot be achieved until individual areas of biology have been treated as authoritatively as Brandon and Sober have treated natural selection (Essay 6). This essay volume, I hope, is another step on that road.

REFERENCES

Beniger, J. R. 1986. *The Control Revolution.* Cambridge, Mass.: Harvard University Press.

Hull, D. 1974. *Philosophy of Biological Science.* Englewood Cilffs, N.J.: Prentice-Hall.

Löw, R. 1980. *Philosophie des Lebendigen. Der Begriff des Organischen bei Kant.* Frankfurt: Suhrkamp.

Mayr, E. 1969. Discussion: footnotes on the philosophy of biology. *Phil. Soc.* 36(2):197–202.

———— 1976. *Evolution and the Diversity of Life.* Cambridge, Mass.: Harvard University Press.

Monod, J. 1971. *Chance and Necessity.* New York: Knopf.

Morgan, T. H. 1932. *The Scientific Basis of Evolution.* New York: Norton.

Rosenberg. A. 1985. *The Structure of Biological Science.* Cambridge: Cambridge University Press.

Ruse, M. 1973. *The Philosophy of Biology.* London: Hutchinson.

Sattler, R. 1986. *Biophilosophy.* New York: Springer.

Smith, C. U. M. 1976. *The Problem of Life.* London: Macmillan.

Waddington, C. F. 1960. *The Ethical Animal.* London: Allen and Unwin.

IS BIOLOGY AN
AUTONOMOUS SCIENCE?

ALL RECENT volumes on the philosophy of biology begin with the question: What is the position of biology in the sciences? "Whether and how biology differs from the other natural sciences . . . is the most prominent, obvious, frequently posed, and controversial issue the philosophy of biology faces," according to Rosenberg (1985:13). This battle over the status of biology has been waged between two distinct camps. One claims that biology does *not* differ in principles and methods from the physical sciences, and that further research, particularly in molecular biology, will in time lead to a reduction of all of biology to physics. Ruse (1973), for example, wondered "whether or not we can look forward to the day when biology as an autonomous discipline will vanish." The other camp claims that biology fully merits status as an autonomous science because it differs fundamentally in its subject matter, conceptual framework, and methodology from the physical sciences (Ayala 1968).

Part of the controversy arose from a different interpretation of the word *autonomy*. If one could plot the domains of the physical and biological sciences on a map, one would find a considerable area of overlap, particularly at the molecular level, where the laws of physics and chemistry dominate. Does this argue against autonomy for biology? For those who define autonomy as a complete separation of the two sciences, this important area of overlap refutes the claim for autonomy. Proponents of the opposing viewpoint, on the other hand, point to the equally important areas *not* overlapped by the physical sciences and insist that only an autonomous science can adequately study them.

This unfortunate controversy is a product of history. When science reawakened after the Middle Ages, in the work of Galileo and Newton and later Lavoisier, it was almost exclusively a movement of the physical sciences. Biology as a discipline was still dormant and did not really come to life until the 1830s and 1840s. For the philosophers, from Bacon and Descartes to Locke and Kant, the physical sciences, and in particular mechanics, were the paradigm of science. The proper way to study the natural world, according to this view, was to define phenomena in terms of movements and forces that obeyed universal laws—that is, laws which were not in any way restricted in time or space nor subject to any exceptions. Such deterministic laws allowed a strict prediction of future events, once the present conditions were understood. The role of chance in natural processes was completely ignored. Consequently, the controlled experiment was considered the only respectable scientific method, whereas observation and comparison were viewed as considerably less scientific.

As everyone was willing to concede, the universality and predictability that seemed to characterize studies of the inanimate world were missing from biology. Because life was restricted to the earth, as far as anyone knew, any statements and generalizations one could make concerning living organisms would seem to be restricted in space and time. To make matters worse, such statements nearly always seemed to have exceptions. Explanations usually were not based on universal laws but rather were pluralistic. In short, the theories of biology violated every canon of "true science," as the philosophers had derived them from the methods and principles of classical physics.

Even after the conceptual framework of physics changed quite fundamentally during the nineteenth and twentieth centuries, a mechanistic approach continued to dominate the philosophy of science. As a result, biology was referred to as a "dirty science," an activity, according to the physicist Ernest Rutherford, not much better than "postage stamp collecting." At best it was a second-class, "provincial" science.

Biologists responded to the claims of the physicists and philosophers in one of three ways. Many of them, particularly those working in physiology and other branches of functional biology, adopted physicalism and attempted to explain all biological processes in terms of movements and forces. Everything was mechanistic, everything was deterministic, and there was no unexplained residue. Jacques Loeb, Carl Ludwig, and Julius Sachs were perhaps the leaders of physicalist biology. As productive as this approach was, particularly in physiology, it left a vast number of phenomena in the living world totally unexplained.

Other biologists, by contrast, felt that a living organism had some constituent that distinguished it from inert matter. These people were customarily lumped together under the term *vitalists*, even though, as we shall see, they held widely differing views of what that constituent might be.

Most biologists, though, simply ignored the philosophical problems of the nature of life and instead concentrated on making new discoveries and elaborating new theories. The result was the unprecedented flowering of evolutionary biology, ecology, ethology, population genetics, cytology, and many other biological disciplines. Each of these fields had its own terminology, methodology, and conceptual framework, and maintained only a minimal contact with the others or with the physical sciences. The worry spread, particularly among philosophers, that science as a whole would be lost, replaced by a large number of independent individual sciences. To counteract this threat, a movement got under way for the unification of science.

But how was unification to be achieved? There seemed to be two broad possibilities:

(1) To bring all sciences down to the common denominator of the physical sciences; in other words, as it was phrased by certain philosophers, "by reducing all sciences to physics."

(2) To adopt a new, broader concept of science that would fit not only the physical sciences but also the life sciences.

It has become quite clear from the discussion of the modern philosophers of science that the validity of the claim of an autonomy of biology depends entirely on the success of the postulated reduction. What we need, then, is an answer to the question: "Can the phenomena, laws, and concepts of biology be successfully reduced to those of the physical sciences?" If such a reduction is impossible, then the autonomy of biology is, so to speak, automatically established.

The 1960s and early 1970s saw quite a few uncompromising reductionists (Schaffner 1967a,b; Ruse 1973), but their number has dwindled in the last ten years. Only one strict reductionist has come to my attention since 1980. One problem in the reductionist camp was that the term *reduction* was being used by different authors in very different senses. One can distinguish three major kinds of reduction (Mayr 1982:59–63; Ayala 1974; 1977):

(1) The term *constitutive reduction* has been applied to any dissection of phenomena, events, and processes into the constituents of which they are composed. Such analysis is not opposed by the modern biologist,

since he does not question that all organic processes can ultimately be reduced to or explained by physico-chemical processes. None of the events and processes encountered in the world of living organisms is in any conflict with a physico-chemical explanation at the level of atoms and molecules. What is controversial are two other kinds of reduction, explanatory reduction and theory reduction.

(2) *Explanatory reduction* claims that all the phenomena and processes at higher hierarchical levels can be explained in terms of the actions and interactions of the components at the lowest hierarchical level. Organicists, by contrast, claim that new properties and capacities *emerge* at higher hierarchical levels and can be explained only in terms of the constituents at those levels. For instance, it would be futile to try to explain the flow of air over the wing of an airplane in terms of elementary particles. Almost any phenomenon studied by a biologist relates to a highly complex system, the components of which are usually several hierarchical levels above the level studied by physical scientists.

(3) Finally, there is *theory reduction*, which postulates that the theories and laws formulated in biology are only special cases of theories and laws formulated in the physical sciences, and that such biological theories can thus be reduced to physical theories. All authors in recent years who have studied this claim, including even several former reductionists, have come to the conclusion that such theory reduction is virtually never successful (Mayr 1982). As a matter of fact, theory reduction has been only partially successful even within the physical sciences, and has been singularly unsuccessful within the biological sciences. Indeed, none of the more complex biological laws has ever been reduced to and explained in terms of the composing single processes (Mainx 1955).

The splendid successes of molecular biology are sometimes cited as evidence for successful reduction, but these cases concern constitutive reduction, and furthermore they are limited almost exclusively to functional biology. Ernest Nagel (1961) was the chief proponent of theory reduction, but most other philosophers of science (Feyerabend, Kuhn, and Kitcher) have vigorously opposed his arguments.

I think it is fair to say that the attempt to unify science by reducing biology to physics has been a failure, as pointed out by Popper (1974), Beckner (1974), Kitcher (1984), and others. Fortunately, changes have taken place in the last several decades in both physics and biology that will greatly facilitate an eventual unification of the two sciences on a very different, much broader basis.

The changes in the physical sciences involve, among other things, a rejection of the strict determinism of classical physics (Mayr 1985). Scientists now recognize that most physical laws are not universal but are rather statistical in nature, and that prediction therefore can only be probabilistic in most cases. They have also realized that stochastic processes operate throughout the universe, at every level, from subatomic particles to weather systems, to ocean currents, to galaxies. In the study of these processes, observation has been elevated to the status of a valid scientific method wherever the experiment is difficult or impossible to perform, as in meteorology and cosmology. And finally, physicists are beginning to recognize that the development of *concepts* can be as powerful a tool as the formulation of laws in understanding physical phenomena.

The changes in biology were, if anything, even more drastic (Mayr 1985). Physiology lost its position as the exclusive paradigm of biology in 1859 when Darwin established evolutionary biology. When behavioral biology, ecology, population biology, and other branches of modern biology developed, it became even more evident how unsuitable mechanics was as the paradigm of biological science. At the same time that an exclusively physicalist approach to organisms was being questioned, the influence of the vitalists was also diminishing, as more and more biologists recognized that all processes in living organisms are consistent with the laws of physics and chemistry, and that the differences which do exist between inanimate matter and living organisms are due not to a difference in substrate but rather to a different organization of matter in living systems.

In the eighteenth and nineteenth centuries, the label *vitalist* was attached to anyone who did not accept the mechanist dogma that matter in motion is an adequate explanatory basis for all aspects of life, and that organisms are simply machines. All those who rejected this characterization were united in their belief that a living organism has some sort of constituent by which it can clearly be distinguished from inert matter. Where a controversy arose, however, was in the interpretation of this constituent.

The classical vitalist ascribed life to the organism's possession of a tangible thing, a real object, whether called a vital fluid, life force, or *Entelechie*. He believed that this vital force was outside the realm of physico-chemical laws; in fact, it had a rather metaphysical flavor in the writings of some vitalists. All attempts to substantiate the existence of this force failed, and the need for it became obsolete when the phenomena it had tried to explain were eventually accounted for by other means, for example, the genetic program.

Other biologists agreed with the classical vitalists that organisms have some unique property that exists in every part of the body, one that is extinguished by death. They attributed to it everything that distinguishes living bodies from inert matter, particularly the form-giving processes of ontogeny. But these authors rejected the idea that this was a nonmaterial force; rather, they viewed life as an organizational property of certain material systems. In the absence of an appropriate term, some of these authors, like the famous physiologist Johannes Müller, referred to these life-giving properties as *Lebenskraft*, but as Delbrück (1971) pointed out, there is a remarkably close analogy between the postulated properties of the *Lebenskraft* of many authors from Aristotle on and the actual properties of the genetic program (DNA).

This second group of biologists might be best referred to as organicists. In any case, it is quite misleading to attach the label *vitalist* to them. Anyone who does this and insists on the strict matter-in-motion definition of organisms will have to call everybody a vitalist who acknowledges the genetic program. Vitalism has become so disreputable a belief in the last fifty years that no biologist alive today would want to be classified as a vitalist. Still, the remnants of vitalist thinking can be found in the work of Alistair Hardy, Sewall Wright, and Charles Birch, who seem to believe in some sort of nonmaterial principle in organisms.

Vitalistic ideas, curiously, were widespread among certain nonbiologists whose simplistic ideas about the nature of physico-chemical systems forced them into vitalism (Crick 1966). Some of the leaders of quantum mechanics, such as Bohr, Schroedinger, Heisenberg, and Pauli, postulated that someday someone would discover physical laws in organisms that were different from those which operate in inert matter. Indeed, when Max Delbrück switched from physics to biology, one of his original objectives was to discover such laws (Kay 1985).

The Emancipation of Biology

Establishing and substantiating the autonomy of biology has been a slow and painful process. It has meant getting rid not only of standard concepts of physicalism, such as essentialism and determinism, but also of some metaphysical concepts favored by certain biologists who intuitively felt the separate status of biology but ascribed it to such metaphysical factors as vitalism or teleology. Even today, many attacks against the notion that biology is an independent science concentrate on refuting vitalism, as though this was still part of the conceptual framework of modern biology. That some early autonomists, like Bertalanffy (1949), supported their position with such vague arguments as dynamics, energy gradients, formative movements, and so on did not enhance the credibility of the new movement. Despite these handicaps, the evidence in support of the autonomy of biology has grown exponentially in recent years.

Let me now describe, one by one, some of the fundamental differences between organisms and inert matter.

THE COMPLEXITY OF LIVING SYSTEMS

Living systems are characterized by a remarkably complex organization which endows them with the capacity to respond to external stimuli, to bind or release energy (metabolism), to grow, to differentiate, and to replicate. Biological systems have the further remarkable property that they are open systems, which maintain a steady-state balance in spite of much input and output. This homeostasis is made possible by elaborate feedback mechanisms, unknown in their precision in any inanimate system.

Such complexity has often been singled out as the most characteristic feature of living systems. Actually, complexity in and of itself is not a fundamental difference between organic and inorganic systems. The world's weather system or any galaxy is also a highly complex system. On the average, however, organic systems are more complex by several orders of magnitude than those of inanimate objects. Even at the molecular level, the macromolecules that characterize living beings do not differ in principle from the lower-molecular-weight molecules that are the regular constituents of inanimate nature, but they are much larger and more complex. This complexity endows them with extraordinary properties not found in inert matter.

The complexity of living systems exists at every hierarchical level, from the nucleus, to the cell, to any organ system (kidney, liver, brain), to the individual, to the species, the ecosystem, the society. The hierarchical structure within an individual organism arises from the fact that the entities at one level are compounded into new entities at the next higher level—cells into tissues, tissues into organs, and organs into functional systems.

To be sure, hierarchical organization is not altogether absent in the inanimate world, where elementary particles form atoms, which in turn form molecules, and then crystals, and so on. But order in the inanimate realm is several levels of magnitude below the order of ontogenetic development, as illustrated by the growth of the peacock's tail or the organization of the central nervous system.

Systems at each hierarchical level have two properties. They act as wholes (as though they were a homogeneous entity), and their characteristics cannot be deduced (even in theory) from the most complete knowledge of the components, taken separately or in other combinations. In other words, when such a system is assembled from its components, new characteristics of the whole emerge that could not have been predicted from a knowledge of the constituents. Such *emergence* of new properties occurs also throughout the inanimate world, but only organisms show such dramatic emergence of new characteristics at every hierarchical level of the system. Indeed, in hierarchically organized biological systems one may even encounter downward causation (Campbell 1974).

ORGANIZATION INTO POPULATIONS

Western thinking for more than 2,000 years after Plato was dominated by essentialism. For Plato and his followers, variable classes of entities consist of imperfect reflections of a fixed number of constant, discontinuous *eide* or essences. This is vividly illustrated by Plato's allegory of the shadows on the cave wall. This concept fits classes of inanimate objects, say the class of chairs or the class of lakes—objects that have no special relation with each other except that they share the same definition (see Essay 20).

In 1859 Darwin introduced the entirely new concept of variable populations composed of unique individuals. For those who have accepted population thinking, the variation from individual to individual within the population is the reality of nature, whereas the mean value (the "type") is only a statistical abstraction. Biopopulations differ fundamentally from classes of inanimate objects not only in their propensity for variation but also in their internal cohesion and their spatio-temporal restriction. There is nothing in inanimate nature that corresponds to biopopulations, and this perhaps explains why philosophers whose background is in mathematics or physics seem to have such a difficult time understanding this concept (see Essay 20). The ability to make the switch from essentialist thinking to population thinking is what made the theory of evolution through natural selection possible.

The concept of biopopulations also made possible the recognition that there are, in nature, two entirely different kinds of evolution, designated by Lewontin (1983) as developmental (transformational) evolution and variational evolution. Any change in an object or system simply as a result of its intrinsic potential, such as the change of a white star to a red star, is developmental evolution. It is entirely due to the action of teleomatic (physical) processes. By contrast, the evolution of organisms is variational evolution, and is due to the selection of certain entities from highly variable populations of unique individuals, and the production of new variation in every generation.

To say that all members of a population are unique does not mean that they differ from one another in every respect. On the contrary, they may agree with one another in most respects, as do conspecific individuals, for instance. Yet each member of a species has a unique constellation of characteristics, some of which are found in no other individual.

Although highly characteristic of the living world, uniqueness is not exclusive to it. Each mountain is unique; so is each weather system, and each planet and star. However, such uniqueness in the inanimate world is limited to complex systems, while the basic building blocks of these systems (elementary particles, atoms, molecules, and crystals) consist of identical components. In the living world, uniqueness is seen even at the molecular level, in the form of DNA or RNA.

POSSESSION OF A GENETIC PROGRAM

Organisms are unique at the molecular level because they have a mechanism for the storage of historically acquired information, while inanimate matter does not. Perhaps there was an intermediate condition at the time of the origin of life, but for the last three billion years or more this distinction between living and nonliving matter has been complete. All organisms possess a historically evolved genetic program, coded in the DNA of the nucleus (or RNA in some viruses). Nothing comparable exists in the inanimate world, except in man-made machines. The presence of this program gives organisms a peculiar duality, consisting of a genotype and a phenotype. The genotype (unchanged in its components except for occasional mutations) is handed on from generation to generation, but, owing to recombination (Essay 6), in ever new variations. In interaction with the environment, the genotype controls the production of the phenotype, that is, the visible organism which we encounter and study.

The genotype (genetic program) is the product of a history that goes back to the origin of life, and thus it incorporates the "experiences" of all ancestors, as Delbrück (1949) said so rightly. It is this which makes organisms historical phenomena. The genotype also endows them with the capacity for goal-directed (teleonomic) processes and activities, a capacity totally absent in the inanimate world.

Since each genetic program is a unique combination of thousands of different genes, the differences among them cannot be expressed in quantitative but only qualitative terms. Thus, quality becomes one of the dominant aspects of living organisms and their characteristics. Qualitative differences are particularly obvious when one compares properties and activities of different species, be it their courtship displays, pheromones, niche occupation, or whatever else may characterize a particular species.

COMPARATIVE VERSUS EXPERIMENTAL METHOD

The experiment has traditionally been the primary means of investigation in the physical sciences, and some philosophers have claimed that it is the only legitimate method of science. In fact, since the days of Copernicus and Kepler, observation and comparison have been exceedingly successful methods in such physical sciences as astronomy, geology, oceanography, and meteorology. And in biology, where observation and comparison have always been of paramount importance, experimental methods have been incorporated into the methodological repertory of many originally observational disciplines, including ecology and ethology.

The roles of the experimental and comparative methods in biology can be understood only if one realizes that biology actually consists of two rather different major fields of study. The first is the biology of proximate causations (broadly, functional biology), and the second is the biology of ultimate causations (evolutionary biology; see Essay 2).

There is nothing in the physical sciences that corresponds to the biology of ultimate causations. The claims that the evolution of galaxies or radioactive decay correspond to biological processes are quite erroneous. Evolution in galaxies is transformational, not variational, evolution (Lewontin 1983), and radioactive decay, controlled by physical laws, is a teleomatic process, not a teleonomic one, as claimed by Nagel (1977).

Early in the century there was virtually no communication between the two biologies of proximate and ultimate causations. As we have seen, the functional biologists tended to be physicalists and inductionists, accepting only the experiment as the method of science. The evolutionary biologists tended to have an opposite point of view, dependent as they were on observation and comparison. Since then, biologists have realized that functional and evolutionary questions are equally legitimate, even though they may require very different approaches. No biological phenomenon can be fully explained until both sets of causations have been explored.

Broadly speaking, functional biology deals with the decoding of the genetic program and with the reactions of an organism to its surrounding world from the moment of fertilization to the moment of death. Evolutionary biology, on the other hand, deals with the history of genetic programs and the changes that they have undergone since the origin of life. A philosopher who fails to recognize both of these two very important and very different aspects of biology will arrive at conclusions that are at best incomplete, but more likely wrong.   .

CONCEPTS IN BIOLOGY
The conceptual framework of biology is entirely different from that of the physical sciences and cannot be reduced to it. The role that such biological processes as meiosis, gastrulation, and predation play in the life of an organism cannot be described by reference only to physical laws or chemical reactions, even though physico-chemical principles are operant. The broader processes that these biological concepts describe simply do not exist outside the domain of the living world. Thus, the same event may have entirely different meanings in several different conceptual domains. The courtship of a male animal, for instance, can be described in the language and conceptual framework of the physical sciences (locomotion, energy turnover, metabolic processes, and so on), but it can also be described in the framework of behavioral and reproductive biology. And the latter description and explanation cannot be reduced to theories of the physical sciences. Such biological phenomena as species, competition, mimicry, territory, migration, and hibernation are among the thousands of examples of organismic phenomena for which a purely physical description is at best incomplete if not irrelevant (Mayr 1982:62–63). For a long time concepts were rather neglected in the physical sciences. Their importance, under the name of *themata*, has recently been emphasized by Holton (1973).

LAWS VERSUS THEORIES
There is perhaps no better way to demonstrate the epistemological differences between the physical sciences and organismic biology than to point to the different roles of laws in the two sciences. In classical physics, laws were considered universal, and Popper's falsifiability principle was based on this conception. Up to the end of the nineteenth century, biologists also tended to explain all phenomena and processes as being due to the operation of laws. Darwin's *Origin of Species* refers to laws controlling certain biological processes no fewer than 106 times in 490 pages.

Today, the word *law* is used sparingly, if at all, in most writings about evolution. Generalizations in modern biology tend to be statistical and probabilistic and often have numerous exceptions. Moreover, biological generalizations tend to apply to geographical or otherwise restricted domains. One can generalize from the study of birds, tropical forests, freshwater plankton, or the central nervous system but most of these generalizations have so limited an application that the use of the word *law*, in the sense of the laws of physics, is questionable.

At the same time, some very comprehensive biological *theories* have been formulated concerning the mechanisms of inheritance, the basic processes of evolutionary change, and certain physiological phenomena from the molecular level up to that of organs. These theories of biology "appear comparable in scope, explanatory power, and evidential support to those of the physical sciences," according to Munson (1975:433). Yet every student of biology is impressed by the fact that there is hardly a theory in biology for which some exceptions are not known.

The so-called laws of biology are not the universal laws of classical physics but are simply high-level generalizations. Hence, as Kitcher (ms.) has stated: "There are a number of sciences that proceed extraordinarily well without employing any statements which can uncontroversially be called laws."

In the physical sciences it is axiomatic that a given process or condition must be explained by a single law or theory. In the life sciences, by contrast, various forms of pluralism are frequent. For instance, a particular adaptation may have been produced by several different evolutionary pathways (Bock 1959). A condition of adaptedness of the phenotype of an individual may have been due to a particular response by the norm of reaction—or it may have been strictly determined by the genotype. The response of a complex system is virtually never a strict response to a single extrinsic factor but rather the balanced response to several factors, and the end result of an evolutionary process may be a compromise between several selection forces. In the study of causations the biologist must always be aware of this potential pluralism.

PREDICTION

A belief in universal, deterministic laws implies a belief in absolute prediction. The ability to predict was therefore the classical test of the goodness of an explanation in physics. In biology, the pluralism of causations and solutions makes prediction probabilistic, if it is possible at all. Prediction in the vernacular sense, that is, the foretelling of future events, is as precarious in biology as it is in meteorology and other physical sciences dealing with complex systems (Mayr 1985:49–50). As Scriven (1959) has pointed out, the ability to predict is not a requirement for the validity of a biological theory.

TELEOLOGY

Since the Greeks, philosophers and theologians have been impressed by the frequency of seemingly end-directed processes in living matter—the growth of an organism from egg to adult, the annual migrations of animals, the perfection of the eye and other organs. The belief that there is a purpose, a predetermined end, in the processes of nature has been referred to as *teleology*. Actually, the term has been applied to four entirely different and independent phenomena, and this has led to considerable confusion (see Essay 3).

Natural selection is not a teleological but a strictly *a posteriori* process (see Essay 6). Adaptedness, as the result of a process of selection, is a condition unknown in the inanimate world. More smoothing and rounding does not make a pebble better adapted for its existence in a river bed. Snow is not an adaptation of water to cold temperature. But many arctic animals (ptarmigans, snowshoe hares) have adaptations that prevent their feet from sinking into the snow (Mayr 1982:47–52, 69–72). Since adaptedness is a result of the past and not an anticipation of the future, it does not qualify for the epithet "teleological."

The Autonomy of Biology and the Unification of Science

The preceding list of biology's unique characteristics as a science explains why attempts to reduce biology and its theories to physics have been a failure. Does this mean that a unification of science is impossible? Not in the least. All it means is that such a unification cannot be achieved by reducing biology to physics. Rather, we have to search for a new foundation for such a unification.

What should it be? G. G. Simpson (1964) has proposed a somewhat extreme interpretation:

> Insistence that the study of organisms requires principles additional to those of the physical sciences does not imply a dualistic or vitalistic view of nature. Life . . . is not thereby necessarily considered as nonphysical or nonmaterial. It is just that living things have been affected for . . . billions of years by historical processes. . . . The results of those processes are

systems different in kind from any nonliving systems and almost incomparably more complicated. They are not for that reason any less material or less physical in nature. The point is that *all* known material processes and explanatory principles apply to organisms, while only a limited number of them apply to nonliving systems. Biology, then, is the science that stands at the center of all science, and it is here, in the field where all the principles of all the sciences are embodied, that science can truly become unified.

We may not need to accept all these sweeping claims. However, Simpson has clearly indicated the direction in which we have to move. I believe that a unification of science is indeed possible if we are willing to expand the concept of science to include the basic principles and concepts of not only the physical but also the biological sciences. Such a new philosophy of science will need to adopt a greatly enlarged vocabulary—one that includes such words as biopopulation, teleonomy, and program. It will have to abandon its loyalty to a rigid essentialism and determinism in favor of a broader recognition of stochastic processes, a pluralism of causes and effects, the hierarchical organization of much of nature, the emergence of unanticipated properties at higher hierarchical levels, the internal cohesion of complex systems, and many other concepts absent from—or at least neglected by—the classical philosophy of science.

Twenty-nine years ago the physicist C. P. Snow vividly described the unbridgeable gap between the physical sciences and the humanities. If biologists, physicists, and philosophers working together can construct a broad-based, unified science that incorporates both the living and the nonliving world, we will have a better base from which to build bridges to the humanities, and some hope of reducing this unfortunate rift in our culture. Paradoxical as it may seem, recognizing the autonomy of biology is the first step toward such a unification and reconciliation.

N O T E

This essay was adapted from a Messenger Lecture presented at Cornell University, Ithaca, New York, October 15, 1985. Previously unpublished.

R E F E R E N C E S

Ayala, F. J. 1968. Biology as an autonomous science. *Amer. Sci.* 56:207–221.
———— 1974. Preface. In Ayala and Dobzhansky (1974), pp. vii–xvii.
———— 1977. Philosophical issues. In Th. Dobzhansky et al. *Evolution*, pp. 497–504. San Francisco: Freeman.

Ayala, F. J., and Th. Dobzhansky, eds. 1974. *Studies in the Philosophy of Biology: Reduction and Related Problems*. Berkeley: University of California Press.

Beckner, M. 1974. Reduction, hierarchies, and organicism. In Ayala and Dobzhansky (1974), pp. 163–177.

Bertalanffy, L. V. 1949. *Das biologische Weltbild: Die Stellung des Lebens in Natur und Wissenschaft*. Vol. 1. Bern: Francke. Eng. trans.: *Problems of Life*. New York: Wiley, 1952.

Bock, W. 1959. Preadaptation and multiple evolutionary pathways. *Evolution* 13:194–211.

Campbell, D. 1974. 'Downward causation' in hierarchically organized biological systems. In Ayala and Dobzhansky (1974), pp. 179–186.

Crick, F. 1966. *Of Molecules and Men*. Seattle: University of Washington Press.

Delbrück, M. 1949. A physicist looks at biology. *Trans. Conn. Acad. Arts Sci.* 38:173–190.

———— 1971. Aristotle-totle-totle. In J. Monod, and E. Borek, eds., *Of Microbes and Life*, pp. 50–55. New York: Columbia University Press.

Holton, G. 1973. *Thematic Origins of Scientific Thought: Kepler to Einstein*. Cambridge, Mass.: Harvard University Press.

Kay, L. E. 1985. Conceptual models and analytical tools: the biology of the physicist Max Delbrück. *J. Hist. Biol.* 18:207–246.

Kitcher, Philip. 1984. 1953 and all that: a tale of two sciences. *Phil. Rev.* 93:335–374.

Lewontin, R. C. 1983. The organism as the subject and object of evolution. *Scientia* 118:63–82.

Mainx, F. 1955. Foundations of Biology. *Int. Encycl. Unified Sci.* 1(9):1–86. Chicago: University of Chicago Press.

Mayr, E. 1976. *Evolution and the Diversity of Life*. Cambridge, Mass.: Harvard University Press.

———— 1982. *The Growth of Biological Thought*. Cambridge, Mass.: Harvard University Press.

———— 1985. How biology differs from the physical sciences. In D. J. Depew and B. H. Weber, eds., *Evolution at a Crossroads: The New Biology and the New Philosophy of Science*, pp. 43–63. Cambridge, Mass.: MIT Press.

Munson, R. 1975. Is biology a provincial science? *Phil. Sci.* 42:428–447.

Nagel, E. 1961. *The Structure of Science*. New York: Harcourt, Brace and World.

———— 1977. Teleology revisited: goal-directed processes in biology. *J. Phil.* 74:261–301.

Popper, K. R. 1974. Scientific reduction and the essential incompleteness of all science. In Ayala and Dobzhansky (1974), pp. 259–284.

Rosenberg, A. 1985. *The Structure of Biological Science*. Cambridge: Cambridge University Press.

Ruse, M. 1973. *The Philosophy of Biology*. London: Hutchinson.

Schaffner, K. 1967a. Approaches to reduction. *Phil. Sci.* 34:137–147.

———— 1967b. Antireductionism and molecular biology. *Science* 157:644.

Scriven, M. 1959. Explanation and prediction in evolutionary theory. *Science* 130:477–482.

Simpson, G. G. 1964. *This View of Life*. New York: Harcourt, Brace and World.

Snow, C. P. 1959. *The Two Cultures and the Scientific Revolution*. Cambridge: Cambridge University Press.

CAUSE AND EFFECT
IN BIOLOGY

BEING a practicing biologist, I feel that I cannot attempt the kind of analysis of cause and effect in biological phenomena that a logician would undertake. I would instead like to concentrate on the special difficulties presented by the classical concept of causality in biology. From the first attempts to achieve a unitary concept of cause, the student of causality has been bedeviled by these difficulties. Descartes's grossly mechanistic interpretation of life, and the logical extreme to which his ideas were carried by Holbach and de la Mettrie, inevitably provoked a reaction leading to vitalistic theories which have been in vogue, off and on, to the present day. I have only to mention names like Driesch (entelechy), Bergson (élan vital), and Lecomte du Noüy, among the more prominent authors of the recent past. Though these authors may differ in particulars, they all agree in claiming that living beings and life processes cannot be causally explained in terms of physical and chemical phenomena. It is our task to ask whether this assertion is justified, and if we answer this question with "no," to determine the source of the misunderstanding.

Causality, no matter how it is defined in terms of logic, is believed to contain three elements: (1) an explanation of past events ("a posteriori causality"); (2) prediction of future events; and (3) interpretation of teleological—that is, "goal-directed"—phenomena.

The three aspects of causality (explanation, prediction, and teleology) must be the cardinal points in any discussion of causality and were quite rightly singled out as such by Nagel (1961). Biology can make a significant contribution to all three of them. But before I can discuss this contribution in detail, I must say a few words about biology as a science.

Two Fields

The word *biology* suggests a uniform and unified science. Yet recent developments have made it increasingly clear that biology is a most complex area—indeed, that the word *biology* is a label for two largely separate fields which differ greatly in method, *Fragestellung*, and basic concepts. As soon as one goes beyond the level of purely descriptive structural biology, one finds two very different areas, which may be designated functional biology and evolutionary biology. To be sure, the two fields have many points of contact and overlap. Any biologist working in one of these fields must have a knowledge and appreciation of the other field if he wants to avoid the label of a narrow-minded specialist. Yet in his own research he will be occupied with problems of either one or the other field. We cannot discuss cause and effect in biology without first having characterized these two fields.

FUNCTIONAL BIOLOGY
The functional biologist is vitally concerned with the operation and interaction of structural elements, from molecules up to organs and whole individuals. His ever-repeated question is "How?" How does something operate, how does it function? The functional anatomist who studies an articulation shares this method and approach with the molecular biologist who studies the function of a DNA molecule in the transfer of genetic information. The functional biologist attempts to isolate the particular component he studies, and in any given study he usually deals with a single individual, a single organ, a single cell, or a single part of a cell. He attempts to eliminate, or control, all variables, and he repeats his experiments under constant or varying conditions until he believes he has clarified the function of the element he studies.

The chief technique of the functional biologist is the experiment, and his approach is essentially the same as that of the physicist and the chemist. Indeed, by isolating the studied phenomenon sufficiently from the complexities of the organism, he may achieve the ideal of a purely physical or chemical experiment. In spite of certain limitations of this method, one must agree with the functional biologist that such a simplified approach is an absolute necessity for achieving his particular objectives. The spectacular success of biochemical and biophysical ressearch justifies this direct, although distinctly simplistic, approach.

EVOLUTIONARY BIOLOGY
The evolutionary biologist differs in his method and in the problems in which he is interested. His basic question is "Why?" When we say "why" we must always be aware of the ambiguity of this term. It may mean "How come?" but it may also mean the finalistic "What for?" When the evolutionist asks "Why?" he or she always has in mind the historical

"How come?" Every organism, as an individual and as a member of a species, is the product of a long history, a history which indeed dates back more than 3,000 million years. As Max Delbrück (1949) has said, "A mature physicist, acquainting himself for the first time with the problems of biology, is puzzled by the circumstance that there are no 'absolute phenomena' in biology. Everything is time-bound and space-bound. The animal or plant or micro-organism he is working with is but a link in an evolutionary chain of changing forms, none of which has any permanent validity." There is hardly any structure or function in an organism that can be fully understood unless it is studied against this historical background. To find the causes for the existing characteristics, and particularly adaptations, of organisms is the main preoccupation of the evolutionary biologist. He is impressed by the enormous diversity of the organic world. He wants to know the reasons for this diversity as well as the pathways by which it has been achieved. He studies the forces that bring about changes in faunas and floras (as in part documented by paleontology), and he studies the steps by which the miraculous adaptations so characteristic of every aspect of the organic world have evolved.

We can use the language of information theory to attempt still another characterization of these two fields of biology. The functional biologist deals with all aspects of the decoding of the programmed information contained in the DNA of the fertilized zygote. The evolutionary biologist, on the other hand, is interested in the history of these programs of information and in the laws that control the changes of these programs from generation to generation. In other words, he is interested in the causes of these changes.

But let us not have an erroneous concept of these programs. It is characteristic of them that the programming is only in part rigid. Such phenomena as learning, memory, nongenetic structural modification, and regeneration show how "open" these programs are. Yet, even here there is great specificity, for instance with respect to what can be "learned," at what stage in the life cycle "learning" takes place, and how long a memory engram is retained. The program, then, may be in part quite unspecific, and yet the range of possible variation is itself included in the specifications of the program. The programs, therefore, are in some respects highly specific; in other respects they merely specify "reaction norms" or general capacities and potentialities.

Let me illustrate this duality of programs by the difference between two kinds of birds with respect to "species recognition." The young cowbird is raised by foster parents—let us say, in the nest of a song sparrow or warbler. As soon as it becomes independent of its foster parents, it seeks the company of other young cowbirds, even though it has never seen a cowbird before! In contrast, after hatching from the egg, a young goose will accept as its parent the first moving (and preferably also calling) object it can follow and become "imprinted" to. What is programmed is, in one case, a definite "gestalt," in the other, merely the capacity to become imprinted to a "gestalt." Similar differences in the specificity of the inherited program are universal throughout the organic world.

The Problem of Causation

Let us now get back to our main topic and ask: Is *cause* the same thing in functional and evolutionary biology?

Max Delbrück (1949), again, has reminded us that as recently as 1870 Helmholtz postulated "that the behavior of living cells should be account-able in terms of motions of molecules acting under certain fixed force laws." Now, says Delbrück correctly, we cannot even account for the behavior of a single hydrogen atom. As he also says, "Any living cell carries with it the experiences of a billion years of experimentation by its ancestors."

Let me illustrate the difficulties of the concept of causality in biology by an example. Let us ask: What is the cause of bird migration? Or more specifically: Why did the warbler on my summer place in New Hampshire start his southward migration on the night of the 25th of August?

I can list four equally legitimate causes for this migration:

(1) *An ecological cause.* The warbler, being an insect eater, must mi-grate, because it would starve to death if it should try to winter in New Hampshire.

(2) *A genetic cause.* The warbler has acquired a genetic constitution in the course of the evolutionary history of its species which induces it to respond appropriately to the proper stimuli from the environment. On the other hand, the screech owl, nesting right next to it, lacks this constitution and does not respond to these stimuli. As a result, it is sedentary.

(3) *An intrinsic physiological cause.* The warbler flew south because its migration is tied in with photoperiodicity. It responds to the decrease in day length and is ready to migrate as soon as the number of hours of daylight have dropped below a certain level.

(4) *An extrinsic physiological cause.* Finally, the warbler migrated on the 25th of August because a cold air mass, with northerly winds, passed over our area on that day. The sudden drop in temperature and the associated weather conditions affected the bird, already in a general physiological readiness for migration, so that it actually took off on that particular day.

Now, if we look over the four causations of the migration of this bird once more, we can readily see that there is an immediate set of causes of the migration, consisting of the physiological condition of the bird interacting with photoperiodicity and drop in temperature. We might call these the *proximate* causes of migration. The other two causes, the lack of food during winter and the genetic disposition of the bird, are the *ultimate* causes. These are causes that have a history and that have been incorporated into the system through many thousands of generations of natural selection. It is evident that the functional biologist would be concerned with analysis of the proximate causes, while the evolutionary biologist would be concerned with analysis of the ultimate causes. This is the case with almost any biological phenomenon we might want to study. There is always a proximate set of causes and an ultimate set of causes; both have to be explained and interpreted for a complete understanding of the given phenomenon.

Still another way to express these differences would be to say that proximate causes govern the responses of the individual (and his organs) to immediate factors of the environment, while ultimate causes are responsible for the evolution of the particular DNA program of information with which every individual of every species is endowed. The logician will, presumably, be little concerned with these distinctions. Yet, the biologist knows that many heated arguments about the "cause" of a certain biological phenomenon could have been avoided if the two opponents had realized that one of them was concerned with proximate and the other with ultimate causes. I might illustrate this by a quotation from Loeb (1916): "The earlier writers explained the growth of the legs in the tadpole of the frog or toad as a case of adaptation to life on land. We know through Gudernatsch that the growth of the legs can be produced at any time even in the youngest tadpole, which is unable to live on land, by feeding the animal with the thyroid gland."

Let us now get back to the definition of "cause" in formal philosophy and see how it fits with the usual explanatory "cause" of functional and evolutionary biology. We might, for instance, define cause as a nonsufficient condition without which an event would not have happened, or as a member of a set of jointly sufficient reasons without which the event would not happen. Definitions such as these describe causal relations quite adequately in certain branches of biology, particularly in those which deal with chemical and physical unit phenomena. In a strictly formal sense they are also applicable to more complex phenomena, and yet they seem to have little operational value in those branches of biology that deal with complex systems. I doubt that there is a scientist who would question the ultimate causality of all biological phenomena—that is, that a causal explanation can be given for past biological events. Yet such an explanation will often have to be so unspecific and so purely formal that its explanatory value can certainly be challenged. In dealing with a complex system, an explanation can hardly be considered very illuminating that states: "Phenomenon *A* is caused by a complex set of interacting factors, one of which is *b*." Yet often this is about all one can say. We will have to come back to this difficulty in connection with the problem of prediction. However, let us first consider the problem of teleology.

The Problem of Teleology

No discussion of causality is complete which does not come to grips with the problem of teleology. This problem had its beginning with Aristotle's classification of causes, one of the categories being the "final" causes. This category is based on the observation of the orderly and purposive development of the individual from the egg to the "final" stage of the adult. Final cause has been defined as "the cause responsible for the orderly reaching of a preconceived ultimate goal." All goal-seeking behavior has been classified as "teleological," but so have many other phenomena that are not necessarily goal-seeking in nature.

Aristotelian scholars have rightly emphasized that Aristotle—by training and interest—was first and foremost a biologist, and that it was his preoccupation with biological phenomena which dominated his ideas on causes and induced him to postulate final causes in addition to the material, formal, and efficient causes. Thinkers from Aristotle to the present have been challenged by the apparent contradiction between a mechanistic interpretation of natural processes and the seemingly purposive sequence of events in organic growth, reproduction, and animal behavior. Such a rational thinker as Bernard (1885) has stated the paradox in these words.

There is, so to speak, a preestablished design of each being and of each organ of such a kind that each phenomenon by itself depends upon the general forces of nature, but when taken in connection with the others it seems directed by some invisible guide on the road it follows and led to the place it occupies.

We admit that the life phenomena are attached to physicochemical manifestations, but it is true that the essential is not explained thereby; for no fortuitous coming together of physicochemical phenomena constructs each organism after a plan and a fixed design (which are foreseen in advance) and arouses the admirable subordination and harmonious agreement of the acts of life . . . Determinism can never be [anything] but physicochemical determinism. The vital force and life belong to the metaphysical world.

What is the x, this seemingly purposive agent, this "vital force," in organic phenomena? It is only in our lifetime that explanations have been advanced which deal adequately with this paradox.

The many dualistic, finalistic, and vitalistic philosophies of the past merely replaced the unknown x by a different unknown, y or z, for calling an unknown factor *entelechia* or *élan vital* is not an explanation. I shall not waste time showing how wrong most of these past attempts were. Even though some of the underlying observations of these conceptual schemes are quite correct, the supernaturalistic conclusions drawn from these observations are altogether misleading.

Where, then, is it legitimate to speak of purpose and purposiveness in nature, and where it is not? To this question we can now give a firm and unambiguous answer. An individual who—to use the language of the computer—has been "programmed" can act purposefully. Historical processes, however, *cannot* act purposefully. A bird that starts its migration, an insect that selects its host plant, an animal that avoids a predator, a male that displays to a female—they all act purposefully because they have been programmed to do so. When I speak of the programmed "individual," I do so in a broad sense. A programmed computer itself is an "individual" in this sense, but so is, during reproduction, a pair of birds whose instinctive and learned actions and interactions obey, so to speak, a single program.

The completely individualistic and yet also species-specific DNA program of every zygote (fertilized egg cell), which controls the development of the central and peripheral nervous systems, of the sense organs, of the hormones, of physiology and morphology, is the *program* for the behavior computer of this individual.

Natural selection does its best to favor the production of programs guaranteeing behavior that increases fitness. A behavior program that guarantees instantaneous correct reaction to a potential food source, to a potential enemy, or to a potential mate will certainly give greater fitness in the Darwinian sense than a program that lacks these properties. Again, a behavior program that allows for appropriate learning and the improvement of behavior reactions by various types of feedback gives greater likelihood of survival than a program that lacks these properties.

The purposive action of an individual, insofar as it is based on the properties of its genetic code, therefore is no more nor less purposive than the actions of a computer that has been programmed to respond appropriately to various inputs. It is, if I may say so, a purely mechanistic purposiveness.

We biologists have long felt that it is ambiguous to designate such programmed, goal-directed behavior "teleological," because the word *teleological* has also been used in a very different sense, for the final stage in evolutionary adaptive processes (see Essay 3).

The development or behavior of an individual is purposive; natural selection is definitely not. When MacLeod (1957) stated, "What is most challenging about Darwin, however, is his re-introduction of purpose into the natural world," he chose the wrong word. The word *purpose* is singularly inapplicable to evolutionary change, which is, after all, what Darwin was considering. If an organism is well adapted, if it shows superior fitness, this is not due to any purpose of its ancestors or of an outside agency, such as "Nature" or "God," that created a superior design or plan. Darwin "has swept out such finalistic teleology by the front door," as Simpson (1960) has rightly said.

We can summarize this discussion by stating that there is no conflict between causality and teleonomy, but that scientific biology has not found any evidence that would support teleology in the sense of various vitalistic or finalistic theories (Simpson 1960; 1950; Koch 1957). All the so-called teleological systems which Nagel discusses (1961) are actually illustrations of teleonomy.

The Problem of Prediction

The third great problem of causality in biology is that of prediction. In the classical theory of causality the touchstone of the goodness of a causal explanation was its predictive value. This view is still maintained in Bunge's modern classic (1959): "A theory can predict to the extent to which it can describe and explain." It is evident that Bunge is a physicist; no biologist would have made such a statement. The theory of natural selection can describe and explain phenomena with considerable precision, but it cannot make reliable predictions, except through such trivial and meaningless circular statements as, for instance: "The fitter individuals will on the average leave more offspring." Scriven (1959) has emphasized quite correctly that one of the most important contributions to philosophy made by the evolutionary theory is that it has demonstrated the independence of explanation and prediction.[1]

Although prediction is not an inseparable concomitant of causality, every scientist is nevertheless happy if his causal explanations simultaneously have high predictive value. We can distinguish many categories of prediction in biological explanation. Indeed, it is even doubtful how to define "prediction" in biology. A competent zoogeographer can predict with high accuracy what animals will be found on a previously unexplored mountain range or island. A paleontologist likewise can predict with high probability what kind of fossils can be expected in a newly accessible geological horizon. Is such correct guessing of the results of past events genuine prediction? A similar doubt pertains to taxonomic predictions, as discussed in the next paragraph. The term *prediction* is, however, surely legitimately used for future events. Let me give four examples to illustrate the range of predictability.

(1) Prediction in classification. If I have identified a fruit fly as an individual of *Drosophila melanogaster* on the basis of bristle pattern and the proportions of face and eye, I can "predict" numerous structural and behavioral characteristics which I will find if I study other aspects of this individual. If I find new species with the diagnostic key characters of the genus *Drosophila*, I can at once "predict" a whole set of biological properties.

(2) Prediction of most physicochemical phenomena on the molecular level. Predictions of very high accuracy can be made with respect to most biochemical unit processes in organisms, such as metabolic pathways, and with respect to biophysical phenomena in simple systems, such as the action of light, heat, and electricity in physiology.

In examples 1 and 2 the predictive value of causal statements is usually very high. Yet there are numerous other generalizations or causal statements in biology that have low predictive values. The following examples are of this kind.

(3) Prediction of the outcome of complex ecological interactions. The statement, "An abandoned pasture in southern New England will be replaced by a stand of grey birch (*Betula populifolia*) and white pine (*Pinus strobus*)" is often correct. Even more often, however, the replacement may be an almost solid stand of *P. strobus*, or *P. strobus* may be missing altogether

and in its stead will be cherry (*Prunus*), red cedar (*Juniperus virginianus*), maples, sumac, and several other species.

Another example also illustrates this unpredictability. When two species of flour beetles (*Tribolium confusum* and *T. castaneum*) are brought together in a uniform environment (sifted wheat flour), one of the two species will always displace the other. At high temperatures and humidities, *T. castaneum* will win out; at low temperatures and humidities, *T. confusum* will be the victor. Under intermediate conditions the outcome is indeterminate and hence unpredictable (Park 1954).

(4) *Prediction of evolutionary events.* Probably nothing in biology is less predictable than the future course of evolution. Looking at the Permian reptiles, who would have predicted that most of the more flourishing groups would become extinct (many rather rapidly), and that one of the most undistinguished branches would give rise to the mammals? Which student of the Cambrian fauna would have predicted the revolutionary changes in the marine life of the subsequent geological eras? Unpredictability also characterizes small-scale evolution. Breeders and students of natural selection have discovered again and again that independent parallel lines exposed to the same selection pressure will respond at different rates and with different correlated effects, none of them predictable.

As is true in many other branches of science, the validity of predictions for biological phenomena (except for a few chemical or physical unit processes) is nearly always statistical. We can predict with high accuracy that slightly more than 500 of the next 1,000 newborns will be boys. We cannot predict the sex of a particular child prior to conception.

Reasons for Indeterminacy in Biology

Without claiming to exhaust all the possible reasons for indeterminacy, I can list four classes. Although they somewhat overlap each other, each deserves to be treated separately.

(1) *Randomness of an event with respect to the significance of the event.* Spontaneous mutation, caused by an "error" in DNA replication, illustrates this cause for indeterminacy very well. The occurrence of a given mutation is in no way related to the evolutionary needs of the particular organism or of the population to which it belongs. The precise results of a given selection pressure are unpredictable because mutation, recombination, and developmental homeostasis are making indeterminate contributions to the response to this pressure. All the steps in the determination

of the genetic contents of a zygote contain a large component of this type of randomness. What we have described for mutation is also true for crossing over, chromosomal segregation, gametic selection, mate selection, and early survival of the zygotes. Neither underlying molecular phenomena nor the mechanical motions responsible for this randomness are related to their biological effects.

(2) *Uniqueness of all entities at the higher levels of biological integration.* In the uniqueness of biological entities and phenomena lies one of the major differences between biology and the physical sciences. Physicists and chemists often have genuine difficulty in understanding the biologist's stress on the unique, although such an understanding has been greatly facilitated by developments in modern physics. If a physicist says "ice floats on water," his statement is true for any piece of ice and any body of water. The members of a class usually lack the individuality that is so characteristic of the organic world, where each individual is unique; each stage in the life cycle is unique; each population is unique; each species and higher category is unique; each interindividual contact is unique; each natural association of species is unique; and each evolutionary event is unique. Where these statements are applicable to man, their validity is self-evident. However, they are equally valid for all sexually reproducing animals and plants. Uniqueness, of course, does not entirely preclude prediction. We can make many valid statements about the attributes and behavior of man, and the same is true for other organisms. But most of these statements (except for those pertaining to taxonomy) have purely statistical validity. Uniqueness is particularly characteristic for evolutionary biology. It is quite impossible to have for unique phenomena general laws like those that exist in classical mechanics.

(3) *Extreme complexity.* The physicist Elsässer stated in a recent symposium: "[An] outstanding feature of all organisms is their well-nigh unlimited structural and dynamical complexity." This is true. Every organic system is so rich in feedbacks, homeostatic devices, and potential multiple pathways that a complete description is quite impossible. Furthermore, the analysis of such a system would require its destruction and would thus be futile.

(4) *Emergence of new qualities at higher levels of integration.* It would lead too far to discuss in this context the thorny problem of "emergence." All I can do here is to state its principle dogmatically: "When two entities are combined at a higher level of integration, not all the properties of the new entity are necessarily a logical or predictable consequence of the properties of the components." This difficulty is by no means confined to

biology, but it is certainly one of the major sources of indeterminacy in biology. Let us remember that indeterminacy does not mean lack of cause, but merely unpredictability.

All four causes of indeterminacy, individually and combined, reduce the precision of prediction.

One may raise the question at this point whether predictability in classical mechanics and unpredictability in biology are due to a difference of degree or of kind. There is much to suggest that the difference is, in considerable part, merely a matter of degree. Classical mechanics is, so to speak, at one end of a continuous spectrum, and biology is at the other. Let us take the traditional example of the gas laws. Essentially they are only statistically true, but the population of molecules in a gas obeying the gas laws is so enormous that the actions of individual molecules become integrated into a predictable—one might say "absolute"—result. Samples of 5 or 20 molecules would show definite individuality. The difference in the size of the studied "populations" certainly contributes to the difference between the physical sciences and biology.

Conclusions

Let us now return to our initial question and try to summarize some of our conclusions on the nature of the cause-and-effect relations in biology.

(1) Causality in biology is a far cry from causality in classical mechanics.

(2) Explanations of all but the simplest biological phenomena usually consist of sets of causes. This is particularly true for those biological phenomena that can be understood only if their evolutionary history is also considered. Each set is like a pair of brackets which contains much that is unanalyzed and much that can presumably never be analyzed completely.

(3) In view of the high number of multiple pathways possible for most biological processes (except for the purely physicochemical ones) and in view of the randomness of many of the biological processes, particularly on the molecular level (as well as for other reasons), causality in biological systems is not predictive, or at best is only statistically predictive.

(4) The existence of complex programs of information in the DNA of the germ plasm permits teleonomic purposiveness. On the other hand, evolutionary research has found no evidence whatsoever for a "goal-

seeking" of evolutionary lines, as postulated in that kind of teleology which sees "plan and design" in nature. The harmony of the living universe, so far as it exists, is an a posteriori product of natural selection.

Finally, causality in biology is not in real conflict with the causality of classical mechanics. As modern physics has also demonstrated, the causality of classical mechanics is only a very simple, special case of causality. Predictability, for instance, is not a necessary component of causality. The complexities of biological causality do not justify embracing nonscientific ideologies, such as vitalism or finalism, but should encourage all those who have been trying to give a broader basis to the concept of causality.

NOTES

This essay is reprinted from *Science* 134:1501–1506; copyright 1961 by the American Association for the Advancement of Science.

1. Various philosophers have published seeming refutations of Scriven's claims. No doubt my views on prediction at that time were rather simplistic. I have since revised them considerably (Mayr 1982:57–59). The philosophical problems of prediction have been well stated by Grünbaum (1963:114–149).

REFERENCES

Bernard, C. 1885. *Leçons sur les phenomènes de la vie*. Vol. 1.

Bunge, M. 1959. *Causality*. Cambridge, Mass.: Harvard University Press.

Delbrück, M. 1949. A physicist looks at biology. *Trans. Conn. Acad. Arts Sci.* 33:173–190.

Grünbaum, A. 1963. *Induction: Some Current Issues*. Middletown, Conn.: Wesleyan University Press.

Huxley, J. 1960. The openbill's open bill: a teleonomic enquiry. *Zool. Jahrb. Abt. Anat. u. Ontog. Tiere*. 88:9–30.

Koch, L. F. 1957. Vitalistic-mechanistic controversy. *Sci. Monthly* 85(5):245–255.

Loeb, J. 1916. *The Organism as a Whole*. New York: Putnam.

MacLeod, R. B. 1957. Teleology and theory of human behavior. *Science* 125:477–480.

Mayr, E. 1982. *The Growth of Biological Thought*. Cambridge, Mass.: Harvard University Press.

Nagel, E. 1961. *The Structure of Science*. New York: Harcourt, Brace and World.

Park, T. 1954. Experimental studies of interspecies competition II. *Physiol. Zool.* 27:177–238.

Pittendrigh, C. S. 1958. Adaptation, natural selection, and behavior. In A. Roe and G. G. Simpson, eds., *Behavior and Evolution*, pp. 390–416. New Haven: Yale University Press.

Scriven, M. 1959. Explanation and prediction in evolutionary theory. *Science* 130:477–482.

Simpson, G. G. 1950. Evolutionary determinism and the fossil record. *Sci. Monthly* 71(4):262–267.

―――― 1960. The world into which Darwin led us. *Science* 131:966–974.

essay three 🙦

THE MULTIPLE MEANINGS
OF TELEOLOGICAL

TELEOLOGICAL language is frequently used in biology in order to make statements about the functions of organs, about physiological processes, and about the behavior and actions of species and individuals. Such language is characterized by the use of the words *function, purpose*, and *goal*, as well as by statements that something exists or is done *in order to*. Typical statements of this sort are: "One of the functions of the kidneys is to eliminate the end products of protein metabolism," or "Birds migrate to warm climates in order to escape the low temperatures and food shortages of winter." In spite of the long-standing misgivings of physical scientists, philosophers, and logicians, many biologists have continued to insist not only that such teleological statements are objective and free of metaphysical content, but also that they express something important which is lost when teleological language is eliminated from such statements. Recent reviews of the problem in the philosophical literature (Nagel 1961; Beckner 1969; Hull 1973; to cite only a few of a large selection of such publications) concede the legitimacy of some teleological statements but still display considerable divergence of opinion as to the actual meaning of the word *teleological* and the relations between teleology and causality.

This confusion is nothing new and goes back at least as far as Aristotle, who invoked final causes not only for individual life processes (such as development from the egg to the adult) but also for the universe as a whole. To him, as a biologist, the form-giving of the specific life process was the primary paradigm of a finalistic process, but for his epigones the order of the universe and the trend toward its perfection became completely dominant. The existence of a form-giving, finalistic principle in the universe was rightly rejected by Bacon and Descartes, but this, they thought, necessitated the eradication of any and all teleological language, even for biological processes, such as growth and behavior, or in the discussion of adaptive structures.

The history of the biological sciences from the seventeenth to the nineteenth centuries is characterized by a constant battle between extreme mechanists, who explained everything purely in terms of movements and forces, and their opponents, who often went to the opposite extreme of vitalism. After vitalism had been completely routed by the beginning of the twentieth century, biologists could afford to be less self-conscious in their language and, as Pittendrigh (1958) has expressed it, were again willing to say "A turtle came ashore to lay her eggs," instead of saying "She came ashore and laid her eggs." There is now complete consensus among biologists that the teleological phrasing of such a statement does not imply any conflict with physicochemical causality.

Yet, the very fact that teleological statements have again become respectable has helped to bring out uncertainties. The vast literature on teleology is eloquent evidence for the unusual difficulties connected with this subject. This impression is reenforced when one finds how often various authors dealing with this subject have reached opposite conclusions (Braithwaite 1954; Beckner 1969; Canfield 1966; Hull 1973; Nagel 1961). They differ from each other in multiple ways, but most importantly in answering the question: What kind of teleological statements are legitimate and what others are not? Or, what is the relation between Darwin and teleology? David Hull (1973) has recently stated that "evolutionary theory did away with teleology, and that is that," yet, a few years earlier MacLeod (1957) had pronounced "what is most challenging about Darwin, is his reintroduction of purpose into the natural world." Obviously the two authors must mean very different things.

Purely logical analysis helped remarkably little to clear up the confusion. What finally produced a breakthrough in our thinking about teleology was the introduction of new concepts from the fields of cybernetics and new terminologies from the language of information theory. The result was the development of a new teleological language, which claims to be able to take advantage of the heuristic merits of teleological phraseology without being vulnerable to the traditional objections.

Traditional Objections to the Use of Teleological Language

Criticism of the use of teleological language is traditionally based on one or several of the following objections. In order to be acceptable teleological language must be immune to these objections.

(1) *Teological statements and explanations imply the endorsement of unverifiable theological or metaphysical doctrines in science.* This criticism was indeed valid in former times, as for instance when natural theology operated extensively with a strictly metaphysical teleology. Physiological processes, adaptations to the environment, and all forms of seemingly purposive behavior tended to be interpreted as being due to nonmaterial vital forces. This interpretation was widely accepted among Greek philosophers, including Aristotle, who discerned an active soul everywhere in nature. Bergson's (1907) *élan vital* and Driesch's (1909) *Entelechie* are relatively recent examples of such metaphysical teleology. Contemporary philosophers reject such teleology almost unanimously. Likewise, the employment of teleological language among modern biologists does not imply adoption of such metaphysical concepts (see below).

(2) *The belief that acceptance of explanations for biological phenomena that are not equally applicable to inanimate nature constitutes rejection of a physicochemical explanation.* Ever since the age of Galileo and Newton it has been the endeavor of the "natural scientists" to explain everything in nature in terms of the laws of physics. To accept special explanations for teleological phenomena in living organisms implied for these critics a capitulation to mysticism and a belief in the supernatural. They ignored the fact that nothing exists in inanimate nature (except for man-made machines) which corresponds to DNA programs or to goal-directed activities. As a matter of fact, the acceptance of a teleonomic explanation (see below) is in no way in conflict with the laws of physics and chemistry. It is neither in opposition to a causal interpretation, nor does it imply an acceptance of supernatural forces in any way whatsoever.

(3) *The assumption that future goals were the cause of current events seemed in complete conflict with any concept of causality.* Braithwaite (1954) stated the conflict as follows: "In a [normal] causal explanation the explicandum is explained in terms of a cause which either precedes it or is simultaneous with it; in a teleological explanation the explicandum is explained as being causally related either to a particular goal in the future or to a biological end which is as much future as present or past." This is why some logicians up to the present distinguish between causal explanations and teleological explanations.

(4) *Teleological language seemed to represent objectionable anthropomorphism.* The use of terms like *purposive* or *goal-directed* seemed to imply the transfer of human qualities, such as intent, purpose, planning, deliberation, or consciousness, to organic structures and to subhuman forms of life.

Intentional, purposeful human behavior, is, almost by definition, teleological. Yet I shall exclude it from further discussion because use of the words *intentional* or *consciously premeditated*, which are usually employed in connection with such behavior, runs the risk of getting us involved in complex controversies over psychological theory, even though much of human behavior does not differ in kind from animal behavior. The latter, although usually described in terms of stimulus and response, is also highly "intentional," as when a predator stalks his prey or when the prey flees from the pursuing predator. Yet, seemingly "purposive," that is, goal-directed behavior in animals can be discussed and analyzed in operationally definable terms, without recourse to anthropomorphic terms like *intentional* or *consciously*.

As a result of these and other objections, teleological explanations were widely believed to be a form of obscurantism, an evasion of the need for a causal explanation. Indeed some authors went so far as to make statements such as "Teleological notions are among the main obstacles to theory formation in biology" (Lagerspetz 1959:65). Yet, biologists insisted on continuing to use teleological language.

The teleological dilemma, then, consists in the fact that numerous and seemingly weighty objections against the use of teleological language have been raised by various critics, and yet biologists have insisted that they would lose a great deal, methodologically and heuristically, if they were prevented from using such language. It is my endeavor to resolve this dilemma by a new analysis, and particularly by a new classification of the various phenomena that have been traditionally designated as teleological.

The Heterogeneity of Teleological Phenomena

One of the greatest shortcomings of most recent discussions of the teleology problem has been the heterogeneity of the phenomena designated as teleological by different authors. To me it would seem quite futile to arrive at rigorous definitions until the medley of phenomena designated as teleological is separated into more or less homogeneous classes. To accomplish this objective will be my first task.

Furthermore, it only confuses the issue to mingle a discussion of teleology with such extraneous problems as vitalism, holism, or reductionism. Teleological statements and phenomena can be analyzed without reference to major philosophical systems.

By and large all the phenomena that have been designated in the literature as teleological can be grouped into three classes:

(1) Unidirectional evolutionary sequences (progressionism, orthogenesis).

(2) Seemingly or genuinely goal-directed processes.

(3) Teleological systems.

The ensuing discussion will serve to bring out the great differences between these three classes of phenomena.

UNIDIRECTIONAL EVOLUTIONARY SEQUENCES

Already with Aristotle and other Greek philosophers, but increasingly so in the eighteenth century, there was a belief in an upward or forward progression in the arrangement of natural objects. This was expressed most concretely in the concept of the *scala naturae*, the scale of perfection (Lovejoy 1936). Originally conceived as something static (or even descending, owing to a process of degradation), the Ladder of Perfection was temporalized in the eighteenth century and merged almost unnoticeably into evolutionary theories such as that of Lamarck. Progressionist theories were proposed in two somewhat different forms. The steady advance toward perfection was either directed by a supernatural force (a wise creator) or, rather vaguely, by a built-in drive toward perfection. During the flowering of natural theology the "interventionist" concept dominated, but after 1859 it was replaced by so-called orthogenetic theories widely held by biologists and philosophers (see Lagerspetz 1959:11–12 for a short survey). Simpson (1949) refuted the possibility of orthogenesis with particularly decisive arguments. Actually, as Weismann had said long ago (1909), the principle of natural selection solves the origin of progressive adaptation without any recourse to goal-determining forces.

It is somewhat surprising how many philosophers, physical scientists, and occasionally even biologists still flirt with the concept of a teleological determination of evolution. Teilhard de Chardin's (1955) entire dogma is built on such a teleology and so are, as Monod (1971) has stressed quite rightly, almost all of the more important ideologies of the past and present. Even some serious evolutionists play, in my opinion, rather dangerously with teleological language. For instance Ayala (1970:11) says,

> the overall process of evolution cannot be said to be teleological in the sense of directed towards the production of specified DNA codes of information, i.e. organisms. But it is my contention that it can be said to be teleological

in the sense of being directed toward the production of DNA codes of information which improve the reproductive fitness of a population in the environments where it lives. The process of evolution can also be said to be teleological in that it has the potentiality of producing end-directed DNA codes of information, and has in fact resulted in teleologically oriented structures, patterns of behavior, and regulated mechanisms.

To me this seems a serious misinterpretation. If *teleological* means anything, it means *goal-directed*. Yet, natural selection is strictly an a posteriori process which rewards current success but never sets up future goals. No one realized this better than Darwin, who reminded himself "never to use the words higher or lower." Natural selection rewards past events, that is the production of successful recombinations of genes, but it does not plan for the future. This is, precisely, what gives evolution by natural selection its flexibility. With the environment changing incessantly, natural selection—in contradistinction to orthogenesis—never commits itself to a future goal. Natural selection is never goal oriented. It is misleading and quite inadmissible to designate such broadly generalized concepts as survival or reproductive success as definite and specified goals.

The same objection can be raised against similar arguments presented by Waddington (1968:55–56). Like so many other developmental biologists, he is forever looking for analogies between ontogeny and evolution. "I have for some years been urging that quasi-finalistic types of explanations are called for in the theory of evolution as well as in that of development." Natural selection "in itself suffices to determine, to a certain degree, the nature of the end towards which evolution will proceed, it must result in an increase in the efficiency of the biosystem as a whole in finding ways of reproducing itself." He refers here to completely generalized processes, rather than to specific goals. It is rather easy to demonstrate how ludicrous the conclusions are which one reaches by over-extending the concept of goal-direction. For instance, one might say that it is the purpose of every individual to die because this is the end of every individual, or that it is the goal of every evolutionary line to become extinct because this is what has happened to 99.9% of all evolutionary lines that have ever existed. Indeed, one would be forced to consider as teleological even the second law of thermodynamics.

One of Darwin's greatest contributions was to make it clear that teleonomic processes involving only a single individual are of an entirely different nature from evolutionary changes. The latter are controlled by the interplay of the production of variants (new genotypes) and their sorting out by natural selection, a process which is quite decidedly not directed toward a specified distant end. A discussion of legitimately teleological phenomena would be futile unless evolutionary processes are eliminated from consideration.

SEEMINGLY OR GENUINELY GOAL-DIRECTED PROCESSES

Nature (organic and inanimate) abounds in processes and activities that lead to an end. Some authors seem to believe that all such terminating processes are of one kind and "finalistic" in the same manner and to the same degree. Taylor (1950), for instance, if I understand him correctly, claims that all forms of active behavior are of the same kind and that there is no fundamental difference between one kind of movement or purposive action and any other. Waddington (1968) gives a definition of his term *quasi-finalistic* as requiring "that the end state of the process is determined by its properties at the beginning."

Further study indicates, however, that the class of end-directed processes is composed of two entirely different kinds of phenomena. These two types of phenomena may be characterized as follows:

Teleomatic processes in inanimate nature.   Many movements of inanimate objects as well as physicochemical processes are the simple consequences of natural laws. For instance, gravity provides the end-state for a rock which I drop into a well. It will reach its end-state when it has come to rest on the bottom. A red-hot piece of iron reaches its end-state when its temperature and that of its environment are equal. All objects of the physical world are endowed with the capacity to change their state, and these changes follow natural laws. They are end-directed only in a passive, automatic way, regulated by external forces or conditions. Since the end-state of such inanimate objects is automatically achieved, such changes might be designated as *teleomatic*. All teleomatic processes come to an end when the potential is used up (as in the cooling of a heated piece of iron) or when the process is stopped by encountering an external impediment (as a falling stone hitting the ground). Teleomatic processes simply follow natural laws, i.e. lead to a result consequential to concomitant physical forces, and the reaching of their end-state is not controlled by a built-in program. The law of gravity and the second law of thermodynamics are among the natural laws which most frequently govern teleomatic processes.

Teleonomic processes in living nature.   Seemingly goal-directed behavior in organisms is of an entirely different nature from teleomatic processes.

Goal-directed *behavior* (in the widest sense of this word) is extremely widespread in the organic world; for instance, most activity connected with migration, food-getting, courtship, ontogeny, and all phases of reproduction is characterized by such goal orientation. The occurrence of goal-directed processes is perhaps the most characteristic feature of the world of living organisms.

For the last 15 years or so the term *teleonomic* has been used increasingly often for goal-directed processes in organisms. I proposed in 1961 the following definition for this term: "It would seem useful to restrict the term teleonomic rigidly to systems operating on the basis of a program, a code of information" (Mayr 1961). Although I used the term *system* in this definition, I have since become convinced that it permits a better operational definition to consider certain activities, processes (like growth), and active behaviors as the most characteristic illustrations of teleonomic phenomena. I therefore modify my definition, as follows: *A teleonomic process or behavior is one which owes its goal-directedness to the operation of a program.* The term teleonomic implies goal direction. This, in turn, implies a dynamic process rather than a static condition, as represented by a system. The combination of teleonomic with the term system is, thus, rather incongruent (see below).

All teleonomic behavior is characterized by two components. It is guided by a "program," and it depends on the existence of some endpoint, goal, or terminus which is foreseen in the program that regulates the behavior. This endpoint might be a structure, a physiological function, the attainment of a new geographical position, or a "consummatory" (Craig 1918) act in behavior. Each particular program is the result of natural selection, constantly adjusted by the selective value of the achieved endpoint.

My definition of *teleonomic* has been labeled by Hull (1973) as a "historical definition." Such a designation is rather misleading. Although the genetic program (as well as its individually acquired components) originated in the past, this history is completely irrelevant for the functional analysis of a given teleonomic processes. For this it is entirely sufficient to know that a "program" exists which is causally responsible for the teleonomic nature of a goal-directed process. Whether this program had originated through a lucky macromutation (as Richard Goldschmidt had conceived possible) or through a slow process of gradual selection, or even through individual learning or conditioning as in open programs, is quite immaterial for the classification of a process as "teleonomic." On the other hand, a process that does not have a programmed end does not qualify to be designated as teleonomic (see below for a discussion of the concept *program*).

All teleonomic processes are facilitated by specifically selected executive structures. The fleeing of a deer from a predatory carnivore is facilitated by the existence of superlative sense organs and the proper development of muscles and other components of the locomotory apparatus. The proper performing of teleonomic processes at the molecular level is made possible by highly specific properties of complex macromolecules. It would stultify the definition of *teleonomic* if the appropriateness of these facilitating executive structures were made part of it. On the other hand, it is in the nature of a teleonomic program that it does not induce a simple unfolding of some completely preformed gestalt, but that it always controls a more or less complex process which must allow for internal and external disturbances. Teleonomic processes during ontogenetic development, for instance, are constantly in danger of being derailed even if only temporarily. There exist innumerable feedback devices to prevent this or to correct it. Waddington (1957) has quite rightly called attention to the frequency and importance of such homeostatic devices which virtually guarantee the appropriate canalization of development.

We owe a great debt of gratitude to Rosenblueth et al. (1943) for their endeavor to find a new solution for the explanation of teleological phenomena in organisms. They correctly identified two aspects of such phenomena: (1) that they are seemingly purposeful, being directed toward a goal, and (2) that they consist of active behavior. The background of these authors was in the newly developing field of cybernetics, and it is only natural that they should have stressed the fact that goal-directed behavior is characterized by mechanisms which correct errors committed during the goal seeking. They considered the negative feedback loops of such behavior as its most characteristic aspect and stated "teleological behavior thus becomes synonymous with behavior controlled by negative feedback." This statement emphasizes important aspects of teleological behavior, yet it misses the crucial point: *The truly characteristic aspect of goal-seeking behavior is not that mechanisms exist which improve the precision with which a goal is reached, but rather that mechanisms exist which initiate, i.e. "cause" this goal-seeking behavior.* It is not the thermostat which determines the temperature of a house, but the person who sets the thermostat. It is not the torpedo which determines toward what ship it will be shot and at what time, but the naval officer who releases the torpedo. Negative feedback only improves the precision of goal-seeking, but does not determine it.

Feedback devices are only executive mechanisms that operate during the translation of a program.

Therefore it places the emphasis on the wrong point to define teleonomic processes in terms of the presence of feedback devices. They are mediators of the program, but as far as the basic principle of goal achievement is concerned, they are of minor consequence.

Recent usages of the term teleonomic. The term *teleonomic* was introduced into the literature by Pittendrigh (1958:394) in the following paragraph:

> Today the concept of adaptation is beginning to enjoy an improved respectability for several reasons: it is seen as less than perfect; natural selection is better understood; and the engineer-physicist in building end-seeking automata has sanctified the use of teleological jargon. It seems unfortunate that the term 'teleology' should be resurrected and, as I think, abused in this way. The biologists' long-standing confusion would be more fully removed if all end-directed systems were described by some other term, like 'teleonomic', in order to emphasize that the recognition and description of end-directedness does not carry a commitment to Aristotelian teleology as an efficient [sic] causal principle.

It is evident that Pittendrigh had the same phenomena in mind as I do,[1] even though his definition is rather vague and his placing the term *teleonomic* in opposition to Aristotle's *teleology* is unfortunate. As we shall see below, most of Aristotle's references to end-directed processes refer precisely to the same things which Pittendrigh and I would call teleonomic (see also Delbrück 1971; Gotthelf 1976).

Other recent usages of the term that differ from my own definition are the following. B. Davis (1961), believing that the term denotes "the development of valuable structures and mechanisms" as a result of natural selection, uses the term virtually as synonymous with adaptiveness. The same is largely true for Simpson (1958:520–521), who sees in *teleonomic* the description for a system or structure which is the product of evolution and of selective advantage:

> The words 'finalistic' and 'teleological' have, however, had an unfortunate history in philosophy which makes them totally unsuitable for use in modern biology. They have too often been used to mean that evolution as a whole has a predetermined goal, or that the utility of organization in general is with respect to man or to some supernatural scheme of things. Thus these terms may implicitly negate rather than express the biological conclusion that organization in organisms is with respect to utility to each separate species at the time when it occurs, and not with respect to any other species

or any future time. In emphasis of this point of view, Pittendrigh [above] suggests that the new coinage 'teleonomy' be substituted for the debased currency of teleology.

Monod (1971) likewise deals with teleonomy as if the word simply meant adaptation. It is not surprising therefore that Monod considers teleonomy "to be a profoundly ambiguous concept." Furthermore, says Monod, all functional adaptations are "so many aspects or fragments of a unique primary project which is the preservation and multiplication of the species." He finally completes the confusion by choosing "to define the essential teleonomic project as consisting in the transmission from generation to generation of the invariance content characteristic of the species. All these structures, all the performances, all the activities con-tributing to the success of the essential project will hence be called teleonomic."

What Monod calls "teleonomic" I would designate as of "selective value." Under these circumstances it is not surprising when Ayala (1970) claims that the term *teleonomy* had been introduced into the philosophical literature in order "to explain adaptation in nature as the result of natural selection." If this were indeed true, and it is true of Simpson's and Davis's cited definitions, the term would be quite unnecessary. Actually, there is nothing in my 1961 account which would support this interpretation, and I know of no other term that would define a goal-directed activity or behavior that is controlled by a program. Even though Pittendrigh's discussion of *teleonomic* rather confused the issue and has led to the sub-sequent misinterpretations, he evidently had in mind the same processes and phenomena which I denoted as *teleonomic*. It would seem well worth-while to retain the term in the more rigorous definition, which I have now given.

The Meaning of the Word "Program"

The key word in my definition of *teleonomic* is *program*. Someone might claim that the difficulties of an acceptable definition for teleological lan-guage in biology had simply been transferred to the term *program*. This is not a legitimate objection, because it fails to recognize that, regardless of its particular definition, a program is (1) something material, and (2) it exists prior to the initiation of the teleonomic process. Hence, it is consistent with a causal explanation.

Nevertheless, it must be admitted that the concept of a program is so new that the diversity of meanings of the term has not yet been fully explored. The term is taken from the language of information theory. A computer may act purposefully when given appropriate programmed instructions. Tentatively, *program* might be defined as *coded or prearranged information that controls a process (or behavior) leading it toward a given end.* As Raven (1960) has remarked correctly, the program contains not only the blueprint but also the instructions of how to use the information of the blueprint. In the case of a computer program or of the DNA of the cell nucleus, the program is completely separated from the executive machinery. In the case of most man-made automata, the program is part of the total machinery.

My definition of program is deliberately chosen in such a way as to avoid drawing a line between seemingly "purposive" behavior in organisms and in man-made machines. The simplest program is perhaps the weight inserted into loaded dice or attached to a "fixed" number wheel so that they are likely to come to rest at a given number. A clock is constructed and programmed in such a way as to strike at the full hour. Any machine which is programmed to carry out goal-directed activities is capable of doing this "mechanically."

The programs which control teleonomic processes in organisms are either entirely laid down in the DNA of the genotype (closed programs) or are constituted in such a way that they can incorporate additional information (open programs) (Mayr 1964), acquired through learning, conditioning, or other experiences. Most behavior, particularly in higher organisms, is controlled by such open programs.

Open programs are particularly suitable to demonstrate the fact that the mode of acquisition of a program is an entirely different matter from the teleonomic nature of the behavior controlled by the program. Nothing could be more purposive, more teleonomic than much of the escape behavior in many prey species (in birds and mammals). Yet, in many cases the knowledge of which animals are dangerous predators is learned by the young who have an open program for this type of information. In other words, this particular information was not acquired through selection and yet it is clearly in part responsible for teleonomic behavior. Many of the teleonomic components of the reproductive behavior (including mate selection) of species which are imprinted for mate recognition is likewise only partially the result of selection. The history of the acquisition of a program, therefore, cannot be made part of the definition of teleonomic.

The origin of a program is quite irrelevant for the definition. It can be the product of evolution, as are all genetic programs, or it can be the acquired information of an open program, or it can be a man-made device. Anything that does *not* lead to what is at least in principle a predictable goal does not qualify as a program. Even though its current gene pool sets severe limits on a species' future evolution, the course of that evolution is largely controlled by the changing constellation of selection pressures and is therefore not predictable. It is not programmed inside the contemporary gene pool.

The entire concept of a program of information is so new that it has received little attention from philosophers and logicians. My tentative analysis may, therefore, require considerable revision when subjected to further scrutiny.

HOW DOES THE PROGRAM OPERATE?

The philosopher may be willing to accept the assertion of the biologist that a program directs a given teleonomic behavior, but he would also like to know how the program performs this function. Alas, all the biologist can tell him is that the study of the operation of programs is the most difficult area of biology. For instance, the translation of the genetic program into growth processes and into the differentiation of cells, tissues, and organs is at the present time the most challenging problem of developmental biology. The number of qualitatively different cells in a higher organism almost surely exceeds one billion. Even though all (or most) have the same gene complement, they differ from each other owing to differences in the repression and derepression of individual gene loci and owing to differences in their cellular environment. It hardly needs stressing how complex the genetic program must be, to be able to give the appropriate signals to each cell lineage in order to provide it with the mixture of molecules which it needs to carry out its assigned tasks.

Similar problems arise in the analysis of goal-directed behavior. The number of ways in which a program may control a goal-directed behavior activity is legion. It differs from species to species. Sometimes the program is largely acquired by experience; in other cases it may be almost completely genetically fixed. Sometimes the behavior consists of a series of steps, each of which serves as reenforcement for the ensuing steps; in other cases the behavior, once initiated, seems to run its full course without need for any further input. Feedback loops are sometimes important, but their presence cannot be demonstrated in other kinds of behavior. Again, as in developmental biology, much of the contemporary research in behavioral biology is devoted to the nature and the operation of the programs which control behavior and more specifically teleonomic behavior sequences (Hinde and Stevenson 1970). Almost any statement one might make is apt to be challenged by one or the other school of psychologists and geneticists. It is, however, safe to state that the translation of programs into teleonomic behavior is greatly affected both by sensory inputs and by internal physiological (largely hormonal) states.

Teleological Systems

The word *teleological,* in the philosophical literature, is particularly often combined with the term *system.* Is it justified to speak of *teleological systems?* Analysis shows that this combination leads to definitional difficulties.

The Greek word *telos* means end or goal. Teleological means end-directed. To apply the word *teleological* to a goal-directed behavior or process would seem quite legitimate. I am perhaps a purist, but it bothers me to apply the word teleological, that is *end-directed,* to a stationary system. Any phenomenon to which we can refer as teleomatic or teleonomic represents a movement, a behavior, or a process that is goal-directed by having a determinable end. This is the core concept of teleological, the presence of a *telos* (an end) toward which an object or process moves. Rosenblueth et al. (1943) have correctly stressed the same point.

Extending the term teleological to cover also static systems leads to contradictions and illogicalities. A torpedo that has been shot off and moves toward its target is a machine showing teleonomic behavior. But what justifies calling a torpedo a teleological system when, with hundreds of others, it is stored in an ordnance depot? Why should the eye of a sleeping person be called a teleological system? It is not goal-directed at anything. Part of the confusion is due to the fact that the term *teleological* system has been applied to two only partially overlapping phenomena. One comprises systems that are potentially able to perform teleonomic actions, like a torpedo. The other comprises systems that are well adapted, like the eye. To refer to a phenomenon in this second class as teleological, in order to express its adaptive perfection, reflects just enough of the old idea of evolution leading to a steady progression in adaptation and perfection to make me uneasy. What is the telos toward which the teleological system moves?

The source of the conflict seems to be that *goal-directed,* in a more or less straightforward literal sense, is not necessarily the same as purposive. Completely stationary systems can be functional or purposive, but they cannot be goal-directed in any literal sense. A poison on the shelf has the potential of killing somebody, but this inherent property does not make it a goal-directed object. Perhaps this difficulty can be resolved by making a terminological distinction between functional properties of systems and strict goal-directedness, that is, teleonomy of behavioral or other processes. However, since one will be using so-called teleological language in both cases, one might subsume both categories under teleology.

R. Munson (1971) has recently dealt with such adaptive systems. In particular, he studied all those explanations that deal with aspects of adaptation but are often called *teleological*. He designates sentences "adaptational sentences," when they contain the terms *adaptation, adaptive,* or *adapted*. In agreement with the majority opinion of biologists, he concludes that "adaptational sentences do not need [to] involve reference to any purpose, final cause, or other non-empirical notion in order to be meaningful." Adaptational sentences simply express the conclusion that a given trait, whether structural, physiological, or behavioral, is the product of the process of natural selection and thus favors the perpetuation of the genotype responsible for this trait. Furthermore, adaptation is a heuristic concept because it demands an answer to the question, in what way the trait adds to the probability of survival and does so more successfully than an alternate conceivable trait. To me, it is misleading to call adaptational statements teleological. "Adapted" is an *a posteriori* statement and it is only the success (statistically speaking) of the owner of an adaptive trait which proves whether the trait is truly adaptive (= contributes to survival) or is not. Munson summarizes the utility of adaptational language in the sentence: "To show that a trait is adaptive is to present a phenomenon requiring explanation, and to provide the explanation is to display the success of the trait as the outcome of selection" (p. 214). The biologist fully agrees with this conclusion. Adaptive means simply: being the result of natural selection.

Many adaptive systems—for instance, all components of the locomotory and the central nervous systems—are capable of taking part in teleonomic processes or teleonomic behavior. However, it only obscures the issue when one designates a system teleological or teleonomic because it provides executive structures of a teleonomic process. Is an inactive, not-programmed computer a teleological system? What "goal" or "end" is it displaying during this period of inactivity? To repeat, one runs into serious logical difficulties when one applies the term *teleological* to static systems (regardless of their potential) instead of to processes. Nothing is lost and much is to be gained by not using the term teleological too freely and for too many rather diverse phenomena.

It may be necessary to coin a new term for systems which have the potential of displaying teleonomic behavior. The problem is particularly acute for biological organs which are capable of carrying out useful functions, such as pumping by the heart or filtration by the kidney. To some extent this problem exists for any organic structure, all the way down to the macromolecules which are capable of carrying out autonomously certain highly specific functions owing to their uniquely specific structure. It is this which induced Monod (1971) to call them teleonomic systems. Similar considerations have induced some authors, erroneously in my opinion, to designate a hammer as a teleological system, because it is designed to hit a nail (a rock, not having been so designed, but serving the same function not qualifying!).

The philosophical complexity of the logical definition of *teleological* in living systems is obvious. Let me consider a few of the proximate and ultimate causes (Mayr 1961), to bring out some of the difficulties more clearly. The functioning of these systems is the subject matter of regulatory biology, which analyzes proximate causes. Biological systems are complicated steady-state systems, replete with feedback devices. There is a high premium on homeostasis, on the maintenance of the *milieu interieur*. Since most of the processes performed by these systems are programmed, it is legitimate to call them teleonomic processes. They are "end-directed" even though very often the "end" is the maintenance of the status quo. There is nothing metaphysical in any of this because, so far as these processes are accessible to analysis, they represent chains of causally interrelated stimuli and reactions, of inputs and of outputs.

The ultimate causes for the efficiency and seeming purposefulness of these living systems were explained by Darwin in 1859. The adaptiveness of these systems is the result of millions of generations of natural selection. This is the mechanistic explanation of adaptiveness, as was clearly stated by Sigwart (1881).

Proximate and ultimate causes must be carefully separated in the discussion of teleological systems (Mayr 1961). A system is capable of performing teleonomic processes because it was programmed to function in this manner. The origin of the program that is responsible for the adaptiveness of the system is an entirely independent matter. It obscures definitions to combine current functioning and history of origin in a single explanation.

The Heuristic Nature of Teleonomic Language

Teleological language has been employed in the past in many different senses, some of them legitimate and some of them not. When the distinctions outlined in my survey above are made, the teleological *Frages-*

tellung is a most powerful tool in biological analysis. Its heuristic value was appreciated already by Aristotle and Galen, but neither of them fully understood why this approach is so important. Questions which begin with "What?" and "How?" are sufficient for explanation in the physical sciences. In the biological sciences no explanation is complete until a third kind of question has been asked: "Why?" It is Darwin's evolutionary theory which necessitates this question: No feature (or behavioral program) of an organism ordinarily evolves unless this is favored by natural selection. It must play a role in the survival or in the reproductive success of its bearer. Accepting this premise, it is necessary for the completion of causal analysis to ask for any feature, why it exists, that is, what its function and role in the life of the particular organism is.

The philosopher Sigwart (1881) recognized this clearly:

> A teleological analysis implies the demand to follow up causations in all directions by which the purpose [of a structure or behavior] is effected. It represents a heuristic principle because when one assumes that each organism is well adapted it requires that we ask about the operation of each individual part and that we determine the meaning of its form, its structure, and its chemical characteristics. At the same time it leads to an explanation of correlated subsidiary consequences which are not necessarily part of the same purpose but which are inevitable by-products of the same goal-directed process.

The method, of course, was used successfully long before Darwin. It was Harvey's question concerning the reason for the existence of valves in the veins that made a major, if not the most important, contribution to his model of the circulation of blood. The observation that during mitosis the chromatic material is arranged in a single linear thread led Roux (1883) to question why such an elaborate process had evolved rather than a simple division of the nucleus into two halves. He concluded that the elaborate process made sense only if the chromatin consisted of an enormous number of qualitatively different small particles and that their equal division could be guaranteed only by lining them up linearly. The genetic analyses of chromosomal inheritance during the next sixty years were, in a sense, only footnotes to Roux's brilliant hypothesis. These cases demonstrate most convincingly the enormous heuristic value of the teleonomic approach. It is no exaggeration to claim that most of the greatest advances in biology were made possible by asking "Why?" questions. This demands asking for the selective significance of every aspect of the phenotype. The former idea that many if not most characters of organisms are "neutral," that is, that they evolved simply as accidents of evolution, has been refuted again and again by more detailed analysis. It is the question as to the "why?" of such structures and behaviors which initiates such analysis. Students of behavior have used this approach in recent years with great success. It has, for example, led to questions concerning the information content of individual vocal and visual displays (Smith 1969; Hinde 1972).

As soon as one accepts the simple conclusion that the totality of the genotype is the result of past selection, and that the phenotype is a product of the genotype (except for the open portions of the program that are filled in during the lifetime of the individual), it becomes one's task to ask about any and every component of the phenotype what its particular functions and selective advantages are.

It is now quite evident why all past efforts to translate teleonomic statements into purely causal ones were such a failure: A crucial portion of the message of a teleological sentence is invariably lost in the translation. Let us take, for instance the sentence: "The Wood Thrush migrates in the fall into warmer countries *in order to* escape the inclemency of the weather and the food shortages of the northern climates." If we replace the words "in order to" by "and thereby," we leave the important question unanswered as to *why* the Wood Thrush migrates. The teleonomic form of the statement implies that the goal-directed migratory activity is governed by a program. By omitting this important message, the translated sentence is greatly impoverished as far as information content is concerned, without gaining in causal strength. The majority of modern philosophers are fully aware of this and agree that "cleaned-up" sentences are not equivalent to the teleological sentences from which they were derived (Ayala 1970; Beckner 1969).

One can go a step further. Teleonomic statements have often been maligned as stultifying and obscurantist. This is simply not true. Actually the nonteleological translation is invariably a meaningless platitude, while it is the teleonomic statement which leads to biologically interesting inquiries.

Aristotle and Teleology

No other ancient philosopher has been as badly misunderstood and mishandled by posterity as Aristotle. His interests were primarily those of a biologist, and his philosophy is bound to be misunderstood if this fact is ignored. Neither Aristotle nor most of the other ancient philosophers made a sharp distinction between the living world and the inanimate. They saw something like life or soul even in the inorganic world. If one can discern purposiveness and goal direction in the world of organisms, why not consider the order of the Kosmos-as-a-whole also as due to final causes, that is, as due to a built-in teleology? As Ayala (1970) said quite rightly, Aristotle's "error was not that he used teleological explanations in biology, but that he extended the concept of teleology to the nonliving world."[2] Unfortunately, it was this latter teleology which was first encountered during the scientific revolution of the sixteenth and seventeenth centuries (and at that in the badly distorted interpretations of the scholastics). This is one of the reasons for the violent rejection of Aristotle by Bacon, Descartes, and their followers.

Although the philosophers of the last forty years acknowledge quite generally the inspiration which Aristotle derived from the study of living nature, they still express his philosophy in words taken from the vocabulary of Greek dictionaries that are hundreds of years old. The time would seem to have come for the translators and interpreters of Aristotle to use a language appropriate to his thinking, that is, the language of biology, and not that of the sixteenth-century humanists. Delbrück (1971) is entirely right when insisting that it is quite legitimate to employ modern terms like *genetic program* for *eidos* where this helps to elucidate Aristotle's thoughts. One of the reasons why Aristotle has been so consistently misunderstood is that he uses the term *eidos* for his form-giving principle, and everybody took it for granted that he had something in mind similar to Plato's concept of *eidos*. Yet, the context of Aristotle's discussions makes it abundantly clear that *his eidos* is something totally different from Plato's *eidos* (I myself did not understand this until recently). Aristotle saw with extraordinary clarity that it made no more sense to describe living organisms in terms of mere matter than to describe a house as a pile of bricks and mortar. Just as the blueprint used by the builder determines the form of a house, so does the *eidos* (in its Aristotelian definition) give the form to the developing organism, and this *eidos* reflects the terminal *telos* of the full-grown individual. There are numerous discussions in many of Aristotle's works reflecting the same ideas. They can be found in the *Analytika* and in the *Physics* (Book II), but particularly in the *Parts of Animals* and in the *Generation of Animals*. Much of Aristotle's discussion becomes remarkably modern if one inserts modern terms to replace obsolete sixteenth- and seventeenth-century vocabulary. There is, of course, one major difference between Aristotle's interpretation and the modern one. Aristotle could not actually *see* the form-giving principle (which, after all, was not fully understood until 1953) and assumed therefore that it had to be something immaterial. When he said "Now it may be that the Form (*eidos*) of any living creature is soul, or some part of soul, or something that involves soul" (P. A. 641a 18), it must be remembered that Aristotle's psyche (soul) was something quite different from the conception of soul later developed in Christianity. Indeed, the properties of "soul" were to Aristotle something subject to investigation. Since the modern scientist does not actually "see" the genetic program of DNA either, it is for him just as invisible for all practical purposes as it was for Aristotle. Its existence is inferred, as it was by Aristotle.

As Delbrück (1971) points out correctly, Aristotle's principle of the *eidos* being an "unmoved mover" is one of the greatest conceptual innovations. The physicists were particularly opposed to the existence of such a principle by

> having been blinded for 300 years by the Newtonian view of the world. So much so, that anybody who held that the mover had to be in contact with the moved and talked about an 'unmoved mover', collided head on with Newton's dictum: *action equals reaction.* Any statement in conflict with this axiom of Newtonian dynamics could only appear to be muddled nonsense, a leftover from a benighted prescientific past. And yet, 'unmoved mover' perfectly describes DNA: it acts, creates form and development, and is not changed in the process (Delbrück, 1971:55).

As I stated above, the existence of teleonomic programs—unmoved movers—is one of the most profound differences between the living and the inanimate world, and it is Aristotle who first postulated such a causation.

Kant and Cosmic Teleology

From the Greeks on, there was a widespread belief that everything in nature and its processes has a purpose, a predetermined goal. Those who held this conviction saw their views confirmed not only by the scala naturae but also by the complete unity and harmony of nature in its manifold adaptations. Those who opposed such a teleology, the strict mechanists, were only a small minority in the seventeenth and eighteenth centuries, and in fact until 1859. This viewpoint became particularly important as it came to be recognized in the eighteenth century that the world had not been created merely 6,000 years ago and as something perfect, but had evolved gradually. The deists explained this by saying that God had not only created the world but had also given it a set of laws which regulate everything that goes on in the world, and these laws would lead the world to ever-greater perfection. Nowhere else was the belief in such a cosmic teleology as strong as in Germany, from Leibniz to Herder and Kant.

Kant was a strict mechanist with respect to inanimate nature, but he saw teleological forces acting in all processes of the living world. I am sure we are all unanimous in our admiration for Kant. He was one of the very great in the history of humanity. What I admire particularly in Kant is how successfully he kept up with the advances in the sciences in spite of his isolation in Königsberg. Yet his familiarity with the work of Buffon, Haller, Wolff, and Blumenbach could get him only as far as these authors had gone. With the exception of the solution of some problems of physiology and a rather primitive systematics and anatomy, it must be realized that biology at the time when Kant wrote his *Critique of Judgment* (1790) and his *Opus Postumum* was virtually a total terra ignota. The birth of scientific biology falls in the period from 1828 to 1865, characterized by the names von Baer, Schwann, Schleiden, Liebig, Bernard, Virchow, Darwin, and Mendel.

When a philosopher attempts today to refute Darwin by citing Kant's opinions, this is no less an anachronism than someone rejecting Einstein, Bohr, and Heisenberg because their theories are in conflict with those of Galileo and Newton. In Kant's day, virtually none of the information needed to establish a theory of evolution was available. Indeed, I am rather convinced that if he had known all that Darwin knew in 1859, Kant would have been the first to draw the necessary consequences, as Darwin did.

The primitive condition of biology in Kant's day made it impossible to explain the riddles of living nature, yet since Kant was unwilling to exclude life from his philosophy of science, he accepted provisionally a teleological explanation. At that time this was perhaps the only possible solution. However, when now, 200 years later, a philosopher bases his explanation of the phenomena of life on Kant and rejects everything that we have learned since his time, such anachronism can hardly be condoned.

The first step in a modern analysis consists in partitioning seemingly teleological phenomena into the four classes I have described. Three of them, teleomatic and teleonomic processes, as well as adapted systems, can be explained scientifically, as shown above. A fourth category consists of cosmic teleology, that is, the belief that there is a force immanent in the world which guides it toward a final goal or at least to ever greater perfection. In contrast with the other three forms of teleology, the existence of such a cosmic teleology cannot be documented in any way whatsoever. Neither have any natural laws been found which could effect such a teleology nor a proper program that would be able to do it, and for this reason cosmic teleology must be excluded from science. Indeed, I do not know of a single modern scientist who believes in it.

Darwin had solved Kant's great puzzle, who had not dared "to hope that someday a Newton would be born who could explain the production of a stem of grass on the basis of natural laws which no purpose has ordered." Naturally, even today we cannot explain organic development on the basis of purely physical laws such as they were known to Kant. Some philosophers have declined to call Darwin the Newton of biology with the comment that not even a molecular biologist could "produce stems of grass." Anyone who makes that demand forgets that Newton likewise to the very end of his life never succeeded in "producing" suns and planets. All Kant demanded was an explanation and not a production of natural phenomena.

From the time of the Greek philosophers until the middle of the last century, a controversy existed between a teleological and a purely causal mechanical explanation of the world. Sometimes one, sometimes the other, of these two views seemed to be victorious. Or else one could deal with the problem as was done by Kant and be a strict mechanist with respect to inanimate nature but a teleologist in the treatment of the world of life. One of the reasons why Darwin was attacked so vigorously was that his theory of selection made a belief in a cosmic teleology unnecessary. It was their unshakable belief in teleology that induced Karl Ernst von Baer and other of Darwin's contemporaries to attack the theory of selection so temperamentally. Indeed, the belief in a teleological force in nature was so firmly anchored in the thinking of many that even among the evolutionists this belief had more followers in the first 80 years after 1859 than did Darwin's theory of selection.

As Max Delbrück has emphasized so rightly, teleonomic and adaptational phenomena have a history and cannot be explained directly through a strictly causal mechanical explanation, as is possible for processes in inanimate nature.

A comparison of Kant's discussion with our new concepts provides a most informative insight into the role of scientific advances in the formulation of philosophical problems. Equally informative is a comparison of three treatments of Kant's teleology, roughly separated by 50-year intervals (Stadler 1874; Ungerer 1922; and McFarland 1970).

Conclusions

(1) The use of so-called teleological language by biologists is legitimate; it neither implies a rejection of physicochemical explanation nor does it imply noncausal explanation.

(2) The terms *teleology* and *teleological* have been applied to highly diverse

phenomena. An attempt is made by the author to group these into more or less homogeneous classes.

(3) It is illegitimate to describe evolutionary processes or trends as goal-directed (teleological). Selection rewards past phenomena (mutation, recombination, etc.), but does not plan for the future, at least not in any specific way.

(4) Processes (behavior) whose goal-directedness is controlled by a program may be referred to as *teleonomic.*

(5) Processes which reach an end state caused by natural laws (e.g., gravity, first law of thermodynamics) but not by a program may be designated as *teleomatic.*

(6) Programs are in part or entirely the product of natural selection.

(7) The question of the legitimacy of applying the term *teleological* to stationary functional or adaptive systems requires further analysis.

(8) Teleonomic (that is, programmed) behavior occurs only in organisms (and man-made machines) and constitutes a clear-cut difference between the levels of complexity in living and in inanimate nature.

(9) Teleonomic explanations are strictly causal and mechanistic. They give no comfort to adherents of vitalistic concepts.

(10) The heuristic value of the teleological *Fragestellung* makes it a powerful tool in biological analysis, from the study of the structural configuration of macromolecules up to the study of cooperative behavior in social systems.

POSTSCRIPT

There have been numerous developments in the analysis of the meaning of *teleological* since this essay was first published (1974). I will not deal with all those papers and books in the purely philosophical literature in which the four meanings of teleological are still merrily intermingled. Instead, I will call attention to two aspects of the problem which require comment.

Aristotle has been traditionally misinterpreted as a cosmic teleologist. Modern students of Aristotle are in agreement that he was not (Gotthelf 1976; Nussbaum 1978; Sorabji 1980; Balme 1981). As already understood by Delbrück (1971), Aristotle's concept of the *eidos,* in the context of ontogenetic development, is in some respects remarkably similar to the modern concept of the genetic program. What the standard histories of philosophy write about Aristotle's teleology is unfortunately largely wrong, and must be ignored. I myself misinterpreted Aristotle before I became acquainted with the modern literature.

The other aspect to be discussed is the reaction of one philosopher to my use of the concept *program.* In 1977 the late Ernest Nagel, distinguished philosopher at Columbia University, published an essay "Teleology revisited: goal directed processes in biology," the first part of which consisted of a rather adverse critique of my treatment of teleology.

Not surprisingly, Nagel questioned particularly those of my proposals that he considered to be in conflict with the logical-positivistic tradition. This is not the place for a detailed analysis of Nagel's propositions and criticisms, particularly since we both agree in a total rejection of cosmic teleology and of nonempirical explanations. He is worried about the predictability of teleonomic explanations and about the logical structure of sentences as used by evolutionary biologists. Most of all, however, Nagel, who was perhaps the most consistent reductionist among recent philosophers, is critical of the concept *program.* He finds my definition of program unacceptable, because historically evolved genetic programs do not exist in the inanimate world. Their recognition would automatically concede that not all biological phenomena can be reduced (without residue) to physical processes.

In order to invalidate the claim that programs, as defined by me, are a special property of the world of life, Nagel attempts to demonstrate the existence of programs in the inanimate world. He suggests that the radioactive decay of a chunk of uranium could be considered also to be controlled by a program. This claim is simply wrong. Radioactive decay is controlled by laws and not by any particular program; it obeys the same laws any time any where. Programs are highly specific and often unique. The importance of the concept *program* is increasingly being recognized. I refer in particular to Beniger (1986).

The objective of my own analysis (see above) had been to show that such a heterogeneous aggregate of phenomena as was discussed by philosophers under the label "teleological" could never be elucidated simply by logical analysis, as had been tried by so many philosophers (including Nagel 1961). The first step in my analysis had been to sort these phenomena and processes into homogeneous classes. I recognized four of such classes, three of which have a sound empirical basis. By contrast, there is no evidence for the existence of a fourth class, cosmic teleology, as had already been shown by Darwin (see Essay 14). One further class, that of teleomatic processes (see above), is of no special interest to philosophers of biology.

Nagel agrees with the biologist that this leaves two classes of phenomena in the biological realm to which the term *teleological* has been applied: *goal-directed activities,* called by me teleonomic, and *functional activities* of organs or structures, called by me the activities of adapted systems. Unfortunately, in his account, he frequently confounds the two, leaving him at times greatly puzzled. He refers to the "goal" of certain endocrine tissues to maintain levels of blood sugar, and to the "function" of the kidneys to eliminate waste products from the blood.

Actually, the two processes are equivalent, and endocrine tissues, being "systems," have no goal. In his account of the goal of a rabbit's flight from a hound, he gets hopelessly entangled and winds up with the statement: "Survival itself does not appear to have any function." This ignores that in the genetic as well as somatic program of rabbits, numerous subprograms exist dealing with predator thwarting. And if these did not have any survival value, their origin would not have been favored by natural selection.

Nagel's arguments are largely based on the principles of the "received view." He is reluctant to accept programs until they are fully reduced to "the components and structures of DNA molecules." For him, as logician, it is, for instance, apparently important whether or not the words *goal-directed* occur in the explanation. He is trying by every means to avoid the adoption of the concept *program* because through such an avoidance "explanations of goal-directed processes in biology are in principle possible, whose structure is like the structure of explanations in the physical sciences in which teleological notions have no place." In other words, Nagel would translate the sentence, "The turtle swims to the shore to lay her eggs" into the sentence, "The turtle swims to the shore and lays her eggs." Then, we would be back precisely to the point where Pittendrigh (1958) found himself when he introduced the term *teleonomic* in order to restore meaning to a biologically meaningless sentence.

In the end Nagel, ruefully but honestly, comes to the conclusion that "none of these conclusions concerning the character of explanations of goal and function ascriptions shows that the laws and theories of biology are reducible to those of the physical sciences" (p. 300). He also agrees with the nonvitalistic biologists, from Aristotle to the present, "that teleological concepts and teleological explanations [except cosmic teleology—E.M.] do not constitute a species of intellectual constructions that are inherently obscure and should therefore be regarded with suspicion" (p. 301).

There has been one recent development in my thinking that might facilitate a certain degree of rapproachment between the traditional philosopher and the modern evolutionary biologist. It deals with the properties of programs. I recognized two kinds of progams, *closed ones* that are entirely coded in the DNA of the genotype, and *open programs* that can incorporate additional information. Although this is a useful classification for certain purposes, particularly for the discrimination between innate and learned behaviors, it fails to meet the needs of many explanations in developmental biology. Here, it is more informative to speak of *genetic* and *somatic* programs. For instance, when a turkey gobbler displays to a hen, his display movements are not directly controlled by the DNA in his cell nuclei, but rather by a somatic program in his central nervous system. To be sure this neuronal program was laid down during development under the control of instructions from the genetic program. But it is now an independent somatic program.

All adapted systems of an organism can be considered to be somatic programs.

If this were accepted, then one could call the functional activities of adapted systems teleonomic activities.

The recognition of somatic programs is important in behavioral biology, but it is even more important in developmental biology, where many larval or embryonic structures seem to serve as somatic programs for the later stages of development. This has been understood by embryologists since Kleinenberg (1886) and probably earlier. Most of the embryonic structures that have been cited as evidence for recapitulation, like the gill arches of tetrapod embryos, are presumably somatic programs. They cannot be removed by natural selection without seriously interfering with the subsequent development.

As I have said elsewhere in this volume, acceptance of the term *program* from informatics is not anthropomorphism. There is a strict equivalence of the "program" of the information theorist and the genetic and somatic programs of the biologist.

I am rather amused to notice that Nagel's rebuttal of my ideas has been cited with approval in several recent papers in philosophical journals, but not one of these philosophers descended to discuss or even list the paper of the biologist whom Nagel had criticized.

NOTES

This essay is adapted from a paper which first appeared in *Boston Studies in the Philosophy of Science* 14(1974):91–117 under the title "Teleological and teleonomic: a new analysis."

1. This is quite evident from the following explanatory comment I have received from Professor Pittendrigh by letter (dated February 26, 1970):

You ask about the word 'teleonomy'. You are correct that I did introduce the term into biology and, moreover, I invented it. In the course of thinking about that paper which I wrote for the Simpson and Roe book (in which the term is introduced) I was haunted by the famous old quip of Haldane's to the effect that 'Teleology is like a mistress to a biologist: he cannot live without her but he's unwilling to be seen with her in public'. The more I thought about that, it occurred to me that the whole thing was nonsense—that what it was the biologist couldn't live with was not the illegitimacy of the relationship, but the relationship itself. Teleology in its Aristotelian form has, of course, the end as immediate, 'efficient', cause. And that is precisely what the biologist (with the whole history of science since 1500 behind him) cannot accept: it is unacceptable in a world that is always mechanistic (and of course in this I include probabilistic as well as strictly deterministic). What it was the biologist could not escape was the plain fact—or rather the fundamental fact—which he must (as scientist) explain:

that the objects of biological analysis are organizations (he calls them organisms) and, as such, are end-directed. Organization is more than mere order; order lacks end-directedness; organization *is* end-directed. [I recall a wonderful conversation with John von Neumann in which we explored the difference between 'mere order' and 'organization' and *his* insistence (I already believed it) that the concept of organization (as contextually defined in its everyday use) always involved 'purpose' or end-directedness.]

I wanted a word that would allow me (all of us biologists) to describe, stress or simply to allude to—without offense—this end-directedness of a perfectly respectable mechanistic system. Teleology would not do, carrying with it that implication that the end is causally effective in the current operation of the machine. Teleonomic, it is hoped, escapes that plain falsity which is anyhow unnecessary. Haldane was, in this sense wrong (surely a rare event): we can live without teleology.

The crux of the problem lies of course in unconfounding the mechanism of evolutionary change and the physiological mechanism of the organism abstracted from the evolutionary time scale. The most general of all biological 'ends', or 'purposes' is of course perpetuation by reproduction. *That* end [and all its subsidiary 'ends' of feeding, defense and survival generally] is in some sense effective in causing natural selection; in causing evolutionary change; but not in causing itself. In brief, we have failed in the past to unconfound causation in the historial origins of a system and causation in the contemporary working of the system . . . .

You ask in your letter whether or not one of the 'information' people didn't introduce it. They did not, unless you wish to call me an information bloke. It is, however, true that my own thinking about the whole thing was very significantly affected by a paper which was published by Wiener and Bigelow with the intriguing title 'Purposeful machines'. This pointed out that in the then newly-emerging computer period it was possible to design and build machines that had ends or purposes without implying that the purposes were the cause of the immediate operation of the machine.

2. For more recent treatments of Aristotle's teleology, see Nussbaum (1978); Sorabji (1980); Balme (1980); and, most of all, Gotthelf (1976).

REFERENCES

Ayala, F. J. 1970. Teleological explanations in evolutionary biology. *Phil. Sci.* 37:1–5.

Baer, K. E. von. 1876. Über den Zweck in den Vorgängen der Natur. *Studien, etc.*, pp. 49–105, 170–234 St. Petersburg.

Balme, D. M. 1980. Aristotle's biology was not essentialist. *Arch. Gesch. Phil.* 62:1–12.

Beckner, M. 1969. Function and teleology. *J. Hist. Biol.* 2:151–164.

Beniger, J. R. 1986. *The Control Revolution.* Cambridge, Mass.: Harvard University Press.

Bergson, H. 1907. *Evolution Créative.* Paris: Alcan.

Braithwaite, R. D. 1954. *Scientific Explanation.* Cambridge: Cambridge University Press. (Also in Canfield 1966, pp. 27–47.)

Canfield, J. V., ed. 1966. *Purpose in Nature.* New Jersey: Prentice-Hall.

Craig, W. 1918. Appetites and aversions as constituents of instincts. *Biol. Bull.* 34:91–107.

Davis, B. D. 1961. The teleonomic significance of biosynthetic control mechanisms. *Cold Spring Harbor Symposia* 26:1–10.

Delbrück, M. 1971. Aristotle-totle-totle. In J. Monod and E. Borek, eds. *Of Microbes and Life.* New York: Columbia University Press.

Driesch, H. 1909. *Philosophie des Organischen.* Leipzig: Quelle und Meyer.

Gotthelf, A. 1976. Aristotle's conception of final causality. *Rev. Metaphysics* 30:226–254.

Hinde, R. A., ed. 1972. *Non-Verbal Communication.* Cambridge: Cambridge University Press.

Hinde, R. A., and J. G. Stevenson. 1970. Goals and response controls. In L. R. Aronson et al., eds., *Development and Evolution of Behavior.* San Francisco: Freeman.

Hull, D. 1973. *Philosophy of Biological Science.* Foundations of Philosophy Series. New Jersey: Prentice-Hall.

Kant, I. 1790. *Kritik der Urteilskraft.* Zweiter Teil.

Kleinenberg, N. 1886. *Über die Entwicklung durch Substitution von Organen.* *Zeitschr. wiss. Zoologie,* pp. 212–224. Leipzig: Engelmann.

Lagerspetz, K. 1959. Teleological explanations and terms in biology. *Ann. Zool. Soc. Vanamo* 19:1–73.

Lehman, H. 1965. Functional explanation in biology. *Phil. Sci.* 32:1–20.

Lovejoy, A. O. 1936. *The Great Chain of Being.* Cambridge, Mass.: Harvard University Press.

MacLeod, R. B. 1957. Teleology and theory of human behavior. *Science* 125:477.

Mainx, F. 1955. *Foundations of Biology,* Foundations of the Unity of Science 1(9):1–86.

Mayr, E. 1964. The evolution of living systems. *Proc. Nat. Acad. Sci.* 51:934–941.

McFarland, J. D. 1970. *Kant's Concept of Teleology.* Edinburgh: University of Edinburgh Press.

Monod, J. 1971. *Chance and Necessity.* New York: Knopf.

Munson, R. 1971. *Biological Adaptation. Phil. Sci.* 38:200–215.

Nagel, E. 1961. The structure of teleological explanations. *The Structure of Science.* Harcourt, Brace and World.

———— 1977. Teleology revisited: goal directed processes in biology. *J. Phil.* 74:261–301.

Nussbaum, M. C. 1978. *Aristotle's De Motu Animalium.* Princeton: Princeton University Press.

Pittendrigh, C. S. 1958. Adaptation, natural selection and behavior. In Roe and Simpson (1958), pp. 390–416.

Raven, C. P. 1960. The formalization of finality. *Folia Biotheorctica* 5:1–27.

Roe, A., and G. G. Simpson, eds. 1958. *Behavior and Evolution.* New Haven: Yale University Press.

Rosenblueth, H., N. Wiener, and J. Bigelow. 1943. Behavior, purpose, and teleology. *Phil. Sci.* 10:18–24. (Also in Canfield 1966, pp. 6–16.)

Roux, W. 1883. *Über die Bedeutung der Kerntheilungsfiguren. Eine hypothetische Erörterung.* Leipzig.

Sigwart, C. 1881. *Der Kampf gegen den Zweck.* Kleine Schriften. 2. Freiburg: Mohr.

Simpson, G. G. 1949. *The Meaning of Evolution: A Study of the History of Life and of Its Significance for Man.* New Haven: Yale University Press.

Smith, W. John. 1969. Messages of vertebrate communication. *Science* 165:145–150.

Sorabji, R. 1980. *Necessity, Cause, and Blame: Perspectives on Aristotle's Theory.* Ithaca: Cornell University Press.

Sommerhoff, G. 1950. *Analytical Biology.* London: Oxford University Press.

Stadler, H. 1874. *Kant's Teleologie und ihre erkenntnistheoretische Bedeutung.* F. Dümmler.

Taylor, R. 1950. Comments on a mechanistic conception of purposefulness. *Phil. Sci.* 17:310–317. (Also in Canfield 1966, pp. 17–26.)

Teilhard de Chardin, P. 1955. *Le Phénomène Humain.* Paris: Editions de Seuil.

Theiler, W. 1925. *Zur Geschichte der Teleologischen Naturbetrachtung bis Aristoteles.* Zürich und Leipzig.

Ungerer, E. 1922. *Die Teleologie Kants und ihre Bedeutung für die Logik der Biologie.* Berlin: Bornträger.

Waddington, C. H. 1957. *The Strategy of the Genes.* London: Allen and Unwin.

———— 1968. *Towards a Theoretical Biology I.* Edinburgh: Edinburgh University Press.

Weismann, A. 1909. The selection theory. In A. C. Seward, ed. *Darwin and Modern Science.* Cambridge: Cambridge University Press.

Wimsatt, W. C. 1972. Teleology and the logical structure of function statements. *Stud. Hist. Phil. Sci.* 3:1–80.

THE PROBABILITY
OF EXTRATERRESTRIAL
INTELLIGENT LIFE

A NUMBER of very different problems are often confused during discussions of the SETI project: (1) the probability of the *existence* of "life" elsewhere in the universe, (2) the probability of *intelligent* extraterrestrial life, and (3) the chances of being able to *communicate* with such life, if it should exist.[1]

At the present time we have no positive evidence whatsoever that life exists elsewhere, and thus, of course, also of intelligent life. The probabilities of either can be guessed at only by highly indirect inferences.

Life in the Universe

When the Mars missions were being prepared, the astronomer Donald Menzel and I had a $5 bet as to whether or not "life as on earth" [our precise designation] would be discovered on Mars. The physical scientist Menzel said yes, the evolutionary biologist Mayr said no. Who was right is on record. By now it is quite evident that none of the other planets in this solar system is suitable for life.

One negative instance, of course, proves nothing. If all suns in the universe have planets (actually a rather dubious assumption), we would have hundreds of millions of planets. Surely, it is argued, some of these should have spawned life. And I agree, the probability for a multiple origin of a self-replicating nucleic acid-protein aggregate is indeed high.

It has been known for some time that smaller organic molecules, like amino-acids, purines, and pyrimidines, can arise spontaneously in the universe, and that such processes can be duplicated in the laboratory. Nevertheless, for a long time it seemed impossible to explain how the amino acids (and peptides) could get together with nucleic acids to form truly replicating, that is, living, macromolecules. Through the researches of Eigen and his school (Küppers 1983) and the discovery of the enzymatic nature of RNA, there seems to be no longer a difficulty of principle. What is particularly interesting is the important role played by natural selection, even during the pre-biotic phase. The probability of the repeated origin of macromolecular systems with an ability for information storage and replication can no longer be doubted.

What is still entirely uncertain is how often this has happened, where it has happened, and how much evolution might have occurred subsequent to the origin of such life. We who live on the earth do not fully appreciate what an inhospitable place most planets must be. To be able to support life they must be just the right distance from their sun, have the right temperature, a sufficient amount of water, a sufficient density to be able to hold an atmosphere, a protection against damaging ultraviolet radiation, and so forth. Furthermore, every planet changes in the course of its history, and the sequence of changes has to be just right. If, for instance, there were too much free oxygen at an early stage, it would destroy life. The total set of prerequisites for the origin and maintenance of life drastically reduces the number of planets that would have been suitable for the origin of life. There is, indeed, the possibility that the combination and sequence of conditions that permitted the origin of life on earth was not duplicated on a single other planet in the universe. I do not make such a claim, and it would not be science if I did, since it would be impossible ever to refute it. However, measured by the possibility of refutation, the claims of the proponents of extraterrestrial life and intelligence are equally outside the bounds of science. The only thing we know for sure is that of the nine planets of the solar system the earth is the only one that has produced life. Let us assume, however, for the sake of the argument that life has originated on some of the supposedly hundreds of millions of planets in the universe. Since we do not know how many suns have planets, the mentioned figure might be a gross overestimation.

The Existence of Extraterrestrial Intelligence

It is interesting and rather characteristic that almost all the promoters of the thesis of extraterrestrial intelligence are physical scientists. They are joined by a number of molecular and microbial biologists, and by a handful of romantic organismic biologists.

Why are those biologists who have the greatest expertise on evolutionary probabilistics so almost unanimously skeptical of the probability of extra-terrestrial intelligence? It seems to me that this is to a large extent due to the tendency of physical scientists to think deterministically, while organismic biologists know how opportunistic and unpredictable evolution is.

Some 20 years ago when I argued a great deal with the astronomer Donald Menzel about life on Mars, I was forever astonished how certain he was that if life had ever originated on Mars (or been transported to Mars), this would inevitably lead to intelligent humanoids. The production of man was for him like the end product of a chemical reaction chain where the end product can be predicted once you know with what chemicals you had started. He took it virtually for granted that, if there was life on a planet, it would in due time give rise to intelligent life: "Our own Milky Way might contain up to a million planets [favorable to the development of life], all inhabited by intelligent life" (Menzel 1965: 218).

Everybody knows, of course, that determinism is no longer the fashion in modern physics, and yet in conversation with physical scientists I have discovered again and again how strongly they still think along deterministic lines. If organic evolution on earth culminated in intelligence, why should it not have resulted in intelligence on all planets on which life had originated?

By contrast, an evolutionist is impressed by the incredible improbability of intelligent life ever to have evolved, even on earth. To demonstrate this, let us look at the history of life on earth.

Date of origin of kinds of organisms if age of earth (4.5 billion years) is made equivalent to a calendar year:

> Origin of Earth = 1 January
> Life (Prokaryotes) = 27 February
> Eukaryotes = 4 September
> Chordates = 17 November
> Vertebrates = 21 November
> Mammals = 12 December
> Primates = 26 December
> Anthropoids = 30 December, 01:00 a.m.
> Hominid line = 31 December, 10:00 a.m.
> *Homo sapiens* = 31 December, 11:56½ p.m.
> (= 3½ minutes before year's end).

Let us look at the chronology of major evolutionary events on earth. (All cited figures are rough estimates, the upward or downward revision of which would have no effect on the argument; the order of magnitude, however, is right.) Let us assume the earth originated 4.5 billion years ago. There is evidence that life began only about 700 million years (*my*) later. Definite early prokaryote fossils are known from 3.5 billion years ago. What is most remarkable is that for about 2,000 *my* nothing very spectacular happened as far as life on earth was concerned. There was apparently a rich diversification of prokaryotes, but these—although quite successful in their way—are poor potential as progenitors of intelligent life. Nevertheless, they displayed remarkable metabolic diversification, the blue-green bacteria even became phototropic and produced oxygen. Up to that time the earth's atmosphere had been reducing.

Sometime between 1,400 and 1,500 *my* ago a most improbable event took place. According to the most likely explanation, a symbiosis was established between two (or more) kinds of prokaryotes, one of them supplying cytoplasmic organelles, the other one the nucleus of an entirely new type of organism, the first eukaryote. This was apparently such a successful combination that within several 100 *my* four new kingdoms evolved, the protists (one-celled animals and plants), fungi, plants, and animals. All higher organisms are eukaryotes, characterized by the possession of a well-organized nucleus and chromosomes in each cell.

We can see that from the origin of life to the origin of the eukaryotes more than half of the age of the earth had passed by without any noticeable events except for diversification within the prokaryotes. But once the eukaryotes had been "invented," an almost explosive innovative diversification took place. Within each of the four mentioned kingdoms scores of separate evolutionary lines originated, many of them strikingly different from each other. However, in none of these kingdoms, except that of the animals, was there even the beginning of any evolutionary trends toward intelligence.

What about evolution of intelligence among the animals? After the animalian "type" had been invented, different structural types originated with such fertility that one could probably recognize at least 50 different phyla of animals in the Cambrian, including unique types in the Ediacaran and surviving Burgess shale formations. Many of these became extinct rather quickly, and there is no good evidence for the origin of any new phylum after the end of the Cambrian (500 *my* ago). However, the surviving phyla experienced a continuing abundant proliferation into classes, orders, families, and lower taxa. Of the 50 or so original phyla of animals, only one, that of the chordates, eventually gave rise to intelligent life, but the world still had to wait some 500 *my* before this happened. At first, still in Paleozoic, the vertebrates appeared in exceedingly diverse types, formerly all lumped together under the name "Fishes," but it is now realized how different the early vertebrates were from each other. Among this multitude of types, only one gave rise to the amphibians; and among the various types of amphibians, only one to the reptiles. What are called the reptiles are again a highly diverse group of vertebrates including such different organisms as turtles, lizards, snakes, crocodiles, and numerous extinct lineages such as ichthyosaurs, plesiosaurs, pterodactyls, and dinosaurs. Among these numerous types of reptiles, only two, the pseudosuchians (ancestors of birds) and the therapsids (ancestors of mammals), gave rise to descendants to some of whom a reasonable degree of intelligence can be attributed. But with all my bias in favor of birds, I would not say that a raven or parrot has the amount and kind of intelligence to found a civilization. So we have to continue with the mammalian class. It contains such unusual types as the monotremes (for example, the platypus) and marsupials, as well as a rich assortment of placental orders, some still living, many others having become extinct in the course of the Tertiary. Forms with a rather high development of the central nervous system and a good deal of intelligence are quite common among the mammals, but only one of these many orders led to the development of a truly superior intelligent life, the primates. The primates, however, are a rather diversified group, with prosimians (lemurs and so on), New World monkeys, and Old World monkeys, but only the anthropoid apes produced intelligence that clearly surpasses other mammals. Only after 18 of the 25 *my* of the existence of the anthropoid apes, and after a splitting of this major lineage into a number of minor lineages, like the gibbons (and relatives), the orangutan (and relatives), the African apes (chimpanzee and gorilla), and a considerable number of extinct lineages, did the lineage emerge which eventually, less than one-third of a million years ago, led to *Homo sapiens*.

The reason why I have buried you under this mass of tedious detail is to make one point, but an all-important one. In conflict with the thinking of those who see a straight line from the origin of life to intelligent man, I have shown that at each level of this pathway there were scores, if not hundreds, of branching points and independently evolving phyletic lines, with only a single one in each case forming the ancestral lineage that ultimately gave rise to Man.

If evolutionists have learned anything from a detailed analysis of evolution, it is the lesson that the origin of new taxa is largely a chance event. Ninety-nine of 100 newly arising species probably become extinct without giving rise to descendant taxa. And the characteristic of any new taxon is to a large extent determined by such chance factors as the genetic composition of the founding population, the special internal structure of its genotype, and the physical as well as biotic environment that supplies the selection forces of the new species population.

My argument based on the incredibly low probability of intelligence ever having originated on earth must not be misunderstood. I do not claim in the least that an extraterrestrial intelligent "life" must have the slightest anatomical similarity to man. I already mentioned that we get a certain amount of intelligence in other mammals and even in birds. Indeed, an ability to make use of previous experience in subsequent actions, in other words a rudimentary kind of intelligence, is widely distributed in the animal kingdom. Intelligence, on another planet, might reside in a being inconceivably different from any living being on earth. Any devotee of science fiction will have no trouble in coming up with possibilities.

The point I am making is the incredible improbability of genuine intelligence emerging. There were probably more than a billion species of animals on earth, belonging to many millions of separate phyletic lines, all living on this planet earth which is hospitable to intelligence, and yet only a single one of them succeeded in producing intelligence.

Proponents of extraterrestrial intelligence have mentioned the convergent evolution of two such "highly improbable" organs as the eye of the cephalopods and of the vertebrates as an analog to the presumably equally improbable but not at all impossible convergent evolution of intelligence. Those who have thus argued, unfortunately, do not know their biology. The case of the convergent evolution of eyes is, indeed, of decisive importance for the estimation of the probability of convergent evolution of intelligence. The crucial point is that the convergent evolution of eyes in different phyletic lines of animals is not at all improbable. In fact, eyes evolved whenever they were of selective advantage in the animal kingdom. As Salvini-Plawen and I have shown, eyes evolved independently no less than at least 40 times in different groups of animals (Salvini-Plawen and Mayr 1977). This shows that a highly complicated organ can evolve repeatedly and convergently when advantageous, provided such evolution is at all probable. For genuine intelligence this is evidently not the case, as the history of life on earth has shown.

One additional improbability must be mentioned. Somehow, the supporters of SETI naively assume that "intelligence" means developing a technology capable of intragalactic or even intergalactic communication. But such a development is highly improbable. For instance, Neanderthal Man, living 100,000 years ago, had a brain as big as ours. Yet, his "civilization" was utterly rudimentary. The wonderful civilizations of the Greeks, the Chinese, the Mayas, or the Renaissance, although they were created by people who were for all intents and purposes physically identical with us, never developed such a technology, and neither did we until a few years ago. The assumption that any intelligent extraterrestrial life must have the technology and mode of thinking of late twentieth-century Man is unbelievably naive.

Civilizations, as human history demonstrates, are fleeting moments in the history of an intelligent species. For two civilizations to communicate with each other, it is necessary that they flourish simultaneously. Let me illustrate the importance of this point by a little fable. Let me assume there is another high technology civilization in our galaxy. By some extraordinary instrumentation their inhabitants were able to discover the origin of the earth 4.5 billion years ago. At once they began to send signals to the earth and continued to do so for 4.5 billion years. Finally at the time of the birth of Christ they decided that they would terminate their program after another 1900 years, if they had not received any answer by then. When they abandoned their program in the year 1900, they had proven to their own satisfaction that there was no other intelligent life in our galaxy.

I am trying to demonstrate by this fable that even if there were intelligent extraterrestrial life, and even if it had developed a highly sophisticated technology (although if they were truly intelligent they would probably carefully avoid this), the timing of their efforts and those of our engineers would have to coincide to an altogether improbable degree, considering the amounts of astronomical time available. Every aspect of "extraterrestrial intelligence" that we consider confronts us with astronomically low probabilities. If one multiplies these with each other, one comes out so close to zero that it is zero for all practical purposes. This was already pointed out by Simpson in 1964. Those biologists who doubt the probability of ever establishing contact with extraterrestrial intelligent life if it should exist do not "deny categorically the possibility of extraterrestrial intelligence," as they have been accused. How could they? There are no facts that would permit such a categorical denial. Nor have I seen a published statement of such a categorical denial. All they claim is that the probabilities are close to zero. This is why evolutionary biologists, as a group, are so skeptical of the existence of extraterrestrial intelligence, and even more so of any possibility of communicating with it, if it exists.

In my view, SETI is a deplorable waste of taxpayers' money, money that could be spent more usefully for other purposes.[2]

NOTES

This essay is reprinted from E. Regis, Jr., ed., *Extraterrestrials: Science and Alien Intelligence*, pp. 23–30. Cambridge: Cambridge University Press. © Cambridge University Press, 1985.

1. From the time of the ancients, imaginative thinkers have speculated on the possible existence of living beings elsewhere in the universe, and there is a considerable literature on the subject (see Beck 1985; Crowe 1986; Dick 1982; Regis 1985). The book where this essay first appeared (Regis 1985) examines the pros and cons of the project to search for extraterrestrial intelligence (SETI).

2. After completing the manuscript, I reread Simpson's classical paper (1964) on the subject and was struck by how similar his analysis was to mine. Perhaps subconsciously I still remembered his arguments. Be that as it may, I warmly recommend reading his famous essay on the "Nonprevalence of humanoids." It is as pertinent today as it was 20 years ago.

REFERENCES

Bates, D. R. 1978. On making radio contact with extraterrestrial civilizations. *Astrophysics and Space Science* 55:7–13.

Beck, L. W. 1985. Extraterrestrial intelligent life. In Regis (1985), pp. 3–18.

Billingham, J., ed. 1981. *Life in the Universe*. Cambridge, Mass.: MIT Press.

Crowe, M. 1986. *Life in the Universe*. Cambridge: Cambridge University Press.

Dick, S. 1982. *Plurality of Worlds: The Origins of the Extraterrestrial Life Debate from Democritus to Kant*. Cambridge: Cambridge University Press.

Hart, M. H., and B. Zuckerman, eds. 1980. *Extraterrestrials: Where are they?* New York: Pergamon.

Küppers, B. O. 1983. *Molecular Theory of Evolution*. New York: Springer.

Maltove, E. F., et al. 1978. *A Bibliography on the Search for Extraterrestrial Intelligence*. NASA Reference Publication 1021.

Menzel, D. H. 1965. Life in the Universe. *Grad. J.* 7:195–219.

Regis, E., Jr., ed. 1985. *Extraterrestrials: Science and Alien Intelligence*. Cambridge: Cambridge University Press.

Salvini-Plawen, L. v., and E. Mayr. 1977. On the evolution of photoreceptors and eyes. *Evol. Biol.* 10:207–263.

Sammons, V. O. 1982. *Extraterrestrial Life*. Library of Congress Science Tracer Bulletin TB 82-4. Washington, D.C.: Library of Congress.

Simpson, G. G. 1964. The nonprevalence of humanoids. In *This View of Life*. New York: Harcourt, Brace and World.

Tarter, J. 1985. Searching for extraterrestrials. In Regis (1985), pp. 167–190.

THE ORIGINS OF
HUMAN ETHICS

T H E publication in 1859 of the *Origin of Species* signified the end of an automatic acceptance of the god-given nature of human morality. To be sure, philosophers, well before that time, had raised the question of the source of human morality, but their real objective was to find out how one could determine the correct or best ethical norms. In this essay I will not deal with that question. I will not try to establish norms stating what is good, what is bad, what is ethical. I take it for granted that our culture as well as every individual that belongs to it has a rather definite idea as to what is moral, what is good or bad. Rather, I will attempt to discuss the origins of human ethics during the evolution of *Homo sapiens* from his primate ancestors, and I will begin with the problem of altruism.

Behavior is altruistic when it benefits another organism, not closely related, while being apparently detrimental to the organism performing the behavior. The presence of truly altruistic behavior, it is generally believed, distinguishes human beings from all animals. The difference seems so drastic and discontinuous that it was often used in the early anti-Darwinian arguments as evidence for a special origin of mankind. Darwin fully realized how different man was from all animals, by stating, "I fully subscribe to the judgment of those writers who maintain that of all the differences between Man and the lower animals the moral sense or conscience is by far the most important" (1871). Yet, as we shall later see, he presented a fully elaborated theory as to how this could have evolved gradually.

In the last 50 years it has been recognized that a particular form of altruism is quite widespread in animals, primarily in species with parental care or those that form social groups consisting principally of extended families. Here one finds an altruistic defense of the offspring by the mother and sometimes the father, a willingness to defend or warn close relatives and to share food, and other kinds of behavior evidently beneficial to the recipient but at least potentially harmful to the actor. As has been pointed out by Haldane, Hamilton, and numerous sociobiologists, such behavior can be favored by natural selection because it enhances the fitness of the genotype of the altruist. As long as the behavior results in an overall benefit for the genotype of the altruist it is, looked at critically, egotistical rather than altruistic behavior. The literature of sociobiology contains literally hundreds of cases of seemingly altruistic behavior actually directed towards enhancing inclusive fitness.

Inclusive fitness altruism is one of the major bones of contention of the current evolutionary literature. Some authors seem to think that all human ethics is more or less raw inclusive fitness altruism. Other authors think that when genuine human ethics evolved, it altogether replaced inclusive fitness altruism. My own position is somewhat intermediate. I discern many remnants of inclusive fitness altruism in the human species, such as the instinctive love of a mother for her children and the different moral stance we adopt when dealing with strangers or aliens as compared with our own group. The moral norms laid down in the Old Testament are characteristic of this heritage. But I do not believe that inclusive fitness is all there is to human ethics.

Interestingly, Darwin was already fully aware of the existence of inclusive fitness. Speaking of the occurrence of men with superior faculties in the human tribe, he stated, "If such men left children to inherit their mental superiority, the chance of the birth of still more ingenious members would be somewhat better, and in a very small tribe, decidedly better. Even if they left no children, the tribe would still include their blood relations" (1871:161), who, as Darwin explains, have a similar genetic endowment. A selection for the spread of inclusive fitness altruism occurred, of course, not only in primitive men but in all social animals in which extended families are the nucleus of social groups. That social animals have a remarkable ability to recognize their relatives is emphasized by Darwin again and again: "The social instincts never extend to all the individuals of the same species" (1871:85). How well developed this sensing of relationship is in certain animals has been excellently documented experimentally in recent years by Pat Bateson and his group at Cambridge University.

Solitary animals have no opportunity to acquire such altruism. They have no behavior that natural selection could convert into altruism. By contrast, the altruistic tendencies of organisms living in social groups are an excellent basis as the starting point for human morality.

The Emergence of Genuine Ethics

As we shall see, genuine human ethics emerged from the inclusive fitness altruism of our primate ancestors. Although it is not possible to establish in each case a sharp line of demarcation between true ethics and inclusive fitness altruism, in general one can say that ethical behavior is based on conscious thought that leads to the making of deliberate choices. The altruistic behavior of a mother bird is not based on choice; it is instinctive, not ethical.

Simpson (1969) characterized the situation quite well: "The concept of ethics is meaningless unless the following conditions exist:

A) There are alternative modes of action;

B) Man is capable of judging the alternatives in ethical terms;

C) He is free to choose what he judges to be ethically good."

Thus it clearly depends on man's capacity to foresee the results of his actions and includes the acceptance of individual responsibility for their results. This is the basis for the origin and the function of the moral sense.

Ayala (1987) expressed more or less the same thought when saying that humans exhibit ethical behavior because their biological constitution determines the presence in them of the three necessary and jointly sufficient conditions for ethical behavior. These conditions are: (1) the ability to anticipate the consequences of one's own actions; (2) the ability to make value judgments; and (3) the ability to choose between alternative courses of action.

The capacity for ethical behavior thus is closely correlated with the evolution of other characteristically human capacities. The difference between an animal, which acts instinctively, and a human being, who has the capacity for making choices, is the line of demarcation for ethics. I take it for granted that a person *can* make the decision whether or not to rob somebody or to send a bullet through his head. Since a person can predict the outcome of his or her actions, he is fully responsible for the ethical evaluation of the result. Human beings have the capacity to make such judgments because of the reasoning power provided by the evolving human brain. *The shift from an instinctive altruism based on inclusive fitness to an ethics based on decision making was perhaps the most important step in humanization.* Of course brain development coincided with several other evolutionary changes in man, such as a great extension of the period of infancy and youth and thus of parental care, and it is on the whole impossible in these correlated developments to determine what is cause and what is effect. Other correlated factors were a trend toward enlargement of the hominid troups beyond the extended family and the development of tribal traditions and culture (see below).

Because the whole concept of ethics rests on the firm conviction that a person is able, up to a point, to control his or her actions, ethics depends on the existence of free will. And quite rightly the problem of free will has occupied a large space in the analysis of ethics by philosophers. This is not the place to defend the existence of free will; all I want to say is that its acceptance does not deny causality. In retrospect, every action can be causally explained, but it is, so to speak, an a posteriori causality. If I have an option of either one or the other of two responses to an ethical situation, no matter which I choose it will always result in a strictly causal sequence.

One of the most important problems in the origin of human ethics is the growth of the hominid group from an extended family into a larger, more open society. For this enlargement to occur, the altruism which before was reserved for close relatives had to be extended to nonrelatives. We see the rudiments of this behavior among a few primate groups, where some interchange of unrelated individuals also occurs. In the course of human evolution, some hominids must have discovered that an enlarged troup had a better chance of being victorious in a fight with another troup than a small one consisting simply of an extended family. One can assume that a troup in possession of a desirable cave, water hole, or hunting ground would attract outsiders wanting to benefit from these advantages. It was of selective advantage for the troup to be strengthened by such added manpower, even though it required the extension of altruism to nonrelatives, that is, beyond the range of inclusive fitness. In other words, it required the development of genuine ethics. The importance of the shift from kinship morality to ethics is a step in human evolution that has not before been stressed sufficiently.

Even though there is a "difference in kind" between human ethics and the inclusive fitness altruism of animals, it surely did not occur as a saltation but rather evolved gradually. The 4–8 million-year gap between *Homo sapiens* and our nearest anthropoid relatives provided more than sufficient time for all the necessary intermediate stages.

A mutually beneficial interaction among nonrelated individuals has been designated in the literature as *reciprocal altruism* (Trivers 1971). The cleaning fishes which free large predatory fishes of their parasites is a typical example. Actually, the term altruism is here quite inappropriate because the putative altruist always benefits either at once or in the long run. Such reciprocal interactions, particularly among primates, always imply a kind of reasoning: "If I help this individual in his fight, he will help me when I get into a fight." In other words, it is highly egotistical behavior rather than altruistic behavior. This type of reasoning, according to some sociobiologists, is at the root of the majority of human altruistic behaviors. The benefit expected by the altruist is the approval, respect, or even admiration of his fellow citizens (Alexander 1987:102). Opponents of this view have quite rightly pointed out how often altruistic acts are performed (for instance, anonymously) where the altruist does not expect, in fact does not even want, a reward of any sort.

The Human Cultural Group

The evolution of uniquely human ethics was closely correlated with the evolution of human cultural groups. These groups—enlargements of the original family groups—were held together by leadership, dialect, geography, rituals, and cultural traditions. The crucial question to be asked is whether such cultural groups could act as units of selection in the evolution of human ethics. That is, could a cultural group be the target of selection?

Those evolutionists who believe in group selection in its various forms have answered this question affirmatively. Those who reject group selection have asserted that it cannot be applied to man any more than to any other organism. Personally, I disagree with both of these camps. In my view, one must avoid lumping under the term *group selection* entirely different evolutionary phenomena. I agree with Williams, Sober, and others that group selection among animals (as proposed by Wynne-Edwards and others) is not supported by any evidence. Of the three kinds of so-called group selection among animals that I can distinguish, none is supportable by the evidence (see Essay 7). In all of the animal groups, the individual is the target of selection.

But human cultural groups are something quite different. There is a great deal of evidence that human cultural groups, as wholes, can serve as the target of selection. Rather severe selection among such cultural groups has been going on throughout hominid history. Darwin was fully aware of this: "All that we know . . . [shows] that from the remotest times successful tribes have supplanted other tribes" (1871:160). This form of selection is of such special importance because, in contrast with individual selection, cultural group selection may reward altruism and any other virtues that strengthen the group, even at the expense of individuals. As history repeatedly illustrates, those behaviors will be preserved and those norms will have the longest survival that contribute the most to the well-being of a cultural group as a whole.

The ability to apply group norms appropriately is intimately correlated with the evolving reasoning capacity of the human brain. The correlated evolution of a larger brain and a larger social group made two new aspects of ethical behavior possible: (1) a selective reward for certain unselfish traits that benefitted the group, and (2) ethical behavior by deliberate choice, rather than purely by the instinct of inclusive fitness.

These conclusions still leave two major questions unanswered, however:

(1) How does a cultural group develop a set of ethical norms?

(2) How does a given individual acquire knowledge of the norms that he should adopt?

HOW DOES A CULTURAL GROUP ACQUIRE ITS PARTICULAR SET OF ETHICS?

This question has been debated by philosophers from Aristotle, Spinoza, and Kant right up to modern times. The two most widely adopted answers were that moral norms were god-given or else they were the product of human reason. But if reason was the causal factor, one must ask whose reason and on the basis of what criteria? I shall not discuss the scores of answers given to these questions.

After the adoption of evolution, numerous attempts were made to derive moral norms from evolutionary theory. For instance, it was suggested that the "good" or "superior" was that which tends to lift man above the level of an animal. But there is no evidence that such a rational approach was ever used consciously by anybody in order to develop a superior ethical system. Indeed, eugenics was conceived by its founders as a way of lifting humans toward greater perfection. It is sadly ironic that this noble original objective eventually led to some of the most heinous crimes mankind has ever seen.

For Darwin the ethical yardstick consisted of man's relation to society. He considered as moral that which expressed itself as obedience to the "social instincts," as he called them. This solution only sidestepped the question because it led to the follow-up question of where the social instincts originated. Bertrand Russell promoted a similar principle, considering as "objectively right . . . [that] which best serves the interest of the group. A comparison of ethical norms throughout the world shows that those groups were most successful in which the self-interest of the individual was at least to some extent subordinated to the welfare of the community." But Russell's statement comes nearer to a satisfactory answer than Darwin's, because it refers to the relative success of different human cultural groups. Some had moral norms that enhanced the probability of success (the longevity) of the group; others had moral norms that led to rapid extinction.

It is easy to image how a particular value system within a culture might lead to the prosperity and numerical increase of the group, which might, in turn lead to genocidal warfare against neighbors, with the victor taking over the territory of the defeated. Any divisive tendencies within a group would weaken it and in due time lead to its extinction. Thus, the ethical system of each social group or tribe would be modified continuously by trial and error, success and failure, as well as by the occasional modifying influence of certain leaders. Intragroup altruism and any other behaviors that strengthen the group would be rewarded by selection over the course of time. Thus, the moral norms according to which the members of a group decide whether a particular action is either right or wrong are not the result of biological evolution but of cultural evolution.

When we ask whether moral norms are the product of reason or simply the result of random survival among competing groups of those with the most constructive ethics, we encounter diverse answers. The enormous variety in the moral norms of primitive human tribes would indicate that many of the differences are simply due to chance. But when we compare the major religions and philosophies, including those of China and India, we discover ethical codes that are remarkably similar, despite their largely independent histories. This suggests that the philosophers, prophets, or law-givers responsible for these codes must have carefully studied their societies and, using their ability to reason on the basis of these observations, must have decided which norms were beneficial and which others were not. The norms announced by Moses or by Jesus in the Sermon on the Mount were surely to a large extent the product of reason. Once adopted, such norms become part of the cultural tradition and are culturally inherited from generation to generation.

HOW DOES THE INDIVIDUAL PERSON ACQUIRE MORAL NORMS?

The answer to this question has been controversial for generations and has ranged between two extremes. Some sociobiologists—Alexander (1987) and Ruse (1987) tend in that direction—believe that even in man there is little genuine altruism, all seeming altruism being inclusive fitness altruism, ultimately advantageous to the seeming altruist. They also believe that such behavior is largely innate that is, it has a genetic basis. Psychologists who have studied ethical traits, on the other hand, are still rather uncertain about whether they have any basis in the genotype. Curiously, it seems that heritability is more easily demonstrated for bad traits than for good ones. Darwin cited the occurrence of kleptomania through several generations in highly affluent families as evidence for strict heritability. A genetic predisposition can often be inferred also in the case of psychopaths. But, as Darwin rightly said, "If bad tendencies are transmitted, it is probable that good ones are likewise transmitted" (1871:102).

I do not believe that definite genes controlling character traits of high ethical value have ever been demonstrated. Rather, what is inherited are tendencies and capacities. Admittedly, a genetic predisposition for high ethical qualities is difficult to prove, since it will be overlain by much cultural acquisition. But there are abundant indications for innate personality differences that affect ethical behavior. This is not to deny that behavioral tendencies that are ultimately selfish vastly preponderate in the inherited component of our behavioral attitudes; selfish behavior was strongly favored by natural selection in prehuman days. Yet the evidence indicates that the genetic component in human ethics is over all, of minor importance. By far the largest part of the moral values of a human being are individually acquired through interactions with other members of the cultural group. There is impressive evidence for the noninnateness of mankind's ethical goodness. Evidence for this claim is highly diverse, but I might mention some of it:

(1) The drastically different kinds of morality in different ethnic groups and tribes.

(2) The total breakdown of morality under certain political regimes.

(3) The ruthless and quite amoral behavior often displayed against minorities, particularly slaves.

(4) The callous behavior exhibited in war, for instance the uninhibited bombing of civilian population centers.

(5) The warping of the character of a child when deprived of a mother or mother substitute during a critical period in its infancy.

Such observations have led to an equally extreme opposing viewpoint, which is that we are born, so to speak, as a *tabula rasa*, and that *every* aspect of our behavior is learned. Many behaviorists and their followers adopted this view. They deny the existence of any inborn component and believe that all moral behavior is the result of reasoning, based on conditioned responses.

Both of these views, in their one-sidedness, are contradicted by much evidence. The greatest weakness of both approaches is that they deal with the situation rather typologically. Two aspects in particular tend to be neglected. First, as already stated, is the great individual variation in moral propensities. There are some individuals who from childhood on seem to be nasty, ruthless, totally selfish, thoroughly dishonest, and so on. There are others who seem to be little angels from the start—warmhearted, utterly unselfish, always reliable and cooperative, honest to the core. Modern twin and adoption studies document that there is a considerable genetic component in these different tendencies.

Second, and far more important, is the neglect in most modern discussions of the difference between asserting an innateness of definite moral norms and asserting merely an innateness of a *capacity for adopting* ethical behavior. If an individual has such a capacity, he is able to adopt a second set of ethical norms supplementing and in part replacing the biologically inherited norm based on inclusive fitness. This capacity permits the individual to adopt the culturally inherited norms of the society to which he or she belongs. The great importance of the cultural norms is that they counteract the basic selfish tendencies of the individual and impose on him an altruism that benefits the group as a whole but of course ultimately the individual whose well-being is closely connected with the well-being of the group as a whole.

Culturally transmitted values are characterized by considerable plasticity. This is documented not only by the sometimes quite astonishing differences among the moral norms of different human groups but also by drastic changes in human cultures in historical times. Wilson (1975) reminds of the changes of the value system of the Irish during the potato famine (1846–1848) and of the Japanese culture during the American occupation after World War II.

Acknowledging that much of our value system is not biologically but culturally inherited allows us finally to attempt to answer the question how the individual acquires these norms. Numerous studies have concluded that ethical norms are acquired during infancy and youth. To say that these values are learned is not much of a help, because there are so many different kinds of learning. I am rather persuaded of the validity of

Waddington's thesis (1960) that a very special kind of learning is involved, a type of learning akin to the imprinting in animals so well described by the ethologists, illustrated by the attachment of a young gosling to its mother.

Man is distinguished from all other animals by the openness of its behavioral program. By this I mean that many of the objects of behavior and the reactions to these objects are not instinctive, that is, not part of a "closed program," but are acquired in the course of life. Just as in the case of the gosling the gestalt of the mother goose becomes imprinted in the behavior program of the gosling, so in human beings ethical norms and definite values are laid down in the open behavior program of the infant. As Waddington proposed it, "The human infant is born with probably a certain innate capacity to acquire ethical beliefs, but without any specific beliefs in particular" (1960:126). Darwin was fully aware of the power of imprinting in early youth: "It is worthy of remark that a belief constantly inculcated during the early years of life, while the brain is impressable, appears to acquire almost the nature of an instinct" (1871:100). This power of indoctrination, says Darwin, leads not only to the adoption of ethical norms but also to the unquestioned acceptance of observed customs, such as the burning of widows by Hindus or the proscription of Moslim women to expose their faces, or other "absurd rules of conduct," as Darwin called them (1871:99).

Every child psychologist knows how anxious children are to receive new information, including normative rules, and how ready they are on the whole to accept them. The findings of the child psychologists, for instance Kohlberg (1981; 1984), seem to support Waddington's thesis. The value system of a person is largely controlled by what was incorporated in his or her youth into this open behavioral program. It is the vast capacity of this open program in humans that makes ethics possible. And the foundation laid in infancy lasts, under normal circumstances, throughout life. In what part of the brain this information is stored and how it is retrieved under the appropriate circumstances is as yet totally unknown.

The story of the acquisition of ethical norms by the individual is not complete unless two further aspects are mentioned.

(1) Learning psychologists have demonstrated that certain things are learned much more readily than others. An olfactory animal learns smells much more easily than a visual animal, and vice versa. It is distinctly possible that if certain moral norms contributed to the sur-

vival potential of certain groups during hominid history, this would also favor the selection of a structure of the open program facilitating the storage of this behavior norm.

(2) If Waddington's thesis is correct, it follows that ethical education is of the utmost importance. We have just passed through a period in which exaggerated importance was placed on the so-called freedom of the child, allowing him to develop his own goodness. We have made fun of the moralizing in children's books and have tended to remove all traces of moral education from the schools. This causes few problems when parents perform their roles properly. But it may spell disaster when parents fail to do their job. In view of our better understanding of the origin of the morality of the individual, would it not seem time again to place greater stress on moral education?

Are the Traditional Ethical Norms of the Western World Adequate?

The ethical norms of Western culture are those of the Judeo-Christian tradition, that is, they are based on the various commandments and precepts articulated in the Old and New Testaments. As phrased in the sacred texts, such commandments would seem to be absolute, allowing for no deviation. The commandment "Thou shalt not kill" has ordinarily absolute validity. But to withdraw life-support machinery from an intensely suffering terminal patient is an act of mercy, not killing. And a similar flexibility must be applied in the case of abortion. When an unwanted child would have to face a life of misery and neglect or when his mother would be led to total despair, then abortion would certainly seem to be the more ethical option. And it is nonsense to drag the question of life into this argument, for as a biologist I know that every egg and spermatozoon also has life.

There are two reasons why the traditional norms of the West are no longer adequate. The first is their rigidity. The essence of the evolutionary process is variability and change, and ethical norms must be sufficiently flexible and versatile to be able to cope with a change of conditions. The second reason is that mankind has indeed experienced a drastic and accelerating change of conditions. Perhaps the most important component in this change has been the steady enlargement of human groups during the last 10 to 15 thousand years. With the coming of agriculture, a larger group was favored because it could better protect against marauders, and the availability of a good food supply likewise favored population growth.

A change in values—for instance, greater emphasis on property rights—
was inevitable.

Some of the ethical norms adopted by the pastoral people of the Near
East more than 3,000 years ago are altogether inadequate for the modern
urbanized mass society. As Simpson has rightly said: "All ethical systems
that originated under tribal, pastoral, or other primitive conditions . . .
have become, to greater or less degree, inadaptive in the extremely different social and other environmental conditions of the present time"
(1969:136).

What then are some of the ethical problems of our modern anonymous
mass society that are not adequately covered by the traditional ethical
norms? I would like to mention three.

The first is what Singer (1981) refers to as the "expanding circle." Not
only in primitive societies but also in the Old Testament, among the
Greeks, and even among eighteenth- and nineteenth-century Europeans
in Africa and Australia, a totally different ethic was employed toward
members of one's own group and toward outsiders (Singer 1981:111–
117). We still had it in the South in our own country until a few decades
ago, and *Apartheid* is an awful contemporary remnant of such group
selfishness. Even within ethnically homogeneous societies, such as early
twentieth-century Britain, there are or have been differences about minor
virtues, loyalties, and injunctions among religious groups, political parties, professional groups, social organizations, and so on. All this sets up
tensions and conflicts.

The ideal of every great moralist is to establish a universal ethic, as
Julian Huxley stated it: "The conception that ethical principles apply to
all humanity, irrespective of race, language, creed, or station" (1947:117).
Expanding the circle, that is, applying one's intra-group ethics also to
outsiders, leads to a fusion of groups with different norms, and this sets
up an almost insoluble conflict, because each group is convinced of the
superiority of its own moral values. Let us just think of the difference
between a modern American and an Islam fundamentalist with respect to
the rights of women, or, in our own country, the different attitudes
toward abortion shown by certain religious groups versus feminist groups.
In spite of all these difficulties, the ethics of the future must be based on
the unrestricted principle of equality between members of one's own group
and outsiders.

The second great ethical problem of our time is excessive egocentricity
and exclusive attention to the rights of the individual. The expanding
circle in our own society has resulted in a legitimate fight for equality,

particularly by minorities and women, but this also has had some undesirable side effects. In recent decades there has been an excessive stress on
rights and freedoms. Martin Luther King has perhaps been the only
freedom fighter who has reminded his followers that all rights must be
accompanied by obligations. I am not a moralist preacher, but anyone
watching the current scene cannot help but see numerous indications of
excessive narcissism and a cult of the ego. This development has many
roots—our mass society, the teachings of Freud, as well as an overreaction
to the preceding period of *dis*regard for the rights of the individual. It is
even favored by our political system because in a democracy a politician's
success depends on his appeal to the individual voter. Last but not least,
the monotheistic religions have tended to emphasize, if not overemphasize,
individual ethics.

Here it is sometimes forgotten that the principal function of cultural
ethics is to restrain the selfish impulses of the individual and to further
the well-being of the community as a whole by the application of laws
and customs. Inevitably modern man faces some virtually insoluble conflicts such as that between equality and merit, or equality and genetic
diversity. In such cases we should always apply two basic principles. One
is that ethical decisions often, if not always, depend on the context, and
absolute prescriptions are sometimes quite unethical. And second, that,
depending on the circumstances, there is often a pluralism of possible
solutions. Let us remind ourselves that at the very core of human ethics
is the possibility of making a choice and of evaluating the conflicting
factors in order to make the right choice. As much as ethical norms are
part of our culture, the responsibility for executing them rests with the
individual; if they are too rigid, the individual may choose not to adhere.

The third great ethical problem of our day is posed by the discovery of
our responsibility toward nature as a whole. Growth, whether economic
growth, populational growth, or whatever other kind of growth, used to
rank very high in our value system. Even though certain influential people
like the Nobel Prize-winning economist Hayek and the Pope have so far
failed to appreciate the danger of overpopulation, I cannot see how it can
be ignored any longer. Certain of our societies, like those of China and
Singapore, have courageously tackled this problem by a reordering of
ethical values. The sooner other societies follow, the better it will be for
the ultimate good of mankind.

The dilemma we are facing is the conflict between traditional values
and newly discovered values. Let me remind you of the conflict between
man's right to unlimited reproduction and to the unlimited exploitation

of the natural world, as against the needs of human posterity as well as the right to existence of the millions of species of wild animals and plants. Where is the proper balance between personal freedom and regard for the welfare of the natural world?

The concept that mankind has a responsibility toward nature as a whole is an ethical concept that seems to have originated remarkably late. In recent times Aldo Leopold, Rachel Carson, and Garrett Hardin have been particularly articulate in their championship of a conservation ethic or a community ethic. But much of what these modern Americans consider ethically valuable is not for the immediate benefit of the individual and is therefore resisted. And yet if mankind and the world as a whole are to have a future, it will be necessary that we reduce the selfish tendencies in our ethics in favor of a higher regard for the community and for the *whole* of Creation.

Conclusion

What have we learned about the relation between evolution and human ethics?

(1) As already Darwin knew, human ethics is not simply the product of the struggle for existence.

(2) Human ethics differs from whatever instinctive altruism is found in the animal kingdom by involving a deliberate choice between alternatives.

(3) Selection among cultural groups has contributed to the spread of those ethical norms in mankind that have contributed most to the well-being of the group.

(4) The human *individual* acquires his ethical standards during childhood and youth. Once entered into his open behavioral program, these ethical maxims tend to retain great power over the individual for all or much of his life.

(5) And finally, ethical norms must be flexible and adaptable, for the shift from a pastoral or agricultural society to an urban mass society requires considerable adjustments, and so does the shift from a thinly settled world to the modern industrial world with its massive overpopulation. In other words, ethical norms must have the capacity to evolve, in order to stay adaptive.

Is there any particular ethics which an evolutionist should adopt? Ethics is a very private matter, a personal choice. My own ethics is rather close to Julian Huxley's evolutionary humanism. "It is a belief in mankind, a feeling of solidarity with mankind, and a loyalty toward mankind. Man is the result of millions of years of evolution, and our most basic ethical principle should be to do everything toward enhancing the future of mankind. All other ethical norms can be derived from this baseline."

Evolutionary humanism is a demanding ethics, because it tells every individual that somehow he has a responsibility toward mankind, and that this responsibility is or should be just as much part of his ethics as individual ethics. Every generation of mankind is the current caretaker not only of the human gene pool but indeed of all nature on our fragile globe.

Evolution does not give us a complete codified set of ethical norms such as the Ten Commandments, yet an understanding of evolution gives us a world view that can serve as a sound basis for the development of an ethical system that is appropriate for the maintenance of a healthy human society, and that also provides for the future of mankind in a world preserved by the guardianship of man.[1]

NOTES

This essay is adapted from a Messenger Lecture presented at Cornell University, Ithaca, New York, October 17, 1985. Previously unpublished.

1. In addition to quoted literature, I include among the references a number of books and papers which I read with profit and which I consider significant contributions to the subject. I have also added some titles that have come to my attention since the Messenger lectures.

REFERENCES

Alexander, R. D. 1979. *Darwinism and Human Affairs*. Seattle: University of Washington Press.
———— 1987. *The Biology of Moral Systems*. Hawthorne, N.Y.: Aldine de Gruyter.
Ayala, F. J. 1987. The biological roots of morality. *Biol. and Phil.* 2:235–252.
Bateson, P., ed. 1983. *Mate Choice*. Cambridge: Cambridge University Press.

Bonner, J. T., and R. M. May. 1981. Introduction to facsimile edition of Charles Darwin's (1871) *The Descent of Man*. Princeton: Princeton University Press.

Caplan, A. L., and B. Jennings, eds. 1984. *Darwin, Marx, and Freud: Their Influence on Moral Theory*. New York: Plenum.

Darwin, C. 1859. *On the Origin of Species by Means of Natural Selection or the Preservation of Favored Races in the Struggle for Life*. London: John Murray.

———— 1871. *The Descent of Man*. London: John Murray.

Ebling, F. J., ed. 1969. *Biology and Ethics*. Symposia Institute of Biology, no. 18. New York: Academic.

Garcia, J. D. 1971. *The Moral Society*. New York: Julian Press.

Gillespie, N. C. 1979. *Charles Darwin and the Problem of Creation*. Chicago: University of Chicago Press.

Glass, Bentley, 1965. *Science and Ethical Values*. Chapel Hill: University of North Carolina Press.

Greene, John C. 1981. From Huxley to Huxley: transformations of the Darwinian credo. In *Science, Ideology, and World View*, pp. 158–193. Berkeley: University of California Press.

Huxley, J. 1947. *Evolution and Ethics 1893–1943*. London: Pilot Press.

Huxley, T. H. 1893. *Evolution and Ethics*. Romanes Lecture. London: Oxford University Press. Rptd. New York: Appleton, 1898.

Mackie, J. L. 1977. *Ethics: Inventing Right and Wrong*. New York: Penguin.

Mayr, E. 1973. [Comments in a panel on population and behavior.] In *Ethical Issues in Biology and Medicine: A Symposium on the Identity and Dignity of Man*, pp. 115–118. Cambridge, Mass.: Schenkman.

———— 1976. *Evolution and the Diversity of Life*. Cambridge, Mass.: Harvard University Press.

———— 1984. Evolution and ethics. In Caplan and Jennings (1984).

Moore, J. R. 1979. *The Post-Darwinian Controversies: A Study of the Protestant Struggle to Come to Terms with Darwin in Great Britain and America, 1870–1900*. Cambridge: Cambridge University Press.

Richards, R. J. 1986. A defense of evolutionary ethics. *Biol. and Phil.* 1:265–354.

Ruse, M. 1984. The morality of the gene. *Monist* 67:167–199.

———— 1986. *Taking Darwin Seriously*. New York: Basil Blackwell.

Sedgwick, A. 1859. [Review of Darwin's *Origin*.] In D. L. Hull, ed., *Darwin and His Critics*, pp. 155–170. Cambridge, Mass.: Harvard University Press, 1973.

Sherman, P. W., and W. G. Holmes. 1985. Kin recognition: issues and evidence. *Fortschr. Zool.* 31:437–460.

Simpson, G. G. 1969. Biology and ethics. In G. G. Simpson, *Biology and Man*, pp. 130–148. New York: Harcourt, Brace and World.

Singer, P. 1981. *The Expanding Circle*. New York: Farrar, Straus and Giroux.

Trivers, R. 1985. *Social Evolution*. Menlo Park: Benjamin/Cummings.

Vogel, C. 1985. Evolution und Moral. In Maier-Leibnitz, ed., *Zeugen des Wissens*, pp. 467–507. Mainz: Hase und Koehler.

Waddington, C. F. 1960. *The Ethical Animal*. London: Allen and Unwin.

Wilson, E. O. 1975. *Sociobiology*. Cambridge, Mass.: Harvard University Press.

Wynne-Edwards, V. C. 1962. *Animal Dispersion in Relation to Social Behaviour*. Edinburgh: Oliver and Boyd.

part two ❧

NATURAL SELECTION

Introduction

No other concept in evolutionary biology is as important as that of natural selection; and, owing to its novelty, no other concept has encountered as much resistance. It was in conflict not only with some of the tenets of Christianity but also with major axioms of the accepted philosophy of the mid-nineteenth century (Mayr 1982). With few exceptions, natural selection continued to be rejected by most philosophers for the next 100 years or more. And when philosophers finally began to accept it as a legitimate scientific theory, their interpretations encountered unexpected complexities. Is selection a teleological concept? Is the simple principle of natural selection a theory or is it a theory only when applied to concrete cases? What is the target of selection? How deterministic and predictive is selection? What is the material of selection? Is selection a one-step or two-step phenomenon? Can selection produce perfection?

During the evolutionary synthesis, selection was universally accepted by evolutionary biologists as the only cause of inherited adaptation and as the only direction-giving factor in evolutionary change. Nevertheless, it became quite clear during the ensuing 50 years how many misconceptions and uncertainties were still attached to this concept. It is particularly gratifying that various modern philosophers, instead of trying to refute the principle, have attempted to sharpen definitions and eliminate equivocations and confusions. This includes analyses by Brandon (1978; 1982), Mills and Beatty (1979), and Tuomi (1981), the important book by Elliott Sober (see Essay 7), and the introductory discussions in Brandon and Burian (1984).

Interestingly, the major disagreements turned out to be not between biologists and philosophers but among the biologists themselves. There are still opposing opinions concerning the target of selection (gene, individual, group, species), the role of sexual selection, "neutral evolution," the role of development, deterministic vs. probabilistic aspects of selection, and the nature and power of constraints, to mention only a few. It is these largely biological questions that I address in the two ensuing essays.

REFERENCES

Brandon, R. N. 1978. Adaptation and evolutionary theory. *Stud. Hist. Phil. Sci.* 9:181–206.
———— 1982. The levels of selection. *PSA 1982* 1:315–324. East Lansing, Mich.: Philosophy of Science Association.
Brandon, R. N., and R. M. Burian. 1984. *Genes, Organisms, Populations.* Cambridge, Mass.: MIT Press.
Mayr, E. 1982. *The Growth of Biological Thought.* Cambridge, Mass.: Harvard University Press.
Mills, S., and J. Beatty. 1979. The propensity interpretation of fitness. *Phil. Sci.* 46:263–288.
Tuomi, J. 1981. Structure and dynamics of Darwinian evolutionary theory. *Syst. Zool.* 30:22–31.

AN ANALYSIS OF
THE CONCEPT
OF NATURAL SELECTION

PROGRESS in science is achieved in two ways: through new discoveries, such as x-rays, the structure of DNA, and gene splicing, and through the development of new concepts, such as the theories of relativity, of the expanding universe, of plate tectonics, and of common descent. Among all the new scientific concepts, perhaps none has been as revolutionary in its impact on our thinking as Darwin's theory of natural selection. In its unprecedented departure from the time-honored traditional thinking of philosophy and in its antibiblical implications, as well as for various other reasons, it encountered at once the most determined opposition. Furthermore, since it was not the discovery of a concrete object or process that one could examine physically, the concept of natural selection was interpreted in various ways even among its followers. It was as true for this concept as for all new concepts that everyone had to incorporate it into his own conceptual framework and adjust it to that which he already believed. The result was that everyone looked at natural selection in a somewhat different way, and much of the ensuing argument about the validity of natural selection suffered from misunderstandings and semantic difficulties.

For Darwin, the concept was simplicity itself. He said, "The preservation of favorable variations and the rejection of injurious variations, I call Natural Selection" (1859:81). His opponents, however, raised the most diverse objections to this simple formulation. These objections were in part based on ideology, religion, or philosophy, and in part on the claim that Darwin had failed to apply the proper scientific method. I shall not cover this long-lasting controversy in this essay, since I have done so elsewhere (Mayr 1982).

But it must be admitted that Darwin's opponents raised some valid points. First of all, his argument was largely deductive, and he was not able to produce anything that his contemporaries were willing to accept as proof. The term *selection* was a particularly formidable obstacle in the way of acceptance, because it implied something forward-looking, some teleological process, some selecting agency. Yet, was there such an agency? And it if existed, why were its activities so haphazard and unpredictable? This teleological implication was transferred to adaptation, the product of selection; even today an evolutionist may say that a species of desert plants had adapted itself to its arid environment. A long time passed before even evolutionary biologists fully realized that selection is, if I may use this seemingly contradictory terminology, an *a posteriori* process. Selection simply is the fact that in every generation a few individuals among the hundreds, thousands, or millions of offspring of a set of parents survive and are able to reproduce because these individuals happened to have a combination of characteristics that favored them under the constellation of environmental conditions they encountered during their lifetime. I have avoided the somewhat misleading, because seemingly circular, expression "survival of the fittest," which has so often led to the claim that the whole concept of natural selection is based on tautology.[1] That this claim is not valid has been shown by many recent authors.

So much evidence for the reality of the process of natural selection has been produced since the publication of the *Origin* that systematic opposition has more or less died down, at least within science. There is, first of all, abundant experimental evidence for the efficacy of selection under laboratory conditions, as summarized in recent textbooks. But there is also—and perhaps this is more important—abundant evidence for the occurrence of natural selection in nature (Endler 1986). The theory of natural selection has an enormous heuristic value and permits so-called predictions under specified environmental conditions. For instance, if on a certain island there are two species of finches, a small-billed one specializing on small seeds and a large-billed one specializing on large seeds (with a certain amount of overlap of the food niche), there should be a higher mortality of the small-billed species during a drought period that hits particularly hard the plants that produce small seeds. And this indeed is what is found (Grant 1986). I do not think I exaggerate if I say that every year some 20 to 30 papers (perhaps far more) are published in which confirmation is provided for such "predictions" derived from the selection theory (Naylor and Handford, 1985; Fergusson 1976).

It has been argued that the theory of natural selection is not a scientific theory at all, because it cannot be refuted: "Natural selection explains nothing, because it explains everything" (Lewontin 1972:181). Indeed, it must be made clear what the nature of "the" theory of natural selection is. The principle of natural selection is more comprehensive than a specific theory, and it has therefore been referred to as a generic theory or basic general principle (Tuomi and Haukoja 1979; Tuomi 1981; Beatty 1981; Brandon 1981), which as such can neither be refuted nor does it have predictive powers. It becomes a genuine theory, called by Tuomi a theoretical model, only when enriched with specific ancillary assumptions, for instance, with particulate inheritance, the postulate of small mutations, and the occurrence of genetic recombination. The principle can be dissected into individual theories such as that stated in Bergmann's rule. And such specific theories can be tested and refuted.

There are various ways in which definitions used in selection theory can be made more precise—for instance, by expressing fitness not in terms of the actual contribution to the gene pool of the next generation but rather as the propensity to make such a contribution. If two identical twin brothers go out on a walk and one is killed by lightning, it will drastically affect his contribution to the gene pool of the next generation, but not his original propensity for making such a contribution. "The fitness of an organism is its propensity to survive and reproduce in a particular specified environment and population" (Mills and Beatty 1979:42; Brandon 1978).

Actually, for every situation in nature we can imagine innumerable solutions that could *not* be explained by natural selection. Let us take two islands with one species of finch on each. On island A only plants with large heavy-shelled seeds occur, no plants with small seeds, and hardly any insects. On island B, by contrast, only plants with small seeds occur, and an abundance of small insects. Frankly, I would not know how to explain it by natural selection if the finch on island A were thin-billed and the one on island B a grosbeak with a heavy bill. The fact of the matter is that we take adaptation in nature for granted to such an extent that we would consider situations like the fictitious scenario of the two finches as totally absurd. Actually, of course, the normally observed "harmony of living nature" is spectacular evidence for the power of natural selection. Maynard Smith (1972) has pointed this out also, illustrating it with a different example.

A Two-Step Process

One of the favored arguments among Darwin's opponents has always been, "How can natural selection produce favorable variations?" Or the even more absurd comparison, made from Jonathan Swift and J. F. Herschel to contemporary critics, of natural selection to a troup of monkeys. Natural selection, they claimed, has no more chance to produce an adaptation than a troup of monkeys to compose *Hamlet* by aimlessly throwing the letters of the alphabet together. These critics overlooked the fact that natural selection proper is only the second stage of a two-step process. The first step consists of the production of variation in every generation, that is, of suitable genetic or phenotypic variants that can serve as the material of selection, and this will then be exposed to the process of selection. This first step of variation is completely independent of the actual selection process, and yet selection would not be possible without the continuous restoration of variability. Several recent critics have failed to understand the relation of the two steps of the selection process to each other. When an author asks (as several have actually done), "is evolution due to molecular processes or due to selection?" it amounts to asking: "Is evolution a change due to step one or to step two of natural selection?" This question is meaningless since the second step, selection *sensu stricto*, deals with the previously produced variation *(a posteriori)* and is not a process which itself produces variation.

Darwin saw clearly that a fixed type, a constant Platonic essence, cannot evolve. And yet the struggle for existence, the mechanism responsible for change during the second step, has no influence whatsoever on the first step, that is, on the amount and nature of the variation produced prior to a bout of reproduction.

As aware as Darwin was of the importance of variation (he devoted to it three of the first five chapters of the *Origin* and later (1868) an entire two-volume work), neither he nor anyone else could solve the riddle of the source of variation and its inheritance until the rediscovery of Mendel's principles in 1900.

The two points Darwin stressed particularly were that a seemingly inexhaustible supply of variation was available at all times and that the production of any new variation was independent of the needs of the organism. In contrast to the evolutionary theories of some of his opponents, variation for Darwin was nondirected. This is what the evolutionist means when he says new mutations are random. This terminology has often been misunderstood. It does not in the least mean that any variation can occur anywhere, any time. On the contrary, mutations, in a given species, are highly "constrained," which means that only a very restricted range of mutations is possible. An eye of *Drosophila* can vary from white to pink, red, or brown, but never to blue or green. A tetrapod can vary in the modifications of its two pairs of extremities, but the occurrence of one extra pair (as in insects) or two extra pairs (as in spiders) is not part of the possible normal variation of a tetrapod. When it is said that mutation or variation is random, the statement simply means that there is no correlation between the production of new genotypes and the adaptational needs of an organism in the given environment. Owing to numerous constraints, the statement does not mean that every conceivable variation is possible (Mayr 1963:176).

After 1900 it became customary to see the origin of new variants as the result of mutation. This was the thinking not only of the early Mendelians but also of much of the Morgan school, and one could read in biology textbooks until the 1960s or even 70s that evolution is due to "mutation and selection." Throughout much of this period genetic recombination, as produced during meiosis, was rather neglected. Although crossing over had been known since the 1880s (Weismann), it was rarely invoked in evolutionary discussions until C. D. Darlington's "equating of chiasmata and cross-overs led to the recognition of the universality of intrachromosomal recombination in natural species" (White 1981:289). Recombination is the process responsible for the fact that each individual in sexually reproducing species is different from all others.

Evolution is not a smooth, continuous process but consists, in sexually reproducing organisms, of the formation of a brand-new gene pool in every generation. Furthermore, there is a steady alternation of the first step, which consists of the meiotic production of new gametes and their fate prior to fertilization, with the second step, the "struggle" of the new zygotes to reach the reproductive stage and to reproduce successfully. Selection is not a forward-looking process but simply a name for the survival of those few individuals that have successfully outlasted the "struggle for existence."

When essentialism was still reigning, there was much talk about the elimination of degenerations of the type. Unfortunately, even after 1859 many authors continued to see in selection a purely negative process, the weeding out of the unfit, and—by implication—the restoration of the purity of the type. To counteract this kind of thinking, several evolutionists, as J. Huxley, Dobzhansky, and Simpson, have called selection a creative process. This designation is justified because evolution mixes in every generation the genetic endowment of the few survivors during sexual reproduction and thus creates abundant new genotypes, which are then tested in the next generation. Each individual is, so to speak, a new experiment that is tested for its fitness in the struggle for existence. No other process is known in nature that could achieve this in such an extraordinarily successful manner as the combination of the first and second steps of natural selection.

The Target of Selection

In Darwin's time and indeed in the first years of Mendelism ("unit characters"), the difference between genotype and phenotype was not understood. The organism as a whole, that is, its phenotype, was considered the target of selection; and something representing the phenotype was handed from generation to generation. Darwin's theory of pangenesis was very much in this tradition. However, it was also realized that the genotype was not simply a condensed projection of the adult phenotype, as is evident from the theory of recapitulation. But it was not until after 1910 that the question of the relation of genotype and phenotype during development became a problem. At that time V. Haecker, Woltereck, Goldschmidt, and other German geneticists worried considerably over it. In that literature there was a tendency to consider both genotype and phenotype rather holistically, with a stress on genetic interactions. As a result the individual as a whole was treated as the target of selection, as it had been by Darwin. By contrast a rather strong reductionist trend prevailed in the English-language scientific literature, and genes were studied more or less in isolation from each other and from the phenotype. Ever since the rise of mathematical populational genetics, and as a matter of fact already in the thinking of some members of the Morgan school (Sturtevant, Muller), the gene was considered the currency of evolution. Consequently, in the mathematical approaches of R. A. Fisher and J. B. S. Haldane, individual genes were accepted as the units on which calculations were based. Evolution was now defined as a change in gene frequencies, and relatively simple mathematical formulae were believed to permit an adequate representation of what happens during evolution. Accepting the change of gene frequencies as the earmark of evolution, some authors unfortunately began to ignore whether such change was due to genetic drift or to selection. And in the single-minded concentration on genes, it was often forgotten how important individuals, populations, and species are in determining the course of evolution.

It is rather unfortunate that the customary question to ask was what is the "unit of selection"? The term *unit* was never properly defined, and this led to all sorts of misunderstandings (discussed in Essay 7). As a result, a protracted controversy developed as to whether the gene, the genotype, the individual (phenotype), the group, or the species is the "unit of selection," or all of them. The misunderstandings go back to the early post-Darwinian period, when certain essentialists wrote that selection was "for the good of the species." A long controversy ensued, one which saw the geneticists, on the whole, on the side of championing the gene, and the naturalists favoring the individual or the group (Brandon and Burian 1984). I myself have favored the individual since the 1950s, and an increasing number of recent authors have come to the same conclusion.

The insistence that the individual as a whole rather than each separate gene is the target of selection would be unimportant if the genotype were an aggregate of separate, independent genes. However, those who favor the individual consider the genotype to be a well-integrated system analogous to an organism with structure and organs. This view was developed by the Russian evolutionary geneticists, particularly Schmalhausen, and by Dobzhansky and his school. It was forcefully presented in Lerner's (1954) *Genetic Homeostasis* and in my own writings (1963; 1976; see also Essay 24). The controversy between this viewpoint and the reductionist one (both within the Darwinian framework) has not yet been resolved. There is, however, a great deal of recent evidence that the system of interactions among components of the genotype is far more complex than conceived of in classical population genetics, and the system nature of the genotype is strongly corroborated by the recent recognition of the great functional diversity of kinds of DNA (see Essay 24). Much of the recent criticism of "neo-Darwinism" is directed against a reductionist-atomistic conception of the genotype. These critics ignore that other representatives of the synthetic theory, like Schmalhausen, Waddington, Rensch, Lerner, members of Dobzhansky's school, and myself, have for many decades adhered to a far more holistic concept of the genotype.

The clarification of this problem has been helped by Sober's (1984) analysis. He pointed out that much of the confusion was caused by the fact that people did not distinguish between asking "selection of" and "selection for." In the controversy over whether the gene, individual, or group is the unit of selection, the question to be asked is "selection of." And this makes it quite clear why the individual and not the gene must be considered the target of selection. There are many ways of documenting this primary importance of the individual. First of all, it is the individual as a whole that either does or does not have reproductive success. Second, the selective value of a particular gene may vary greatly depending on the genotypic background on which it is placed. Third, since different individuals of the same population differ at many loci, it would be exceedingly difficult to calculate the contributions of each of these loci to the fitness of a given individual. And fourth, accepting the individual as a whole makes it unnecessary to make the confusing distinction (as some authors have) between internal selection (dealing with processes going on during ontogeny) and external selection (dealing with the interaction of the adult with the environment).

When we say that the individual is the target of selection, we mean the individual at all stages of its life cycle, from the moment of fertilization of the maternal egg through all embryonic stages right up to death. There is no justification in separating as internal selection all those components of selection that occur during ontogeny. Furthermore, what is involved during that stage of the life cycle is, for the most part, a removal of defective zygotes (J. Remane 1983), thus elimination rather than selection.

Considering the individual as a whole rather than single genes to be the target of selection clarifies a number of evolutionary puzzles. After the enormous variability of enzyme genes had been discovered in the 1960s through the use of electrophoresis, Kimura and others proposed the theory that most changes in gene frequencies are of no selective significance but are neutral. Not that Kimura (1983) and those who followed him completely denied the existence of natural selection, but they certainly reduced its importance as a factor in the evolutionary change of the genotype.

A great deal of recent research has indeed established that some changes at the molecular level seem to be either entirely neutral or of very small selective significance. Nevertheless serious objections can be raised against the neutrality concept of molecular evolution. First, many of the supposedly neutral mutations have subsequently been found to be subject to selection (Nevo 1983). Also, Kimura's models have been challenged, and it has been asserted that the number of neutral mutations appears to be far smaller if a different method of calculation is employed (Gillespie 1986). Second, the selective value of a given allele is not necessarily constant, and may change in a different environment or in a different constellation of the genotype. Finally, and most importantly, if the individual as a whole is the target of selection, and if certain individuals are favored by selection owing to others of their characteristics, then numerous neutral genes can be carried along as "hitchhikers." Hence, the neutrality of certain mutations is not in the slightest in conflict with the theory of selection, which rewards only the integrated selective value of the individual *as a whole*. DNA sequences believed to be functionless, such as pseudo-genes and certain introns, behave as if selectively neutral and may thus be subject to rapid change, owing to genetic drift and to their being immune to stabilizing selection.

The normal process of mutation is the replacement of an allele by a new allele, or, in molecular terms, the replacement of an amino acid residue or of a base pair in the DNA. However, genetics and molecular biology have discovered numerous other processes that may lead to a change in the genotype and to increased variability. Some of these deserve to be mentioned because they seem to contradict the concept of natural selection. This includes genes for the distortion of segregation (meiotic drive), various modes of production of "selfish DNA," and so-called molecular drive. These molecular processes are not random and might be referred to as "biased variation." It is a mistake, in my opinion, to claim that the phenomena of biased variation prove that genes can serve as the target of selection. This is not the case, because it is always the resulting zygote that is the actual target. Normal selection begins to operate as soon as the biased variation affects the fitness of the phenotype.

There are a number of problems of definition with respect to all those genes that are responsible for "biased variation." But as far as I am concerned, their activity without exception affects the first step of natural selection, that of the generation of variability, while the second step, exposing individuals to the forces of selection, is not affected. There is a continuing selection for unlinked modifiers for the reduction of the intensity of the meiotic drive. To be sure, strong biased variation can be more potent than selection, as in the case of the segregation-distorter genes, but this does not affect the conceptual issues. I believe one is thus justified in stating categorically that the answer to the question, what is the target of selection (*selection of*), is *the individual*.

Although I discuss species selection elsewhere (Essay 26), I want to mention at least one frequently cited case of species selection. If two species of flour beetles (*Tribolium*) are introduced into a container with flour, experience shows that only one of the two species will remain after a number of generations. This has been considered a case of species selection. Actually there is no typological competition between two species but rather each individual of either species is competing with all the coexisting individuals (of both species). Thus this is again a typical case of *selection of* individuals.

The Objectives of Selection

At this point we have to remember once more that one must distinguish between selection *of* and selection *for*. Having discussed "selection *of*" under the heading of "The Target of Selection," we are now discussing "selection for"; we are discussing what kind of traits might be favored by selection.

The success of selection in a given generation can be expressed in terms of the numerical contribution to the gene pool of the next generation. However, successful reproduction—and Darwin saw this much more clearly than any of his contemporaries—might be due to two entirely different causes. Darwin designated as *natural selection* anything that favored survival, whether this was an increase or decrease of body size, a broadening of the niche or its more efficient utilization, better protection against the environment or an increased tolerance of environmental extremes, a superior ability to cope with diseases or to escape enemies. Natural selection would favor anything that would accomplish greater ecological-physiological efficiency or save energy. Any individual favored by such selection would contribute genotypes to the gene pool that were apt to spread in future generations and thus enhance the adaptation of the population as a whole.

However, Darwin saw also—and again much more clearly than any of his contemporaries—that not all selection leads to improved adaptation, as this term is currently understood. An individual might contribute more genes to the next generation not through physiological efficiency or any other component of viability, but simply through being more successful in reproduction. This kind of selection Darwin called *sexual selection*. He further realized that there was a potential conflict between these two kinds of selection, the natural one for improved fitness (in the vernacular meaning of this word) and the sexual one for mere reproductive success of individuals. A purely egotistical selection for reproductive success might favor the evolution of traits that do not add to fitness (as traditionally understood) but might actually make the species more vulnerable. It leads to the gorgeous plumes of the birds of paradise, the extraordinary tail of the peacock, and the gigantic size of the elephant seal bulls.

Darwin devoted almost two thirds of the text of his *The Descent of Man* (1871) to the discussion of sexual selection. Yet this important process was largely ignored during the ensuing 100 years, in part owing to the reductionist definition of the target of selection by the mathematical population geneticists. By making the gene the unit of selection and defining anything as selectively superior that leads to an increased representation in the gene pool of the next generation, the entire difference between survival selection and selection for reproductive success was eliminated. Unfortunately this concealed some of the most interesting aspects of selection.

Even though I called attention in 1963 (pp. 199–201) to the importance of "selection for reproduction success," this phenomenon continued to be neglected until the commemoration of the centennial (1971) of Darwin's *Descent of Man*. Then, within a few years, literally hundreds of cases of behaviors and devices were discovered that tended to favor the reproductive success of an individual without contributing to its survival qualities. This new interest in selection for reproductive success is undoubtedly one of the few major developments in post-synthesis evolutionary biology (Trivers 1985).

The existence of selfish selection for reproductive success poses a dilemma for the evolutionary biologist. Classical natural selection ordinarily resulted in genotypes that were adaptively superior. This was so obvious that it was even suggested to define an adaptation as anything that was a "product of selection." However, this definition does not fit sexual selection at all. Is the gigantic size of the elephant seal bulls of adaptive value for this species? Is the survival of birds of paradise enhanced by the brilliant plumes of the adult males? Surely not. On the contrary, it is quite possible that an excessive development of certain characters favored by sexual selection contributed to the extinction of some species.

Does Selection Lead to Perfection?

The opponents of natural selection have often used the following argument. Since selection by necessity leads to perfection, they say, and since evidence for insufficient perfection and maladaptiveness abound in nature, then obviously natural selection does not operate. But which modern evolutionist has ever claimed that natural selection would lead to perfection? Darwin already had emphasized "natural selection tends only to make each organic being as perfect as, or slightly more perfect than, the other inhabitants of the same country with which it has to struggle for existence" (1859:201, also 472). He cites as proof the rapid extinction of fauna and flora in New Zealand when European competitors were brought to that island.

To be sure, selection is an optimization process, but one of a very special kind. It is neither teleologically programmed nor controlled by any law, but is entirely opportunistic. As many recent authors have pointed out, the process of selection is subject to so many constraints that it could not possibly achieve perfection. The most striking refutation of the frequently made claim that "selection can do anything" is provided by the frequency of extinction. More than 99.9 percent of all species that ever existed on earth have become extinct. This includes even such temporarily so-flourishing groups as the trilobites, ammonites, and dinosaurs. Here it does not matter what factor was responsible for the extinction: competition, a pathogen, a climatic catastrophe, or an asteroid's impact. In each case selection had been unable to find, among the available variants, an appropriate answer to the new situation.

Constraints

Curiously, the existence of constraints was already well known prior to the establishment of Darwinian evolution. Thus, Cuvier said prior to 1820 that no carnivore would ever have horns. The whole concept of the archetypes of the idealistic morphologists was based on the silent acceptance of developmental constraints. The naturalists of the nineteenth century knew very well that each taxon had severe limits to its potential variation. Weismann, who was a pan-selectionist if there ever was one, frequently stressed the constraints encountered by selection (Essay 27).

The importance of such constraints was, however, unfortunately neglected after 1900, when the geneticists thought of evolution as a matter of genes rather than of whole organisms. And for this reason it has been wholesome that authors such as Gould and Lewontin (1979) have again called attention to the power of the constraints on selection.

There are many different kinds of constraints that prevent an optimal response to selection forces. They affect all aspects of the selection process, from the production of variation to the interactions between genotype and environment. I shall discuss some of them. Additional ones are recorded in the literature.[2]

A CAPACITY FOR NONGENETIC MODIFICATION

The more pliable the phenotype (owing to developmental flexibility), the more this reduces the force of an adverse selection pressure. Plants and particularly micoorganisms have a far greater capacity for phenotypic adaptation than higher animals. The existence of this capacity is, however, present even in man, as shown for instance by his capacity for physiological adjustment when he goes from the lowlands to high altitudes. In the course of days and weeks he can become reasonably well adapted to the lowered atmospheric pressure, but even under these circumstances natural selection is not entirely neutralized. First of all, the very capacity for nongenetic adaptation is under strict genetic control; but also when a population shifts to a new, specialized environment, genes will be favored by selection during the ensuing generations that reenforce and eventually largely replace the capacity for nongenetic adaptation.

THE INDIVIDUAL AS TARGET

In order to be favored by natural selection, it is sufficient that an individual is competitively superior to most other individuals of its population (species). This superiority may be achieved by particular features, indeed sometimes by a single gene. In that case natural selection "tolerates" the remainder of the genotype even when some of its components are more or less neutral or even slightly inferior (see above). Genetic linkage therefore can also exert considerable constraints.

AVAILABILITY OF VARIATION

There is a limit to the kind and amount of genetic variation that can be contained in a population, and therefore the kind of genes needed for an appropriate response to new selection pressure may not be present in the gene pool.

Darwin apparently took it for granted that there was always sufficient variability in natural populations to satisfy the demands of natural selection. Many geneticists tended to disagree. De Vries, Bateson, and other Mendelians did not actually deny the existence of variation, but they thought that ordinary continuous variation was evolutionarily irrelevant and that evolution proceeded via occasional more or less drastic mutations. This was contradicted by the work of Chetverikov (1926) and his school and by the ecological geneticists (Dobzhansky 1937; Ford 1964), who showed that abundant natural variation was always available. But, said Muller, and this opinion was widely adopted, this variation simply consists of deleterious recessives that are eliminated as quickly as they become homozygous. Finally, in 1966 Darwin seemed to be vindicated when Lewontin and Harris demonstrated the enormous variability of enzyme genes as revealed by electrophoresis. But that by no means ended the argument, because Kimura and others presented evidence that much of this variation was neutral, that is, not suitable as material for natural selection.

More importantly, the high frequency of extinction, as well as the high frequency of cases where artificial selection was unable to achieve its objective, indicate that more is needed than abundant variability. It must be variability of the needed portions of the genotype. Success in evolution

may depend on highly specific genes or combinations of genes, and these
are sometimes not present when needed.

MULTIPLE PATHWAYS

Several alternate responses are usually possible for any environmental
challenge. This is well illustrated by pelagic swimmers that have developed
either from sessile or benthic (crawling) or actively swimming ancestors
belonging to many different animal phyla and that have become adapted
to the pelagic mode of life through entirely different adaptations. Each
solution is a very different compromise between the new need and the
previously existing structure. This illustrates splendidly Jacob's principle
of opportunistic "tinkering" (Jacob 1977). Whatever solution is chosen
depends on a constellation of circumstances. The adoption of a particular
solution may greatly constrain the possibilities of future evolution, for
instance the acquisition either of an internal or external skeleton.

INTERFERENCE BY EVOLUTIONARY "NOISE"

This refers to all the chance factors that interfere with the superficially so
deterministic process of selection. They include not only the rich array of
chance factors during the first step of selection but also the breaking apart
of favorable gene constellations through crossing over as well as the
numerous accidents to which a zygote is exposed during its maturation
and reproduction. Hence the *propensity* for fitness is rarely fully expressed
in the *realized* fitness. (For a further discussion of this "noise" factor see
also below, under "Chance.")

COHESION OF THE GENOTYPE

As I have shown in a number of recent analyses (see Essay 24), and as is,
of course, not at all unknown to geneticists, genes are frequently so tightly
functionally interconnected with each other that any change is deleterious.
Presumably this is usually due to epistatic interactions during ontogeny.
New information turns up all the time supplying further evidence for the
resistance of the genotype to any but the most superficial changes. More
drastic changes may lead to the production of deleterious phenotypes.

DEVELOPMENTAL CHANNELING

Natural selection does not piece together each organism from scratch. All
it can do is to attempt to modify slightly an already existing highly
complex structure. Evolutionary pathways, for this reason, are largely
channeled. Some developments are possible, others are not. The Hawaiian
finches (Drepanididae) show that there is little channeling with respect to

the shape of the bill. The birds of paradise of New Guinea document that in that family there is a broad leeway with respect to the elaboration of plumes. Yet in most higher taxa there is an extraordinary basic similarity of all members. This is why a good zoologist can usually tell of any animal not only to what phylum but often also to what class, order, or even family it belongs. If one is an entomologist one can tell at once almost any buprestid, cerambycid, weevil, or member of quite a few other families. Curiously, it is rarely emphasized what convincing demonstration of developmental constraint this taxonomic similarity is. All the well-known cases of recapitulation during ontogeny are further illustrations of such developmental constraints.

LIMITS TO THE POTENTIAL OF A BAUPLAN

Certain organisms, through the nature of their Bauplan, as the comparative anatomists now call the morphotype, are preadapted for evolutionary departures, as the lobe-finned coelacanths were for terrestrial life, or the archosaurian ancestors of birds for flight. Others are totally unsuited for a shift into a particular adaptive zone. There is nothing in the structure of a turtle that would permit the origin of flight in its descendants. The needed Bauplan is so drastically different that no amount of natural selection would be able to effect such a shift.

The existing genotype, thus, prescribes definite channels for future evolution. Is this fact in any conflict with the universal statement that natural selection is the only direction-giving factor in evolution? The answer is yes and no. Directional changes in evolution are caused by natural selection, but constrained by the potential of the existing genotype. Indeed, there is much evidence that the cohesion of the genotype is so tight that in most cases there is very little change of the phenotype over millions of years. It needs some special events or processes to destabilize the genotype. This has often happened during domestication (Belyaev 1979), and this is what seems to happen sometimes during speciation in founder populations (Essay 25).

The Genetics of Developmental Constraints and Phylogenetic Channeling

One of the great new insights of the animal systematists of the early nineteenth century was that animals cannot be arranged in a smooth, continuous series from the simplest to the most perfect, as the proponents of the *scala naturae* had claimed. Instead, a limited number of discrete types could be recognized. This new insight was developed by Cuvier, von Baer, Oken, Owen, Agassiz, and their followers. Even after everybody had adopted Darwin's theory of common descent, most major taxa remained discrete entities, each with its own Bauplan.

Atomistic genetics has been incapable of coming up with an explanation for the stability of the Bauplan. The fact that all tetrapods have basically one pair of anterior and one pair of posterior extremities can perhaps be functionally explained. Yet why all insects should have three pairs of extremities, and all spiders four pairs, can be explained only by the conservatism of the developmental system built into their genotypes. The same argument is true for the five-rayed foot (or hand) of terrestrial vertebrates. Even where it is reduced, as in the foot of the horse or in the wings of birds, or supplemented by additional rays, as in certain marine vertebrates, the structure is always laid down in ontogeny with five rays. One could go on and on about such conservative aspects of the Bauplan, either extended into the adult phenotype or visible only during ontogeny, but this does not add to the solution of the problem. There are presumably internal structures of the genotype that reductionist genetics has not even begun to explain (Essay 24). One must ask: What are these structures, and what happens to them during the short evolutionary periods when rather drastic ("saltational") evolutionary innovations are initiated?

No answers can yet be given to these questions. However, the recent advances in molecular genetics justify the hope that we are not too far from a solution. It is now clear that there are many different kinds of DNA, including middle and highly repetitive DNA, mobile genetic elements, and inducible mutation systems. All of them may and probably do have regulatory functions, but classical genetics was unable to work with this heterogeneity of DNA. The classical genes are structural genes and for the most part seem to be involved in the *production* of substrate rather than in its regulation. But the aspect of gene function that is crucial for evolution is how this substrate is used and when. For some 20 years everybody has been proclaiming that regulatory genes are the crucial element in evolution, but so far extremely little is known about which parts of the DNA function as regulatory genes and how these genes, in turn, are being controlled. The only thing that is clear, at least to me, is that the classical beanbag type of genetics is altogether insufficient to provide satisfying evolutionary explanations.

Chance

Selection is often described as a deterministic process, indeed sometimes even as a teleological process, because it seems to result in long-term evolutionary trends. These designations are, however, quite misleading. First of all, a close analysis of long-term evolutionary trends has shown almost invariably that they are actually quite irregular and often even terminated by reversals. Also, how could a process be deterministic in which there is no actual continuity because the genes of a population are returned in each generation to the common gene pool, and are thoroughly reassembled, with an entirely new start being made through the random production of new zygotes.

The importance of chance during evolution has been stressed by certain authors for more than 100 years (Mayr 1963:204). As early as 1871 Gulick insisted that the differences among snail populations on Oahu Island in the Hawaiian Islands were due to random variation and not due to selection. Since that time no one has stressed the role of chance factors more emphatically than Sewall Wright. Chance operates at every level of the process of reproduction, from crossing-over to the survival of newly formed zygotes (Mayr 1962). This includes the locus at which mutations occur, the location of chiasmata involved in crossing-over, the segregation of chromosomes during the reduction division, the survival of the millions or billions of gametes, the meeting of two gametes of opposite sex prior to fertilization, and finally the untold interactions of a zygote with its environment (in the widest sense of the word). There is also genetic drift in all of its forms, particularly significant in small populations, and all the effects of linkage. It is thus evident that a considerable percentage of differential survival and reproduction is not the result of ad hoc selection but rather of chance (Beatty 1984; 1987). Chance is also introduced by the phenomenon of pleiotropy. If a gene has multiple expressions, it will be selected for the most important of these, and other expressions of the gene will be carried along incidentally.

The large number of stochastic processes in populations of finite size, as well as the constraints which operate during selection, prevent selection from ever being a deterministic process. Rather, it must be remembered at all times that selection is probabilistic. This is true even for the success of the zygote. Each individual encounters in its environment numerous unpredictable adverse forces, such as catastrophes, epidemics, and unexpected encounters with enemies in which the outcome is largely probabilistic.

Finally, survival may depend on aspects of population structure. A certain genotype may have a high survival probability in a small founder population, while it would be clearly inferior in a very large, widespread population.

It should be evident from this discussion how misleading the picture of natural selection is which some authors have. A careful reading of Darwin's version shows that his concept of natural selection was far more mature than that of most of his opponents. Nevertheless, even he did not fully appreciate the power of the constraints and of chance.

The modern concept of natural selection has been rather frustrating for essentialist-deterministic philosophers. And yet who could question that the study of natural selection is a legitimate component of science? A philosopher thus has three choices: he can close his eyes to the existence of natural selection, or he can claim that, lacking the attributes of the phenomena of inanimate nature, it has to be excluded from science, or, finally, he can revise his view of what science is and enlarge his vocabulary and inventory of principles. It is the latter choice that has been made by almost all the younger philosophers of biology.

NOTES

This essay is previously unpublished.

1. Curiously, the tautology argument has been raised again and again, even though it must have been refuted by at least 8 or 10 qualified biologists and philosophers, for instance Caplan (1977), Mills and Beatty (1979), Ruse (1973), Stebbins (1977), M. B. Williams (1973), Riddiford and Penny (1984), Brandon (1981), and Hodge (1983).

In fact, it is possible to formulate the theory in a clearly nontautological manner (Lewontin 1972; Brandon 1981). Actually, this had already been done by Darwin. He pointed out that in every generation there is a great overproduction of individuals, only a small percentage of whom can survive and reproduce. Second, all these individuals differ in their genetic endowment, and therefore differ, at least in principle, in their adaptedness to their common environment. And third, the causes of the differences in adaptedness are in part heritable. It follows by simple logic that those with the highest adaptedness have the greatest chance to survive and reproduce.

2. The importance and prevalence of constraints on every aspect of evolution is now universally recognized. Annually, scores of papers are published dealing with various constraints. I will mention here only a few which happened to have come to my attention while I was writing these pages. Let me emphasize: these are only a small fraction of a vast literature: Peters (1985), D. B. Wake (1982), Cole (1985), Cheverud et al. (1985), Gould (1980), Riddiford and Penny (1984), Dawkins (1982). Lists of constraints were previously given by Gould and Le-

wontin (1979), Mayr (1982b), and Reif et al. (1985). Actually, constraints have been cited in the evolutionary literature from Darwin, Weismann, and Whitman on. Nothing is sillier than the question of some anti-Darwinians, "If selection can do everything, why are there maladaptions, extinction, etc.?" Only a person who completely lacks any understanding of evolutionary constraints can ask such a misinformed question.

REFERENCES

Beatty, J. 1981. What's wrong with the received view of evolutionary theory. *PSA 1980* 2:397–426 (Philosophy of Science Association).
——— 1984. Chance and natural selection. *Phil. Sci.* 51:183–211.
——— 1987. Natural selection and the null hypothesis. In J. Dupré, ed., *The Latest on the Best: Essays on Evolution and Optimality*. Cambridge, Mass.: MIT Press.
Belyaev, D. K. 1979. Destabilizing selection as a factor in domestication. *J. Hered.* 70:301–308.
Brandon, R. N. 1978. Adaptation and evolutionary theory. *Stud. Hist. Phil. Sci.* 9:181–206
——— 1981. A structural description of evolutionary theory. *PSA 1980* 2:427–439 (Philosophy of Science Association).
Brandon, R. N., and R. M. Burian, eds. 1984. *Genes, Organisms, Populations: Controversies over the Units of Selection*. Cambridge, Mass.: MIT Press.
Caplan, A. L. 1977. Tautology, circularity, and biological theory. *Amer. Nat.* 111:390–393.
Chetverikov, S. S. 1926. On certain aspects of the evolutionary process. [Russian.] Eng. trans: *Proc. Amer. Phil. Soc.* 105(1961):167–195.
Cheverud, J. M., M. M. Dow, and W. Leutenegger. 1985. The quantitative assessment of phylogenetic constraints in comparative analyses: sexual dimorphism in body weight among primates. *Evolution* 39:1335–1351.
Cole, B. J. 1985. Size and behavior in ants: constraints on complexity. *Proc. Nat. Acad. Sci.* 82:8548–8551.
Darwin, C. 1859. *On the Origin of Species*. London: John Murray. Facs. ed. Cambridge, Mass.: Harvard University Press, 1964.
——— 1868. *The Variation of Animals and Plants under Domestication*. 2 vols. London: John Murray.
——— 1871. *The Descent of Man, and Selection in Relation to Sex*. 2 vols. London: John Murray.
Dawkins, R. 1982. *The Extended Phenotype*. San Francisco: Freeman.
Dobzhansky, Th. 1937. *Genetics and the Origin of Species*. New York: Columbia University Press.

Endler, J. A. 1986. *Natural Selection in the Wild*. Princeton: Princeton University Press.

Ferguson, A. 1976. Can evolutionary theory predict? *Amer. Nat.* 110:1101–1104.

Ford, E. B. 1964. *Ecological Genetics*. London: Methuen.

Gillespie, J. H. 1986. Rates of molecular evolution. *Ann. Rev. Ecol. Syst.* 17:637–655.

Gould, S. J. 1980. The evolutionary biology of constraint. *Daedalus* 109(2):39–52.

Gould, S. J., and R. C. Lewontin. 1979. The spandrels of San Marco and the Panglossian paradigm. *Proc. Roy. Soc. London* B205:581–598.

Grant, P. R. 1986. *Ecology and Evolution of Darwin's Finches*. Princeton: Princeton University Press.

Ho, M.-W., and P. T. Saunders. 1984. *Beyond Neo-Darwinism: An Introduction to the New Evolutionary Paradigm*. London: Academic.

Hodge, J. 1983. The development of Darwin's general biological theorizing. In D. S. Bendall, ed., *Evolution from Molecules to Men*, pp. 43–62. Cambridge: Cambridge University Press.

Jacob, F. 1977. Evolution and tinkering. *Science* 196:1161–1166.

Kimura, M. 1983. *The Neutral Theory of Molecular Evolution*. Cambridge: Cambridge University Press.

Lerner, M. 1954. *Genetic Homeostasis*. Edinburgh: Oliver and Boyd.

Lewontin, R. C. 1972. Testing the theory of natural selection. *Nature* 236:181–182.

Maynard Smith, J. 1972. *On Evolution*. Edinburgh: Edinburgh University Press.

Mayo, O. 1983. *Natural Selection and Its Constraints*. New York: Academic.

Mayr, E. 1962. Accident or design: the paradox of evolution. In *The Evolution of Living Organisms*, pp. 1–14. Symp. Roy. Soc. Victoria, Melbourne 1959. Victoria: Melbourne University Press.

———— 1963. *Animal Species and Evolution*. Cambridge, Mass.: Harvard University Press.

———— 1976. *Evolution and the Diversity of Life*. Cambridge, Mass.: Harvard University Press.

———— 1982. *The Growth of Biological Thought*. Cambridge, Mass.: Harvard University Press.

Mills, S., and J. Beatty. 1979. The propensity interpretation of fitness. *Phil. Sci.* 46:263–286.

Naylor, B. G., and P. Handford. 1985. In defense of Darwin's theory. *Bioscience* 35:478–484.

Nevo, E. 1983. In X. Oxford and X. Robinson, eds., *Adaptive and Taxonomic Significance of Protein Variation*, pp. 239–000. New York:

Peters, D. S. 1985. Mechanical constraints canalizing the evolutionary transformation of tetrapod limbs. *Acta Biotheoretica* 34:157–164.

Reif, W.-E., R. D. K. Thomas, and M. F. Fischer. 1985. Constructive morphology: the analysis of constraints in evolution. *Acta Biotheoretica* 34:233–248.

Remane, J. 1983. Selektion und Evolutionstheorie. *Paläontol. Zeitschr.* 57:205–212.

Riddiford, A., and D. Penny. 1984. The scientific status of modern evolutionary theory. In J. W. Pollard, ed., *Evolutionary Theory: Paths into the Future*. London: Wiley.

Ridley, M. 1985. [Review of Ho and Saunders (1984).] *Nature* 313:823–824.

Ruse, M. 1973. *Philosophy of Biology*. London: Hutchinson.

Sober, E. 1984. *The Nature of Selection: Evolutionary Theory in Philosophical Focus*. Cambridge, Mass.: MIT Press.

Stebbins, G. L. 1977. In defense of evolution: tautology or theory? *Amer. Nat.* 111:386–390.

Trivers, R. 1985. *Social Evolution*. Menlo Park, Cal.: Benjamin/Cummings.

Tuomi, J. 1981. Structure and dynamics of Darwinian evolutionary theory. *Syst. Zool.* 30:22–31.

Tuomi, J., and E. Hankioja. 1979. Predictability of the theory of natural selection: an analysis of the structure of the Darwinian theory. *Savonia* 3:1–8.

Wake, D. B. 1982. Functional and developmental constraints and opportunities in the evolution of feeding systems in urodeles. In Mossakowski and Roth eds., *Environmental Adaptation and Evolution*, pp. 51–66. Stuttgart and New York: Gustav Fischer.

White, M. J. D. 1981. Tales of long ago. *Paleobiology* 7:287–291.

Williams, M. B. 1973. The logical status of the theory of natural selection and other evolutionary controversies. In M. Bunge, ed., *The Methodological Unity of Science*, pp. 84–102. Dordrecht: Reidel.

Wilson, D. S. 1983. The group selection controversy: history and current status. *Ann. Rev. Ecol. Syst.* 14:159–187.

PHILOSOPHICAL ASPECTS
OF NATURAL SELECTION

❧ Many of the controversies in biology, as well as in the philosophy of biology, are due to the equivocal use of certain terms. To eliminate much of this equivocation in relation to the theory of natural selection was one of the successful achievements of E. Sober's *The Nature of Selection*. But other equivocations remain. For instance, considerable confusion surrounds the term *unit of selection*; also few authors have recognized that the term *group* as used in the combination *group selection* is quite equivocal. In this essay I point out that there are four different kinds of groups, each different from the others as potential targets of selection. Since one of these four groups is the kinship group, I explain why there has been so much argument on the question whether kin selection is group selection. Anyone who attacks or defends group selection must, in my opinion, specify which of the four kinds of group selection he is talking about.

IN HIS preface E. Sober admits quite frankly that "until about eight years ago I had only the most cursory impression of what evolutionary theory is about." It must be admired what remarkable understanding of evolutionary biology Sober acquired in this short time. A difference of opinion within evolutionary biology is involved in the few cases where I disagree with Sober's interpretations.

Sober's book is far more than just a work on natural selection. Not only does he attempt to clarify other problems of evolutionary biology, he earns the gratitude of the biologist for his detailed philosophical analysis of such concepts as force, tautology, causation, chance, explanation, and correlation. These are terms nonphilosophers often use but interpret rather carelessly and sometimes wrongly. Sober shows again and again how easily one may slip into error by not defining terms rigorously. For example, he shows that defining group selection in terms of heritable variation in the fitness of groups or in terms of altruism is faulty. It is simply impossible to evaluate group selection properly until the term is rigorously defined.

The book is divided into two parts. The first one, entitled "Fitness, Selection, Adaptation" (pp. 13–211), contains six chapters. The second part, entitled "The Group Above and the Gene Below" covers pages 215–368 in three chapters and deals with the target of selection.

The Nature of Theorizing

What is the task of the philosopher when dealing with the problem of natural selection or, for that matter, with any biological problem? He must think, says Sober, "about causation, chance, explanation, and reduction." This is of course true for all biological theorizing, whether done by scientists or philosophers; but Sober also tried "to contribute what a philosopher can do—to uncover presuppositions and make them explicit." This is perhaps Sober's most important contribution, because hidden assumptions have been the bane of all scientific controversies.

The theory of natural selection is, basically, simplicity itself: If there is heritable variation in fitness in a population, evolution ensues automatically. And yet this seemingly so obvious principle was almost universally resisted for three-quarters of a century. The major reasons are, first, that Darwin was able neither to explain the source of variation nor to prove the actual occurrence of natural selection in nature; second, that a probabilistic theory was unpalatable in an age of determinism; and, third, that just about everybody was a typologist (essentialist), and selection simply makes no sense without population thinking.

Perhaps the most valuable part of Sober's analysis is his critical study of certain terms, like *causation, explanation,* and *correlation,* which by most scientists are accepted as so free of ambiguity that they can be used without any special concern. Sober makes us realize that the establishment of a correlation is not proof of a causation, nor is an explanation the same as a statement of causation. This becomes important when applied to more specific evolutionary terms like *fitness, adaptation,* or *sexual selection.*

In his second chapter Sober discusses the claim that the theory of natural selection, as articulated in the metaphor "the survival of the fittest," is a tautology. Sober's argument is actually somewhat different from that of other opponents of the tautology argument (for example, Beatty, Brandon, Kitcher, M. Williams), paying special attention to the problem of disentangling issues of causation from issues of explanation. He asks, can a theory be valid even though the underlying mechanisms are not yet understood (for example, continental drift without plate tectonics, natural selection without genetics)? What exactly is an explanation? Are there a priori truths?

Adaptation

The fact that, prior to the theory of natural selection, the term *adaptation* had teleological connotations should be completely irrelevant for a Darwinian. Yet, considering how widespread misconceptions concerning adaptation still are, these misunderstandings must be refuted at regular intervals, and Sober rightly devotes chapter 6 to this task. We are fortunate that several other philosophers and biologists, such as Brandon, Bock, Burian, and Wallace, have also recently published valuable discussions of the subject (see Part Three).

At the end Sober arrives at the traditional Darwinian conclusion (contra Lewontin, Williams) that there is no evidence "that selection is insufficient for adaptation" (p. 208). He also presents a philosophically rigorous definition of adaptation (p. 208), a simplified version of it being "adaptedness is that which has resulted from selection." This definition is supposed to cover both natural and sexual selection, but I fear that many results of sexual selection lead to structural or behavioral excesses which I find it difficult to designate as aspects of adaptedness.

Group Selection

Sober states in his introduction that it was G. C. Williams's analysis of group selection (1966) that got him interested in natural selection, and he devotes far more pages of his book to group selection than to any other subject. It is indeed a fascinating topic for a logician and historian of concepts. Can a group function as the target of selection? What is a group? Can a group have a fitness value that is independent of the fitness of the individuals of which it is composed? Can group selection be reduced to individual selection or to genic selection? It is through these and similar questions that Sober advances his analysis.

Ghiselin (1974) has demonstrated how widespread thinking in terms of group selection has been all along. Good Darwinians intuitively rejected it, however, and those who believed in group selection never made a major issue of it until the controversy was brought to a boil by the publication in 1962 of Wynne-Edwards's *Animal Dispersion in Relation to*

Social Behavior. Wynne-Edwards based his thesis on the conviction that animals, particularly social animals, had many characteristics which simply could not be explained by Darwinian individual selection. Some recent writers have narrowed this down to the problem of altruism, but the issue is much broader (Mayr 1963:197–198). Two authors almost immediately opposed Wynne-Edwards's thesis, but with rather different strategies. David Lack (1966, and several earlier publications, all curiously never mentioned by Sober) maintained the Darwinian thesis that all seeming cases of group selection could be explained in terms of individual advantage. G. C. Williams (1966) took over the silent assumption of the mathematical population geneticists that the gene is the unit of selection and that all group selection can be reduced to genic selection. Quite an enormous literature on the subject has since been generated; even a recent reader (Brandon and Burian 1984) is largely devoted to this subject. Seeing how many loose ends are left unresolved in Sober's account, it can be predicted that the issue will keep evolutionists arguing for a long time to come. Indeed, Wynne-Edwards has published a new book on the subject (1986).

As much ingenuity as Sober has invested in his analysis of group selection, his discussion is not altogether satisfactory because he fails to distinguish clearly among four different meanings of the term group, each of which has a different relation to natural selection. I think it helps an understanding of this complex problem to treat each of the four kinds of groups separately.

HAMILTONIAN GROUPS

These are kinship groups, consisting essentially of extended families. They are not ordinarily geographically isolated, although they may temporarily occupy a definite territory. In every generation they usually exchange a few individuals with other such groups. The development of a new altruistic trait in such a group may raise its reproductive potential, and this may raise its competitive success over other groups. Maynard Smith (1964) has correctly pointed out that selection for the enhancement of inclusive fitness—that is, kin selection—is not group selection at all, but rather a subdivision of individual selection. The defense of her young by a mother, or warning cries of an individual that benefit his kin, contribute to the spread of the genotype of these seemingly altruistic individuals, and since the prospective benefit is greater than the harm, such behavior by definition does not qualify as altruistic. All that is essential for kin selection is that close relatives live sufficiently near each other so that an animal's behavior can influence the survival or fecundity of the relatives with whom it shares part of its genotype.

WRIGHTIAN GROUPS

These are the local demes, described by Sewall Wright, that are temporarily isolated from each other but soon fuse again with the main body of the species population or with other Wrightian groups. It is primarily for these demes that group selection has been postulated.

Is it really true that interdemic selection among Wrightian groups can explain evolutionary phenomena inexplicable by individual selection? It would seem to me that this claim is based on several invalid assumptions:

(1) There are well-demarcated demes within species. This is of course a highly controversial subject, but it would seem to me that in the last 15 years so much evidence has been found for copious gene flow among most populations that such a claim cannot be upheld. One of a number of prerequisites for demes to serve as genuine entities is that the rate of gene flow by migration could not be greater than 5% per generation (Levin and Kilmer 1974). I would presume that such a low rate of gene flow would be found only in strongly isolated founder populations.

(2) The deme has an advantageous "group phenotype." Sober makes the peculiarly typological statement "group selection exists . . . when group phenotypes have certain sorts of effect on the fitness values of organisms" (p. 325). What is a group phenotype? Is dark skin a group phenotype of Blacks, or is it simply, as I am sure Sober would be the first to agree, a phenotypic property of every individual Black? Sober quite rightly rejects the procedure of averaging the characters of members of a group, and yet his concept of group phenotype and "group fitness value" seem to be based on exactly this procedure of averaging. Actually, I do not know of any deme in sexually reproducing species in which there is not considerable genetic variability and in which the major struggle for existence is not among members of the deme.

(3) The group as a single entity is the target of selection. Neither Sober nor anyone else has ever provided evidence in favor of this assumption. And when Sober says "the superior group sends out colonists," he is making a mistake that he himself has repeatedly warned against. No, it is not the group but rather the reproductively most successful individuals within the group, whose surplus offspring disperse beyond the group and colonize outside localities.

Sober, following Lewontin, cites the attenuation of the myxoma virus as an instructive example of group selection. Here I agree with Futuyma (1979) that this case can and must be explained entirely in terms of individual selection. The survival of a given rabbit in Australia, after the introduction of the virus, was entirely determined by the degree of virulence of the most virulent strain of the virus population in this rabbit. Less virulent strains within the same rabbit are irrelevant for the process of selection (but not for dispersal). The less virulent the most virulent strain in a rabbit is, the longer will be the survival of the host. This increases the probability of dispersal, and thus effects a continuing selection of increasingly less virulent strains. To refer to a "complete lack of selective advantage of avirulence within demes" is a misleading statement, because the aggregate of viral clones in a rabbit do not form a deme. Only the most virulent virus strain in a rabbit is relevant for selection, and as far as selection is concerned, the exclusive importance of this strain makes it appear as if there were no other strains in the host. Selection, therefore, is based entirely on the degree of virulence of the most virulent strain in a given rabbit. If, in the case of a mixed infection, a less virulent strain was dispersed before the death of the host, it would accelerate the spread of avirulence. By killing their rabbit hosts before they can be dispersed, the more virulent strains of course extinguish themselves.

MAYRIAN (FOUNDER) GROUPS

These are completely isolated and have no competitive relations to other groups, until they have become species, that is, have acquired isolating mechanisms and are able to become parapatric or sympatric. And then species selection is involved.

Wade's (1977) experiments deal with such founder groups. Since his selections are purely arbitrary and no selective competition among groups is involved, such as we would find in nature, I cannot see that exposing isolated founder populations to artificial selection pressures sheds any light on group selection in an evolutionary sense. That one can select for the most improbable characteristics in isolated laboratory populations or in fancy breeds of dogs and pigeons has, of course, long been known. But this has nothing to do with group selection as postulated by Wynne Edwards or other supporters of group selection.

CULTURAL GROUPS IN MAN

These are local assemblages in mankind, held together by cultural characteristics, including tribal customs, supernatural beliefs, and language or dialect, or by leadership. Such groups, particularly in primitive man, usually compete with each other, and may cause each other's extinction, as described already by Darwin (1871). Such cultural groups are found only in man and supply, in my view, the only well-established cases of group selection (Essay 5).

Among animals, group selection (excluding kin selection), if it exists at all, would seem to be an exceedingly rare phenomenon. Under the circumstances I wonder whether it is worth the time and effort so many people have devoted to it, particularly since the same outcomes can be explained in terms of individual selection.

Species Selection

Species are groups of a sort, and so-called species selection is perhaps a logical extension of group selection. Sober recognizes this and includes species selection in his analysis. Unfortunately, it is not a very penetrating study of the subject. The concept of a competition among species, resulting in the success and multiplication of some of them and the extinction of others, goes back to preevolutionary days. Charles Lyell explained the change in faunas in part by the "introduction" of a better adapted species to replace one whose allotted life span had run out. Darwin thought that most extinction of species was due to competition by better adapted species. This coming and going of species fits well with the modern concept of speciation, which considers it a random process, very much like mutation (Mayr 1963:620–621). A process of competition among species can be explained adequately by individual selection (Mayr 1982:169–170), and Sober agrees with this interpretation.

Two other kinds of species selection, recently proposed by Stanley, Gould, and their associates, are, however, not adequately dealt with by Sober. The first is what I call *speciator selection*—the capacity of certain species to establish numerous successful founder populations, that is, the capacity for frequent peripatric speciation. Such a capacity, it is presumed, is of selective advantage for two reasons. First, in periods of mass extinction the gene content of a species has a much greater chance to survive if it has very many daughter and granddaughter species than does the gene content of a species that has not speciated actively. Second, there is a much greater chance of an important evolutionary innovation occurring in a large number of daughter and granddaughter species, and innovative neospecies might initiate a major, highly successful, evolutionary advance. The capacity for rapid speciation thus, in the long run, would seem a great selective advantage and justify the designation *speciation selection* or *speciator selection*.

One can, however, advance some counterarguments. First, a conservative but highly successful widespread species may actually outsurvive all the more localized and therefore more ephemeral daughter species of a close relative. Second, the whole process of speciator selection is strictly based on individual selection. As I have shown elsewhere, peripatric speciation and its success depend on a number of attributes, such as dispersal drive, colonizing ability, and the capacity to speedily acquire new isolating mechanisms, that are the properties of individuals.

Finally, Gould has proposed a third mode of species selection, the process of *biased speciation*. If one were to accept the claim of punctuationism, that evolutionary change can occur only during short bouts of speciation, and if variation is strongly constrained (as indeed it probably is), then a parallel evolutionary trend in a family of related phyletic lines may be due entirely to a series of biased speciation events. Those daughter species would be most favored by selection that lie closest to the direction of the favored evolutionary trend. Whether such species selection occurs is, of course, at the present time, strictly hypothetical, and it will be up to the paleontologists to substantiate it. Owing to all the genetic, ecological, and geographical uncertainties of the process of speciation, it is perhaps rather premature to invest too much thought in species selection. This problem has recently also been considered by Maynard Smith in an analysis that differs considerably from mine, but in the end comes out pretty much the same way. Perhaps I go even a little further than Maynard Smith in claiming that individual selection is sufficient to explain all the phenomena designated as species selection.

Is "Unit of Selection" an Apt Term?

Curiously, Sober nowhere defines the term *unit of selection* rigorously; indeed, he implies repeatedly that it sometimes may mean "selection of" an object and sometimes "selection for" a property. Considering how clearly he himself realizes this ambiguity, it is curious that he nowhere attempted to undertake a terminological distinction. In my own account above, I have consistently referred to the *selection of* an object by the term "the target of selection." It is gratifying that ultimately Sober arrives nevertheless at the obvious conclusion, that to characterize the gene as the unit of replication establishes very little about what the unit of selection is (p. 252). He quite rightly rejects the claim of the genic selectionists that "the unit of selection must have a high degree of per-manence," and that only the gene, the unit of replication, qualifies. Variation, rather than permanence, is the principal prerequisite for selection.

Selection for a property of an individual is, of course, an entirely different matter. Indeed, an individual may be highly favored by selection owing to the possession of a single gene. As Sober rightly points out, the confusion of "of" with "for" is the major basis for the claims of the genic selectionists. Yet it is evident that there is still terminological equivocation in this area. The term *unit* is most often used in the quantification of forces (in physics) and of measurements. It is rather questionable whether it is legitimate to use the term *unit* for something so completely different as the target of selection (see also Ghiselin 1981:280). For me a target is not a unit.

Conclusions

In this analysis I have concentrated on the issues that are still controversial or have remained unresolved by Sober, but let me repeat that Sober has nevertheless given us what is perhaps the most careful and penetrating analysis of the concept of natural selection as it affects the process of evolution. He has exposed the invalidity of many of the arguments with which genic selection and group selection have been promoted, and has laid a foundation for further analyses. When the time comes for the development of a new philosophy of science, which also incorporates the methodology and conceptual framework of evolutionary biology, Sober's analysis will play a prominent role. Working my way through Sober's book made me realize the ambiguity of much of the evolutionary termi-nology. Even if Sober failed to provide a rigorous definition of some of the terms used by him (like *unit of selection*), he nevertheless has succinctly defined most others, and this will reduce future confusion. Both biologists and philosophers will benefit greatly from the study of Sober's careful analysis of the problems of natural selection.

NOTE

This essay is an extract from a review of E. Sober's *The Nature of Selection* (1984) that was first published in *Paleobiology* 12(1986):233–239. Parts of the review duplicated by Essay 6 have been omitted.

REFERENCES

Brandon, R. N., and R. M. Burian, eds. 1984. *Genes, Organisms, Populations.* Cambridge, Mass.: MIT Press.

Darwin, C. 1871. *The Descent of Man.* London: John Murray.

Futuyma, D. 1979. *Evolutionary Biology.* Sunderland, Mass.: Sinauer.

Ghiselin, M. T. 1974. *The Economy of Nature and the Evolution of Sex.* Berkeley: University California Press.

———— 1981. Categories, life, and thinking. *Behav. Brain Sci.* 4:280.

Lack, D. 1966. *Population Studies of Birds.* Oxford: Clarendon Press.

Levin, B. R., and W. L. Kilmer. 1974. Interdemic selection and the evolution of altruism: a computer simulation study. *Evolution* 28:527–545.

Maynard Smith, J. 1964. Group selection and kin selection. *Nature* 201:1145–1147.

Mayr, E. 1982. Adaptation and selection. *Biol. Zentralbl.* 101:161–174.

Wade, M. J. 1977. An experimental study of group selection. *Evolution* 31:134–153.

Williams, G. C. 1966. *Adaptation and Natural Selection.* Princeton: Princeton University Press.

Wynne-Edwards, V. C. 1962. *Animal Dispersion in Relation to Social Behavior.* Edinburgh: Oliver and Boyd.

———— 1986. *Evolution through Group Selection.* Oxford: Blackwell Scientific.

part three ૨ુ

A D A P T A T I O N

Introduction

For the natural theologian, the beautiful harmony of the world was the most convincing proof of its origin by creation. Ascribing this harmony to the action of natural selection meant replacing the hand of God by a game of dice, as Darwin's opponents said accusingly. How can adaptation be achieved by such a haphazard process?

To meet this criticism, Darwinians had to prove that adaptation is indeed a consequence of natural selection. In the ensuing controversy, claims ran the whole gamut from Weismann's slogan of the *Allmacht der Naturzüchtung* to counter claims of a total impotence of natural selection. And this controversy is still going on. Even today there is considerable disagreement on just how much of any genotype or phenotype is the direct result of natural selection.

Since everything created by God was evidently perfect, replacing God by natural selection led, by habit of thought, to the assumption that evolved nature was likewise perfect. Even Darwin thought so in the beginning, but soon realized that this conclusion was unnecessary (Essay 14). Some of his followers, however, succumbed to the teleological flavor of the word *adaptation* and assumed that every characteristic of an organism had to be molded toward perfection, independently of all other characteristics. Many geneticists, assuming the gene to be the target of selection and assuming, furthermore, that at every locus there was always one homozygously best allele (H. J. Muller), concluded likewise that near-perfect adaptation was an attainable goal and that the majority of organisms had indeed come close to it. Such committed adaptationists did not refrain from using a pseudo-teleological language when referring to adaptation.

Some opponents of this extreme position have cited neutral genes and neutral characters as refutations of selectionism. Most of such refutations are flawed because they ignore the fact that the individual as a whole (with its entire genotype) is the target of selection. Molecular genic changes without effect on fitness will be carried along as hitchhikers of selected individuals (Essay 6). This will have no consequences for adaptation, unless the new genotype is in an improved position for further evolution. The same is true for components of the phenotype. Since the phenotype as a whole is the target of selection, those changes that are selectively neutral are irrelevant for evolution. They do not supply an argument against the adaptationist program.

In the last 10–15 years most authors, in order to avoid all possibility of confusion, no longer say that natural selection produces adaptation, but rather that it results in adaptedness. But this does not yet eliminate all ambiguity. There is also considerable confusion about the term *fitness*.

Even though it may be difficult to express it in words, the average person knows exactly what the word *fitness* has meant traditionally. It is the capacity of an organism to survive in a given environment or, in other words, to cope with the conditions of the environment. Unfortunately, when searching for a term by which to express the genetic contribution of an organism (or genotype, or gene) to the next generation, R. A. Fisher chose *fitness,* defining it as "the per capita rate of increase." This gave the term, of course, a very different meaning from the traditional one. But this is not the only weakness of the transferred term. It is also unable to make any distinction between natural and sexual selection. Only natural selection gave fitness, as traditionally defined, but both natural and sexual selection can lead to an increase in fitness as defined by the population geneticists.

Fitness in the new definition measures only actual, "realized," success. However, an individual can be exceedingly well adapted, yet leave no offspring because it is prematurely killed by lightning or some other accident. Then it has zero "realized" fitness. It became necessary therefore to find another term for what was traditionally called fitness, and the term adaptedness came into use for it. But it had to be defined as a dispositional property. Adaptedness, consequently, was defined as "the propensity to survive" (Brandon 1978; Mills and Beatty 1979); or, to include the consequences of sexual selection, as "the propensity to survive and reproduce." (For further discussions of this problem, see Bernstein et al. 1983; Michod 1986.)

There are two conventions about the application of the term *adaptation*. For some authors an adaptation is any trait that not only increases the adaptedness of its possessor but that also was the product of evolution by natural selection (Brandon ms.). Other authors, in view of the difficulty of documenting the evolutionary history of a trait, are satisfied to call any trait an adaptation that increases the adaptedness of its possessor, regardless of its history (Bock 1980).

Adaptation, because it is so easily misunderstood and is a term that has legitimately several meanings, has been thoroughly discussed by a number of recent authors. Perhaps most important are Bock and von Wahlert (1965), Lewontin (1978; 1984), Bock (1980), Brandon (1985), and an excellent review by Burian (1983).

In spite of these analyses, there is still a widespread feeling that the concept and the process of adaptation are not yet fully understood. The geneticist Krimbas (1984) has even gone so far as to claim that the concept is not only superfluous but actually impedes the understanding of Darwinian evolution. This claim was at once rejected by Wallace (1984). It is true, however, that there is perhaps no other term in the Darwinian vocabulary about which there is so much uncertainty and misunderstanding as the term adaptation—and its relation to natural selection. To eliminate some of these uncertainties is the objective of Essay 8.

The problem of the ecological niche has played a considerable role in the thinking of various authors on the topic of adaptation. Some ecologists have rightly pointed out that many potentially available resources of the environment are unused or at least considerably underexploited. This means that the ecological niches actually filled by species leave a considerable residue of unused niche space. To what extent can new genetic variation within a species be utilized to expand the existing niche in order to exploit unused available resources? The fact that invading species sometimes have only a rather slight impact on the status and the abundance of the other members of a biota also documents the existence of unused niche space. Such unused resources are, of course, an invitation for evolution by the present biota and for the acquisition of new adaptations. Hence, the process of adaptation is encouraged not only by a steadily changing environment but also by previously unused resources. There are no empty niches if niche is defined as the requirements of a species. But such a definition ignores the existence of unused resources. The existence of such potential niche space explains, of course, why speciation is sometimes successful.

Considering how widespread and sometimes unexpected many adaptations are, it is not surprising that unexplained features of an organism are a challenge to the searching mind of the evolutionist. It was this challenge which induced Harvey to ask "What is the purpose of valves in the veins of vertebrates?" and thus to discover the circulation of blood. It induced Roux to speculate on the meaning of mitosis and come up with an essentially correct theory, even though he himself later gave it up because it was in conflict with others of his ideas (see also Mayr 1982:49–50).

The search for an unknown functional meaning of a feature of an organism has been called *the adaptationist program*. In the opinion of a few critics, some adaptationists have gone too far in searching for the significance of structures or behaviors, and these efforts have been ridiculed by Gould and Lewontin (1979). Some of the arguments against the adaptationist program are undoubtedly valid, though the attitude of these critics seems, as a whole, to be unnecessarily negative (see Reif 1982). Essay 9 is an attempt to outline the principles of a balanced adaptationist program.

Part of the controversy over adaptation arose from the relatively recent efforts to discover an adaptive significance of differences among populations and species. In the first 50 to 80 years after the *Origin* there was enormous interest in adaptation, but adaptation of the higher taxa. It was rather easy to demonstrate that the characters by which mammals differ from lizards or those which enabled birds to adopt an aerial mode of life were adaptive (no matter how defined). But except for the geographical regularities expressed in Bergmann's and the other climatic rules, little effort was made to demonstrate a selective significance of differences among populations and related species. This new research program was virtually initiated by the Chetverikov school (including Dobzhansky) for genetic and chromosomal characters and by Lack (finch bills) and Cain and Sheppard (snail banding) for phenotypes. Through H. Böker, D. D. Davis, and W. Bock it spread into evolutionary morphology. To look for this type of evidence definitely required the adoption of an adaptationist program. And it is in most instances a very difficult enterprise. No one, for instance, has up to now been able to suggest an adaptive significance for the remarkable geographic variation in coloration found in the Papuan bird genera *Pachycephala, Monarcha, Rhipidura,* and *Myzomela.* When such differences are due to sexual selection, as most of these color differences probably are, they are adaptive only in a rather special sense. Whole classes of characters previously considered neutral, such as specializations in male genitalia, make sense when considered the result of sexual selection (Eberhard 1985).

The adaptationist program would be only a way to produce "just-so stories," if it were not possible to test suggested functions. The hypothesis that valves in veins regulate blood flow can be tested very easily. To merely ask whether a structure has any possible function can never lead to an answer, unless one asks some more specific questions first. And that is what the adaptationist program does, and that is what has been done in a number of recent researches.

The major difficulty in establishing the adaptedness of characters is methodological. It is usually possible only through correlation to establish the selective value of a given character. The connection between a selective value for a given character and optimal fitness is often not at all evident. However, it could be shown for a European flycatcher that tarsus length was closely correlated with survival from fledging until breeding—with clutch size in females, and in males with the ability to attract mates (Alatalo and Lundberg 1986). Tarsus length, of course, is merely an indication of a certain body form, which is what is actually selected for. Leisler (1975; 1977) showed that many species-specific morphological features of European warblers (*Sylvia, Acrocephalus, Locustella*), particularly the proportions of the several components of their extremities, are closely correlated with features of their locomotion and the properties of their environment. This was first shown by Palmgren (1936). One may be at a loss to prove the adaptedness of such species-specific features in the same manner as one proves a mathematical theorem, but the inference is certainly strong.

REFERENCES

Alatalo, R. V., and A. Lundberg. 1986. Heritability and selection on tarsus length in the pied flycatcher. *Evolution* 40:574–583.

Bernstein, H., H. C. Byerly, F. A. Hopf, R. A. Michod, and G. K. Vermulapalli. 1983. The Darwinian dynamic. *Quart. Rev. Biol.* 58:185–207.

Bock, W. J. 1980. The definition and recognition of biological adaptation. *Amer. Zool.* 20:217–227.

Bock, W. J., and G. v Wahlert. 1965. Adaptation and the form–function complex. *Evolution* 19:269–299.

Brandon, R. N. 1978. Adaptation and evolutionary theory. *Stud. Hist. Phil. Sci.* 9:181–206.

———— 1985. Adaptation explanations. In D. J. Depew and B. H. Weber, eds., *Evolution at a Crossroads*, pp. 81–96. Cambridge, Mass.: MIT Press.

Burian, R. M. 1983. Adaptation. In M. Grene, ed., *Dimensions of Darwinism*, pp. 287–314. Cambridge: Cambridge University Press.

Eberhard, W. G. 1985. *Sexual Selection and Animal Genitalia*. Cambridge, Mass.: Harvard University Press.

Gould, S. J., and R. C. Lewontin. 1979. The spandrels of San Marco and the Panglossian paradigm. *Proc. Roy. Soc. London Ser. B.* 205:581–598.

Krimbas, C. B. 1984. On adaptation, neo-Darwinian tautology and population fitness. *Evol. Biol.* 17:1–57.

Leisler, B. 1975. Die Bedeutung der Fussmorphologie für die ökologische Sonderung mitteleuropäischer Rohrsänger (*Acrocephalus*) und Schwirle (*Locustella*). *J. Orn.* 116:117–153.

———— 1977. Ökomorphologische Aspekte von Speziation und adaptiver Radiation bei Vögeln. *Vogelwarte* 29(Sonderheft):136–153.

Lewontin, R. C. 1978. Adaptation. *Sci. Amer.* 239:156–169.

———— 1984. Adaptation. In E. Sober, ed., *Conceptual Issues in Evolutionary Biology,* pp. 235–251. Cambridge, Mass.: MIT Press.

Mayr, E. 1982. *The Growth of Biological Thought.* Cambridge, Mass.: Harvard University Press.

Michod, R. E. 1986. On fitness and adaptedness and their role in evolutionary explanation. *J. Hist. Biol.* 19:289–302.

Mills, S., and J. Beatty. 1979. The propensity interpretation of fitness. *Phil. Sci.* 46:263–286.

Palmgren, P. 1936. Bemerkungen über die ökologische Bedeutung der biologischen Anatomie des Fusses bei einigen Kleinvogelarten. *Ornis Fenn.* 13:53–58.

Reif, W.-E. 1982. Functional morphology on the procrustean bed of the neutralism-selectionism debate. *N. Jb. Geol. Paläont. Abh.* 164(2):46–53.

Wallace, B. 1984. Adaptation, neo-Darwinian tautology, and population fitness: a reply. *Evol. Biol.* 17:59–71.

essay eight 🙠

ADAPTATION AND SELECTION

IN THE first 80 years after the publication of Darwin's *Origin of Species* (1859) almost total disagreement prevailed among evolutionists concerning the causes of evolution. In the pre-Darwinian days of natural theology, adaptation was explained as the product of design by a benign and provident Creator. According to the Darwinians, however, natural selection in an infinitely variable world could achieve any observed adaptation. This explanation was unacceptable to most early evolutionists, who considered it as inconceivable that such a mechanical process as selection could account for the beautiful harmony of nature with its admirable adaptations and co-adaptations. Some of these anti-Darwinians proposed therefore the existence of finalistic mechanisms in nature, such as orthogenesis, nomogenesis, or aristogenesis. Others, beginning with Darwin himself, believed that natural selection was supplemented by an inheritance of acquired characters ("use and disuse"). Still others thought that all major evolutionary novelties owed their existence to macromutations (for example, de Vries, Goldschmidt, Schindewolf). The actual existence of all these postulated processes was thoroughly refuted in the first 40 years of the century and, as a result of the evolutionary synthesis of the 1930s and 40s, selection seemed to have the field all to itself.

Since about 1940 we have been very comfortable with the thought that all adaptation is the result of natural selection and that natural selection would prevent any maladaptive developments. There has been occasional dissent and endeavours to call attention to chance components in evolution, but these were generally considered to be of minor, if not negligible, importance. The classical view of Darwinian evolution that natural selection will lead to ever greater fitness until every organism is perfectly adapted to its living and inanimate environment, seemed well substantiated. As Darwin said in the *Origin* (p. 489), "As natural selection works solely by and for the good of each being, all corporeal and mental endowments will tend to progress towards perfection."

This situation has changed rather significantly in the last 15 years. Several authors have emphasized the chance element in evolution far more strongly than had been done previously and have attempted to document a considerable impotence of natural selection. This has gone so far that some extremists have questioned the evolutionary significance of selection altogether, and even less extreme authors have ascribed a major part of evolution to various forms of so-called Non-Darwinian evolution, effected by neutral genes and neutral characters. Since these authors erroneously define Neo-Darwinism as the evolutionary modification of gene frequencies (Eldredge 1980), they claim that the newer findings had refuted Neo-Darwinism. These claims have caused considerable confusion, particularly among younger students, but have raised for all of us the important question: Is what we have believed up to now, really wrong? My own answer is no. In fact, some of the argument is a purely semantic one, caused by misconceptions and the misuse of terms. It seems to me that the time has come for a renewed analysis of the major concepts of evolutionary biology, particularly the terms adaptation and selection. This is the reason for the choice of my topic.

Adaptation

The ambiguous meaning of the term adaptation is particularly disturbing. For instance, it is used both for the process of becoming adapted and for the end stage of having achieved adaptedness. It is, on one side, used for the somatic or nongenetic adaptation of an individual, as for instance for the adaptation of a person to low oxygen pressure at high altitudes and, in the case of bacteria, to the presence or absence of certain nutrients in the substrate, or else it is used in a strictly genetic sense to denote the reconstruction of the genotype owing to continued selection pressure over many generations. Some authors (Bock 1980) would restrict the term adaptation to single components of the phenotype, while most others speak of the adaptation of an individual in an overall sense, considering the term as synonymous with fitness. Among recent treatments of the problems of adaptation, I particularly like that of Lewontin (1978), although I do not agree with him in every detail. Sophisticated definitions of adaptation have been proposed in which adaptation is defined in terms of the saving of energy. However, since no method is available by which anyone has yet succeeded in measuring such postulated savings of energy, I prefer a simple descriptive definition. Adaptedness is the morphological, physiological, and behavioral equipment of a species or of a member of a species that permits it to compete successfully with other members of its own species or with individuals of other species and that permits it to tolerate the extant physical environment. Adaptation is greater ecological-physiological efficiency than is achieved by other members of the population. Improved adaptedness may be due to a particular component of the phenotype, or to a single gene, or to the total genotype. One can ask some very disturbing questions about adaptation. Surely every fish is adapted for its life in water and yet in the history of the vertebrates ten thousands of species of fish have become extinct, either because they were not sufficiently well adapted to some component of the environment or (what may be the same) because they lost in the competition with some "better adapted" other species. The same is true for individuals within a species. All have the same species-specific adaptations and yet only a small minority will survive into the next generation. Were these better adapted than the ones that lost in the struggle for existence? Thus, it is evident that we have "adapted" and "better adapted." This is precisely the process of natural selection, which on the average, favors those that are "better adapted." For more detailed analyses of the problem of adaptation see Bock (1980), Bock and von Wahlert (1965), Dobzhansky (1956), Lewontin (1978), Muller (1949), Stern (1970).

In order to provide a somewhat richer concept of adaptation, one might consider *for what* an animal or plant is adapted. To this question an ecologist might give the quick answer: to its *niche*. When one analyzes the situation in more detail, one finds that this is a rather unsatisfactory answer. Let us take for instance a species of ducks. Is it adapted to the air-niche or to the water-niche or, if to both, is there an air-water-niche? To cope with these difficulties evolutionary ecologists have proposed a number of finer distinctions. G. G. Simpson has proposed the term *adaptive zone* for any major kind of environment to which an organism can become adapted. The shift from an aquatic to a terrestrial mode of living or from a carnivorous to a vegetarian diet (or vice versa) is a shift into a different adaptive zone. Niches describe the partitioning of resources within an adaptive zone. Purely descriptively, a niche is the outward projection of the needs of an organism, or, in G. E. Hutchinson's definition, it is a multidimensional resource space.

Whether niches exist that are *not* occupied by any species is a question the answer to which depends on the definition of niche. That there are resources of nature that are not fully utilized, at least not by the higher components of food chains, can hardly be questioned. The tropical forests of Borneo and Sumatra, for instance, provide resources for 28 species of woodpeckers. By contrast there are no woodpeckers what-so-ever in the exceedingly similar forests of New Guinea, and on that island hardly any other species of birds make use of the woodpecker niche. The existence of insufficiently utilized resources is further documented by successful cases of colonization or invasion which do not result in a visible decline of any previously existing species. These observations confirm the inability of selection to produce perfection, for in a perfectly designed world all resources ought to be utilized with optimal efficiency.

How Good Is Selection?

Some, perhaps even many, evolutionists assumed that adaptation must be "nearly perfect." This assumption is particularly evident from the accounts of those geneticists who believed in a constant fitness value of any given gene and in the existence of one particular allele that would be superior to all others. On the basis of such assumptions, one particular genotype should have not only the greatest actual fitness but also the greatest possible fitness (except for the presence of a few recently mutated and not yet eliminated deleterious recessives). An equally ultraselectionist viewpoint was also adopted by some naturalists who believed that every component of the phenotype was adapted to a particular mode of life and that none of it was a remnant of an ancestral heritage, tolerated by natural selection but not specially shaped for life in a particular adaptive zone.

Such selectionist extremism that looks for an *ad hoc* adaptive value for every detail of organic structure and behavior has been attacked since the days of Darwin (Dobzhansky 1956). The Darwinian botanist Thiselton-Dyer severely criticized in 1883 "facile speculation based on natural selection." Poulton (1896), likewise, admonished evolutionists to take "the necessary precautions" when invoking natural selection.

By far the majority of recent evolutionists have been quite cautious about ascribing the conformation of specific structures to selection, because they realize that much of the phenotype is a byproduct of the evolutionary past, tolerated by natural selection but not necessarily produced under current conditions. (See Mayr 1963, chaps. 8, 9, 10, for constraints on the workings of natural selection.)

Adaptation through Selection

When we study a particular structure or other so-called adaptation, it is, unfortunately, nearly always impossible to prove that this feature is the direct result of natural selection. Already Darwin encountered this difficulty. Almost all of his arguments in the *Origin* in favor of natural selection were based on deductive inferences as follows. If there is heritable variability in a population and if only a small fraction of the offspring survives, then those individuals that have certain attributes facilitating survival and reproduction will obviously have a better chance to contribute to the gene pool of the next generation than other individuals. This is a perfectly logical conclusion, but where was Darwin's proof? Darwin did not have a single conclusive one.

Now, 120 years later, every textbook of evolution has at least one chapter devoted to proofs of selection. Many of them concern the selective value of individual genes, gene combinations, and gene arrangements, as revealed through population cages or other experimental procedures. Sometimes it has also been possible to establish in the field a correlation between gene frequencies and certain environmental factors. The effect of selection would be far more convincing if one could demonstrate a selective value of specific components of the phenotype. And in fact this has been demonstrated by Kettlewell and others for the cryptic color of certain moths, also for fishes, rodents, and other animals.

Darwin was overjoyed when in 1863 Bates published his discovery of *mimicry*. The similarity of models and mimics permits the making of definite predictions, and these predictions were confirmed, as Bates was able to show. The inference that an adaptive change, observed in nature, is due to selection is often very strong. For instance, when the loss of a species from a certain habitat results in the take-over of its niche by another species, and when this species responds to the niche enlargement by an appropriate morphological change, for instance in a bird in an increase of bill size, one is permitted to conclude—in absence of any other reasonable explanation—that this change was due to natural selection. Lack (1947) was the first to establish such a correlation for the variation of certain species of finches in the Galapagos Islands. Since that time numerous cases of character divergence in the area of overlap of competing species have been found.

A strong inference on selection can also be made in the numerous cases where several unrelated species have convergently acquired a similar adaptation because they occur in the same habitat or occupy a similar niche.

Desert coloration or whiteness in the arctic are well-known cases. Others are the numerous similarities among diving birds or among raptorial birds. Actually, this category includes most biological adaptations recorded in the literature. Almost all the characters formerly attributed to an inheritance of acquired characters belong in this category.

The claim, sometimes made, that such convergent similarities might be due to chance is refuted by statistical analysis. If desert coloration were due to chance, we should find in deserts at least as many species with other kinds of coloration, but this we do not find. The same refutation of chance can be made for any of the vast number of adaptations believed to be due to selection. We must remember, however, that often there are multiple pathways to adaptation and that adaptation is sometimes expressed in diversity rather than in convergent uniformity. This is true, for instance, for flower structure, whereby different species of plants facilitate pollination by specific pollinators.

When a pollinating insect, in turn, is beautifully adapted to a specific species of plants, it would be difficult to refute the validity of the thesis of adaptation through natural selection.

I go so far as to claim that adaptation through selection can sometimes be invoked even when the adaptive significance of a structure is not clear. For instance, the specialists still argue whether the conspicuous plates along the backbone of the dinosaur *Stegosaurus* played a role in defense, or in courtship, or in thermoregulation. However, when one views this energy-costly structure, one cannot escape the conclusion that it could not have evolved except through natural selection because it provided some benefit to the possessors of this structure.

The pecten in the eyes of birds is another conspicuous structure, the precise function of which is still controversial. Yet, every selectionist will rightly insist that this structure would not have such a wide distribution among birds if its spread had not been favored by natural selection.

Opponents of natural selection, from 1859 to modern times, have taken great delight in describing various characteristics of organisms for which a selective value could not be demonstrated. Literally hundreds of examples of seemingly "neutral" or even deleterious characters were listed in anti-Darwinian treatises. The most frequent response of the Darwinians was to say that the selective significance of the particular character had not yet been discovered. Up to a point, this was, indeed, a legitimate argument, since the true meaning of numerous structures, such as the lateral line in fishes, was not discovered until many years after 1859. I still remember the long arguments I had with Dobzhansky and Epling whether or not chromosomal polymorphism in Drosophila (Mayr 1945) had selective value. It was Dobzhansky himself who eventually provided the most decisive evidence in favor of selection.

A second argument of the Darwinians was the suggestion that such seemingly neutral or deleterious features are the pleiotropic byproduct of genes selected for other contributions to fitness. Subsequent analysis has, indeed, sometimes demonstrated the occurrence of pleiotropic effects, but what proportion of so-called neutral characters can be explained by pleiotropy is still uncertain.

At this point we are up against a fundamental difficulty. Any adaptation in an organism is the endproduct of a long historical sequence of consecutive steps. In the case of such a *historical narrative*, as the modern philosopher calls it, it is never possible to prove the causation of each step retroactively. Historical causation is fundamentally different from causation as demanded by the laws of the physical sciences. In the absence of a possibility of proof, all we can claim is that our explanation is consistent with all the known facts or that it explains these facts better than any other suggested explanation.

Selectionists, for a long time, were handicapped by a major fault of conceptualization, namely by adopting the wrong target for the action of natural selection, the gene. Those who did so were unable to explain the fixation of neutral or slightly deleterious genes, except by invoking errors of sampling in small populations. Much of this problem disappears, however, as soon as one assumes the individual as a whole to be the target of selection. If the phenotype as a whole has superior fitness, it can carry along many quasi-neutral or slightly deleterious genes and gene combinations, particularly if there is reasonably tight linkage. An organism is an integrated whole and not a collection of independent genes. An atomism that gives every gene a separate fitness value or that dissects the phenotype into the greatest possible number of separate features, each supposedly having evolved for the sake of its own optimal adaptation, is an entirely inappropriate approach toward the understanding of natural selection.

Evolution by Random Walk

The critics of superselectionism deserve credit, however, for having emphasized the evolutionary role of chance. The idea that chance may be important in evolution is, of course, not new. Gulick (1873) was the first author to develop a theory of evolution based on random variation. He claimed that the patternless geographical variation of the *Achatinella* snails on Oahu, Hawaii, could not possibly be explained in terms of natural selection. Such evolution by random variation was supported by the Hagedoorns (1921) and other naturalists, but received its most concrete formulation in the work of Sewall Wright (1931; 1932). Wright postulated that changes of gene frequencies might occur through errors of sampling in small populations and lead to evolutionary change without participation of natural selection. Such populations would go through an "inadaptive phase" during which they would experience "genetic drift." Both Dobzhansky (1937) and G. G. Simpson (1944) adopted these stochastic perturbations as an important evolutionary process. Opponents, even though admitting the temporary occurrence of errors of sampling, called attention to the effect of recurrent mutations and of gene flow that would soon reverse temporary depletions of local gene pools, and would tend to counteract the effects of stochastic processes. Even though the effects of genetic drift are now interpreted somewhat differently than in the 1930s, nevertheless Wright's formulation of a second legitimate factor in evolutionary change resulted in a more balanced approach to the causal interpretation of evolutionary change. Unfortunately the discovery of the legitimacy of random processes led, in the writings of some authors, to the posing of absolute alternatives: Is evolution due to selection or due to "random walk"—stochastic processes? Such strict alternatives produce more heat than light. Given the evidence now available, long-term evolution without natural selection is inconceivable. The mere fact of the vast reproductive surplus in each generation, together with the genetic uniqueness of each individual in sexually reproducing species, makes the importance of selection inescapable. This conclusion, however, does not in the least exclude the probability that random events also affect chances of survival and of the successful reproduction of an individual. The modern theory thus permits the inclusion of random events among the causes of evolutionary change. Such a pluralistic approach is surely more realistic than any onesided extremism.

The theory of evolution by random walk received its greatest boost through the study of enzyme polymorphism. Calculations were offered by Kimura, Lewontin, King and Jukes, and others, purporting to show that much of the enormous variability of enzyme genes could not possibly be due to heterozygote superiority because otherwise the genetic load due to deleterious homozygotes would be too great. Hence, these authors concluded that this variation must be due to selectively neutral mutations. These authors refer to the stochastic processes causing changes in the frequency of neutral genes as Non-Darwinian evolution. Counterarguments have been produced by Ayala, Richmond, Powell, and others, and the controversy is still going on. My own view is that some of this variation is indeed "neutral," producing nothing but evolutionary noise; however, a considerable part is nevertheless due to selection. Neutrality is, of course, particularly true for those mutations that consist of base-pair substitutions which, owing to the degeneracy of the code, do not affect amino-acid production.

The arguments of the proponents of Non-Darwinian evolution are based almost entirely on the variability of enzyme genes as revealed by electrophoresis. The latest findings of molecular biology suggest, however, that this particular class of genes may be of far smaller evolutionary significance than the remainder of the nuclear DNA. Our knowledge of that part of the DNA, that is not involved in the production of soluble enzymes is still very rudimentary. Yet it appears increasingly possible that precisely this part of the genetic material is the most important for natural selection. At the present time this suggestion is merely a hunch, but considering the great contemporary activity in the study of middle repetitive DNA and of transposing genes, I am confident that definitive answers will be available within the next five or ten years.

Species Selection and Extinction

One particular objection made by the opponents of selectionism most frequently was that the gradual replacement of one allele by another and the gradual change of gene frequencies in a population could not possibly explain such major evolutionary innovations as the origin of the eukaryotes, of the metazoans, and of all drastic changes of adaptive zones such as the origin of terrestrial tetrapods or of aerial birds. Evolutionary events of such magnitude, said the opponents of Darwin, cannot possibly be explained by gradual selectionist evolution. Up to a point, these opponents of gradualism were right, as we shall presently see. The escape from this dilemma, however, was already outlined by Darwin in his discussion of extinction.

Let us first look at this problem in its historical context. When the concept of a struggle for existence originated in the eighteenth century, it was primarily seen as a struggle between species, let us say between a predator (wolf) and its prey (sheep), or, to a lesser extent, as a struggle of an organism against the forces of the inanimate environment (cold, drought, and so on). However, in the eyes of natural theologians this struggle was viewed in rather benign terms as a method of adjusting to a fluctuating balance. When the reality of extinction was finally realized (Blumenbach, Cuvier), something Lamarck still had refused to admit, the existing explanatory model of the benign balance was no longer adequate. Ideas were now rediscovered which in their nucleus went all the way back to the Greeks. Perhaps a species has an allotted lifespan, it was said, and would last only till its time had run out, to be replaced by a newly created or otherwise newly originating species. To a considerable extent this was Charles Lyell's explanation for extinction, and Darwin accepted it for some time (as shown in his notebooks of 1837–1838). Lyell, however, added one crucial new thought: Perhaps a species may become extinct because it is less well adapted than some other "more recently introduced" species which is competitively superior. Competition among similarly adapted species, thus, became an added component of the struggle for existence. This provided a potential explanation for many previously puzzling cases of extinction.

Introduction of the concept of competition among species resulted in a considerable expansion of the concept of natural selection as originally conceived by Darwin. The term natural selection, as it is almost universally adopted, refers to an *intra*populational process, because it consists of competition among individuals of the same population. However, potentially there are also two kinds of *inter*populational selection. One of these, called *group selection*, refers to competition among different populations of one and the same species. Wynne-Edwards (1962) and his followers have invoked group selection to explain the origin and spread of traits, let us say aberrant sex ratios, which they believe could not be gradually built up within a population through individual selection. However, much of the analysis of evolutionary biology in the last 20 years has resulted in refuting one after the other of the reputed cases of group selection. Even where it seems to occur, as in social species, it can usually be explained as selection for greater inclusive fitness, that is, by natural selection. (For a discussion of group selection see Essay 7.)

The other kind of interpopulational selection is *species selection*. This refers to competition among whole species. In a mild version, species selection leads to what Darwin called character divergence; in more drastic cases it leads to the extinction of the inferior competitor.

Species selection owing to competition, however, has a rather typological flavor; and after Darwin in 1838 had developed the concept of natural selection based on competition among individuals, competition among species was rather pushed into the background in Darwin's thinking. He took it up again when he developed the concept of character divergence, which was based strictly on competition among species. The concept, however, disappeared again from the consciousness of the evolutionists when geneticists began to dominate evolutionary thinking. For them, evolution was simply the gradual modification of gene pools, and competitive interaction among species did not enter their thinking to any degree.

This all changed when the naturalists, beginning with Lack in 1944, reemphasized the ecological interaction among species. It was now realized that selection, although always involving individuals, is reflected at several levels, not only that of genes in gene pools, but also that of species in natural communities. It led to the recognition that competition among species is one of the major evolutionary forces. Such competition is importantly involved in the turnover of floras and faunas, on islands as well as on continents. A large proportion of extinctions is apparently due to the arrival of a superior competitor.

In order to avoid misunderstandings, it must be emphasized that species selection is not in conflict with individual selection. In species selection the actual process of selection is also carried out through the success or failure of individuals, except that in this case the reproductively more successful individuals tend to belong to one species and the less successful ones to the species which is the loser in the competition. As a result, the gene pool of the losing species continues to dwindle until the species as a whole is extinct.

The Origin of New Species

This still leaves the question unanswered: where did the superior species come from, the species that won the competition owing to the possession of some new faculty, structure, or behavior? How did this species originate? As long as the attention of evolutionists, particularly of paleontologists and of geneticists, was exclusively directed toward the vertical component of evolution, no answer was possible. Haldane and others had shown how evolutionarily inert species are that are widespread and very populous. The replacement of inferior genes in such species is so slow, and connected with such high evolutionary cost, that not much will happen in a million years, even allowing for the existence of stochastic processes. Paleontologists had long recognized, as no one emphasized more than Simpson, that shifts into new adaptive zones are of crucial importance in evolutionary advance. But they did not investigate where the species came from that invaded the new adaptive zones. Nor did they explain how the shift into the new adaptive zone occurred. Usually, when they attempted an explanation they tended to propose models of sympatric speciation. This was also recently done by Lewontin (1978) when he stated that "as the environment changes, the single adaptive peak may become two distinct peaks, and two populations diverge to form distinct species." Actually, in the cases of actual speciation studied by me, I have been unable to find any instances that would be consistent with this explanatory model. What actually seems to happen almost universally is that the first step is a split of the population into two, owing to the establishment of a founder population beyond the species border or owing to a splitting of the population following the origin of a new geographical barrier. The change of the environment mentioned by Lewontin is a secondary consequence of the establishment of a new population.

Why Is the Shift into a New Adaptive Zone Such a Difficult Problem?

Long before anyone thought of evolution, it was understood that the relation between an organism and its environment was largely a steady-state balance. The species population of each organism constantly removes all deviants, in order to preserve its type. It will, of course, attempt to cope with the variations of the physical and biotic environment, but this kind of steady-state adaptation to conditions is essentially conservative and quite unable to produce major evolutionary novelties or drastic departures from existing conditions. Van Valen (1973) has expressed this in his "Red Queen Hypothesis," based on the metaphor that, in a continuously changing world, one must forever run in order to be able to stay at the same place.

But, if this is true, and there is every reason to believe that it is true, how can one explain major evolutionary shifts and the origin of evolutionary novelties? A possible solution occurred to me during the study of the distribution of new species and genera. As early as 1942 (pp. 284–285) I called attention to the occasional origin of new genera in peripherally isolated populations. By 1954 I was ready to propose the theory that major evolutionary shifts are most likely to occur in peripherally isolated founder populations. I pointed out that such incipient species are ideally structured to incorporate chromosomal rearrangements and to shift into new adaptive zones. I call this mode of speciation "peripatric speciation."

The production of such peripherally isolated new species seems to proceed at a high rate, but most of these experiments of nature are unsuccessful. When a new species is competitively inferior either to the parental species or to one of its new sister species, it will become extinct without ever leaving a trace in the fossil record. Occasionally, however, the genetic revolution in the founder population may lead to such a loosening up of constraints that the neospecies can enter a new adaptive zone or make some other kind of evolutionary innovation. This can have two (more or less overlapping) effects: either the new adaptation removes the neospecies from too serious a competition with already existing species, thus leading to an enrichment of the biota (and an increase in diversity) or the neospecies becomes a particularly successful competitor and causes the extinction of one or several existing species.

Any new conquest of a major adaptive zone, let us say the origin of flight among the ancestors of birds, produces such a rapid chain-effect of improvements and extinctions that the new "type" will soon be drastically discontinuous with its ancestors. Since the connecting links are of very short duration, and of limited geographic range, they are not likely ever to be found in the fossil record. Furthermore, they succeed each other with such rapidity that they appear to be discontinuous as far as the fossil record is concerned, and paleontologists thus usually refer to them as saltations. Actually, such adaptive shifts and productions of evolutionary novelties occur in populations and are gradual in a strictly Darwinian sense in spite of the rapidity with which they occur.

It is now clear why species selection is of such fundamental significance in the process of macroevolution. First, because it explains most cases of extinction that would otherwise not be explicable. Second, because it is an important driving force in macroevolution. All the indications are, although this is not accepted universally, that slow changes in gradually evolving lines are of a rather limited evolutionary significance. Apparently they rarely lead to the origin of a major evolutionary novelty or to a switch into a new niche or adaptive zone. By contrast, shifts of macroevolutionary significance occur in peripherally isolated founder populations and are perhaps restricted to them. The discovery of this process provided the answer to all those Darwinian critics who had claimed that macroevolutionary events could not be explained by selection and gradual evolution. The crucial aspect of the new explanatory model is a shift in emphasis from the purely genetic component (change in gene frequencies, mutation) to the phenomena of natural history such as population, isolation, geographic location of populations, competition, and behavioral shifts. The new interpretation of macroevolution is yet another manifestation of the successful synthesis of the thought of several biological disciplines.

Summary

(1) Adaptation, as measured by evolutionary success, consists of a greater ecological-physiological efficiency of an individual than is achieved by most other members of the population or at least by the average.

(2) Adaptation is achieved by the greater survival or higher reproductive success of certain individuals owing to the fact that they possess ecological-physiological traits not, or only partially, shared by other individuals of their population, traits that are useful in the struggle for existence.

(3) Since the individual as a whole is the target of selection, many neutral or even slightly deleterious genes may be carried in a population as "hitch hikers" of favorable gene combinations.

(4) As a result, not every character of an individual or of a species is optimally adapted. Indeed, it is methodologically very difficult to prove the selective value of many characters.

(5) The frequency of genes in populations, particularly in small populations, is subject to stochastic perturbations. Changes in gene frequencies due to genetic drift, however, are not an alternate to evolution by natural selection, because the two processes proceed simultaneously.

(6) There is a struggle for existence not only among individuals of the same species, but sometimes also among individuals of different species. If such competition leads to the extinction of one of the species, this process is referred to as species selection.

(7) None of the new insights of the last 50 years necessitate any essential correction of the basic Neo-Darwinian theory. Adaptations, indeed, are on the whole the result of natural selection. Stochastic processes affecting gene frequencies can occur simultaneously, but are not an alternate mechanism that would lead to greater adaptedness.

NOTE

This essay is an abridged version of one which first appeared in *Biologisches Zentralblatt* 101 (1982): 161–174.

REFERENCES

Bock, W. J. 1980. The definition and recognition of biological adaptation. *Amer. Zool.* 20:217–227.

Bock, W. J., and G. von Wahlert. 1965. Adaptation and the form-function complex. *Evolution* 19:269–299.

Dobzhansky, Th. 1937. *Genetics and the Origin of Species*. New York: Columbia University Press.

———— 1956. What is an adaptive trait? *Amer. Nat.* 90:337–347.

Eldredge, N. 1980. [Response by Niles Eldredge.] 54:642.

Gould, S. J., and R. C. Lewontin. 1979. The spandrels of San Marco and the Panglossian paradigm: a critique of the adaptationist programme. *Proc. Roy. Soc. London* B205:581–598.

Gulick, J. T. 1873. On diversity of evolution under one set of external conditions. London: *Linn. Soc. J., Zool.* 11:496–505.

Hagedoorn, A. L., and A. C. Hagedoorn. 1921. *The Relative Value of the Processes Causing Evolution*. The Hague: Martinus Nijhoff.

Lack, D. 1947. *Darwin's Finches*. Cambridge: Cambridge University Press.

Lewontin, R. C. 1978. Adaptation. *Sci. Amer.* 239:212–230.

Mayr, E. 1942. *Systematics and the Origin of Species*. New York: Columbia University Press.

———— 1945. Symposium on age of the distribution pattern of gene arrangements in *Drosophila pseudoobscura*. *Lloydia* 8:69–83.

———— 1954. Change of genetic environment and evolution. In J. Huxley, A. Hardy, and E.B. Ford, eds., *Evolution as a Process*, pp. 157–180. London: Allen and Unwin.

———— 1963. *Animal Species and Evolution*. Cambridge, Mass.: Harvard University Press.

Muller, H. J., et al. 1949. Natural selection and adaptation. *Proc. Amer. Phil. Soc.* 93:459–519.

Poulton, E. B. 1896. *Charles Darwin and the Theory of Natural Selection*. London: Cassell.

Simpson, G. G. 1944. *Tempo and Mode in Evolution*. New York: Columbia University Press.

Stern, J. T. 1970. The meaning of "adaptation" and its relation to the phenomenon of natural selection. *Evol. Biol.* 4:39–66.

Thiselton-Dyer, W. 1883. Deductive biology. *Nature* 27:554–555.

Van Valen, L. 1973. A new evolutionary law. *Evol. Theory* 1:1–30.

Wright, S. 1931. Evolution in Mendelian populations. *Genetics* 16:97–159.

———— 1932. The roles of mutation, inbreeding, crossbreeding, and selection in evolution. *Proc. VI. Intern. Congr. Genet.* 1:356–366.

Wynne-Edwards, V. C. 1962. *Animal Dispersion in Relation to Social Behavior*. Edinburgh: Oliver and Boyd.

HOW TO CARRY OUT THE ADAPTATIONIST PROGRAM?

TO HAVE been able to provide a scientific explanation of adaptation was perhaps the greatest triumph of the Darwinian theory of natural selection. After 1859 it was no longer necessary to invoke design, a supernatural agency, to explain the adaptation of organisms to their environment. It was the daily, indeed hourly, scrutiny of natural selection, as Darwin had said, that inevitably led to ever greater perfection. Ever since then it has been considered one of the major tasks of the evolutionist to demonstrate that organisms are indeed reasonably well adapted, and that this adaptation could be caused by no other agency than natural selection. Nevertheless, beginning with Darwin himself (remember his comments on the evolution of the eye), evolutionists have continued to worry about how valid this explanation is. The more generally natural selection was accepted after the 1930s, and the more clearly the complexity of the genotype was recognized, particularly after the 1960s, the more often the question was raised as to the meaning of the word *adaptation*.[1] The one thing about which modern authors are unanimous is that adaptation is not teleological, but refers to something produced in the past by natural selection. However, since various forms of selfish selection (for example, many aspects of sexual selection) may produce changes in the phenotype that could hardly be classified as "adaptations," the definition of adaptation must include some reference to the selection forces effected by the inanimate and living environment. It surely cannot have been anything but a lapse when Gould wanted to deny the designation "adaptation" to certain evolutionary innovations in clams, with this justification: "The first clam that fused its mantle margins or retained its byssus to adulthood may have gained a conventional adaptive benefit in its local environment. But it surely didn't know that its invention would set the stage for future increases in diversity" (Gould and Calloway 1980:395).

Considering the strictly a posteriori nature of an adaptation, its potential for the future is completely irrelevant, as far as the definition of the term adaptation is concerned.

A program of research intended to demonstrate the adaptedness of individuals and their characteristics is referred to by Gould and Lewontin (1979) as an "adaptationist program." A far more extreme definition of this term was suggested by Lewontin (1979:6) to whom the adaptationist program "assumes without further proof that all aspects of the morphology, physiology and behavior of organisms are adaptive optimal solutions to problems." Needless to say, in the ensuing discussion I am not defending such a sweeping ideological proposition.

When asking whether or not the adaptationist program is a legitimate scientific approach, one must realize that the method of evolutionary biology is in some ways quite different from that of the physical sciences. Although evolutionary phenomena are subject to universal laws, as are most phenomena in the physical sciences, the explanation of the history of a particular evolutionary phenomenon can be given only as a "historical narrative." Consequently, when one attempts to explain the features of something that is the product of evolution, one must attempt to reconstruct the evolutionary history of this feature. This can be done only by inference. The most helpful procedure in an analysis of historical narratives is to ask "why" questions; that is, questions (to translate this into modern evolutionary language) which ask what is or might have been the selective advantage that is responsible for the presence of a particular feature.

The adaptationist program has recently been vigorously attacked by Gould and Lewontin (1979) in an analysis which in many ways greatly pleases me, not only because they attack the same things that I questioned in my "bean bag genetics" paper (Mayr 1959), but also because they emphasize the holistic aspects of the genotype as I did repeatedly in discussions of the unity of the genotype (Mayr 1970, chap. 10; Essay 24). Yet I consider their analysis incomplete because they fail to make a clear distinction between the pitfalls of the adaptationist program as such and those resulting from a reductionist or atomistic approach in its implementation. I will try to show that basically there is nothing wrong with the adaptationist program, if properly executed, and that the weaknesses and deficiencies quite rightly pointed out by Gould and Lewontin are the result of atomistic and deterministic approaches.

In the period after 1859 only five major factors were seriously considered as the causes of evolutionary change, or, as they are sometimes called, the agents of evolution. By the time of the evolutionary synthesis (by the

1940s), three of these factors had been so thoroughly discredited and falsified that they are now no longer considered seriously by evolutionists. These three factors are: inheritance of acquired characters, intrinsic directive forces (orthogenesis, and so on), and saltational evolution (de Vriesian mutations, hopeful monsters, and the like). This left only two evolutionary mechanisms as possible causes of evolutionary change (including adaptation): chance and selection forces. The identification of these two factors as the principal causes of evolutionary change by no means completed the task of the evolutionist. As is the case with most scientific problems, this initial solution represented only the first orientation. For completion it requires a second stage, a fine-grained analysis of these two factors: What are the respective roles of chance and of natural selection, and how can this be analyzed?

Let me begin with chance. Evolutionary change in every generation is a two-step process: the production of genetically unique new individuals and the selection of the progenitors of the next generation. The important role of chance at the first step, the production of variability, is universally acknowledged (Mayr 1962), but the second step, natural selection, is on the whole viewed rather deterministically: Selection is a nonchance process. What is usually forgotten is the important role chance plays even during the process of selection. In a group of sibs it is by no means necessarily only those with the most superior genotypes that will reproduce. Predators mostly take weak or sick prey individuals but not exclusively, nor do localized natural catastrophes (storms, avalanches, floods) kill only inferior individuals. Every founder population is largely a chance aggregate of individuals, and the outcome of genetic revolutions, initiating new evolutionary departures, may depend on chance constellations of genetic factors. There is a large element of chance in every successful colonization. When multiple pathways toward the acquisition of a new adaptive trait are possible, it is often a matter of a momentary constellation of chance factors as to which one will be taken (Bock 1959).

When one attempts to determine for a given trait whether it is the result of natural selection or of chance (the incidental byproduct of stochastic processes), one is faced by an epistemological dilemma. Almost any change in the course of evolution might have resulted by chance. Can one ever prove this? Probably never. By contrast, can one deduce the probability of causation by selection? Yes, by showing that possession of the respective feature would be favored by selection. It is this consideration which determines the approach of the evolutionist. He must first attempt to explain biological phenomena and processes as the product of natural selection. Only after all attempts to do so have failed is he justified in designating the unexplained residue tentatively as a product of chance.

The evaluation of the impact of selection is a very difficult task. It has been demonstrated by numerous experiments that selection is not a phantom. That it also operates in nature is a conclusion that has often been confirmed (Endler 1986; Grant 1986). Very convincing was Bates' demonstration that the geographic variation of mimics parallels exactly that of their distasteful or poisonous models. The agreement of desert animals with the variously colored substrate also strongly supports the power of selection. In other cases the adaptive value of a trait is by no means immediately apparent.

As a consequence of the adaptationist dilemma, when one selectionist explanation of a feature has been discredited, the evolutionist must test other possible adaptationist solutions before he can resign and say: This phenomenon must be a product of chance. Gould and Lewontin ridicule the research strategy: "If one adaptive argument fails, try another one." Yet the strategy to try another hypothesis when the first fails is a traditional methodology in all branches of science. It is the standard in physics, chemistry, physiology, and archeology. Let me merely mention the field of avian orientation in which sun compass, sun map, star navigation, Coriolis force, magnetism, olfactory clues, and several other factors were investigated sequentially in order to explain as yet unexplained aspects of orientation and homing. What is wrong in using the same methodology in evolution research?

At this point it may be useful to look at the concept of adaptation from a historical point of view. When Darwin introduced natural selection as the agent of adaptation, he did so as a replacement for supernatural design. Design, as conceived by the natural theologians, had to be perfect, for it was unthinkable that God would make something that was less than perfect. It was on the basis of this tradition that the concept of natural selection originated. Darwin gave up this perfectionist concept of natural selection long before he wrote the *Origin*. Here he wrote, "Natural selection tends only to make each organic being as perfect as, or slightly more perfect than, the other inhabitants of the same country with which it has to struggle for existence. And we see that this is the degree of perfection attained under nature" (1859:201). He illustrated this with the biota of New Zealand, the members of which "are perfect . . . compared with another" (p. 201), but "rapidly yielding" (p. 201) to recent colonists and invaders. After Darwin, some evolutionists forgot the modesty of Darwin's claims, but other evolutionists remained fully aware that selection cannot give perfection, by observing the ubiquity of extinction and of physiological and morphological insufficiencies. However, the existence of some perfectionists has served Gould and Lewontin as the reason for making the adaptationist program the butt of their ridicule and for calling it a Panglossian paradigm. Here I dissent vigorously. To imply that the adaptationist program is one and the same as the argument from design (satirized by Voltaire in *Candide*) is highly misleading. When *Candide* was written (in 1759), a concept of evolution did not yet exist and those who believed in a benign creator had no choice but to believe that everything "had to be for the best." This is the Panglossian paradigm, the invalidity of which has been evident ever since the demise of natural theology. The adaptationist program, a direct consequence of the theory of natural selection, is something fundamentally different. Parenthetically one might add that Voltaire misrepresented Leibniz rather viciously. Leibniz had not claimed that this is the best possible world, but only that it is the best of the possible worlds. Curiously one can place an equivalent limitation on selection (see below). Selection does not produce perfect genotypes, but it favors the best which the numerous constraints upon it allow. That such constraints exist was ignored by those evolutionists who interpreted every trait of an organism as an ad hoc adaptation.

The attack directed by Gould and Lewontin against unsupported adaptationist explanations in the literature is fully justified. But the most absurd among these claims were made several generations ago, not by modern evolutionists. Gould and Lewontin rightly point out that some traits, for instance the gill arches of mammalian embryos, had been acquired as adaptations of remote ancestors but, even though they no longer serve their original function, they are not eliminated because they have become integral components of a developmental system. Most so-called vestigial organs are in this category. Finally, it would indeed be absurd to atomize an organism into smaller and smaller traits and to continue to search for the ad hoc adaptation of each smallest component. But I do not think that this is the research program of the majority of evolutionists. Dobzhansky well expressed the proper attitude when saying: "It cannot be stressed too often that natural selection does not operate with separate 'traits.' Selection favors genotypes . . . The reproductive success of a genotype is determined by the totality of the traits and qualities which it produces in a given environment" (1956:340). What Dobzhansky described reflects what I consider to be the concept of the adaptationist program accepted by most evolutionists, and I doubt that the characterization assigned to the adaptationist program by Gould and

Lewontin, "An organism is atomized into traits and these traits are explained as structures optimally designed by natural selection for their functions" (p. 585), represents the thinking of the average evolutionist.

By choosing this atomistic definition of the adaptationist program and by their additional insistence that the adaptive control of every trait must be "immediate," Gould and Lewontin present a picture of the adaptationist program that is indeed easy to ridicule. The objections cited by them are all based on their reductionist definition. Of course, it is highly probable that not all secondary byproducts of relative growth are "under immediate adaptive control." In the case of multiple pathways it is, of course, not necessary that every morphological detail in a convergently acquired adaptation be an ad hoc adaptation. This is true, for instance, in the case cited by them, of the adaptive complex for a rapid turnover of generations that evolved at least three times independently in the evolution of the arthropods. Evolution is opportunistic, and natural selection makes use of whatever variation it encounters. As Jacob (1977) has said so rightly: "Natural selection does not work like an engineer. It works like a tinkerer."

Considering the evident dangers of applying the adaptationist program incorrectly, why are the Darwinians nevertheless so intent on applying it? The principal reason for this is its great heuristic value. The adaptationist question, "What is the function of a given structure or organ?" has been for centuries the basis for every advance in physiology. If it had not been for the adaptationist program, we probably would still not yet know the functions of thymus, spleen, pituitary, and pineal. If one answer turned out to be wrong, the adaptationist program demanded another answer until the true meaning of the structure was established or until it could be shown that this feature was merely an incidental byproduct of the total genotype. It would seem to me that there is nothing wrong with the adaptationist program, provided it is properly applied.

Consistent with the modern theory of science, adaptationist hypotheses allow a falsification in most cases. For instance, there are numerous ways to test the thesis that the differences in beak dimensions of a pair of species of Darwin's finches on a given island in the Galapagos is the result of competition (Darwin's character divergence). One can correlate size of preferred seeds with bill size and study how competition among different assortments of sympatric species of finches affects bill size. Finally, one can correlate available food resources on different islands with population size (Boag and Grant 1981). As a result of such studies, the adaptationist program leads in this case to a far better understanding of the ecosystem.

The case of the beak differences of competing species of finches is one of many examples in which it is possible, indeed necessary, to investigate the adaptive significance of individual traits. I emphasize this because someone might conclude from the preceding discussion that a dissection of the phenotype into individual characters is inappropriate in principle. To think so would be a mistake. A more holistic approach is appropriate only when the analysis of individual traits fails to reveal an adaptive significance.

What has been rather neglected in the existing literature is the elaboration of an appropriate methodology to establish adaptive significance. In this respect a recent analysis by Traub (1980) on adaptive modifications in fleas is exemplary. Fleas are adorned with a rich equipment of hairs, bristles, and spines, some of which are modified into highly specialized organs. What Traub (and various authors before him) found is that unrelated genera and species of fleas often acquire convergent specializations on the same mammalian or avian hosts. The stiffness, length, and other qualities of the mammalian hair are species specific and evidently require special adaptations that are independently acquired by unrelated lineages of fleas. "The overall association [between bristles and host hair] is so profound that it is now possible to merely glance at a new genus or species of flea and make correct statements about some characteristic attributes of its host" (Traub 1980:64). Basically, the methodology consists in establishing a tentative correlation between a trait and a feature of the environment, and then to analyze in a comparative study, other organisms exposed to the same feature of the environment and see whether they have acquired the same specialization. There are two possible explanations for a failure of confirmation of the correlation. Either the studied feature is not the result of a selection force or there are multiple pathways for achieving adaptedness.

When the expanded comparative study results in a falsification of the tentative hypothesis, and when other hypotheses lead to ambiguous results, it is time to think of experimental tests. Such tests are not only often possible but indeed are now being made increasingly often, as the current literature reveals (Clarke 1979). Only when all such specific analyses to determine the possible adaptive value of the respective trait have failed is it time to adopt a more holistic approach and to start thinking about the possible adaptive significance of a larger portion of the phenotype, indeed possibly of the Bauplan as a whole.

Thus, the student of adaptation has to sail a perilous course between a pseudoexplanatory reductionist atomism and stultifying nonexplanatory holism. When we study the literature, we find almost invariably that those who were opposed to nonexplanatory holism went too far in adopting atomism of the kind so rightly stigmatized by Gould and Lewontin, while those who were appalled by the simplistic and often glaringly invalid pseudoexplanations of the atomists usually took refuge in an agnostic holism and abandoned all further effort at explanation by invoking best possible compromise, or integral component of Bauplan, or incidental byproduct of the genotype. Obviously neither approach, if exclusively adopted, is an appropriate solution. How do Gould and Lewontin propose to escape from this dilemma?

While castigating the adaptationist program as a Panglossian paradigm, Gould and Lewontin exhort the evolutionists to follow Darwin's example by adopting a pluralism of explanations. As much as I have favored pluralism all my life, I cannot follow Darwin in this case and, as a matter of fact, neither do Gould and Lewontin themselves. For Darwin's pluralism, as is well known to the historians of science, consisted of accepting several mechanisms of evolution as alternatives to natural selection, in particular the effects of use and disuse and the direct action of external conditions on organisms. Since both of these subsidiary mechanisms of Darwin's are now thoroughly refuted, we have no choice but to fall back on the selectionist explanation.

Indeed, when we look at Gould and Lewontin's "alternatives to immediate adaptation," we find that all of them are ultimately based on natural selection, properly conceived. It is thus evident that the target of their criticism should have been neither natural selection nor the adaptationist program as such, but rather a faulty interpretation of natural selection and an improperly conducted adaptationist program. Gould and Lewontin's proposals (1979:590–593) are not "alternatives to the adaptationist program," but simply legitimate forms of it. Such an improved adaptationist program has long been the favored methodology of most evolutionists. There is a middle course available between a pseudoexplanatory reductionist atomism and an agnostic nonexplanatory holism. Dobzhansky's (1956) stress on the total developmental system and adjustment to a variable environment and my own emphasis on the holistic nature of the genotype (1963, chap. 10; 1970, chap. 10 [considerably revised]; 1975) have been attempts to steer such a middle course, to mention only two of numerous authors who adopted this approach. They all chose an adaptationist program, but not an extreme atomistic one.

Much of the recent work in evolutionary morphology is based on such a middle-course adaptationist program, for instance Bock's (1959) analysis of multiple pathways and my own work on the origin of evolutionary novelties (1960). A semiholistic adaptationist program often permits the explanation of seemingly counter-intuitive results of selection. For instance, the large species of albatrosses (*Diomedea*) have only a single young every second year and do not start breeding until they are 6 to 8 years old. How could natural selection have led to such an extraordinarily low fertility for a bird? However, it could be shown that in the stormy waters of the south temperate and subantarctic zones only the most experienced birds have reproductive success and this in turn affects all other aspects of the life cycle. Under the circumstances the extraordinary reduction of fertility is favored by selection forces and hence is an adaptation (Lack 1968).

The critique of Gould and Lewontin would be legitimate to its full extent if one were to adopt (1) their narrow reductionist definition of the adaptationist program as exclusively "breaking an organism into unitary traits and proposing an adaptive story for each considered separately" (p. 581) and (2) their characterization of natural selection, in the spirit of natural theology, as a mechanism that must produce perfection.

Since only a few of today's evolutionists subscribe to such a narrow concept of the adaptationist program, Gould and Lewontin are breaking in open doors. To be sure, it is probable that many evolutionists have a far too simplistic concept of natural selection: They are neither fully aware of the numerous constraints to which natural selection is subjected, nor do they necessarily understand what the target of selection really is, nor, and this is perhaps the most important point, do they appreciate the importance of stochastic processes, as is rightly emphasized by Gould and Lewontin.

Darwin, as mentioned above, was aware of the fact that the perfecting of adaptations needs to be brought only to the point where an individual is "as perfect as, or slightly more perfect than" any of its competitors. And this point might be far from potentially possible perfection. What could not be seen as clearly in Darwin's day as it is by the modern evolutionist is that there are numerous factors in the genetics, developmental physiology, demography, and ecology of an organism that makes the achievement of a more perfect adaptation simply impossible. Gould and Lewontin (1979) and Lewontin (1978) have enumerated such constraints and so have I, based in part on independent analysis (see essays 6 and 8).

There are chance components in all these constraints, but it must be stated emphatically that selection and chance are not two mutually exclusive alternatives, as was maintained by many authors from the days of

Darwin and the earlier writings of Sewall Wright to the arguments of some anti-Darwinians today. Actually there are stochastic perturbations ("chance events") during every stage of the selection process.

The question whether or not the adaptationist program ought to be abandoned because of presumptive faults can now be answered. It would seem obvious that little is wrong with the adaptationist program as such, contrary to the claims of Gould and Lewontin, but that it should not be applied in an exclusively atomistic manner. There is no better evidence for this conclusion than that which Gould and Lewontin themselves have presented. Aristotelian "why" questions are quite legitimate in the study of adaptations, provided one has a realistic conception of natural selection and understands that the individual-as-a-whole is a complex genetic and developmental system and that it will lead to ludicrous answers if one smashes this system and analyzes the pieces of the wreckage one by one.

A partially holistic approach that asks appropriate questions about integrated components of the system needs to be neither stultifying nor agnostic. Such an approach may be able to avoid the Scylla and Charybdis of an extreme atomistic or an extreme holistic approach.

NOTES

This essay is reprinted from *The American Naturalist* 121(March 1983):324–333. © 1983 by The University of Chicago.

1. The difficulty of the concept of adaptation is best documented by the incessant efforts of authors to analyze it, describe it, and define it. Since I can do no better myself, I refer to a sample of such efforts (Bock and von Wahlert 1965; Bock 1980; Brandon 1978; Burian 1983; Dobzhansky 1956, 1968; Lewontin 1978, 1979; Muller 1949; Munson 1971; Stern 1970; Williams 1966; Wright 1949).

REFERENCES

Boag, P. T., and P. R. Grant. 1981. Intense natural selection in a population of Darwin's finches (Geospizinae) in the Galapagos. *Science* 214:82–85.
Bock, W. J. 1959. Preadaptation and multiple evolutionary pathways. *Evolution* 13:194–211.

———— 1980. The definition and recognition of biological adaptation. *Amer. Zool.* 20:217–227.

Bock, W., and G. von Wahlert. 1965. Adaptation and the form-function complex. *Evolution* 19:269–299.

Brandon, R. N. 1978. Adaptation and evolutionary theory. *Stud. Hist. Phil. Sci.* 9:191–206.

Clarke, B. C. 1979. The evolution of genetic diversity. *Proc. Roy. Soc. London* B205:453–474.

Clutton-Brock, T. H., and P. H. Harvey. 1979. Comparison and adaptation. *Proc. Roy. Soc. London* B205:547–565.

Darwin, C. 1859. *On the Origin of Species by Means of Natural Selection or the Preservation of Favored Races in the Struggle for Life.* London: John Murray.

Dobzhansky, Th. 1956. What is an adaptive trait? *Amer. Nat.* 90:337–347.

———— 1968. Adaptedness and fitness. In R. Lewontin, ed., *Population Biology and Evolution.* New York: Syracuse University Press.

Endler, J. A. 1986. *Natural Selection in the Wild.* Princeton: Princeton University Press.

Gould, S. J., and C. B. Calloway. 1980. Clams and brachiopods—ships that pass in the night. *Paleobiology* 6(4):383–396.

Gould, S. J., and R. Lewontin. 1979. The spandrels of San Marco and the Panglossian paradigm: a critique of the adaptationist programme. *Proc. Roy. Soc. London* B205:581–598.

Grant, P. R. 1986. *Ecology and Evolution of Darwin's Finches.* Princeton: Princeton University Press.

Jacob, F. 1977. Evolution and tinkering. *Science* 196:1161–1166.

Lack, D. 1968. *Ecological Adaptations for Breeding in Birds.* London: Methuen.

Lewontin, R. C. 1978. Adaptation. *Sci. Amer.* 239:156–169.

———— 1979. Sociobiology as an adaptationist program. *Behav. Sci.* 24:5–14.

Mayr, E. 1959. Where are we? *Cold Spring Harbor Symp. Quant. Biol.* 24:409–440.

———— 1960. The emergence of evolutionary novelties. In S. Tax, ed. *The Evolution of Life.* Chicago: University of Chicago Press.

———— 1962. Accident or design, the paradox of evolution. In *The Evolution of Living Organisms: A Symposium of the Royal Society of Victoria Held in Melbourne, December 1959.* Melbourne: Melbourne University Press.

———— 1963. *Animal Species and Evolution.* Cambridge, Mass.: Harvard University Press.

———— 1970. *Population, Species, and Evolution.* Cambridge, Mass.: Harvard University Press.

Muller, H. J., et al. 1949. Natural selection and adaptation. *Proc. Amer. Phil. Soc.* 93:459–519.

Munson, R. 1971. Biological adaptation. *Phil. Sci.* 38:200–215.

Roux, W. 1881. *Kampf der Theile im Organismus.* Leipzig: Jena.

Stern, J. T. 1970. The meaning of "adaptation" and its relation to the phenomenon of natural selection. *Evol. Biol.* 4:39–66.

Traub, R. 1980. Some adaptive modifications in fleas. In R. Traub and H.
Starcke, eds., *Fleas*. Rotterdam: A. A. Baldema.

Williams, G. C. 1966. *Adaptation and Natural Selection.* New Jersey: Princeton
University Press.

Wright, S. 1949. Adaptation and selection. In G. Jepsen, E. Mayr, and G. G.
Simpson, eds., *Genetics, Paleontology and Evolution.* New Jersey: Princeton
University Press.

part four 🎔

DARWIN

Introduction

The philosopher John Passmore (1983) recently pointed out that only one intellectual revolution—the one brought about by Darwin—has been dignified by the suffix *-ism*. Hence we have Darwinism, but not Newtonianism, Maxwellism, Planckism, Einsteinism, or Heisenbergism. This exceptional status is justified, for it would be difficult to refute the claim that the Darwinian revolution was the greatest of all intellectual revolutions in the history of mankind.

Its importance lies in the fact that Darwin caused the overthrow of some of the most basic beliefs of his age. Furthermore, and this is only now fully realized, Darwin established the basis for entirely new approaches in philosophy.

Darwin's conceptual framework was so alien to his age and to the generations that followed him that, not surprisingly, at first his greatness was not at all recognized. Although he was celebrated as an outstanding naturalist and his work soon led to the acceptance of evolution and the theory of common descent, his deeper philosophical ideas were ignored, rejected, or entirely misunderstood.

A new period of Darwin appreciation began with the centenary of the publication of the *Origin* in 1959 and culminated in the centenary of Darwin's death in 1982. I rather suspect that more books and papers dealing with Darwin and his work were published in the 25 years after 1959 than in the 100 years before that date. This activity was greatly furthered by the discovery of the "Darwin Papers," a rich treasure of letters, manuscripts, notebooks, and other unpublished material, most of it now in the Library of Cambridge University. This material facilitates reconstructing the growth of Darwin's thought in the most extraordinary manner. I believe that nothing quite as complete exists for any other of the great thinkers.

There are good reasons for the vastness and diversity of the Darwin literature. As an intellectual pioneer, Darwin himself was sometimes uncertain and then used ambiguous language. Workers with different backgrounds, such as taxonomists, paleontologists, morphologists, or geneticists, have interpreted his writings differently, and some of the controversies thus engendered are still with us. Fortunately, the recent Darwin revival coincided with a considerable clarification of evolutionary thought, as reflected in modern texts by Mayr (1963; 1970), Dobzhansky (1970), Futuyma (1979), Grant (1977), and several others. This resulted in the elimination of at least some of the controversies.

Although I have been reading Darwin's work since my university years, my more active preoccupation with him began with the centenary of Darwin's birth (Mayr 1959a,b,c,d; 1960; 1962a). I studied him even more intensely when I prepared an introduction to a facsimile edition of the *Origin* (1964b). Curiously, most editions of the work then available were of the much revised sixth edition. In the few available reprintings of the first edition, the pagination had been changed, making it impossible to provide exact page references to the original. The 1964 publication of a very inexpensive facsimile print of the first edition made a wide distribution of the *Origin* possible for the first time since its original publication.

In the ensuing years I devoted myself with increasing intensity to the study of Darwin's work, culminating in a comprehensive treatment of Darwin's evolutionary concepts in *Growth of Biological Thought* (1982:394–525). It was, however, impossible in this overview to present detailed analyses of some particular aspects of Darwin's work, and I treated these in a number of separate papers, mostly presented at Darwin commemorative celebrations. The essays included in this chapter represent an endeavor of mine to focus on such previously neglected aspects of Darwin's work and to clarify confusing issues.

An almost embarrassing number of invitations were extended to me in the centennial year 1982 to participate in Darwin celebrations. Even though I accepted only a few, it was impossible to avoid saying things at one symposium that I had also said at another. I have edited the six Darwin essays in this volume to eliminate as far as possible any overlap in their contents.

Curiously, even such a seemingly traditional concept as evolution has been greatly clarified in recent years, or, more correctly, it has been recognized that what commonly was called evolution was not what Darwin had considered.

For instance, I pointed out a decade ago (1977:40) that "the reductionist definition, so widely adopted in recent decades—'Evolution is a change in gene frequencies in populations'—is not only *not* explanatory, but is in fact misleading. Far more revealing is the definition: 'Evolution is change in the adaptation *and* in the diversity of populations of organisms.' The important emphasis here is on the dual nature of evolution. It deals, so to speak, both with the 'vertical' phenomenon of adaptative change and with the 'horizontal' phenomenon of populations, incipient species, and new species." To point this out was important because the diversity-making (taxic) component of evolution was almost consistently ignored by geneticists and paleontologists, as is described in other essays in this volume.

The term evolution was first employed by biologists in embryology to describe the unfolding of the preformed capacities of the germ (Bonnet). Lamarck's concept of evolution was rather similar, designating likewise a completely gradual process, a change due to a trend toward perfection or adjustment to the environment. Darwin's concept of biological evolution was fundamentally different. As Lewontin said: "Before Darwin theories of historical change were all transformational. That is, systems were seen as undergoing change in time because each element in the system undergoes an individual transformation during its life history. Lamarck's theory of evolution was transformational, for it regarded species as changing because each individual organism within the species underwent the same change . . . An example of a transformational theory in modern natural science is the evolution in the cosmos . . . The evolution of the universe is the evolution of every star within it. Theories of human history are all transformational . . .

"In contrast to these transformational theories of change, Darwin proposed a variational principle. Different individual members of the ensemble differ from each other in some properties, and the system evolves by a change in the proportions of the different types . . . so that the nature of the ensemble as a whole changes without any successive changes of individual members. Thus, variation of one kind, variation between objects in space, becomes transformed qualitatively into temporal variation" (Lewontin 1983:65–66).

Darwinian evolution, thus, is not a steady progression, as is Lamarckian transformational evolution; rather, it proceeds by generational steps. Indeed, it is a two-step phenomenon, the first step in each generation being responsible for the production of variation (Mayr 1962a), while this variation is then sorted ("ordered") in the second step, selection proper (Essay 6).

Thus, strictly speaking, Darwinian evolution is discontinuous because a new start is made in every generation when a new set of individuals is produced from the contents of the gene pool. When evolution nevertheless appears to be totally gradual, this is because it is populational. The nature of sexual reproduction and diploidy necessitates the gradual change of populations. No matter how polymorphic certain populations might be, their change is by necessity gradual.

In view of the fact that Darwin's ideas have been so badly misrepresented in some of the more polemic Darwin literature, I wish to call attention to two recent papers in which detailed quotations from Darwin's work are brought together (A. Huxley 1982; Rhodes 1983). These quotations deal particularly with Darwin's ideas on rates of evolution, on stasis, on gradualism (or its absence), and on pluralism. They make it evident how much closer Darwin was to current thinking than is usually assumed.

Needless to say, when Darwin wrote the *Origin* he did not know anything about the laws of genetics. But by observation he did know a number of things. He knew that in asexual reproduction the offspring is identical with the parent, while in sexual reproduction each offspring is different. Furthermore, he knew that each offspring had a mixture of the characters of both parents. This "blending" of the characters had, as Darwin saw it, a beneficial and a detrimental aspect. The beneficial aspect was that any undesirable deviation from the type of the species would be obliterated by blending. The detrimental aspect was that, of course, the same would happen to any desirable novelty. And in this Darwin saw the advantage of isolation on an island, where the inbreeding population was too small for a valuable new character to be obliterated. Thus, even though based on the erroneous theory of blending inheritance, Darwin correctly pinpointed the importance of isolated populations as the cradle of evolutionary innovation.

In spite of all the work on Darwin that has been done in recent years, some unresolved questions remain. It is still not evident why Darwin thought his principle of character divergence was so important—indeed, as he said, of equal importance with his theory of natural selection. In my opinion Darwin saw in it the mechanism that made sympatric speciation in plants possible, but there are other interpretations (Kohn 1985).

The role in Darwin's thinking played by chance has not yet been analyzed properly. The spirit of science, at his time, was still highly deterministic, and yet Darwin's descriptions of the course of evolution made it quite clear how strong a role was played by chance. This incensed John Herschel to such a degree that he referred to the theory of natural selection as the "theory of the higgledy-piggledy." Actually Darwin made a fine distinction, in the case of variations, between their being accidental as far as their "purpose" or selective value is concerned, and their being "accidental as to their cause or origin" (F. Darwin 1887 1:314). A similar train of thought is expressed in the *Variation of Animals and Plants* (1868 2:431). It is evident that Darwin accepted the strict working of what he called natural laws at the physiological level but was aware of stochastic processes at the organismic level.

Authors not fully familiar with Darwin's work, particularly some philosophers, often speak of *the* Darwinian theory, in the singular. There is no such thing as *the* Darwinian theory, since Darwin's evolutionary paradigm is highly composite. I endeavor to point this out in Essay 12.

Far too many critiques of Darwinism are still being published that reveal a total misunderstanding of Darwin's thought and of the conclusions of the evolutionary synthesis. Much as one might wish to ignore such effusions of ignorance, one cannot do so, lest the critics assume that they have produced some unanswerable objections to Darwinism, and lest some readers accept these statements as the last word. It is particularly disappointing to encounter a remarkable lack of understanding of Darwin's theories in the writings of a distinguished historian of ideas who has been interested in evolutionism since his college days. I am referring to John C. Greene. In his case one cannot ascribe the miscomprehension to ignorance, for Greene is exceptionally well acquainted with the Darwinian literature. Rather, he fails to understand Darwin because his own ideology is completely incompatible with Darwin's thought. Since we all believe in the freedom of thought, no one can object if someone holds ideas that are different from those of a Darwinian. What can be objected to, and what calls for correction, is the misrepresenting of Darwin's thought, and this is unfortunately what Greene has been doing. It is the objective of Essay 15 to make such a correction.

Other recent critics of Darwin are far less well informed. In a book review (Mayr 1984) and in a more detailed analysis (Mayr 1986a) I have discussed the profound misunderstanding which Ho and Saunders display in their book *Beyond Neo-Darwinism* (1984). See also Brace (1985) and Futuyma (1984). Their image of neo-Darwinism is a caricature of the actual thought of contemporary Darwinians. The same is, unfortunately, true for Reid (1985), whose volume I recently reviewed (Mayr 1986b). Frankly, it puzzles me why publishers continue to bring out such ill-informed volumes. With all the excellent recent books on Darwin's thought now available, there is no excuse for the publication of such literature.

REFERENCES

Brace, C. L. 1985. [Review of Ho and Saunders (1984).] *Amer. J. Phys. Anthropol.* 68:446–448.

Darwin, C. 1868. *Variations of Animals and Plants.* 2 vols. London: John Murray.

Darwin, F. 1887. *The Life and Letters of Charles Darwin.* 3 vols. London: John Murray.

Dobzhansky, Th. 1970. *Genetics of the Evolutionary Process.* New York: Columbia University Press.

Futuyma, D. 1979. *Evolutionary Biology.* Sunderland, Mass.: Sinauer.

———— 1984. [Review of Ho and Saunders (1984).] *Science* 226:532–533.

Grant, V. 1977. *Organismic Evolution.* San Francisco: Freeman.

Ho, M. W., and P. T. Saunders. 1984. *Beyond Neo-Darwinism: An Introduction to the New Evolutionary Paradigm.* London: Academic.

Huxley, Andrew 1982. [Presidential address.] *Proc. Roy. Soc. London* A379:ix–xvii.

Kohn, D. 1985. Darwin's principle of divergence as internal dialogue. In D. Kohn, ed., *The Darwinian Heritage*, pp. 245–257. Princeton: Princeton University Press.

Lewontin, R. C. 1983. The organism as the subject and object of evolution. *Scientia* 118:63–82.

Mayr, E. 1959a. Isolation as an evolutionary factor. *Proc. Amer. Phil. Soc.* 103(2):221–230.

———— 1959b. Darwin and the evolutionary theory in biology. In B. J. Meggers, ed., *Evolution and Anthropology: A Centennial Appraisal*, pp. 1–10. Washington, D.C.: Anthropological Society.

———— 1959c. Agassiz, Darwin, and evolution. *Harvard Lib. Bull.* 13(2):165–194.

———— 1959d. [Review of Gertrude Himmelfarb, *Darwin and the Darwinian Revolution.*] *Sci. Amer.* 201:209–216.

———— 1960. The emergence of evolutionary novelties. In S. Tax, ed., *Evolution after Darwin.* Vol. 1: *The Evolution of Life: Its Origin, History, and Future.* pp. 349–380. Chicago: University of Chicago Press.

———— 1962a. Accident or design, the paradox of evolution. In *The Evolution of Living Organisms*, pp. 1–14. Symp. Roy. Soc. Victoria, Melbourne 1959. Victoria: Melbourne University Press.

———— 1962b. [Review of G. Wichler, *Charles Darwin.*] *Science* 134:607.

———— 1963. *Animal Species and Evolution.* Cambridge, Mass.: Harvard University Press.

———— 1964a. Evolutionary theory today. In *McGraw-Hill Yearbook of Science and Technology*, pp. 86–95. New York: McGraw-Hill.

———— 1964b. *On the Origin of Species.* Facs. 1st. ed., introd. Ernst Mayr. Cambridge, Mass.: Harvard University Press.

———— 1970. *Evolution and the Diversity of Life*. Cambridge, Mass.: Harvard University Press.

———— 1971. Open problems of Darwin research. *Stud. Hist. Phil. Sci.* 2(3):273–280.

———— 1972a. The nature of the Darwinian revolution. *Science* 176:981–989.

———— 1972b. Sexual selection and natural selection. In B. Campbell, ed., *Sexual Selection and the Descent of Man*, pp. 87–104. Chicago: Aldine.

———— 1973. The descent of man and sexual selection. In *Atti del Colloquio Internazionale sul tema: "L'Origine dell'Uomo" (28–30 Oct. 1971)*, pp. 33–48. Rome: Accademia Nationale dei Lincei.

———— 1977. The study of evolution, historically viewed. In C. E. Goulden, ed., *The Changing Scenes in the Natural Sciences, 1776–1976*, pp. 39–58. Philadelphia: Fulton.

———— 1978. Evolution. *Sci. Amer.* 239(3):47–55.

———— 1981. [Review of Paul H. Barrett, Donald J. Weinshank, and Timothy T. Gottleber, eds., *Concordance to Darwin's Origin of Species*, 1st ed.] *Isis* 73:476–477.

———— 1982a. Foreword. In M. Ruse, *Darwinism Defended*, pp. xi–xii. Reading, Mass.: Addison-Wesley.

———— 1982b. Darwinistische Missverständnisse. In K. Bayertz, B. Heidtmann, and Hans-Jörg Rheinberger, eds., *Dialektik 5, Darwin und die Evolutionstheorie*. Köln: Pahl-Rugenstein.

———— 1984. The triumph of evolutionary synthesis. *Times Literary Supplement* 4257:1261–1262.

———— 1985. Darwin and the definition of phylogeny. *Syst. Zool.* 34(1):97–98.

———— 1986a. What is Darwinism today? *PSA 1984* 2:145–156.

———— 1986b. [Review of R. G. B. Reid, *Evolutionary Theory*.] *Isis* 77:358–359.

Meggers, B. J., ed. 1959. *Evolution and Anthropology: A Centennial Appraisal*. Washington, D.C.: Anthropological Society.

Passmore, J. 1983. The mysterious case of Charles Darwin. In D. S. Bendall, ed., *Evolution from Molecules to Men*, pp. 569–575. Cambridge: Cambridge University Press.

Reid, R. G. B. 1985. *Evolutionary Theory: The Unfinished Synthesis*. Ithaca: Cornell University Press.

Rhodes, F. H. T. 1983. Gradualism, punctuated equilibrium, and the Origin of Species. *Nature* 305:269–272.

DARWIN, INTELLECTUAL REVOLUTIONARY

WE LIVE in a time of short memories. Only few papers nowadays cited in the scientific literature are more than three years old. And many of the younger generation are regrettably unaware of the revolutionary contributions made by the great men of the past. One can hardly find a better illustration for this than the work of Charles Darwin. Nearly all of his great innovations have become to such an extent an integral component of Western thinking that only the historians appreciate Darwin's pioneering role.

We speak rather glibly of the Darwinian revolution, but if one challenges the speaker to specify precisely just what he means by this term, one will invariably get an answer that is, at best, incomplete but more often partly wrong. I admit that it is almost impossible for a modern person to project himself back to the early half of the last century, to reconstruct the thinking of the pre-Darwinian period, that is, the framework of ideas that was so thoroughly destroyed by Darwin. What we must, therefore, do is to ask what were the widely or universally held views that were challenged by Darwin in 1859 in the *Origin?* I will endeavour to show that the intellectual revolution generated by Darwin produced a far more fundamental and far-reaching change in the thinking of the Western World than is usually appreciated. And I shall also make it clear that this was not merely an incidental by-product of some biological speculations, but that Darwin had a comprehensive programme, to review and, where necessary, to revolutionize traditional ideas. I shall show that Darwin was a bold and often quite radical revolutionary.

Let me preface my analysis by the comment that I shall try to avoid as much as possible analyzing by what pathway Darwin reached his novel ideas. Rather, I will begin by asking whether there was a prevailing world view in the West prior to 1859. The answer is no. Indeed, the prevailing thinking in England, France, or Germany was perhaps more different from that of the other two countries in the first half of the nineteenth century than in any period before or since. Natural theology was still dominant in England, but already obsolete in France and Germany. The conservative and rational France of Cuvier, as well as the romantic Germany of the *Naturphilosophen* were worlds apart from the England of Lyell, Sedgwick, and John Herschel. Darwin's intellectual milieu, of course, was that of England, and his arguments were directed against his peers and their thinking.

Let me now take up the major concepts or ideas which Darwin encountered and which he tried to modify or replace. By far the most comprehensive of these, in fact an all-pervading ideology, was that of creationism. By creationism is understood the belief that the world must be interpreted as reflecting the mind of the Creator. But physicists and naturalists interpreted the implementation of the mind of the Creator in different ways. The order and harmony of the universe made the physical scientists search for laws, for wise institutions in the running of the universe installed by the Creator. Everything in nature was caused, but the causes were secondary causes, regulated by the laws instituted by the primary cause, the Creator. To serve his Creator best, a physicist studied His laws and their working. Most naturalists, by contrast, concentrated on the wonderful adaptations of living creatures. These could not be explained readily as the result of general laws such as those of gravity, heat, light, or movements. Nearly all the marvellous adaptations of living creatures are so unique that it seemed vacuous to claim that they were due to "laws." It rather seemed that these aspects of nature were so special and unique that they could be interpreted only as caused by the direct intervention of the Creator, by His specific design. Consequently the functioning of organisms, their instincts, and their manifold interactions provided the naturalist with abundant evidence for design, and seemed to constitute irrefutable proof for the existence of a Creator. Darwin, even though a naturalist, tried valiantly to apply the thinking of the physicists, by looking for laws.

It was the task of natural theology to study the design of creation, and natural theology was thus as much science as it was theology. The two endeavours, theology and science, were indeed inseparable. Consequently, most of the greater scientific works of this period, as exemplified by Lyell's *Principles of Geology* (1830–33) or Louis Agassiz's *Essay on Classification* (1857), were simultaneously treatises of natural theology. Science and theology were fused into a single system and, as is obvious with hindsight, there could not be any truly objective and uncommitted science until science and theology had been cleanly and completely divorced from each other. The publication of Darwin's *Origin of Species* was to a greater extent responsible for this divorce than anything else. Although Darwin was not yet a complete agnostic in 1859, he argued throughout the *Origin* against creationism, the intrusion of theology into science. And when Darwin said "This whole volume is one long argument," he was clearly referring to his endeavour to "get theology out of science," as Gillespie (1979) has put it. In this effort Darwin was completely successful, and the historians accept the year 1859 as the end of creditable natural theology.

I used to believe that it was a loss of Christian faith that induced Darwin to seek a complete autonomy of science from theology (Essay 13). However, the recent analyses of Gillespie (1979), Ospovat (1981), and others have persuaded me that scientific findings had been primarily responsible for Darwin's change of mind. In particular it was Darwin's realization of the invalidity of three prominent doctrines among the numerous beliefs of creationism that was of crucial importance for Darwin's change of mind:

(1) that of an unchanging world of short duration,

(2) that of the constancy and sharp delimitation of created species, and

(3) that of a perfect world explicable only by the postulate of an omnipotent and beneficent Creator.

As we shall presently discuss, Darwin's conclusions were reinforced by his conviction of the invalidity of several other tenets of creationism, such as the special creation of man, but the three stated dogmas were clearly of primary importance.

Even though Bishop Ussher's calculation that the world had been created as recently as 4004 years B.C. was still widely accepted among the pious, the men of science had long become aware of the great age of the earth. The researches of the geologists, in particular, left no doubt of the immense age of the earth, thus providing all the time needed for abundant organic evolution. Lamarck was the first to draw the necessary conclusions from this.

Among the various discoveries of geology the one that was most important and most disturbing for the creationist was the discovery of extinction. Already in the eighteenth century Blumenbach and others had accepted extinction of formerly existing types like ammonites, belemnites, and trilobites, and of entire faunas, but it was not until Cuvier had worked out the extinction of a whole sequence of mammalian faunas in the Tertiary of the Paris Basin that the acceptance of extinction became inevitable. The ultimate proof for it was the discovery of fossil mastodons and mammoths, animals so huge that any living survivors could not possibly have remained undiscovered in some remote part of the globe.

Three explanations for extinction were offered. According to Lamarck, no organism ever became extinct; there simply was such drastic transformation that the formerly existing types had changed beyond recognition. According to the progressionists, like Miller, Murchison, and Agassiz, each former fauna had become extinct as a whole and was replaced by a newly created, more progressive fauna. Such catastrophism was unpalatable to Charles Lyell, who produced a third theory consistent with his uniformitarianism. He believed that individual species became extinct one by one as conditions were changing and that the gaps thus created in nature were filled by the introduction of new species. Lyell's theory was an attempt at a reconciliation between those who recognized a changing world of long duration and those who supported the tenets of creationism.

The question of how these new species were "introduced" was left unanswered by Lyell. He bequeathed this problem to Darwin, who in due time made it his most important research program. Darwin thus approached the problem of evolution in an entirely different manner from Lamarck. For Lamarck, evolution was a strictly vertical phenomenon, more or less a temporalization of the *scala naturae,* proceeding in a single dimension, that of time. Evolution for him was a movement from less perfect to more perfect, an endeavour to establish a continuity among the major types of organisms, from the most primitive infusorians up to the mammals and Man. Lamarck's *Philosophie Zoologique* was the paradigm of what I designate as vertical evolutionism. Species played no role in Lamarck's thinking. New species originated all the time by spontaneous generation from inanimate matter, but this produced only the simplest infusorians. Each such newly established evolutionary line gradually moved up to ever greater perfection. Lamarck rightly called his work a "philosophy," because he did not present testable scientific theories.

Darwin was unable to build on this foundation but rather started from the fundamental question that Lyell had bequeathed to him. Although Lyell had appealed to "intermediate causes" as the source of the new species, his description of the process was that of special creation. "Species may have been created in succession at such times and at such places as to enable them to multiply and endure for an appointed period and occupy an appointed space on the globe." For Lyell, each creation was a carefully planned event. The reason why Lyell, like Henslow, Sedgwick, and all the others of Darwin's scientific friends and correspondents in the middle of the 1830s, accepted the unalterable constancy of species was ultimately a philosophical one. The constancy of species was the one piece of the old dogma of a created world that remained inviolate after the concepts of the recency and constancy of the physical world had been abandoned.

Another ideology which prevented Lyell from recognizing the change of species was *essentialism*, which had dominated Western thinking for more than two thousand years after Plato. According to the essentialist, the changing variety of things in nature is a reflection of a limited number of constant and sharply delimited underlying *eide*, or essences. Variation is merely the manifestation of imperfect reflections of the constant essences. For Lyell all nature consisted of constant types, each created at a definite time. "There are fixed limits beyond which the descendants from common parents can never deviate from a certain type." And he added emphatically: "It is idle . . . to dispute about the abstract possibility of the conversion of one species into another, when there are known causes, so much more active in their nature, which must always intervene and prevent the actual accomplishment of such conversions" (Lyell 1835:162). For an essentialist there can be no evolution, there can only be a sudden origin of a new essence by a major mutation or saltation.

To the historian it is quite evident that no genuine and testable theory of evolution could develop until the possibility was accepted that species have the capacity to change, to become transformed into new species, and to multiply into several species. For Darwin to accept this possibility required a fundamental break with Lyell's thinking. The question which we must ask ourselves is how Darwin was able to emancipate himself from Lyell's thinking, and what observations or conceptual changes permitted Darwin to adopt the theory of a transforming capacity of species.

As Darwin tells us in his autobiography, he encountered many phenomena during his visit to South America on the *Beagle* which any modern biologist would unhesitatingly explain as clear evidence for evolution. Furthermore, when sorting his collections on the homeward voyage, his observations in the Galapagos Islands made him pen the memorable sentences (approximately July 1836): "When I see these islands in sight of each other and possessed of but a scanty stock of animals, tenanted by these birds but slightly differing in structure and filling the same place in nature, I must suspect they are varieties . . . if there is the slightest foundation for these remarks the zoology of the archipelagoes will be well worth examining: for such facts would undermine the stability of species"

(Barlow 1963). Yet, it is evident that Darwin at that date had not yet consciously abandoned the concept of constant species. This Darwin apparently did in two stages. The discovery of a second, smaller species of *Rhea* (South American ostrich) led him to the theory, consistent with essentialism, that an existing species could give rise to a new species, by a sudden leap. Such an origin of new species had been postulated scores of times before (Osborn 1894), from the Greeks to Robinet and Maupertuis. Typological new origins, however, are not evolution (Mayr 1982:301–42). The diagnostic criterion of evolutionary transformation is gradualness. The concept of gradualism, the second step in Darwin's conversion, was apparently first adopted by Darwin, as Sulloway (1982) has demonstrated so convincingly, when the ornithologist John Gould, who prepared the scientific report on Darwin's bird collections, pointed out to him that there were three different endemic species of mockingbirds on three different islands in the Galapagos. Darwin had thought they were only varieties.

The mockingbird episode was of particular importance to Darwin for two reasons. The Galapagos endemics were quite similar to a species of mockingbirds on the South American mainland and clearly derived from it. What was important was that the Galapagos birds were not the result of a single saltation, as Darwin had postulated for the new species of *Rhea* in Patagonia, but had gradually evolved into three different although very similar species on three different islands. This fact helped to convert Darwin to the concept of gradual evolution. Even more important was the fact that these three different species had branched off from a single parental species, the mainland mockingbird, an observation which gave Darwin the solution to the problem of the multiplication of species.

The problem of the introduction of new species posed by Lyell was thus solved by Darwin. New species can originate by what we now call geographical or allopatric speciation. This theory of speciation says that new species may originate by the gradual genetic transformation of isolated populations. By this thought Darwin founded a branch of evolutionism which, for short, we might designate as horizontal evolutionism, in contrast with the strictly vertical evolutionism of Lamarck. The two kinds of evolutionism deal with two entirely different aspects of evolution even though the processes responsible for these aspects proceed simultaneously. Vertical evolutionism deals with adaptive changes in the time dimension, while horizontal evolutionism deals with the origin of new diversity, that is, with the origin of new populations, incipient species, and new species, which enrich the diversity of the organic world and which are the potential founders of new evolutionary departures, of new higher taxa, of the occupants of new adaptive zones.

From 1837 on, when Darwin first recognized and solved the problem of the origin of diversity, this duality of the evolutionary process has been with us. Unfortunately, there were only few authors with the breadth of thought and experience of Darwin to deal simultaneously with both aspects of evolution. As it was, paleontologists and geneticists concentrated on or devoted themselves exclusively to vertical evolution, while the majority of the naturalists studied the origin of diversity as reflected in the process of speciation and the origin of higher taxa.

The first author to have proposed geographic speciation was the geologist L. von Buch (1825) in a short statement which he failed to develop any further. M. Wagner (1841) and Alfred Russel Wallace (1855) also proposed it independently of Darwin. The discovery of the divergence of contemporary, geographically isolated populations made it possible to incorporate the origin of organic diversity within the compass of evolution. For Darwin this was of crucial importance, because horizontal thinking permitted the solution of three important evolutionary problems:

(1) The problem of the multiplication of species.

(2) The resolution of the seeming conflict between the observed discontinuities in nature and the concept of gradual evolution.

(3) The problem of the evolution of the higher taxa owing to common descent.

Perhaps the most decisive consequence of the discovery of geographic speciation was that it led Darwin automatically to a branching concept of evolution. This is why branching entered Darwin's notebooks at such an early stage.

How Are Relatives Connected?

For those who accepted the concept of the *scala naturae*—and in the eighteenth century this included most naturalists to a lesser or greater extent—all organisms were part of a single linear scale of ever-growing perfection. Lamarck still adhered, in principle, to this concept even though he allowed for some branching in his classification of the major phyla. Pallas and others had also published branching diagrams, but it required the categorical rejection of the *scala naturae* by Cuvier in the first and second decades of the nineteenth century before the need for a new way to represent organic diversity became crucial. The Quinarians experimented with indicating relationship by osculating circles, but their diagrams did not fit reality at all well. The archetypes of Owen and the Naturphilosophen strengthened the recognition of discrete groups in nature, but the use of the term *affinity* in relation to these groups remained meaningless prior to the acceptance of the theory of evolution.

Apparently very soon after Darwin had understood that a single species of South American mockingbird had given rise to three daughter species in the Galapagos Islands, he seemed to have realized that such a process of multiplication of species, combined with their continuing divergence, could give rise in due time to different genera and still higher categories. The members of a higher taxon then would be united by descent from a common ancestor. The best way to represent such common descent would be a branching diagram. Already in the summer of 1837 Darwin clearly stated that "organized beings represent an irregularly branched tree" (*Notebook B*, p. 21), and he drew several tree diagrams in which he even distinguished living and extinct species by different symbols. By the time he wrote the *Origin,* the theory of common descent had become the backbone of his evolutionary theory, not surprisingly so because it had extraordinary explanatory powers. Indeed, the manifestations of common descent as revealed by comparative anatomy, comparative embryology, systematics ("natural system"), and biogeography became the main evidence for the occurrence of evolution in the years after 1859. Reciprocally the stated biological disciplines, which up to 1859 had been primarily descriptive, now became causal sciences, with common descent providing explanation for nearly everything that had previously been puzzling.

In these studies the comparative method played an important role. To be sure, the practitioners of idealistic morphology and the Naturphilosophen had also practised comparison with excellent results. But the archetypes which they had reconstructed had no causal explanation until they were reinterpreted by Darwin as reflecting the putative common ancestors.

The theory of common descent, once proposed, is so simple and so obvious that it is hard to believe Darwin was the first to have adopted it consistently. Its importance was not only that it had such great explanatory powers but also that it provided for the living world a unity which had been previously missing. Up to 1859 people had been impressed primarily by the enormous diversity from the lowest plants to the highest vertebrates, but this diversity took on an entirely different complexion when it was realized that it all could be traced back to a common origin. The final proof of this, of course, was not supplied until our time, when it was demonstrated that even the prokaryotes have the same genetic code as animals and plants.

Perhaps the most important consequence of the theory of common descent was the change in the position of man. For theologians and philosophers alike man was a creature apart from all other living nature. Aristotle, Descartes, and Kant agreed in this, no matter how much they disagreed in other aspects of their philosophies. Darwin, in the *Origin,* confined himself to the cautiously cryptic remark, "Light will be thrown on the origin of Man and his history" (p. 488). But Ernst Haeckel, T. H. Huxley, and in 1871 Darwin himself demonstrated conclusively that humans must have evolved from an ape-like ancestor, thus putting him right into the phylogenetic tree of the animal kingdom. This was the end of the traditional anthropocentrism of the Bible and of the philosophers. To be sure, the claim that "man is nothing but an animal" was quite rightly rejected by all the more perceptive students of the human species, yet one cannot question that Darwin was responsible for a fundamental reevaluation of the nature of man and his role in the universe.

Gradualism

Darwin's solution for the problem of the multiplication of species and his discovery of the theory of common descent were accompanied by a number of other conceptual shifts. The most important one was his partial abandoning of essentialism in favour of gradualism and population thinking. The literature on the history of evolutionary biology lists numerous authors as forerunners of Darwin by having adopted evolutionism. When more closely examined, nearly all of these claims turn out to be invalid. Changes in the living world were ascribed to new origins by these earlier authors. But the sudden origin of a new species or still higher taxon is not evolution. Indeed, as has been rightly said by Reiser, if one is an essentialist, one cannot conceive of gradual evolution. Since an essence is constant and sharply delimited against other essences, it cannot possibly evolve. Change for an essentialist occurs only through the introduction of new essences. This was precisely Charles Lyell's interpretation of the introduction of new species throughout geological history, or Darwin's early explanation of the origin of the lesser ostrich of Patagonia.

The observation of nature seemed to give powerful support to the essentialist's claims. Wherever one looked one saw discontinuities, between species, between genera, between orders and even higher taxa. Quite naturally, the gaps between the higher taxa, like birds and mam-

mals, or beetles and butterflies, were mentioned particularly often by
Darwin's critics. And yet, as far back as Aristotle and his principle of
plenitude (Lovejoy 1936), there had been an opposing trend. It was
expressed in the *scala naturae,* and even such an arch-essentialist as Linnaeus
stated that the orders of plants were touching each other like countries
on a map.

Nevertheless Lamarck was the first person to apply the principle of
gradualism to explain the changes in organic life that could be inferred
from the geological record. But there is no evidence that Darwin derived
his gradualistic thinking from Lamarck. References to gradual changes are
scattered through Darwin's notebooks from an early time (Kohn 1980).
Yet it is still somewhat uncertain what the exact sources of Darwin's
gradualistic thinking were. One of the intellectual sources surely was
Lyell's uniformitarianism, which Darwin had adopted quite early. There
is also the fact that Darwin considered the changes of organisms either to
be produced directly by the environment or to be at least an answer to
the changes in the environment. Hence, "the changes in species must be
very slow, owing to physical changes slow" (*Notebook C,* p. 17). Gradualness was also favored by Darwin's conclusion that changes in habit or
behavior may precede changes in structure (*Notebook C,* pp. 57, 199). At
that time Darwin still believed in a principle called Yarrell's Law, according to which it takes many generations of impact for the effects of the
environment or of use and disuse to become strongly hereditary. As Darwin
stated, "Variety when long in blood, gets stronger and stronger" (*Notebook
C,* p. 136). Various other sources for Darwin's gradualistic thinking have
been suggested in the recent literature, such as the writings of J. B.
Sumner (Gruber 1974:125), or Leibniz's principle of plenitude (Stanley
1981), but to me it seems more likely that Darwin had arrived at his
gradualism empirically (see Essay 12).

At least three observations may have been influential. First, the slightness of the differences among the mockingbird populations on the three
Galapagos islands and the South American mainland, as well as a similarly
slight difference among many varieties and species of animals. Second,
the barnacle researches, where Darwin complained constantly to what
great extent species and varieties were intergrading. And third, Darwin's
work with the races of domestic pigeons, where he convinced himself that
even the most extreme races which, if found in nature, would be unhesitatingly placed by taxonomists in different genera were nevertheless the
product of painstaking, long-continued, gradual, artificial selection. In
his *Essay* of 1844, Darwin argues in favor of gradual evolution by analogy

with what is found in domesticated animals and plants. And he postulates therefore that "there must have existed intermediate forms between all the species of the same group, not differing more than recognized varieties differ" (p. 157). The abundant evidence which Darwin adduces in support of gradualism was to a considerable extent instrumental in the weakening and eventual refutation of essentialism. It was only by adopting a theory of gradual evolution that Darwin was able to escape from the constraints of essentialism. Yet evolutionary gradualism has remained a controversial subject to the present day (see essays 12, 23, and 26).

Natural Selection

The concept for which Darwin is better known than for any other is that of natural selection. When we speak of Darwinism today, we mean evolution by natural selection. The meaning of natural selection, its limits, and the processes by which it achieves its effects are now the most active areas of evolutionary research. Our present concept of natural selection is presented in Essay 6. For Darwin's views on natural selection see also essays 12 and 13.

The fundamental importance of the theory of natural selection when proposed in 1858/59 was that it challenged two of the most universally held ideologies of the period, that of cosmic teleology, and that of natural theology (Mayr 1982). According to the teleologists, the world either was established toward an end or was still moving toward an objective, either by the guiding hand of a Creator or by secondary causes, that is, by laws that were guiding the course of events toward an ultimate goal. All this was categorically denied by Darwin, for whom each evolutionary change was a singular event controlled by the temporary constellation of selection forces. Indeed, Darwin even stated that there would not be any change at all when there was no change in the environment. A nondeterministic process like natural selection was quite unintelligible for any philosopher thinking in terms of Newtonian laws.

Devastating as the denial of teleology was for many of Darwin's contemporaries such as Sedgwick and von Baer, the denial of design was even more sweeping. To explain all the beautiful adaptations of organisms, their adjustment to each other, their well-organized interdependence, and indeed the whole harmony of nature, as being the result of such a capricious process as natural selection was quite unacceptable to almost all of Darwin's contemporaries. The theory of natural selection amounted to the proposal to replace the hand of the Creator by a purely material and mechanical process, at that by one not deterministic and not predictive. As one critic put it, it dethroned God. Although an accommodation between religion and Darwinism was eventually reached, their mutual relation had first to go through a rather traumatic period. The current renaissance of the creationist movement shows that this relationship is still a precarious one. For a more detailed treatment of the ideological resistance to natural selection, see Mayr (1988).

Critical Dates

A historian likes to cite dates for events in intellectual history as much as for political history. When a pioneer of intellectual history like Darwin introduces revolutionary new ideas, one would like to pinpoint the moment when he formulated the new concepts. The fixing of a definite date often helps in the determination of the factors that were instrumental in the development of the new ideas. In line with this reasoning, we have come to say that Darwin "became an evolutionist" in March 1837 when Gould persuaded him that the varieties of mockingbirds on the Galapagos were good species, and we say that Darwin discovered the principle of natural selection on 28 September 1838 when reading Malthus. In both cases this dating is strongly reinforced by assertions made by Darwin himself.

A thorough study of the literature has convinced me that a major shift, virtually amounting to a conversion, did indeed occur on the mentioned dates. However, in both cases the crucial new information that brought about the conversion struck a "prepared mind." Darwin had already been vaguely thinking of a non-Lyellian change in species sometime before March 1837, a thought foreshadowed by his famous "islands in sight of each other" statement, penned a year earlier while sailing from the Cape of Good Hope to Ascension (Sulloway 1982) and by his saltational theory of the origin of Darwin's *Rhea*.

The same is true for natural selection; although clearly 28 September 1838 was decisive, yet Darwin, even prior to this date, made in his notebooks a number of "proto-selectionist" statements, as Kohn (1980) calls them. Most interesting among these is Darwin's reference to sexual selection (*Notebook C*, p. 61). "Whether species may not be made by a little more vigour being given to the chance offspring who have any slight peculiarity of structure. Hence seals take victorious seals, hence deer victorious deer, hence males armed and pugnacious all order; cocks all war-like." And after the date of the Malthus episode his understanding of natural selection continued to mature (Hodge 1983; Ospovat 1981).

To repeat, even though the shifts both to gradual evolutionism and to natural selection were clearly rather sudden and drastic processes, nevertheless there are indications for a prior erosion of opposing views, and a preparing of the mind for the final, decisive shift.

Darwin's Uncertainties

Darwin, like all great pioneers, removed old uncertainties and created new ones. He was unable to find satisfying solutions for several of the problems he tried to solve. Some of the questions he posed are still unsolved or at least still the subject of controversy. There was no other subject on which he was as uncertain as he was on variation. The science of genetics has removed some of these uncertainties, but the recent discovery of different kinds of DNA has reopened the subject.

Uncertainty still exists concerning speciation. At first Darwin was entirely convinced of the importance of isolation, even though apparently for the wrong reasons. Later, particularly after he thought that varieties of plants were as much incipient species as were varieties of animals, he downgraded the importance of isolation and attempted to replace it by his principle of character divergence. As the current writings on sympatric speciation, parapatric speciation, and stasipatric speciation show, the question of the importance of geographical isolation is still controversial today.

Darwin often commented on the steady action of natural selection, and it is usually assumed that he favored a steady evolutionary progression. Actually, Darwin was fully aware of the potential for evolutionary stasis. In his *Essay* of 1844 he said, "There is no necessity of modification in a species, when it reaches a new and isolated country. If it be able to survive and if slight variations better adapted to the new conditions are not selected, it might retain . . . its old form for an indefinite time" (p. 202). He also said, "As we see some species at present adapted to a wide range of conditions, so we may suppose that such species would survive unchanged and unexterminated for a long time" (p. 165; Rhodes 1983).

Darwin was remarkably well aware that the size of a population might be an important evolutionary factor. However, since he considered variation to be an intermittent phenomenon, occurring mostly under special circumstances, he thought that the larger a population was, the greater the chance that it would produce a variation that would be favored by selection. This opinion was reinforced by his observation that species from

large continents were usually competitively superior to species from restricted areas. The genetic discoveries of the last 80 years make it possible
to claim by contradistinction that it is precisely very small founder populations that have the best opportunity to initiate new evolutionary departures. However, this also is still a controversial area in which the last
word has not yet been said.

Again and again when we examine a modern controversy carefully, we
find that the problem was already known to Darwin and that he had
already made tentative suggestions as to its solution. There seem to have
been no limits to the fertility of his mind.

Darwin's Life Work

Darwin would be remembered as an outstanding scientist even if he had
never written a word about evolution. Indeed, J. B. S. Haldane has gone
so far as to say, "In my opinion, Darwin's most original contribution to
biology is not the theory of evolution, but his great series of books on
experimental botany published in the latter part of his life" (Haldane
1959:358). This achievement is little known among nonbiologists, nor is
Darwin's equally outstanding work on the adaptation of flowers and on
animal psychology, or his competent work on the barnacles and his
imaginative work on earthworms. In all these areas Darwin was a pioneer,
and although in some areas it took more than half a century before others
continued to build on the foundations which Darwin had laid, it is now
clear that he had attacked important problems with extraordinary originality, thereby becoming the founder of several now well-recognized separate disciplines. Darwin was the first to work out a sound theory of
classification, one which is still adopted by the majority of taxonomists.
His approach to biogeography, in which so much emphasis was placed on
the behavior and the ecology of organisms as factors of distribution, is
much closer to modern biogeography than the purely descriptive–geographical approach that dominated biogeography for more than half a
century after Darwin's death.

Darwin was a pioneer in other areas. I have already mentioned his
contribution to the end of anthropomorphism, and to the recognition of
population thinking. It was clearly Darwin who established the philosophical foundations of historical biology, a branch of science which in my
opinion is as scientific as are the physical sciences, even though dominated
by rather different methods of discovery and explanation from those
considered valid and necessary in the physical sciences. This includes

limits to prediction, the importance of the observational–comparative method (as compared with the experimental one), the role of historical narratives, and much else. I am mentioning this only in order to illustrate the extraordinary richness and originality of Darwin's thought and the impact of his writings right up to the present day.

Conclusion

Most students of the history of ideas believe that the Darwinian revolution was the most fundamental of all intellectual revolutions in the history of mankind. While such revolutions as those brought about by Copernicus, Newton, Lavoisier, or Einstein affected only one particular branch of science, or the methodology of science as such, the Darwinian revolution affected every thinking man. A world view developed by anyone after 1859 was by necessity quite different from any world view formed prior to 1859. It is therefore eminently proper that we pay tribute to the memory of this great man. But how can we explain Darwin's greatness?

That he was a genius is hardly any longer questioned, some of his earlier detractors notwithstanding. But there must have been a score of other biologists of equal intelligence who failed to match Darwin's achievement. What then is it that distinguishes Darwin from all the others?

Perhaps we can answer this question by investigating what kind of a scientist Darwin was. As he himself has said, he was first and foremost a naturalist. He was a splendid observer, and like all other naturalists he was interested in organic diversity and in adaptation. Naturalists are, on the whole, describers and particularists, but Darwin was also a great theoretician, something only very few naturalists have ever been. In that respect Darwin resembles much more some of the leading physical scientists. But Darwin differed from the run-of-the-mill naturalists also in another way. He was not only an observer but also a gifted and indefatigable experimenter whenever he dealt with a problem whose solution could be advanced by an experiment.

I think this suggests some of the sources of Darwin's greatness. The universality of his talents and interests had preadapted him to become a bridge builder between fields. It enabled him to use his background as a naturalist to theorize about some of the most challenging problems with which man's curiosity is faced. And contrary to widespread beliefs, Darwin was utterly bold in his theorizing.

There may be other ingredients to Darwin's greatness, but those that are apparent are these:

A brilliant mind, great intellectual boldness, and an ability to combine the best qualities of a naturalist-observer, of a philosophical theoretician, and of an experimentalist. The world has so far seen such a combination only once, and this accounts for Darwin's unique greatness.

NOTE

This essay is an abridged version of one which first appeared in D. S. Bendall, ed., *Evolution from Molecules to Men*, pp. 23–41. © Cambridge University Press, 1983.

REFERENCES

Barlow, N., ed. 1963. Darwin's ornithological notes. *Bull. Brit. Mus. (Nat. Hist.)*, Hist. Ser., 2:201–278.
Bowler, P. J. 1976. *Fossils and Progress*. New York: Science History Publications.
Buch, L. von. 1825. *Physicalische Beschreibung der Canarischen Inseln*, pp. 132–133. Berlin: Kgl. Akad. Wiss.
Darwin, C. 1844. Essay. In F. Darwin, ed., *The Foundations of the Origin of Species*. Cambridge: Cambridge University Press, 1909.
———— 1859. *On the Origin of Species*. Ed. E. Mayr. Facs. ed. Cambridge, Mass.: Harvard University Press, 1964.
Gillespie, N. C. 1979. *Charles Darwin and the Problem of Creation*. Chicago: University of Chicago Press.
Gruber, H. E. 1974. *Darwin on Man*. New York: Dutton.
Haldane, J. B. S. 1959. An Indian perspective of Darwin. *Cent. Rev. Arts Sci., Mich. State Univ.* 3:357.
Hodge, M. J. S. 1983. The development of Darwin's general biological theorizing. In D. S. Bendall, ed., *Evolution from Molecules to Men*, pp. 163–67. Cambridge: Cambridge University Press.
Kohn, D. 1980. Theories to work by: rejected theories, reproduction, and Darwin's path to natural selection. *Stud. Hist. Biol.* 4:67–170.
Lamarck, J.-B. 1809. *Philosophie Zoologique*. Paris.
Lovejoy, A. O. 1936. *The Great Chain of Being*. Cambridge, Mass.: Harvard University Press.
Lyell, C. 1830–1833. *Principles of Geology*. 3 vols. London: John Murray.
———— 1835. *Principles of Geology*. 4th ed. Vol. 3, p. 162. London: John Murray.
Maynard Smith, J. 1975. *The Theory of Evolution*. 3rd ed. Harmondsworth: Penguin.

Mayr, E. 1982. *The Growth of Biological Thought.* Cambridge, Mass.: Harvard University Press.

———— 1988. *Der ideologische Widerstand gegen Darwin's Selektionstheory.* Munich: Siemens Stiftung.

Osborn, H. F. 1894. *From the Greeks to Darwin.* New York: Columbia University Press.

Ospovat, D. 1981. *The Development of Darwin's Theory.* Cambridge: Cambridge University Press.

Rhodes, F. H. T. 1983. Gradualism, punctuated equilibrium, and the Origin of Species. *Nature* 305:269–272.

Stanley, S. M. 1981. *The New Evolutionary Timetable.* New York: Basic.

Sulloway, F. J. 1982. Darwin and his finches: the evolution of a legend. *J. Hist. Biol.* 15:1–53.

Wagner, M. 1841. *Reisen in der Regentschaft Algier in den Jahren 1836, 1837 und 1838.* Leipzig: Leopold Voss.

Wallace, A. R. 1855. On the law which has regulated the introduction of new species. *Annals and Magazine of Nat. Hist.,* ser. 2, 16:184–196.

THE CHALLENGE OF DARWINISM

ONE OF the most striking, indeed almost puzzling, aspects of Darwin's thought is the difference in reaction that it has induced in different constituencies. For a biologist, the concept of evolution by common descent caused by natural selection resulted in a complete reorientation of his thinking, for "nothing in biology makes sense except in the light of evolution," as Dobzhansky has said so rightly. Intellectual historians likewise have recognized and emphasized the profound impact of the Darwinian revolution. Yet, at the same time some historians have brushed off Darwin as a weak intellect, and numerous authors have tended to agree with Louis Agassiz's judgment, that the Darwinian theory was a "scientific mistake, untrue in its facts, unscientific in its methods, and mischievous in its tendency" (1860:154).

When a book with the title *At the Deathbed of Darwinism* came into the hands of V. Kellogg in 1907, he wrote that "ever since there has been Darwinism there have been occasional 'deathbeds of Darwinism' on the title pages of pamphlets, addresses and sermons." Curiously, this situation has not materially changed in the ensuing 75 years even though the theories of Darwin have been strengthened in the meantime through the researches of genetics, systematics, biogeography, and population biology. In this respect Darwinism is unique among major scientific theories. Thus, it is necessary to explain why Darwinism has encountered a so much greater and more broadly based opposition than any other well-established scientific theory. Let me try, therefore, to analyze historically the reasons for the opposition to Darwin and to present the current consensus of leading evolutionists concerning the validity of Darwin's ideas.

It is rarely noticed how recent the Darwinian revolution is in our intellectual history. The *Origin* was published only 45 years before I was born. And yet Europe had been in the throes of a continuous intellectual upheaval during the three preceding centuries, culminating in the Scientific Revolution of the sixteenth and seventeenth centuries and in the Enlightenment of the eighteenth century. Why did it take so long for evolution to be seriously proposed? And why did Darwinism face such an uphill battle after it was proposed? The reason, I contend, is that Darwin challenged some of the basic beliefs of his age. Let me enumerate the more important ones.

(1) *A belief in a constant world.* In spite of Lamarck and the Naturphilosophen, it was still widely, if not almost universally, accepted in 1859 that except for minor perturbations (floods, volcanism, mountain building) the world had not changed materially since creation. And in spite of Buffon, Kant, Hutton, Lyell, and the ice age theory, the prevailing opinion was still that of a rather recently created world.

(2) *A belief in a created world.* Species and other taxa were believed to be unchanging, and therefore the existing diversity of the living world could only be due to an act of creation. This was a single creation as believed by the orthodox Christians or repeated creations, either of whole biota as believed by the so-called progressionists (for example L. Agassiz), or of individual species as proposed by Charles Lyell.

(3) *A belief in a world designed by a wise and benign Creator.* Even though the world had its imperfections, it was the best of the *possible* worlds (Leibniz). The adaptation of organisms to their physical and living environment was perfect because it had been designed by an omnipotent Creator. (For fuller discussion, see Essay 14.)

(4) *A belief in a cosmic teleology.* See Essay 3.

(5) *A belief in the philosophy of essentialism.* The variable phenomena of this world, according to Plato and his followers, are the reflections (or outward manifestations) of constant, sharply delimited essences. These essences (natures, *eide*) are identical for all the members of a class or species. They are unchanging, all deviations being "accidents." This philosophy, of course, made evolution impossible. The only seriously opposing philosophy, that of nominalism (names bracketing individuals into classes), was only slightly less incompatible with evolution. According to Lamarck, all individuals exposed to the same environment would show identical reactions.

(6) *A belief in an interpretation of the causal processes of nature as they had been elaborated by the physicists.* During the Scientific Revolution, as a result of the work of Galileo, Descartes, Newton, and their associates

and followers, a theory of the world had been elaborated, dominated by a belief in laws, determinism, rigid prediction, classes, and a reductionist approach. This, as Darwin saw clearly, was quite unsuited to explain historical processes such as those which have caused the diversity and adaptedness of the living world.

(7) *A belief in the unique position of man in the Creation.* This was an anthropocentric world in the eyes of the Christian religion as well as in that of the foremost philosophers. Man had a soul, something animals did not have. There was no possible transition from animal to man.

How Did Darwin Challenge These Traditional Views?

The theory of evolution proposed by Darwin challenged all seven of these traditional and well entrenched views. Actually, Darwin's comprehensive theory of evolution was an entire set of theories dealing with both the "vertical" phenomenon of adaptive change and with the "horizontal" phenomenon of the diversity of populations, incipient species, and new species.

The origin of adaptation and the origin of diversity are two entirely separate problems, and their study represents two separate traditions in biology. Lamarck consistently concentrated on vertical evolution. Darwin, particularly early in his career, was far more interested in the origin of diversity, that is, in horizontal evolution. The two fathers of evolution thereby established two different traditions that are still with us, and that to a considerable extent have lived side by side, often almost without any contact. The leaders of the new systematics, for instance, were almost entirely concerned with the origin of diversity, while the paleontologists and geneticists concerned themselves almost entirely with phyletic evolution and its adaptive trends and shifts.

The comprehensiveness of Darwin's theory of evolution, dealing both with vertical and horizontal evolution, is made apparent when we dissect it into five major components or subtheories.

(1) Evolution as such

(2) Evolution by common descent

(3) The origin of diversity (speciation)

(4) Gradualness

(5) Natural selection

As I have shown elsewhere (1982:505–510), this package of theories is not an indivisible whole, since most evolutionists after 1859, including so-called Darwinians, accepted some but rejected others of these five theories. A more detailed treatment of the five theories is presented in Essay 12.

Who Were Darwin's Opponents?

When a new scientific theory is proposed, it usually has to battle one or two competing theories in the same scientific field. Although ideological commitments are often at the bottom of the arguments (see for instance Roe 1981 for the controversy of preformation *v.* epigenesis), nevertheless only a single front is usually involved. By proposing his five-pronged theory of evolution, Darwin had opened a battle on about half a dozen fronts. Several of these are still raging. The camps of his opponents may be characterized (quite tentatively) as follows.

(1) Orthodox Christians.   This group corresponds more or less to what we would now call fundamentalists, that is, those who reject anything that is in conflict with the literal interpretation of the Bible, as far as Creation and a constant world of short duration is concerned.

(2) Natural theologians.   Many of them were reasonable, liberal deists who nevertheless continued to believe in the argument from design and who saw the evidence of teleology everywhere in nature. This group included some of the outstanding scientists in Darwin's day, like Sedgwick in England and K. E. von Baer on the Continent.

(3) Lay-persons.   Darwin's theories seemed to be contradicted by the evidence encountered by any lay-person. How can one believe in gradual speciation when every species is sharply separated from every other one in our gardens, fields and woods? How could the gaps between birds, mammals, and reptiles ever be bridged, not to mention those between animals and plants, or any of the higher taxa of organisms? Intermediate stages between these types are quite unthinkable, it was said. How can selection lead to perfection when there has been so much extinction in the earth's history? Many similar questions were raised in the post-Darwinian battles where 'common sense' clearly seemed to contradict Darwin.

(4) The philosophers.   With a few exceptions (Baldwin, Dewey, Goudge, and a currently active group of young philosophers), philosophers have been essentialists and physicalists. In spite of sincere efforts by some of them, they have been quite unable to adopt population thinking, the concept of historical narratives, the distinction between proximate and

ultimate causations (Essay 2), the absence or at least irrelevance of laws (as defined by the physicists) in evolutionary biology, the invalidity of much of reductionism, the necessity to partition teleology, and various other basic concepts of evolutionary biology. The anti-Darwinian arguments of certain philosophers (including some contemporary ones) are to such a degree beside the point that, to a competent evolutionist, it is almost embarrassing to read them.

(5) *Physical scientists.*   Most physicists believed that all phenomena in nature, whether living or inanimate, have to obey the same laws and have to be investigated by the same methods. All of them were essentialists and more or less strict determinists. They strongly inclined to an atomistic reductionism and considered experimentation to be the only true scientific method. At the time of Darwin the physical scientists considered themselves the only true natural philosophers and were convinced that they had the necessary expertise to pass judgment on anything in science. The usually appallingly ignorant reviews of the *Origin* by physical scientists and mathematicians are eloquent testimony of "the arrogance of the physicists" (Hull 1973). Yet, their prestige and authority was so great that in any argument between a physical scientist and an evolutionist, as in that between Lord Kelvin and Darwin on the age of the earth, everyone automatically assumed that the physicist (being a "true scientist") had to be right.

(6) *Non-Darwinian biologists.*   Although the biologists accept evolution and common descent almost unanimously, most of them had reservations with respect to natural selection. Here, one must realize that there were two classes of opponents. A radical minority denied that natural selection played any constructive role whatsoever in evolution, while the majority of evolutionists did not deny the existence of selection but denied (as had Darwin himself) that selection alone could account for all adaptations and evolutionary changes. All these partial selectionists (Plate called them the "old Darwinians," to distinguish them from the neo-Darwinians) invoked some other evolutionary force, be it inheritance of acquired characters, finalistic forces, or saltations. Only the neo-Darwinians, beginning with Weismann, rejected any or all auxiliary agents and relied exclusively on natural selection.

None of the opposition to Darwin was as serious as that coming from his own profession. In the 80 years after 1859, the non-Darwinian biologists were decidedly in the majority. In fact in some countries, France for instance, there were virtually no true Darwinians. But even in the English- and German-speaking worlds, the flowering of Mendelism after

1900 seemed to spell the defeat of Darwinism to such an extent that books on the "death of Darwinism" could be published.

The prestige of Darwinism was perhaps never lower than in the first quarter of the twentieth century. Far more than 50% of the biologists, in fact even of the evolutionary biologists, subscribed to neo-Lamarckism, to various orthogenetic theories, or to saltationism. And yet it was in this period that the foundation was laid for a synthesis of seemingly conflicting evolutionary schools, a synthesis that took place in the 1930s and 40s. The delay was due to the fact that the two groups of biologists who supplied the information and the set of concepts that made the synthesis possible were hardly communicating with each other, and each had a one-sided approach.

On one side were the experimental geneticists concerned with phenomena at the level of the gene, preoccupied with the study of single gene pools, strongly reductionist in approach, concerned only with the vertical components of evolution (adaptation), and dealing with discontinuous entities (genes, chromosomes). In spite of these handicaps, the geneticists supplied the answer to the problem that had always eluded Darwin, that is, the nature and origin of the genetic variation that is necessary for natural selection to be successful. Also, the geneticists had a major share in refuting soft inheritance (inheritance of acquired characters). In addition, they were unable to find any evidence for the existence of any orthogenetic processes.

The other group consisted of the naturalists-systematists who worked with populations, who were interested in horizontal evolution (populations, incipient species), who thoroughly believed in gradual evolution and who appreciated that the individual is the true target of natural selection. Their major contribution was the realization that evolutionary processes and phenomena involve not only mechanisms at the level of the gene but also whole individuals and, what is most important, that the divergence of populations (geographic variation and geographic speciation) and of species (macroevolution) were major components of evolution. Only by including populations and species in the evolutionary analysis was it possible to explain the major problems of evolution, such as the multiplication of species, the origin of higher taxa, the origin of evolutionary novelties, and other manifestations of evolution in nature. These phenomena can not be explained at the level of genes. The synthesis occurred when actual students of evolution, like Huxley, Rensch, Simpson, Mayr, and Stebbins, agreed that the evolutionary phenomena studied by them are fully explicable through gradual evolution by means of natural selection (Mayr and Provine 1980). The synthesis and what happened afterwards was in many respects (except for the rejection of soft inheritance) a return to a purer Darwinism. For Darwin, it had always been clear that the individual and not the gene is the target of selection and that evolution has two components, the vertical one (adaptation) and the horizontal one (diversity).

Post-Synthesis Developments

Neither the discovery of numerous new facts relating to evolution nor the development of new concepts of speciation and genetic variation have required any essential revision of the picture of evolution as developed during the evolutionary synthesis. I emphatically deny the claims of various authors that these recent developments have led to an end of Darwinism, or of neo-Darwinism, or of the evolutionary synthesis. They are simply a filling-in of missing pieces in the edifice that the evolutionary synthesis had constructed. If one wanted to characterize the post-synthesis developments in a few sentences, one might mention the following items:

The reductionist concept introduced by the mathematical population geneticists, that evolution is a change in gene frequencies in populations, was rejected, and Darwin's old concept that evolution is a change in the components of the phenotypes, of whole individuals, populations, species, and higher taxa, was restored. A new theory of speciation was proposed (Mayr 1954) in which the role of peripherally isolated founder populations was emphasized. This formulation called attention to the fact that the rate of evolutionary change is often inversely correlated with population size. A better understanding of the chemical and structural nature of the genetic material was achieved through the work of molecular genetics. And finally, while the evolutionary synthesis was preoccupied with phenomena at the level of populations and species, much work in the post-synthesis era has been devoted to an understanding of the phenomena of macroevolution. The results of these studies considerably expand the framework laid down during the evolutionary synthesis by Rensch and Simpson. None of these developments have led, however, to a rejection of the basic principles established during the evolutionary synthesis. (For a somewhat different consideration of the post-synthesis developments, see Essay 28.)

Darwin's Theories in the Light of Current Thinking

In conclusion, I would like to analyse Darwin's views on evolution in the light of current thinking. I am careful not to say "to analyse where Darwin was right and where he was wrong," because there are still sufficient uncertainties concerning certain aspects of the evolutionary process to make it perilous to assert too dogmatically what *is* right and what *is* wrong.

The basic theory of evolution has been confirmed so completely that most modern biologists consider evolution simply a fact. How else except by the word *evolution* can we designate the sequence of faunas and floras in precisely dated geological strata? And evolutionary change is also simply a fact owing to the changes in the content of gene pools from generation to generation.

Darwin's theory of common descent has also been gloriously confirmed by all researches since 1859. Everything we have learned about the physiology and chemistry of organisms supports Darwin's daring speculation that "all the organic beings which have ever lived on this earth have descended from some one primordial form, into which life was first breathed" (1859:484). The discovery that the prokaryotes have the same genetic code as the higher organisms was the most decisive confirmation of Darwin's hypothesis.

Darwin's theory of gradualism, unpalatable even to his close friends Huxley and Galton, has triumphed decisively and makes the more sense, the more clearly we realize that evolution is a process involving populations. The only apparent exceptions are the occasional abandonment of sexual reproduction and certain chromosomal processes such as polyploidy.

Darwin's great emphasis on the development of diversity as an important component of the evolutionary process, undeservedly neglected during the first third of this century, is again at the forefront of interest, particularly in palaeontology and ecology. As far as speciation is concerned— the process that serves as the source of new diversity—Darwin was somewhat confused (Kottler 1978; Sulloway 1979). Although supporting geographic speciation on islands, Darwin believed in a widespread occurrence of sympatric speciation on continents (see Part 7).

The greatest triumph of Darwinism is that the theory of natural selection, for 80 years after 1859 a minority opinion, is now the prevailing explanation of evolutionary change. It must be admitted, however, that it has achieved this position less by the amount of irrefutable proofs it has been able to present than by the default of all the opposing theories.

It must further be admitted that the modern theory is not quite the same as that of Darwin.

Darwin took it for granted that a nearly unlimited amount of variation was at all times available to provide material for natural selection. He had no idea as to the source of this variation and supported several genetic theories (soft inheritance, pangenesis, blending inheritance) that have since been refuted. Nevertheless, advances in genetics continue to strengthen rather than weaken the theory of natural selection. Recent developments in our thinking about natural selection are discussed in Essay 6.

When reviewing all the additions to our knowledge and all the changes in our concepts of the last 50 years, I do not have the feeling that they constitute any decisive change in the overall view of evolution as composed during the evolutionary synthesis. The numerous claims to the contrary in the recent literature were made without any supporting evidence (Stebbins and Ayala 1981; see also Essay 28).

Darwin's Impact

There is perhaps no better way to characterize Darwin's impact on our thinking than by listing the concepts that he refuted and those that he newly introduced into our thinking, or at least to whose acceptance he contributed materially.

CONCEPTS REFUTED BY DARWIN

Creation.   Darwin refuted the belief that the diversity of organisms found on earth was the result of creation. (See also Gillespie 1979.)

Newness of the earth.   In the battle with Lord Kelvin, Darwin clearly emerged as the victor, and his estimate of organic life being several thousand million years old is now universally accepted.

Cosmic teleology.   All the phenomena that previously had been ascribed to design or to finalistic causes Darwin was able to explain in terms of natural selection.

Anthropocentrism.   Darwin and his followers showed conclusively that man is not a separate creation but the product of common descent.

NEW CONCEPTS INTRODUCED OR BOLSTERED IN A
DECISIVE FASHION BY DARWIN

Population thinking.   Living nature does not consist of types but of variable populations in which each individual is unique.

Natural selection.   Owing to the vast reproductive surplus, there is an intense competition among individuals of the same population for survival and reproduction. Certain genotypes have a greater probability of leaving offspring than others.

Geographic speciation.   Different populations undergo different genetic modifications, and if such populations are isolated, the genetic changes may be compounded into species differences.

Evolutionary progress.   Since there is no cosmic teleology, there is no necessary organic progress. Whatever seeming progress we find between the first origin of life and the existing biota is due to competition among species and to character divergence. There are numerous processes, such as change of function of a structure, that facilitate the acquisition of new capabilities of advantage in species competition.

It was through Darwin that we have come to realize that every biological phenomenon and process requires at least two explanations, a purely functional one and an evolutionary one, and that evolution has two major components, changes in adaptation and changes in diversity. But Darwin's work has had an impact far transgressing the domain of biology. No other scientific theory has challenged and, in fact, refuted so many commonly held beliefs as Darwin's theory of evolution by natural selection. But more than that: Darwin introduced entirely new ways of thinking and of carrying out scientific research. In short, no other philosopher or scientist has had as great an impact on the thinking of modern man as Darwin.

NOTE

This essay is an abridged and revised version of one which first appeared under the title "Epilogue" in the *Biological Journal of the Linnean Society, London* 17(1982):115–125.

REFERENCES

Agassiz, L. 1860. Prof. Agassiz on the origin of species. *Amer. J. Sci. Arts*, 2nd ser., 30:154.

Berry, R. J. 1979. Genetical factors in animal population dynamics. In R. M. Anderson, B. D. Turner, and L. R. Taylor, eds., *Population Dynamics*, pp. 53–80. Oxford: Blackwell Scientific.

Darwin, C. 1859. *On the Origin of Species*. London: John Murray.

Gillespie, N. C. 1979. *Charles Darwin and the Problem of Creation.* Chicago: University of Chicago Press.

Gould, S. J., and R. Lewontin. 1979. The spandrels of San Marco and the Panglossian paradigm: a critique of the adaptationist program. *Proc. Roy. Soc. London* B205:581–598.

Haldane, J. B. S. 1957. The cost of natural selection. *J. Genet.* 55:511–524.

Hull, D. 1973. *Darwin and His Critics.* Cambridge, Mass.: Harvard University Press.

Kellogg, V. L. 1907. *Darwinism To-day.* New York: Henry Holt.

Kottler, M. 1978. Charles Darwin's biological species concept and theory of geographic speciation. *Amer. Sci.* 35:275–297.

Mayr, E. 1954. Change of genetic environment and evolution. In J. Huxley, A. C. Hardy, and E. B. Ford, eds., *Evolution as a Process,* pp. 157–180. London: Allen and Unwin.

———— 1962. Accident or design, the paradox of evolution. In G. W. Leeper, ed., *The Evolution of Living Organisms,* pp. 1–14. Melbourne: Melbourne University Press.

———— 1982. *The Growth of Biological Thought.* Cambridge, Mass.: Harvard University Press.

Mayr, E., and W. Provine, eds. 1980. *The Evolutionary Synthesis.* Cambridge, Mass.: Harvard University Press.

Roe, S. 1981. *Matter, Life, and Generation.* Cambridge: Cambridge University Press.

Stebbins, G. L., and F. J. Ayala. 1981. Is a new evolutionary synthesis necessary? *Science* 213:967–971.

Sulloway, F. 1979. Geographic isolation in Darwin's thinking: the vicissitudes of a crucial idea. *Stud. Hist. Biol.* 3:23–65.

WHAT IS DARWINISM?

IN RECENT controversies on evolution one frequently finds references to "Darwin's theory of evolution," as though it were a unitary entity. In reality, Darwin's "theory" of evolution was a whole bundle of theories, and it is impossible to discuss Darwin's evolutionary thought constructively if one does not distinguish its various components. But quite aside from the fact that it helps understanding of the structure of evolutionary theory to carry the analysis to the level of the subtheories Darwin adopted, it is important to call attention to the composite nature of the Darwinian theory for three very specific additional reasons.

The first is that this knowledge is very important for the proper understanding of the term *Darwinism*. This term has numerous meanings, depending on who has used the term and at what period. At first, in Darwin's day, Darwinism to most people simply meant a belief in evolution, and perhaps man's descent from the apes. By contrast, if a modern biologist uses the term, his emphasis is entirely on natural selection. He would define Darwinism as the theory that attributes evolutionary change to selection forces. Nonbiologists, however, have often used the term in a much broader sense. I need refer only to the term *Social Darwinism,* an ideology that had more to do with Spencer than with Darwin. Indeed, in the last third of the nineteenth century the term Darwinism, often used with a distinctly derogatory connotation, was applied to a materialisticatheistic Weltanschauung (Greene 1981), which as a matter of fact had little to do with Darwin's own thinking. To repeat, an understanding of the meaning of the term Darwinism will be helped considerably by a discrimination among the various evolutionary theories held by Darwin.

The second reason for such discrimination is that one cannot answer the question correctly of how and when "Darwinism" was accepted in different countries throughout the world unless one focuses on the various Darwinian theories separately. If we look at the contributions made by several authors to the volume by Glick (1974) or the book on the fate of

Darwinism in France by Conry (1974), we find that the treatment suffers from a failure to deal with the different Darwinian ideas individually. As we now know, what Darwin presented in 1859 in the *Origin* was a compound theory, and the five subtheories I shall single out had very different fates in the eighty years after Darwin.

A third reason to determine clearly and unambiguously what Darwinism is, is the frequently made recent claim that the Darwinian theory is obsolete, that neo-Darwinism has been refuted, or even that the evolutionary synthesis is dated if not refuted. It is quite impossible to test the validity of these claims until one has clearly determined the meaning of the terms *Darwinism* and *neo-Darwinism*, that is, has determined of what theories they consist, and how they are constructed.

I want to issue a warning at the outset. Darwin was a great pioneer, a person with an exceptionally fertile mind, but like other fertile thinkers, he had considerable trouble sticking to a consistent "party line." On almost any subject he dealt with—and this includes almost all of his own theories—he not infrequently reversed himself. For instance, he might say that only slight variations are evolutionarily important, but then on another page he might talk about rather strikingly different varieties, like the ancon sheep and the turnspit dog, both of them extremely short-legged. Darwin's pluralism has recently been emphasized by Gould and Lewontin (1979). In addition to natural selection, for instance, Darwin allowed also for use and disuse and occasionally even for a direct influence of the environment. Although Darwin at first fully supported geographic speciation, eventually he also allowed considerable scope to various forms of sympatric speciation. I could quote many more examples of his pluralism. It was this that led one of Darwin's twentieth-century critics to assert: "Darwin's hedging and self-contradiction—enabled any unscrupulous reader to choose his text from the *Origin of Species* or the *Descent of Man* with almost the same ease of accommodation to his purpose as if he had chosen from the Bible" (Barzun 1958:75). Evidently, then, it is not legitimate to refute the validity of one of Darwin's multiple choices and then claim this is a refutation of Darwinism.

Some historians (for example, Kohn, Ospovat, Hodge) have referred to the combination of theories Darwin held at various times as his "unified theory" of that period. I will not argue against this, if this is the historians' practice. But it must not be forgotten that each of these "unified" theories consisted of a very heterogeneous set of components, each of them a full theory in its own right. There is one particularly cogent reason why Darwinism cannot be a single homogeneous theory, which is that organic evolution consists of two essentially independent processes, transformation in time and diversification in (ecological and geographical) space. The two processes require a minimum of two entirely independent and very different theories. That writers on Darwin have nevertheless almost invariably spoken of the combination of these various theories as "Darwin's theory" in the singular, is in part Darwin's own doing. He not only referred to the theory of evolution itself as "my theory," but he also called the theory of descent by natural selection "my theory," as if common descent and natural selection were a single theory.

The discrimination among his various theories was not helped by the fact that he treated speciation under natural selection in chapter 4 of the *Origin* and that he ascribed many phenomena, particularly those of geographic distribution, to natural selection when they were really the consequences of common descent. Under the circumstances I consider it urgently necessary to dissect Darwin's conceptual framework of evolution into a number of major theories that formed the basis of his evolutionary thinking. For the sake of convenience I have partitioned Darwin's evolutionary paradigm into five theories, but of course others might prefer a different division. The selected theories are by no means all of Darwin's evolutionary theories; others were, for instance, sexual selection, pangenesis, effect of use and disuse, and character divergence. However, when later authors referred to Darwin's theory they invariably had a combination of some of the following five theories in mind. For Darwin himself these five theories were apparently much more a unity than they appear to a person who analyzes them with modern hindsight. The five theories were: (1) evolution as such, (2) common descent, (3) gradualism, (4) multiplication of species, and (5) natural selection. Someone might claim that indeed these five theories are a logically inseparable package and that Darwin was quite correct in treating them as such. This claim, however, is refuted by the fact, as I have demonstrated elsewhere (1982:506), that most evolutionists in the immediate post-1859 period—that is, authors who had accepted the first theory—rejected one or several of Darwin's other four theories. This demonstrates that the five theories are not one indivisible whole.

Evolution as Such

This is the theory that the world is neither constant nor perpetually cycling but rather is steadily and perhaps directionally changing, and that organisms are being transformed in time. It is difficult for a modern to visualize how widespread the belief still was in the first half of the last century, particularly in England, that the world is essentially constant and of short duration. Even the majority of those who, like Charles Lyell, were fully aware of the great age of the earth and of the steady march of extinction refused to believe in a transformation of species.

I must stress at this point the word *transformation* as representing a process guaranteeing continuity. This was overlooked by Osborn in his *From the Greeks to Darwin* (1894), where he lists scores if not hundreds of authors whom he designated as forerunners of Darwin. To be sure, these authors proposed "new origins," that is, the production of new species or new types, but invariably by discontinuous saltations. This was inevitably so because all these authors had been essentialists, and an essence cannot evolve. Any change must be due to the production of new essences. Accordingly, even Lyell explained the steady change of faunas as due to extinction and a mysterious "introduction of new species," a discontinuous process.

When Darwin began to break away from Lyell's thinking, he did so in stages. When he discovered the second species of *Rhea* (South American "ostriches") occurring (in certain districts) side by side with the better-known large Rhea, he explained it not as a new "introduction" (à la Lyell) to fill a vacant ecological niche in nature, but rather as a derivation from the older species by a saltation. Darwin thus adopted transmutation, but transmutation essentialistically conceived.

To the best of my knowledge Lamarck was the first author to propose a consistent theory of gradual transformation. After 1800, but before 1859, the idea of gradual evolution was accepted by a considerable number of authors on the Continent (Mayr 1982), but none of these authors developed the idea of evolution into a consistent and well-documented theory. This is what Darwin achieved in the *Origin*.

Evolution as such is no longer a theory for a modern author. It is as much a fact as that the earth revolves around the sun rather than the reverse. The changes documented by the fossil record in precisely dated geological strata are a fact that we designate as evolution. It is the factual basis on which the other four evolutionary theories rest. For instance, all the phenomena explained by common descent would make no sense if evolution were not a fact.

Common Descent

The case of the species of Galapagos mockingbirds provided Darwin with an important new insight. The three species had clearly descended from a single ancestral species on the South American continent. From here it was only a small step to postulate that all mockingbirds were derived from a common ancestor—indeed, that every group of organisms descended from an ancestral species. This is Darwin's theory of common descent.

It must be emphasized that the terms *common descent* and *branching* describe exactly the same phenomenon for an evolutionist. Common descent reflects a backward-looking view and branching a forward-looking view. As an aside it might be remarked that the concept of common descent runs into difficulties in the relatively rare cases of "reticulate evolution," that is, when a phyletic lineage is the product of a merger (owing to hybridization or symbiosis) of two previously separate lineages.

With branching being inseparably connected to descent in the mind of an evolutionist, it seems at first puzzling that the term *branching* was used so widely long before 1859. Just exactly what did branching mean to an author who did not believe in evolution? Pallas expressed relationship in the form of branching trees, and Cuvier called his major phyla "embranchements." For Agassiz and Milne-Edwards, branching reflected a divergence in ontogeny, so that the adult forms were far more different than the earlier embryonic stages. From all these examples it is evident that static branching diagrams of nonevolutionists are no more indications of evolutionary thinking than branching flow charts in business or branching diagrams in administrative hierarchies. The concept of common descent was, however, not entirely original with Darwin. Buffon had already considered it for close relatives, such as horses and asses; but not accepting evolution, he had not extended this thought systematically. There are occasional suggestions of common descent in a number of other pre-Darwinian writers, but historians so far have not made a careful search for early adherents of common ancestry. It is a theory that was definitely not upheld by Lamarck, who, although he proposed the occasional splitting of "masses" (higher taxa), never thought in terms of a splitting of species and regular branching. He derived diversity from spontaneous generation and the vertical transformation of each line separately into stages of higher perfection. For him descent was linear descent within each phyletic line, and the concept of common descent was alien to him.

The concept of branching occurred to Darwin quite early in his evolutionary speculating, and the rough sketches of branching trees in his *Notebooks* have often been described (Gruber and Barrett 1974:142–143). Branching, by necessity, means divergence; and every modern student of phylogeny, regardless of which school he belongs to, takes it for granted that evolutionary lines, once they have become completely separated from each other, will steadily diverge to a greater or lesser degree. And the evidence indicates that Darwin made the same assumption in his earlier branching diagrams. Yet in the 1850s Darwin discovered his "principle of divergence," which he considered, together "with Natural Selection—[to be] the keystone of my book" (Darwin 1903 1:109). Ever after he discovered this principle, sometime between 1854 and 1857, he referred to it always with great excitement, as if it had been a major departure from his previous thinking (for example, 1958:120–121). His wording implies that earlier he had thought phyletic lines would remain parallel after completing the process of speciation. His earlier understanding that insular phyletic lines could diverge drastically was apparently completely forgotten. In some respects one has the impression that the principle emerged as a result of a shift in Darwin's concept of speciation from insular speciation by geographical varieties to continental speciation by ecological varieties as described by botanists. But Ospovat (1981) makes a strong case for considering the principle an outcome of Darwin's thinking about classification. Actually in the years 1844 to 1858 Darwin's concepts underwent such a strong change in several important respects that it would probably not be correct to point to a single factor as the source for the principle of divergence. It seems to me that the conceptual connections in Darwin's intellectual development between speciation, common descent, and character divergence are not yet fully understood, and even Kohn's (1985) analysis is not the last word.

None of Darwin's theories was accepted as enthusiastically as common descent; it is probably correct to say that no other of Darwin's theories had such enormous immediate explanatory powers. Everything that had seemed to be arbitrary or chaotic in natural history up to that point now began to make sense. The archetypes of Owen and of the comparative anatomists could now be explained as the heritage from a common ancestor. The entire Linnaean hierarchy suddenly became quite logical, because it was now apparent that each higher taxon consisted of the descendants of a still more remote ancestor. Patterns of distribution that previously had seemed capricious could now be explained in terms of the dispersal of ancestors. Virtually all the proofs for evolution listed by Darwin in the *Origin* actually consist of evidence for common descent. To establish the line of descent of isolated or aberrant types became the most popular research program of the post-*Origin* period, and has largely remained the research program of comparative anatomists and paleontologists almost up to the present day. To shed light on common ancestors also became the program of comparative embryology. Even those who did not believe in strict recapitulation often discovered similarities in embryos that were obliterated in the adults. These similarities, such as the chorda in tunicates and vertebrates, or the gill arches in fishes and terrestrial tetrapods, had been totally mystifying until they were interpreted as vestiges of a common past.

Nothing helped the rapid adoption of evolution more than the explanatory power of the theory of common descent. Soon it was demonstrated that even animals and plants, seemingly so different from each other, could be derived from a common, one-celled ancestor. This Darwin had already predicted, when he suggested that "all our plants and animals [have descended] from some one form, into which life was first breathed" (1975:248). The studies of cytology (meiosis, chromosomal inheritance) and biochemistry fully confirmed the evidence from morphology and systematics for a common origin. It was one of the triumphs of molecular biology to be able to establish that eukaryotes and prokaryotes have the identical genetic code, thus leaving little doubt about the common origin even of these groups. Though there are still a number of connections among higher taxa to be established, particularly among the phyla of plants and invertebrates, there is probably no biologist left today who would question that all organisms now found on the earth have descended from a single origin of life.

There was only one area in which the application of the theory of common descent encountered vigorous resistance: the inclusion of man into the total line of descent. To judge from contemporary cartoons, none of the Darwinian theories was less acceptable to the Victorians than the derivation of man from the other primates. Yet at the present time this derivation is not only remarkably well substantiated by the fossil record, but the biochemical and chromosomal similarity of man and the African apes is so great that it is quite puzzling why they are so relatively different in morphology and brain development.

Gradualism

Darwin's third theory was that evolutionary transformation always proceeds gradually, never in jumps. One will never understand Darwin's insistence on the gradualism of evolution, nor the strong opposition to this theory, unless one realizes that virtually everyone at that time was an essentialist. The occurrence of new species, documented by the fossil record, could take place only by new origins, that is, by saltations. Since the new species were perfectly adapted, however, and since there was no evidence for the frequent production of maladapted species, Darwin saw only two alternatives. Either the perfect new species had been specially created by an all-powerful and all-wise Creator, or else—if such a supernatural process were unacceptable—the new species had evolved gradually from pre-existing species by a slow process, at each stage of which they maintained their adaptation. It was this second alternative that Darwin adopted.

This theory of gradualism was a drastic departure from tradition. Theories of a saltational origin of new species had existed from the pre-Socratics to Maupertuis and the progressionists among the so-called catastrophist geologists. These saltationist theories were consistent with essentialism. Perhaps one can distinguish three kinds of such saltationist theories:

(1) Extinct species are replaced by newly created ones that are more or less at the same level as those that they replace (Lyell 1830–1833).

(2) Extinct species are replaced by new creations at a higher level of organization (progressionists, such as Buckland, Sedgwick, Hugh Miller, L. Agassiz).

(3) New species originate through saltations of pre-existing species (E. Geoffroy Saint-Hilaire, Darwin in Patagonia ["Petise"], Galton, Goldschmidt).

Darwin's totally gradualist theory of evolution—not only species but also higher taxa arise through gradual transformation—immediately encountered strong opposition. Even Darwin's closest friends were unhappy about it. T. H. Huxley wrote to Darwin on the day before the publication of the *Origin:* "You have loaded yourself with an unnecessary difficulty in adopting *Natura non facit saltum* so unreservedly" (Huxley 1900 2:27). In spite of the urgings of Huxley, Galton, Kölliker, and other contemporaries, Darwin insisted almost obstinately on the gradualness of evolution, even though he was fully aware of the revolutionary nature of this concept. With the exceptions of Lamarck and Geoffroy, almost everybody else who had ever thought about changes in the organic world had been an essentialist and had resorted to saltations.

While still on the *Beagle* Darwin had accepted Lyell's sudden introduction of new species, and even when he derived new species from pre-existing species, like the Petise "ostrich," he contended it occurred through a saltation. Even though Darwin continued to use the term *transmutation* for many more years, after March 1837 he held that evolutionary change was more or less gradual, and it had become transformation instead of transmutation. Furthermore his adherence to gradualism became stronger with time; eventually (after the 1867 critique by F. Jenkin) he minimized even more the evolutionary role of "sports" (drastic variations).

The source of Darwin's strong belief in gradualism is not quite clear. The problem has not yet been analyzed adequately. Gruber (Gruber and Barrett 1974) thinks Darwin was influenced by the theologian Sumner (1824:20), who suggested that sudden occurrences are indications of an intervention by the Creator and thus of supernatural origin. By stressing continuity and gradualness Darwin was stressing natural causation. Stanley (1981) thinks that Darwin applied the principle of plenitude. Most likely gradualism is the extension of Lyell's uniformitarianism from geology to the organic world. Lyell's failure to do so had rightly been criticized by Bronn. Darwin, of course, also had strictly empirical reasons for his insistence on gradualism. His work with domestic races, particularly his work with pigeons and his conversations with animal breeders, convinced him how strikingly different the end products of slow, gradual selection could be. This fitted well with his observations on the Galapagos mockingbirds and tortoises, which were best explained as the result of gradual transformation.

Finally Darwin had didactic reasons for insisting on the slow accumulation of rather small steps. He answered the argument of his opponents that one should be able to "observe" evolutionary change owing to natural selection by saying: "As natural selection acts solely by accumulating slight successive favorable variations, it can produce no great or sudden modifications; it can act only by very short and slow steps" (1859:471). There is little doubt that the general emergence of population thinking in Darwin strengthened his adherence to gradualism. As soon as one adopts the concept that evolution occurs in populations and slowly transforms them—and this is what Darwin increasingly believed—one is automatically forced also to adopt gradualism. Gradualism and population thinking probably were originally independent strands in Darwin's conceptual framework, but eventually they reinforced each other powerfully.

After Darwin's death the concept of gradualism became even less popular than it had been in Darwin's own time. This began with Bateson's 1894 book and reached a climax with the mutationist theories of the Mendelians. Both Bateson and De Vries missed no opportunity to make fun of Darwin's belief in gradual evolution and upheld instead evolution by macromutations (Mayr and Provine 1980; Essay 25). A mild popularity of saltationist theories continued right through the evolutionary synthesis (Goldschmidt 1940; Willis 1940; Schindewolf 1950).

The naturalists were the main supporters of gradual evolution, which they encountered everywhere in the form of geographic variation. Eventually geneticists arrived at the same conclusion through the discovery of ever slighter mutations, of polygeny, and of pleiotropy. The result was that gradualism was able to celebrate a complete victory during the evolutionary synthesis in spite of the continuing opposition by Goldschmidt and Schindewolf.

Now, some forty years later, the argument has flared up again in the wake of the theory of punctuated equilibria. But the current argument simply boils down to the question: What is gradual, and how is it to be defined? Punctuated evolution for Goldschmidt and Schindewolf, as well as for Bateson and De Vries, was the production of a new species or higher taxon through the origin of a single individual that had experienced a complete genetic reorganization. Gradual evolution, as now defined, is populational evolution, during which the genetic changes are effected in the course of a series of generations. This does not preclude the possibility that the reorganization of the phenotype at the end of a series of generations may be rather dramatic, as suggested in my theory of genetic revolutions during peripatric speciation (1954; Essay 25).

Defining gradualism as populational evolution—and this is what Darwin basically had in mind—permits us to say that in spite of all the opposition to him, Darwin ultimately prevailed even with his third evolutionary theory. The only exceptions to gradualism that are clearly established are cases of stabilized hybrids that can reproduce without crossing (like allotetraploids).

Nothing is said in the theory of gradualism about the rate at which the change may occur. Darwin was aware of the fact that evolution could sometimes progress quite rapidly, but, as Andrew Huxley (1981) has recently quite rightly pointed out, it could also contain periods of complete stasis "during which these same species remained without undergoing any change." In his well-known diagram in the *Origin* (opposite p. 117), Darwin lets one species (F) continue unchanged through 14,000 generations or even through a whole series of geological strata (p. 124). The understanding of the independence of gradualness and evolutionary rate is important for the evaluation of the theory of punctuated equilibria.

The Multiplication of Species

This theory of Darwin's dealt with the explanation of the origin of the enormous organic diversity. It is estimated that there are 5 to 10 million species of animals and 1 to 2 million species of plants on earth. Even though in Darwin's day only a fraction of this number was known, the problem of why there are so many species and how they originated was already present. Lamarck had ignored the possibility of a multiplication of species in his *Philosophie Zoologique* (1809). For him diversity was produced by differential adaptation. New evolutionary lines originated by spontaneous generation, he thought. In Lyell's steady-state world, species number was constant and new species were introduced to replace those that had become extinct. Any thought of the splitting of a species into several daughter species was absent among these earlier authors.

To find the solution to the problem of species diversification required an entirely new approach, and only the naturalists were in the position to find it. L. von Buch in the Canary Islands, Darwin in the Galapagos, Wagner in North Africa, and Wallace in Amazonia and the Malay Archipelago were the pioneers in this endeavor. By adding a horizontal (geographic) dimension to the vertical dimension that had previously monopolized evolutionary thought, they all were able to discover geographically representative (allopatric) species or incipient species. But more than that, these naturalists found numerous allopatric populations that were in all conceivable intermediate stages of species formation. The sharp discontinuity between species that had so impressed John Ray, Carl Linnaeus, and other students of the nondimensional situation (the local naturalists) was now supplemented by a continuity among species owing to the incorporation of the geographical dimension.

At first the situation was not understood as clearly as it is now. This is well reflected in Wallace's statement in 1855: "Every species has come into existence coincident both in time and space with a pre-existing closely allied species," and more specifically "the most closely allied species are found in the same locality or in closely adjoining localities" (1855:6, 5). It reflects the typological species concept still prevailing at that time. There is no reference to geographically representative populations.

If one defines species simply as morphologically different types, one evades the real issue. A more realistic formulation of the problem of speciation did not occur until the development of the biological species concept (K. Jordan, Poulton, Stresemann, Mayr). Only then was it seen that the real problem is not the acquisition of difference but of "distinctness." The problem is thus the acquisition of reproductive isolation in relation to other contemporary species. Transformation of a phyletic line in the time dimension (gradual phyletic evolution, as it was later designated) sheds, of course, no light on the origin of diversity.

Darwin struggled with the problem of the multiplication of species all his life. When he discovered in Patagonia a second species of *Rhea* (which he called "ostrich"), Darwin modified only slightly Lyell's concept of a sudden introduction of new species. In contrast with Lyell, he derived it from an existing species (the northern Rhea); but he did so by an essentialistic saltation. Only after he had discovered the three new species of mockingbirds on different islands in the Galapagos did Darwin develop a fully consistent concept of geographic speciation. At once he found additional examples of apparent geographic speciation, as stated in his B *Notebook* (see Darwin 1960–1967): "Galapagos tortoises, . . . Falkland fox, Chiloe fox.—English and Irish Hare" (*B* 7). It was at this time that Darwin provided species definitions which basically agree with the modern biological species concept (*B* 24: "repugnance to intermarriage"; see also *B* 122, *B* 213; 1967 C:161). His thinking, at that period, seems to have been derived exclusively from the zoological literature. Even though Darwin considered isolation on islands as the principal speciation mechanism, he seems to have had difficulties in explaining speciation on continents. At one time, to account for the rich species diversity in South Africa, he postulated large-scale geological changes, up and down movements of the crust, during which South Africa was temporarily converted into an archipelago, setting the stage for abundant speciation.

But it was not until Darwin, with the help of Hooker, became better acquainted with the botanical literature that he began seriously to speculate on sympatric speciation on continents. There is still some uncertainty about the development of his thoughts on this subject, in spite of extensive clarification by Kottler (1978) and Sulloway (1979). Part of Darwin's difficulty was caused by the ambiguous use of the term *variety* in the taxonomic literature. Zoologists tended to use the term for geographic races (subspecies) and these, when isolated, were incipient species. Hence, in the 1840s, varieties for Darwin were incipient species.

In the botanical literature, however, varieties more often than not were individual variants ("morphs") within a population. Applying his previously established axiom "varieties are incipient species" to such coexisting varieties required a theory of sympatric speciation. This Darwin established in *Natural Selection* with numerous examples, most of them actually zones of secondary hybridization (pp. 252–261). The same thesis, but in a rather abbreviated form, is presented in the *Origin* (p. 103). It was this claim of sympatric speciation that led to his controversy with M. Wagner. Evidently Darwin was not aware of some of the difficulties for sympatric speciation to which modern authors have called attention.

Although Darwin deserves credit, together with Wallace, for having posed concretely for the first time the problem of the multiplication of species, the pluralism of his proposed solution led to a history of continuous controversy that is not ended to this very day. At first, from the 1870s to the 1940s, sympatric speciation was perhaps the more popular theory of speciation, although some authors, particularly ornithologists and specialists of other groups displaying strong geographic variation, insisted on exclusive geographic speciation. The majority of entomologists, however, and likewise most botanists, even though admitting the occurrence of geographic speciation, considered sympatric speciation to be the more common and thus more important form of speciation. After 1942 allopatric speciation was more or less victorious for some twenty-five years, while now the controversy is again in full swing (White 1978; Essay 21).

Paleontologists, on the whole, completely ignored the problem of the multiplication of species. For instance, one finds no discussion of it in the work of G. G. Simpson. Indeed, the material of the paleontologists is not suitable for an analysis of the speciation process. When paleontologists finally incorporated speciation into their theories (Eldredge and Gould 1972), their conclusions were based on the speciation research of those who study living organisms.

There are three reasons why speciation is still an open problem 125 years after the publication of *Origin*. The first is that, as in so much of evolutionary research, the evolutionist analyzes the results of past evolutionary processes and is thus obliged to reach conclusions by inference. Consequently, one encounters all the well-known difficulties met in the reconstruction of historical sequences. The second difficulty is that in spite of all the advances of genetics, we are still almost entirely ignorant as to what happens genetically during speciation. And finally, there are reasons to believe that rather different genetic mechanisms may be involved in the speciation of different kinds of organisms and under different circumstances. Yet Darwin's model of speciation, as first developed on the basis of the Galapagos mockingbirds, is still very much alive, and presumably essentially correct.

For many years I have extolled Darwin's introduction of population thinking into biology. Its importance cannot be questioned. The whole Malthus episode of 28 September 1838 would not have made sense if Darwin had not suddenly appreciated on that day that the struggle for existence was among members of the same population and that selection could operate successfully only if there were individual differences among the members of the population. Even earlier, when in March 1837 Darwin had come to the conclusion that the mockingbirds on each of three islands in the Galapagos had undergone gradual change, he could not have interpreted this in terms of essentialism; it required a populational explanation. The great stress in the B *Notebook* on generation as the source of variation is also a populational explanation. Owing to the enormous explanatory power of the populational approach, one would think that from 1838 on Darwin would have employed it for all his explanations, as the modern evolutionist does; but this is not the case. In *Natural Selection* and the *Origin* many explanations are given in a strictly typological language. No historian has yet made a detailed analysis of all of Darwin's references to variation to determine what proportion of them was based on essentialistic or on populational thinking. This much is certain, however; the development of the theory of natural selection would have been impossible without the prerequisite of populational thinking. Nor could any essentialist have come to terms with gradualism.

Natural Selection

Darwin's theory of natural selection was his most daring, most novel theory. It dealt with the mechanism of evolutionary change and, more particularly, how this mechanism could account for the seeming harmony and adaptation of the organic world. It attempted to provide a natural explanation in place of the supernatural one of natural theology. His theory for the natural mechanism that would be able to direct evolutionary change was unique. There was nothing like it in the whole philosophical literature from the pre-Socratics to Descartes, Leibniz, or Kant. It replaced teleology in nature with an essentially mechanical explanation. How Darwin came to develop his theory is described in Essay 13. To judge from his writings, Darwin had a much simpler concept of natural selection than the modern evolutionist. For him there was a steady production of individuals, generation after generation, with those that were "superior" having a reproductive advantage. It seemed essentially to be a single-step process, the conveying of reproductive success. The modern evolutionist agrees with Darwin that the individual is the target of selection; but we now also know that the production of a new individual is an exceedingly complex process. Indeed, we realize that natural selection is actually a two-step process, the first one consisting of the production of genetically differing individuals, while in the second step the survival and reproductive success of these individuals is determined. All aspects of natural selection, as we now understand this process, are discussed in Essay 6.

The intellectual prehistory of the concept has been dealt with elsewhere (Mayr 1982).

Although I call the theory of natural selection Darwin's *fifth* theory, it is actually, in turn, a small package of theories. This includes the theory of the perpetual existence of a reproductive surplus (superfecundity), the theory of the heritability of individual differences, the discreteness of the determinants of heredity, and several others. Many of these were not explicitly stated by Darwin but are implicit in his model as a whole.

Evolutionists from Darwin on have always emphasized the continuity of populational evolution, in contrast with the discontinuous character of saltational evolution by way of reproductively isolated individuals. The fact is invariably ignored, however, that even continuous evolution is mildly discontinuous owing to the sequence of generations. In each generation an entirely new gene pool is reconstituted from which the new individuals are drawn that are the target of selection in that generation.

The theory of natural selection was the most bitterly resisted of all of Darwin's theories. If it were true, as some sociologists have claimed, that the theory was the inevitable consequence of the Zeitgeist of early nineteenth-century Britain, of the industrial revolution, of Adam Smith and the various ideologies of the period, one would think that the theory of natural selection would have been embraced at once by almost everybody. Exactly the opposite is true: the theory was almost universally rejected. Only a few naturalists, like Wallace, Bates, Hooker, and Fritz Müller, could be called consistent selectionists in the 1860s. Lyell never had any use for natural selection, and even T. H. Huxley, defending it in public, was obviously uncomfortable with it and probably did not really believe in it (Poulton 1896; Kottler 1985). Before 1900 not a single experimental biologist either in Britain or elsewhere adopted the theory (Weismann was basically a naturalist). Of course Darwin himself was not a total selectionist, since he always allowed for the effects of use and disuse and an occasional direct influence of the environment. The most determined resistance came from those who had been raised under the ideology of natural theology. They were quite unable to abandon the idea of a world designed by God and to accept a mechanical process instead. More importantly, a consistent application of the theory of natural selection meant a rejection of any and all cosmic teleology. Sedgwick and K. E. von Baer were particularly articulate in resisting the elimination of teleology.

Natural selection represents not only the rejection of any finalistic causes that may have a supernatural origin, but it rejects any and all determinism in the organic world. Natural selection is utterly "opportunistic," as G. G. Simpson has called it; it is a "tinkerer" (Jacob 1977). It starts, so to speak, from scratch in every generation. Throughout the nineteenth century the physical scientists were still deterministic in their outlook, and so indeterministic a process as natural selection was simply not acceptable to them. One has only to read the critiques of the *Origin* written by some of the best-known physicists of the period (Hull 1973) to see how strongly the physicists objected to Darwin's "law of the higgledy-piggledy" (see Huxley 1900 2:240). From the Greeks to the present day there has been a never-ending argument as to whether the events of nature are due to chance or due to necessity (Monod 1970). Curiously, in the controversies over natural selection, the process has been described sometimes as "pure chance" (Herschel and many other opponents of natural selection) or as a strictly deterministic optimization process. Both classes of claimants overlook the two-step nature of natural selection and the fact that in the first step chance phenomena prevail, while the second step is decidedly of an anti-chance nature. As Sewall Wright has so correctly said: "The Darwinian process of continued interplay of a random and a selective process is not intermediate between pure chance and pure determination, but in its consequences qualitatively utterly different from either" (1967:117; see Essay 14).

When Darwin first developed his theory of natural selection, he was still inclined to think that it was able to produce near-perfect adaptation, in the spirit of natural theology (Ospovat 1981). More thinking and the realization of the numerous deficiencies in the structure and function of organisms—perhaps particularly the incompatibility of a perfection-producing mechanism with extinction—led Darwin to reduce his claims for selection, so that all he demanded in the *Origin* was that "natural selection tends only to make each organism, each organic being, as perfect as, or slightly more perfect than, the other inhabitants of the same country with which it has to struggle for existence" (p. 201). Today we are even more conscious of the numerous constraints that make it impossible for natural selection to achieve perfection, or, to state it perhaps more realistically, to come even anywhere near perfection (Essay 6).

The Varying Fates of Darwin's Five Theories

We can now summarize the subsequent fate of each of the five theories of Darwin which I have discussed above. Evolution as such, as well as the theory of common descent, were adopted very quickly. Within fifteen years of the publication of the *Origin* hardly a qualified biologist was left who had not become an evolutionist. Gradualism, by contrast, had to struggle, and populational thinking was a concept that was apparently very difficult to adopt for anyone who was not a naturalist. Even today, in the discussions of punctuated equilibria, statements are made indicating that some people still do not understand the core of population thinking. What counts is not the size of the individual mutation but only whether the introduction of evolutionary novelties proceeds through their incorporation into populations or through the productions of single new individuals that are the progenitors of new species or higher taxa.

That a theory of the multiplication of species is an essential, in fact integral, component of evolutionary theory, as first pronounced by Wallace and Darwin, is now taken for granted. How this multiplication proceeds is still controversial (Essay 21).

Finally, the importance of natural selection, the theory that is usually meant by the modern biologist when speaking of Darwinism, is firmly accepted by nearly everyone. Rival theories—like finalistic theories, neo-Lamarckism, and saltationism—are so thoroughly refuted that they are no longer seriously discussed. Where the modern biologist perhaps differs from Darwin most is in assigning a far greater role to stochastic processes than did Darwin or the early neo-Darwinians. Chance plays a role not only during the first step of natural selection, the production of new, genetically unique individuals, but also during the probabilistic process of the determination of reproductive success of these individuals. Yet when one looks at all the modifications that have been made in the Darwinian theories between 1859 and 1984, one finds that none of these changes affects the basic structure of the Darwinian theories. There is no justification whatsoever for the claim that the Darwinian paradigm has been refuted and has to be replaced by something new. It strikes me as almost miraculous that Darwin in 1859 came so close to what would be considered valid 125 years later.

NOTE

This essay is an abridged version of one which first appeared under the title "Darwin's five theories of evolution," in D. Kohn, ed., *The Darwinian Heritage,* pp. 755–772. Princeton: Princeton University Press, 1985.

REFERENCES

Barzun, J. 1958. *Darwin, Marx and Wagner: Critique of a Heritage.* 2nd ed. Garden City, N.Y.: Anchor.
Conry, Y. 1974. *L'introduction du darwinisme en France au XIXe siècle.* Paris: Vrin.

Darwin, C. 1859. *On the Origin of Species.* London: John Murray.

———— 1903. *More Letters of Charles Darwin.* Ed. F. Darwin and A. C. Seward. 2 vols. London: John Murray.

———— 1958. *The Autobiography of Charles Darwin.* Ed. Nora Barlow. London: Collins.

———— 1960–1967. Darwin's notebooks on transmutation of species. Ed. G. de Beer. *Bull. Brit. Mus. (Nat. Hist.)* 2:27–300, 3:129—176. (Notebooks B, C, D, E.)

———— 1967. Darwin's notebooks. Pages excised by Darwin. Ed. G. de Beer, M. J. Rowlands, and B. M. Skramovsky. *Bull. Brit. Mus. (Nat. Hist.)* 3:129– 176.

———— 1975. *Natural Selection.* Ed. R. C. Stauffer. Cambridge: Cambridge University Press.

Darwin, F. 1887. *The Life and Letters of Charles Darwin.* 3 vols. London: John Murray. Rptd. New York: Johnson Reprint Corp., 1969.

Eldredge, N., and S. J. Gould. 1972. Punctuated equilibria: an alternative to phyletic gradualism. In T. J. M. Schopf, ed., *Models in Paleobiology,* pp. 82– 115. San Francisco: Freeman.

Glick, T. F., ed. 1974. *The Comparative Reception of Darwinism.* Austin: University of Texas Press.

Goldschmidt, R. 1940. *The Material Basis of Evolution.* New Haven: Yale University Press.

Gould, S. J., and R. C. Lewontin. 1979. The spandrels of San Marco and the Panglossian paradigm: a critique of the adaptationist programme. *Proc. Roy. Soc. London* B205:581–598.

Greene, J. C. 1981. Darwinism as a world-view. In J. C. Greene, *Science, Ideology, and World-view: Essays in the History of Evolutionary Ideas,* pp. 128–157. Berkeley: University of California Press.

Gruber, H. E., and P. H. Barrett. 1974. *Darwin on Man: A Psychological Study of Scientific Creativity.* Together with Darwin's *Early and Unpublished Notebooks.* New York: Dutton.

Hull, D. L. 1973. *Darwin and His Critics: The Reception of Darwin's Theory of Evolution by the Scientific Community.* Cambridge, Mass.: Harvard University Press.

Huxley, A. 1981. Anniversary address of the president. Suppl. *Roy. Soc. News* 12:i–vii.

Huxley, T. H. 1900. *Life and Letters of Thomas Henry Huxley, by His Son Leonard Huxley.* 2 vols. London: Macmillan.

Jacob, F. 1977. Evolution and tinkering. *Science* 196:1161–1166.

Kohn, D. 1985. Darwin's principle of divergence as internal dialogue. In D. Kohn, ed., *The Darwinian Heritage,* pp. 245–257. Princeton: Princeton University Press.

Kottler, M. J. 1985. Charles Darwin and Alfred Russel Wallace: two decades of debate over natural selection. In D. Kohn, ed., *The Darwinian Heritage*, pp. 367–432. Princeton: Princeton University Press.

Lyell, C. 1830–1833. *Principles of Geology, being an attempt to explain the former changes of the earth's surface, by reference to causes now in operation.* 3 vols. London: John Murray.

———— 1881. *The Life, Letters and Journals of Sir Charles Lyell.* Ed. K. M. Lyell. 2 vols. London: John Murray.

Mayr, E. 1954. Change of genetic environment and evolution. In J. Huxley, A. C. Hardy, and E. B. Ford, eds., *Evolution as a Process*, pp. 157–180. London: Allen and Unwin.

———— 1982. *The Growth of Biological Thought.* Cambridge, Mass.: Harvard University Press.

Mayr, E., and W. B. Provine, eds. 1980. *The Evolutionary Synthesis: Perspectives on the Unification of Biology.* Cambridge, Mass.: Harvard University Press.

Monod, J. 1970. *Le hasard et la nécessité.* Paris: Èditions du Sueil. Trans., *Chance and Necessity: An Essay on the Natural Philosophy of Modern Biology.* New York: Knopf, 1971.

Osborn, H. F. 1894. *From the Greeks to Darwin: An Outline of the Development of the Evolution Idea.* New York: Columbia University Press.

Ospovat, D. 1981. *The Development of Darwin's Theory: Natural History, Natural Theology, and Natural Selection, 1838–1859.* Cambridge: Cambridge University Press.

Poulton, E. B. 1896. *Charles Darwin and the Theory of Natural Selection.* London: Cassell.

Schindewolf, O. H. 1950. *Grundfragen der Paläontologie.* Stuttgart: Schweizerbart.

Stanley, S. 1981. *The New Evolutionary Timetable.* New York: Basic.

Sulloway, F. J. 1979. Geographic isolation in Darwin's thinking: the vicissitudes of a crucial idea. *Stud. Hist. Biol.* 3:23–65.

Sumner, J. B. 1824. *The Evidence of Christianity Derived from Its Nature and Reception.* London: Hatchford.

Wallace, A. R. 1855. On the law which has regulated the introduction of new species. *Ann. Mag. Nat. Hist.* 16:184–196. Reptd. in *Natural Selection and Tropical Nature*, pp. 3–19. London: Macmillan, 1891.

White, M. J. D. 1978. *Modes of Speciation.* San Francisco: Freeman.

Willis, J. C. 1940. *The Course of Evolution by Differentiation or Divergent Evolution Rather than by Selection.* Cambridge: Cambridge University Press.

Wright, S. 1967. Comments. In P. S. Moorhead and M. M. Kaplan, eds., *Mathematical Challenges to the Neo-Darwinian Interpretation of Evolution*, pp. 117–120. Philadelphia: Wistar Institute Press.

DARWIN AND NATURAL SELECTION

T H E Q U E S T I O N concerning the conceptual sources of Darwin's theory of natural selection is still highly controversial. The favorite interpretation of the historians has always been that it was a manifestation of the thinking of upper-class England in the first half of the nineteenth century (empiricism, mercantilism, industrial revolution, poor laws, etc.). Darwin's admission that reading Malthus had given him the crucial insight seemed to provide a powerful confirmation of this interpretation. The evolutionists, by contrast, favored an interpretation based on "internal causation," relying on Darwin's insistence that his familiarity with the practices of the animal breeders had provided him with the decisive evidence. The rediscovery of Darwin's notebooks covering the one and a half years prior to the date of his "conversion" has provided us with a great amount of new information, but—although narrowing down our options—it still permits conflicting interpretations. What I present here is anything but the last word in a still ongoing controversy. It will require further research before the remaining disagreements can be removed.

Darwin had returned to England from the voyage of the *Beagle* in October 1836. While working on his bird collections, and particularly through discussions with the ornithologist John Gould, Darwin became an evolutionist, apparently in March 1837. Certainly by July 1837 he had firmly accepted evolution by common descent. I like to designate his new interpretation of the world as the *first Darwinian revolution.* It consisted not only in replacing a static or steady-state world by an evolving one but also, and more important, in depriving man of his unique position in the universe and placing him into the stream of animal evolution. Darwin, after this date, never questioned the fact of evolution even though he continued for another twenty years to collect supporting evidence. Yet, the causes of evolution were at first a complete mystery to him.

For a year and a half Darwin speculated incessantly, developing and then again rejecting one theory after the other,[1] until he finally had the decisive illumination on 28 September 1838. In his autobiography he describes it as follows (p. 120):[2]

> Fifteen months after I had begun my systematic enquiry, I happened to read for amusement Malthus on Population, and being well prepared to appreciate the struggle for existence which everywhere goes on, from long-continued observation of the habits of animals and plants, it at once struck me that under these circumstances favorable variations would tend to be preserved, and unfavorable ones to be destroyed. The result of this would be the formation of new species. Here, then, I had at last got a theory by which to work.

It was the theory later called by Darwin the theory of natural selection. It was a most daring innovation, since it proposed to explain by natural causes, so to speak mechanically, all the wonderful adaptations of living nature hitherto attributed to "design." I would like to call this the *second Darwinian revolution,* since it introduced an entirely new way of explaining our world.

Darwin makes it sound as if the concept of natural selection was simplicity itself. But his memory deceived him. His autobiography was written almost 40 years later (in 1876), largely for the benefit of his grandchildren and replete with characteristically Victorian self-denigrations, and Darwin had forgotten what a complex shift in four or five major concepts had been required to arrive at the new theory. He probably never fully realized himself how unprecedented his new concept was and how totally opposed to many traditional assumptions. Everyone now agrees that it was one of the most novel and most daring new conceptualizations in the history of ideas.

Indeed, the concept of natural selection was so strange to Darwin's contemporaries when proposed in the *Origin of Species* in 1859 that only a handful adopted it. It took nearly three generations until it became universally accepted even among biologists. Among nonbiologists the idea is still rather unpopular, and even those who pay lip service to it often reveal by their comments that they do not fully understand the working of natural selection.[3] It is only when one is aware of the complete unorthodoxy of this idea that one can appreciate Darwin's revolutionary intellectual achievement. And this poses a powerful riddle: How could Darwin have arrived at an idea which not only was totally at variance with the thinking of his own time but which was so complex that even now, 139 years later, it is widely misunderstood in spite of our vastly greater understanding of the processes of variation and inheritance? This is the question which the historian of ideas would like to see answered.[4]

Darwin's own version (in his autobiography) was that contemplation of the success of animal breeders in producing new breeds had provided him with the clue for the mechanism of evolution and was thus the basis for his theory of natural selection. We know that this is a vast oversimplification—a revision of our thinking which we owe to the rediscovery of Darwin's notebooks. In July 1837 he had started to write down (in a series of notebooks) all the facts as well as his own brain waves and speculations "which bore in any way on the variation of animals and plants under domestication and [in] nature." Even though he later cut out occasional pages, to use them for his book manuscripts, Darwin never discarded these notebooks, and they were rediscovered in the 1950s among the Darwin papers at the Cambridge University Library.[5] Darwin's day-by-day records throw an entirely new light on the development and the changes in his thought during the period from July 1837 to 28 September 1838, when the theory of natural selection was conceived.

One fact, the importance of which has not been reduced by the recent discoveries, is the impact of Darwin's reading of Malthus. The interpretation of the Malthus episode, however, has become the subject of considerable controversy among the Darwin scholars. According to some of them—de Beer and S. Smith, and, to a lesser extent, H. Gruber and myself—it was merely the culmination in the gradual development of Darwin's thinking, a little nudge that pushed Darwin across a threshold at which he was already standing. According to others—C. Limoges and D. Kohn, for example,—it constituted a rather drastic break, almost equivalent to a religious conversion. Which of these two interpretations is nearer to the truth?

There are essentially two methods by which we can try to find an answer. Either we can attempt to analyze all the entries in the notebooks, in a chronological sequence, or we can try to reconstruct Darwin's explanatory model of natural selection and then study separately the history of each of its individual components. My own choice is in favor of the second method, placed in a chronological framework, although both methods are necessary for a full understanding. The first method attempts to reconstruct the trials and errors of Darwin's gradual approach, as reflected in successive entries in the Notebooks. It also examines tentative ideas that he later rejected. Gruber, Kohn, and, in part, Limoges have favored this method.

What were the components of Darwin's explanatory model? For my analysis I found it most convenient to recognize five facts and three inferences, as diagrammed in Figure 13.1. I shall attempt to determine, first, at what time Darwin became aware of these five facts, then at what time he made the three inferences, and whether or not these inferences had already been made previously and could be found in the literature.

When we consider the five facts in Figure 13.1 carefully, it becomes apparent that all of them were widely known. Not only had they been in Darwin's own hands well before the Malthus episode, but they had been available to Darwin's contemporaries, only a single one of whom, A. R. Wallace, used them in exactly the same way as Darwin. Merely having these facts obviously was not enough. They had to be related to each other in a meaningful manner; that is, they had to be placed in an appropriate conceptual background. In other words, Darwin had to be intellectually prepared to see the connections among these facts.

This leads us to the most interesting but also the most difficult question: What had been going on in Darwin's mind in the one and a half years prior to the Malthus episode? All the indications are that it was a period of unprecedented intellectual activity in Darwin's life. Precisely what the changes in Darwin's thinking were and how they were connected with each other has not yet been investigated nearly as fully as it deserves. Gruber and Kohn have examined this problem more fully than anyone else, but the Darwin correspondence of that period and other manuscript materials that have not yet been analyzed are bound to provide new insights. My own tentative conclusions may, therefore, turn out to be incorrect. However, my reading suggests to me that Darwin's beliefs changed moderately or drastically in four areas, which I shall simply list here and then discuss in the context of Darwin's model.

(1) *The gradual substitution of the concept of the uniqueness of individuals for one that assumes all individuals of a species to be essentially alike.*[6] This is the belief, ultimately going back to Plato (*eidos*), that the observed variability of phenomena reflects a limited number of constant, discontinuous essences—the antonym being "population thinking," a belief in the importance of individual differences and the reality of the variation within a population. Most of Darwin's earlier statements on species and varieties were strictly typological. It is my impression that they became more "populational" as Darwin delved deeper into the literature of the animal breeders.

(2) *A shift from soft toward hard inheritance.* In his earlier statements Darwin seemed to assume that most, if not all, inheritance was "soft"—

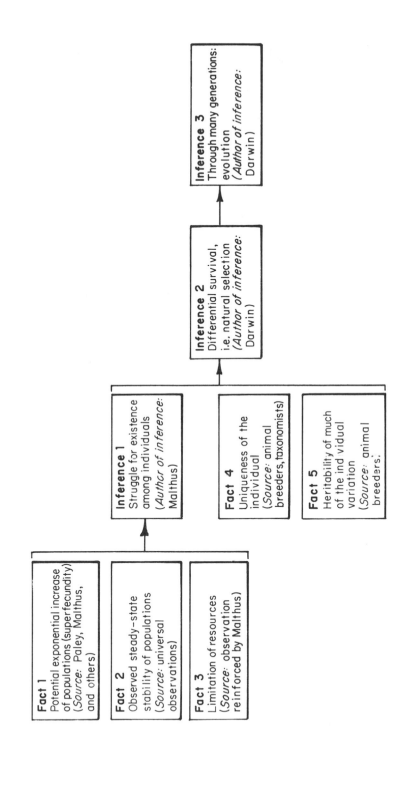

Fact 1
Potential exponential increase of populations (superfecundity)
(*Source:* Paley, Malthus, and others)

Fact 2
Observed steady-state stability of populations
(*Source:* universal observations)

Fact 3
Limitation of resources
(*Source:* observation reinforced by Malthus)

Inference 1
Struggle for existence among individuals
(*Author of inference:* Malthus)

Fact 4
Uniqueness of the individual
(*Source:* animal breeders, taxonomists)

Fact 5
Heritability of much of the individual variation
(*Source:* animal breeders)

Inference 2
Differential survival, i.e. natural selection
(*Author of inference:* Darwin)

Inference 3
Through many generations: evolution
(*Author of inference:* Darwin)

that is, he assumed that the material basis of inheritance is not unchange-
ably constant but can be modified alike by the inheritance of acquired
characters ("use and disuse,"), by physiological activities of the body, by
a direct influence of the environment on the genetic material, or by an
inherent tendency to progress and to become more perfect. With the
growth of his population thinking, there is an increasing stress on genetic
differences among individuals, indicative of a growing awareness of the
need to postulate "hard" inheritance.

(3) *A changing attitude toward the balance of nature.*   This affected two
aspects; whether the balance is purely static or, rather, dynamic, and
whether the balance is maintained by benign adjustments or by constant
war.

(4) *A gradual loss of his Christian faith.*   It is apparent that Darwin lost
his faith in the years 1836–1839, much of it clearly prior to the reading
of Malthus.[7] In order not to hurt the feelings of his friends and of his
wife, Darwin often used deistic language in his publications, but much
in his notebooks indicates that by this time he had become a "materialist"
(more or less equivalent to an atheist).

These four changes in Darwin's thinking are to some extent interconnected. Since they were largely unconscious, they are usually reflected in
Darwin's notebooks only in subtle changes of wording, and there is
considerable leeway in possible interpretation. However, keeping these
four points in mind will sharpen our awareness of possible changes in
Darwin's thinking in the years prior to reading Malthus, while we make
a point-by-point analysis of Darwin's explanatory model (Figure 13.1).

The Struggle for Existence

When recording his reaction to reading Malthus on 28 September 1838,
Darwin makes it quite clear that it was not Malthus's general attitude
that had acted as a catalyst on his thoughts, but one particular sentence,
for he says that "yet until the one sentence of Malthus no-one clearly
perceived the great check amongst men."[8] De Beer succeeded in determining what Malthus's crucial sentence was: "It may safely be pronounced,
therefore, that the population, when unchecked, goes on doubling itself
every twenty-five years, or increases in a geometrical ratio."[9] From then
on, Darwin stressed that it was Malthus's demonstration of the exponential
increase of populations that was decisive in his discovery of the importance
of natural selection ("Fact 1").

Yet there is a puzzling difficulty. Why did it take Darwin so long to recognize the evolutionary significance of the Malthusian principle? The prodigious fertility of animals and plants had been pointed out by many of Darwin's favorite and most frequently read authors, like Erasmus Darwin, Charles Lyell, Alexander von Humboldt, and William Paley. Furthermore, Malthus's principle was widely discussed in the essay literature of the period. Why then did this suddenly impress Darwin so profoundly on 28 September 1838?

Three reasons might be suggested—the first one being, as pointed out by Gruber,[10] that Darwin had learned on the three preceding days (between September 25 and 27) of the unbelievable fertility of protozoans by reading Ehrenberg's work on the subject. This quite likely primed Darwin's receptivity for Malthus's thesis. The second reason is that when Malthus applied the principle to man, a species with relatively few offspring, Darwin suddenly realized that a potentially exponential increase of a population was entirely independent of the actual number of offspring of a given pair. The third reason is that the Malthus episode came at a time when population thinking had begun to mature in Darwin's mind.

The second fact in Figure 13.1—population stability—was not in the slightest controversial. No one questioned that the number of species and, aside from temporary fluctuations, the number of individuals in every species maintained a steady-state stability. This is implicit in the concept of plenitude of the Leibnizians and in the harmony-of-nature concept of the natural theologians. If there is any extinction, it is balanced by speciation, and if there is high fertility it must be counterbalanced by mortality. In the end everything adds up to a steady-state stability.

The third fact—limitation of resources—again, was not at all controversial, being very much part of the balance-of-nature concept of natural theology, so dominant in England in the first half of the nineteenth century.

Darwin's first great inference, derived from these three facts, was that exponential population growth combined with a fixed amount of resources would result in a fierce struggle for existence. We must ask, Was this inference original with Darwin, and if so, what part of it did he owe to Malthus? This is perhaps the most controversial question raised by the analysis of the selection theory. The main difficulty is that the term "struggle for existence" and similar synonymous terms were used in different senses by different authors.

Before we can analyze them, we must deal with one other concept, the idea of a perfect balance of nature, an idea prevalent in the eighteenth century: nothing in nature is too much, nothing too little, everything is designed to fit with everything else. Rabbits and hares have lots of young because food must be available for foxes and other carnivores. The whole economy of nature forms a harmonious whole that can in no way be disturbed. This is why Lamarck, who was very much an adherent of this concept, could not conceive of extinction. Cuvier likewise had adopted it, as shown in correspondence with his friend Pfaff. He transferred the same concept to the structure of an organism, which he visualized as a "harmonious type" in which nothing could be changed. Everything in such a complex system is so perfect that any change would lead to deterioration.

This type of thinking was still dominant in Darwin's day, not only among the natural theologians of England but also on the Continent. Indeed, one can find a number of entries in Darwin's notebooks which seem to reflect this kind of thinking. But was Darwin still a wholehearted supporter of the concept of a harmonious balance of a benign nature? This is very much a question, and it is a very important question because it affects the interpretation of what Darwin understood under the term *struggle for existence.*

For us moderns the term means a rather fierce fight with no holds barred. But for the natural theologians the struggle for existence was a beneficial feedback device, the function of which was to maintain the balance of nature. It is, as Herder called it, "the balance of forces which brings peace to the creation."[11] Linnaeus devoted an entire essay to the "Police of Nature" and emphasized that "those laws of nature by which the number of species in the natural kingdoms is preserved undestroyed, and their relative proportions kept in proper bounds are objects extremely worthy of our attentive pursuit and researches."[12] Lamarck expressed similar sentiments.

Was this benign interpretation of the struggle for existence unanimous? Unfortunately, even today we have no reliable analysis that would give us an answer to this question. My impression is, however, that as the interaction of predators and prey, of parasites and their victims, the frequency of extinction, and the struggles of competing species became better known, the struggle for existence was more and more recognized as a "war" or fight, a struggle for survival, "red in tooth and claw" as Tennyson later expressed it. Bonnet (1781) and de Candolle (1820)[13] emphasized that this war among species consisted not merely of a predator-prey relationship but of competition for any and all resources.[14] However, it was not at all appreciated how fierce this struggle is, and Darwin admits that "even the energetic language of de Candolle does not convey the warring of the species as [convincingly as does the] inference from Malthus."[15]

Nevertheless, it is highly probable that Darwin had been gradually conditioned by his reading to a far less benign interpretation of the struggle for existence than that held by the natural theologians. The mere fact that Darwin had adopted evolution must have made him aware of the frequency of extinction and of the unbalances and adaptational lags caused by evolutionary changes. From Aristotle to the natural theologians it was considered axiomatic that a belief in a harmonious universe and "perfect adaptation" in nature, or in a creator continuously active in correcting imperfections and imbalances, was incompatible with a belief in evolution. By necessity, accepting evolutionary thinking undermined a continued adherence to a belief in a harmonious universe.[16]

Struggle among Species or Individuals?

Of far greater importance is a second question: Between whom does the struggle for existence take place? This question allows for two drastically different answers. In the entire essentialistic literature the struggle is considered to take place among species. The balance of nature is maintained by this struggle, even if it occasionally causes the extinction of a species. This is the interpretation of the struggle for existence in the literature of natural theology, up to de Candolle and Lyell, and is the major emphasis of Darwin's notebooks up to the Malthus reading. The main function of this struggle is to correct disturbances in the balance of nature, but it can never lead to changes; on the contrary, it is a device to preserve a steady-state condition. As such it continued even after 1838 to be an important component of Darwin's thinking, particularly in his biogeographic discussions (such as determination of species borders).

It is only when one applies population thinking to the struggle for existence that one can make the crucial conceptual shift to recognizing a struggle for existence among individuals of a single population. This, as Sandra Herbert was the first to recognize clearly, was Darwin's decisive new insight resulting from his reading of Malthus.[17] If most individuals of every species are unsuccessful in every generation, then there must be a colossal competitive struggle for existence among them. It was this conclusion which made Darwin think at once of various other facts that had been slumbering in his subconscious but for which, up to that moment, he had had no use.

Darwin's reading of Malthus was dramatic and climactic, and it does not matter whether one interprets it as a complete reversal of Darwin's thinking or whether one believes that "the evidence suggests that the change in choice of unit was a protracted process, stretched over a year or more, and linked to other aspects of his thought."[18] I myself hold with the latter view, because the capacity to be able to interpret the Malthus statement on exponential growth of populations and to apply it to individuals requires population thinking, and this Darwin had been gradually acquiring during the preceding year and a half. That everything came to a dramatic climax on 28 September 1838, however, is beyond question.

Uniqueness of the Individual

The whole concept of competition among individuals would be irrelevant if all these individuals were typologically identical—if they all had the same essence. Variability does not become meaningful in an evolutionary sense until a concept has developed that allows for differences among the individuals of the same population. According to this concept, each individual may differ in the ability to tolerate climate, to find food and a place in which to live, to find a mate, and to raise young successfully. The recognition of the role of individuality is not only of the utmost importance for an understanding of the history of biology but it is one of the most drastic conceptual revolutions in Western thought ("Fact 4"). We call the concept which emphasizes the uniqueness of every individual *population thinking.*

There is little doubt that Darwin's population thinking received an enormous boost through reading Malthus at that right moment. Yet, curiously, when we go through Malthus's writings we find no trace of population thinking. There is nothing whatsoever even faintly relating to the subject in those early chapters of Malthus that gave Darwin the idea of exponential growth. There is, however, a reference to animal breeding in chapter 9, but here the subject is introduced to prove exactly the opposite point. After referring to the claims of the animal breeders, Malthus states, "I am told that it is a maxim among the improvers of cattle that you may breed to any degree of nicety you please, and they found this maxim upon another, which is, that some of the offspring will possess the desirable qualities of the parents in a greater degree." He then produces all sorts of facts and reasons why this cannot possibly be true, leading him to the conclusion that "it cannot be true, therefore, that among animals, some of the offspring will possess the desirable qualities of the parents in a greater degree; or that animals are indefinitely affect-able."[19]

Where, then, did Darwin get his population thinking, since he evidently did not get it from Malthus? In his autobiography and in various letters, Darwin emphasized again and again that he had been mentally prepared for the Malthus principle by studying the literature of animal breeding. Recent commentators have insisted that this must be a lapse of Darwin's memory because there is nothing about selection and very little about animal breeding in Darwin's notebooks until about three months after the Malthus reading. For myself, I am rather convinced that Darwin's own presentation is nevertheless essentially correct.

If we ask ourselves what Darwin would be likely to enter in his notebooks, we would certainly say new facts or new ideas. Hence, since it was not a new subject, animal breeding surely would not qualify. Darwin's best friends at Cambridge University were the sons of country squires and of owners of estates. As Himmelfarb called them, they were the "horsy set" of Cambridge University, riding or hunting with dogs on every occasion. One can hardly doubt that all of them to a greater or lesser degree were intensely interested in animal breeding. They must have argued a great deal among themselves about Bakewell and Sebright and the best methods of breeding and improving dogs, horses, and livestock.

How else—other than that it had a great interest for him—can one explain that Darwin, in the excessively busy period after the return of the *Beagle,* devoted so much of his time to studying the literature of the animal breeders? To be sure, Darwin's primary interest was in the origin of variation, but in the course of his reading Darwin could not help absorbing the important lesson from the breeders—that every individual in the herd was different from every other one, and what extreme care had to be used in selecting the sires and dams from which to breed the next generation. I am quite convinced that it is no coincidence that Darwin studied the literature of the animal breeders so assiduously exactly during the six months before reading Malthus. In this I agree entirely with Michael Ruse.[20]

It was not the process of selection but the fact of the differences among individuals which Darwin remembered when suddenly becoming aware of the competition among individuals, of the struggle for existence among individuals. Here we have the fortuitous coming together of two important concepts—excessive fertility and individuality—which jointly provide the basis for an entirely new conceptualization.

Variation can be of evolutionary significance—that is, it can be se-lected—only if at least part of it is heritable ("Fact 5"). Like the animal breeders from whom he got so much of his information, Darwin took this heritability completely for granted.

And this assumption can be held quite independently of the assumptions concerning the nature of the genetic material and of the origin of new genetic factors. Darwin's ideas on these subjects were quite confused, but fortunately Darwin treated genetic variation as a "black box." As a natu-ralist and reader of the animal breeding literature, he knew that variation was always present, and this is all he *had* to know. He was also convinced that the supply of variation was renewed in every generation and thus always abundantly available as material for natural selection. In other words, a correct theory of genetics was *not* a prerequisite for the estab-lishment of the theory of natural selection.

Natural Selection

The next question we have to answer is how Darwin arrived at the actual concept of natural selection on the basis of the stated five facts and his first inference (Figure 13.1). In his autobiography (pp. 118–120) Darwin stresses that he "collected facts on a wholesale scale, more especially with respect to domesticated productions, by printed enquiries, by conversation with skillful breeders and gardeners, and by reading . . . I soon perceived that selection was the key-stone of man's success in making useful races of animals and plants. But how selection could be applied to organisms living in a state of nature remained for some time a mystery to me."[21] In 1859 he wrote to Wallace, "I came to the conclusion that selection was the principle of change from the study of domestic productions; and then, reading Malthus, I saw at once how to apply this principle."[22] To Lyell he wrote, with reference to Wallace's theory, "We differ only [in] that I was led to my views from what artificial selection had done for domestic animals." Traditionally, these statements were accepted by the Darwin students as a correct representation of the facts.

This interpretation, however, has been challenged in recent years in the wake of the discovery of Darwin's notebooks. Limoges and Herbert point out that in the first three notebooks Darwin nowhere refers to selection or to the selective activities of animal breeders, particularly in the pro-duction of new domestic races. They claim that Darwin was interested in domestic animals only because he hoped to find evidence concerning the occurrence of variations and the mechanisms of their production, matters that are difficult to study in wild populations.

It is true that the term *selection* does not occur in Darwin's notebooks; it is first found in his 1842 sketch in the words "natural means of selection."[23] Darwin here refers to artificial selection by the term "human selection." Actually, in the notebooks Darwin not infrequently refers to the process of selecting, but he uses a different term—*picking.*

I am willing to grant to the recent critics that there is no evidence in the notebooks of a simple application to the evolutionary process of the analogy between selection by man and selection by nature. This is quite evident when one reads the crucial entry in the notebooks on 28 September 1838 (here reproduced in the original telegraph style):

> Take Europe on an average every species must have same number killed year with year by hawks, by cold etc.—even one species of hawk decreasing in number must affect instantaneously all the rest.—The final cause of all this wedging, must be to sort out proper structure, and adapt it to changes—to do that for form, which Malthus shows is the final effect (by means however of volition) of this populousness on the energy of man. One may say there is a force like a hundred thousand wedges rrying [to] force every kind of adapted structure into the gaps in the economy of nature, or rather forming gaps by thrusting out the weaker ones.[24]

The simile here is "wedging," not "selecting." Thus it appears that the arguments of the critics have considerable validity. However, the analogy between artificial selection and natural selection is not necessary for Darwin's conclusions. Inference 1 and Fact 4 automatically result in Inference 2 (natural selection). It is quite likely that Darwin did not see the obvious analogy between artificial and natural selection until some time after the Malthus reading. Yet, I have little doubt that the copious reading which Darwin had done in the field of animal breeding had prepared his mind to appreciate the role of the individual and its heritable qualities. Indeed I am convinced, and in this I agree with Ruse, that the many years during which Darwin had been exposed to the ideas of the animal breeders had preconditioned his mind to appreciate the importance of the Malthus principle. This dormant knowledge was actualized under the impact of reading Malthus.[25]

The natural selection of individuals with particular heritable qualities, continued over many generations, automatically leads to evolution, as in Inference 3 (Figure 13.1). In fact this process is sometimes used as the definition of evolution. In this connection it must be emphasized once more that Darwin's inference is exactly the opposite of that of Malthus, who had denied that "some of the offspring will possess the desirable qualities of the parents in a greater degree." Indeed, Malthus used his entire argument as a refutation of the thesis of Condorcet and Godwin of human perfectability. The Malthusian principle, dealing with populations of essentialistically identical individuals, causes only quantitative, not qualitative, changes in populations.[26]

How Great Was Darwin's Debt to Malthus?

That it was the Malthus reading which acted as a catalyst in Darwin's mind in producing the theory of natural selection cannot be disputed and was emphasized by Darwin himself again and again. However, when we analyze the components of the theory, as we have just done, we find that it is primarily the insight that competition is among individuals rather than among species that is clearly a Malthusian contribution. To be sure, this in turn led Darwin to a reevaluation of other phenomena, such as the nature of the struggle for existence, but only as second-order consequences. I agree with those who think that the Malthusian thesis of exponential growth was the capstone of Darwin's theory. "The one sentence of Malthus" acted like a crystal dropped into a supercooled fluid.

There is, however, also a second and more subtle Malthusian impact. The world of the natural theologians was an optimistic world: everything that is happening is for the common good and helps to maintain the perfect harmony of the world. The world of Malthus was a pessimistic world: there are ever-repeated catastrophes, an unending, fierce struggle for existence, yet the world essentially remains the same. However much Darwin might have begun to question the benign nature of the struggle for existence, he clearly did not appreciate the fierceness of this struggle until reading Malthus. And it permitted him to combine the best elements of Malthus and of natural theology: it brought him to the belief that *the struggle for existence is not a hopeless steady-state condition à la Malthus but the very means by which the harmony of the world is achieved and maintained. Adaptation is the result of the struggle for existence.*

The events on 28 September 1838 are of great interest to students of theory formation. Seeing to what an extent Darwin was in the possession of all the other pieces of his theory prior to this date, it becomes clear that in the case of a complexly structured theory, it is not sufficient to have most of the pieces; no, one must have them all. Even a small deficiency, like defining the word "variety" typologically instead of populationally, might be sufficient to prevent the correct piecing together of the components. Equally important is the general ideological attitude of the theory-builder. A person like Blyth might have had in his possession the very same components of the theory as Darwin but would have been unable to piece them together correctly owing to an incompatible ideological commitment. Nothing illustrates better how important the general attitude and conceptual framework of the maker of a theory is than the simultaneous, independent proposal of the theory of natural selection by A. R. Wallace. He was one of the few people, perhaps the only one, who had had a similar set of past experiences: a life dedicated to natural history, years of collecting on tropical islands, and the experience of reading Malthus.

Preparation of Darwin's Mind

It is obvious that we have learned a great deal in recent years about the ingredients that went into Darwin's theory of natural selection, but we are still to a considerable extent in the dark as to what went on in Darwin's mind in the year and a half between his becoming an evolutionist (March 1837) and the day of the Malthus reading on 28 September 1838. What has become clear is that it required a shift in at least four major concepts, permitting Darwin to see five well-known facts in a new light.

One of these shifts has been rather consistently sidestepped by all those who have occupied themselves with the history of the theory of natural selection. It is the question of the extent that Darwin's loss of Christian faith affected the conceptual framework on which the theory of natural selection rests. They have overlooked the fact that the discussion of his religious beliefs is an important chapter in Darwin's autobiography.[27] From this account it becomes evident that the years between October 1836 and January 1839 were the years in which Darwin lost much of his faith.[28] Adopting natural selection rather than the hand of God as the active factor responsible for all that was formerly considered evidence for design was, of course, the last step. However, even the acceptance of evolution was already a fatal undermining of natural theology. That Darwin lost his faith in the same crucial years in which he began to adopt population thinking (and "hard" inheritance) is a synergistic constellation, the importance of which should not be underestimated.

The traditional interpretation, promulgated by Darwin himself, that his theory was the result of a simple interaction between a knowledge of the importance of artificial selection and the realization of the exponential growth of populations is, however, quite misleading. The theory is far too complex for such a simple origin.

And it is this complexity which is responsible for the enormous resistance the theory encountered in subsequent generations. It led, still in Darwin's lifetime, to the proposal of many competing explanations of evolution, among which saltationism, orthogenesis, and neo-Larmarckism are best known. To report on the struggle between Darwinism and these other explanatory theories would be a separate story. Evolutionary biologists finally came to complete agreement in the 1930s and 1940s, during the so-called "evolutionary synthesis." Yet, the resistance to selectionism among nonbiologists and lay people has still not been overcome entirely even in our day.

NOTES

This essay first appeared in *American Scientist* 65(1977):321–327. This interpretation of Darwin's pathway to the discovery of the theory of natural selection was made before much of the unpublished Darwin material had become available or analyzed. Since then a number of other accounts have been published which come to somewhat different conclusions. It would have been confusing for the historian if I had revised this reprint in order to take the more recent literature into consideration. My 1977 essay, therefore, is reprinted unchanged. To reach a balanced view I suggest reading also the literature listed in the References at the end of the essay.

    1. Darwin's earlier theories of evolution, prior to the Malthus episode, have been perceptively analyzed by H. E. Gruber. 1974. Darwin on Man. New York: Dutton; C. Limoges. 1970. *La sélection naturelle.* Paris: Presses Universitaires de France; and E. D. Kohn. 1975. *Charles Darwin's Path to Natural Selection.* University of Massachusetts diss.
    2. Nora Barlow, ed. 1958. *The Autobiography of Charles Darwin, 1809–1882.* London: Collins.
    3. For instance J. Monod (1971) in the last two chapters of *Chance and Necessity.* New York: Knopf.
    4. In addition to Kohn, Gruber, and Limoges (note 1), those who have discussed this problem in recent years are: S. Smith. 1960. The origin of "The Origin." *Advancement of Science* 64:391–401. G. de Beer. 1965. *Charles Darwin: A Scientific Biography.* New York: Doubleday Anchor. M. T. Ghiselin. 1969. *The Triumph of the Darwinian Method.* Berkeley: University of California Press. R. M. Young. 1969. Malthus and the evolutionists. *Past and Present* 43:109–141. 1971.

Darwin's metaphor: does nature select? *Monist* 55:443–503. P. Vorzimmer. 1969. Darwin, Malthus, and the theory of natural selection. *J. Hist. Ideas* 30:527–542. S. Herbert. 1971. Darwin, Malthus, and selection. *J. Hist. Biol.* 4:209–217. P. Bowler. 1974. Darwin's concepts of variation. *J. Hist. Med. and Allied Sci.* 29:196–212. 1976. Malthus, Darwin, and the concept of struggle. *J. Hist. Ideas* 37:631–650. See also J. C. Greene. 1975. Reflections on the progress of Darwin studies. *J. Hist. Biol.* 8:243–273.

5. G. de Beer, ed. 1960–61. Darwin's notebooks on transmutation of species, pts. 1–5. *Bull. Brit. Mus. (Nat. Hist.)*, Hist. Series, vol. 2, nos. 2–6 (abbrev. NBT). G. de Beer, M. J. Rowlands, and B. M. Skramovsky. 1967. Darwin's notebooks on transmutation of species, pt. 6., vol. 3, no. 5.

6. For indications of a shift in Darwin toward population thinking, see also Ghiselin (note 4), pp. 57–59.

7. See also Gruber (note 1), pp. 208–214.

8. NBT, 5 (Notebook D, p. 135).

9. Found by Darwin in the 6th edition of Malthus 's *Essay on the Principle of Population* (1836), vol. 1, p. 6.

10. Gruber (note 1), pp. 161–163.

11. J. G. Herder. 1784. *Ideen zur Philosophie der Geschichte der Menschheit,* II, 3, p. 89.

12. C. Linnaeus. [1781]. *Politia Naturae [Amoen. Academicae]*, trans. F. J. Brand, pp. 131–32. London.

13. A. S. de Candolle, 1820. *Essai Elementaire de Geographie Botanique.*

14. Even Lyell emphasized this, and his lengthy list of the causes of extinction is strikingly in conflict with the concept of a perfectly adapted, well-balanced world.

15. NBT, 5 (Notebook D, pp. 134–135).

16. See also Limoges (note 1), pp. 40, 44–45, 70, and 79–80.

17. S. Herbert (note 4), pp. 209–217. See also Ghiselin (note 4), p. 59.

18. Gruber (note 1), p. 165.

19. Malthus. 1798 [see note 9]. Chap. 9, p. 163.

20. Michael Ruse. 1975. Charles Darwin and artificial selection. *J. Hist. Ideas* 36:339–350.

21. *Autobiography* (see note 2), pp. 118–120.

22. Darwin to Wallace, 6 April 1859. A. R. Wallace. *Letters and Reminiscences,* vol. 1, p. 136.

23. Darwin. 1842. Sketch. In F. Darwin. 1909. *The Foundations of the Origin of Species,* p. 46. Cambridge: Cambridge University Press.

24. NBT, 5 (Notebook D, pp. 134–135).

25. Gruber (note 1), p. 118.

26. Limoges (note 1), p. 80; also de Beer and others.

27. *Autobiography* (see note 2), pp. 85–96.

28. Gruber (note 1), pp. 208–214.

REFERENCES

Gillespie, N. C. 1979. *Charles Darwin and the Problem of Creation*. Chicago: Chicago University Press.

Hodge, M. J. S. 1983. The development of Darwin's general biological theorizing. In D. S. Bendall, ed., *Evolution from Molecules to Men*, pp. 43–62. Cambridge: Cambridge University Press.

—————— 1987. Natural selection as a causal, empirical, and probabilistic theory. In L. Krüger et al., eds., *The Probabilistic Revolution*, pp. 233–270. 2 vols. Cambridge, Mass.: MIT Press.

Hodge, M. J. S., and D. Kohn. 1985. The immediate origins of natural selection. In D. Kohn, ed., *The Darwinian Heritage*, pp. 185–206. Princeton: Princeton University Press.

Kohn, D. 1980. Theories to work by: rejected theories, reproduction, and Darwin's path to natural selection. *Stud. Hist. Biol.* 4:67–170.

Ospovat, D. 1981. *The Development of Darwin's Theory: Natural History, Natural Theology, and Natural Selection, 1838–1859*. Cambridge: Cambridge University Press.

Ruse, M. 1979. *The Darwinian Revolution*. Chicago: University of Chicago Press.

Sulloway, F. 1982. Darwin's conversion: the Beagle voyage and its aftermath. *J. Hist. Biol.* 15:327–398.

THE CONCEPT OF FINALITY
IN DARWIN AND AFTER DARWIN

WHEN THE modern reader studies the works of seventeenth- and eighteenth-century thinkers, he finds it almost incomprehensible to what extent God was part of every explanation. Arguments which in the nineteenth and twentieth centuries would have been scientific or purely philosophical were theological arguments in the eighteenth and earlier centuries, and most often were frankly anthropomorphic. Such a God-centered world view was by no means peculiar to Christianity but is reflected also in the writings of Plato and Cicero. There were a number of different versions of such a world view, as we shall see, but they all agreed that this world, and everything in it, had to have a meaning or purpose because, as Aristotle had said it, "nature does nothing in vain," and neither does God, a Christian would say. God, the almighty, would not make anything that is not perfect, or at least as perfect as is possible within the network of the laws of nature. If there is, or has been, any change in the world, this change was due to final causes, moving the particular object or phenomenon toward an ultimate goal. It was universally believed that things in the empirical world "strive" to attain ends.

A finalistic world view is largely if not entirely deterministic. To a large extent it was the opposite of the view of Democritus, who insisted that everything in the world was due to chance or necessity, with chance being the prevailing cause. That something in the world could be due to chance was unacceptable for a theist, and chance was therefore largely ignored in the writings of the theistic thinkers. What arguments there were about the interpretation of creation concerned almost entirely the "necessity" component.

The meaning of the word *necessity* changed according to different concepts of creation. Creation, of course, always meant the implementation of God's thought, and God was the final cause of everything created. However, as the understanding of the phenomena of this world improved from the seventeenth century on, a number of different versions of creation developed, each with a somewhat different concept of final causes.[1]

(1) A world of short duration, remaining constant from its beginning a few thousand years ago until the day of judgement soon to come. There is no noticeable change, and the finalistic element is simply the thought of God as expressed in His creation.

(2) The thinkers of the "scientific revolution" were fascinated by motion. They worked out the laws of a falling body or of the motion of the planets around the sun; for them the world was a world of motion controlled by eternal laws. God had instituted them at the time of creation but from that point on it was these laws that kept the world moving. God was the final cause of everything but He ruled the world through His laws and not by continuous intervention. Descartes was one of the chief spokesmen for this strictly mechanistic world view, but it was more or less adopted even by the naturalist Buffon and carried to the most extreme consequences by Holbach. Even those who adopted this concept of a mechanized world had certain misgivings about applying it to the living world. Buffon, for instance, was fully aware of the conflicts between the mechanized world picture and many phenomena encountered in the study of organisms. Yet any alternative view was unacceptable to him.

(3) Those who were dissatisfied with a strictly mechanistic world, entirely run by laws, developed a different explanatory framework. They credited God with a much larger role in designing the world down to the last detail and in effecting the changes that had taken place since creation. It was distasteful to them to remove God from the running of His world and to replace Him by the efficient causes of His laws. Not only that, but they also found it inconceivable that the observed harmony of nature and all the mutual adaptations of organisms to each other could be due simply to efficient causes. Their answer was to stress the elaborateness of the original design of the world to a far greater degree than done by the mechanists. No matter where you looked in nature, they claimed, you would find evidence for the infinite wisdom of the Creator. Anyone who would study His work (nature) was as legitimate a theologian as he who would study His word (the Bible). Beginning with John Ray (1691) and William Derham (1706), the study of nature became physicotheology or natural theology.

There were two further developments that strengthened the belief in final causes. One was the increasingly strong belief that God had created the world for the sake of man. This was foreshadowed by Aristotle's statement (Politics I, 8, 1256a, b), "Now if nature makes nothing incomplete and nothing in vain, the inference must be that she has made all animals for the sake of man." It was made legitimate by corresponding statements in Genesis. The other reenforcement of the belief in final causes came from manifold observations indicating ongoing changes in the world. This led to a new concept of creation. Creation was no longer seen as something that had happened instantaneously (or in 6 days), but as a gradual and slow process, directed by final causes, culminating in the production of man. Consistent with this modified concept of creation Leibniz and Herder temporalized the scala naturae. Furthermore, the Chain of Being was more and more considered a scale of perfection. One of the foremost objectives of the writings of the physicotheologians was to demonstrate how perfectly everything in the world was designed. Since God could not have created anything that was not perfect, the world was considered the "best of all possible worlds." This was a dominant theme of that vast literature from Ray and Derham to Paley and the Bridgewater Treatises. It dominated even Darwin's early thinking and certainly that of most of his contemporaries. As long as evolution was not accepted there was no conceivable alternate to chance but "necessity," that is, God's design.

Much of the literature of natural theology is quite admirable. Boyle (1688) for instance understood perfectly well that the explanation of the mechanical workings of a structure is an entirely independent endeavor from the explanation of the reason why the organ exists and what its role in the life of the organism is. Thus he made quite clearly a distinction between proximate and ultimate causations. Proximate causations could be explained mechanistically, by physical laws, but one could not do without postulating final causes for the explanation of ultimate causations (Lennox 1983).

Though the beginnings of natural theology go back to the Greeks and even old Egypt, its period of true dominance, at least in England, lasted from the last quarter of the seventeenth century to 1859. Even Kant (1790) believed that numerous phenomena of nature could not be explained purely mechanistically. He therefore postulated final causes. Although we must come to the conclusion, he says, that we cannot consider nature in all of its admirable arrangements as anything but the product of reason, nevertheless theoretical analysis of nature will never reveal to us whether this reason had had a final goal. Actually it made little difference whether an author believed that everything in the world was governed by laws or was specifically regulated by God, because in either case God was either directly or indirectly responsible. He was the final cause for everything.

A belief in cosmic teleology fitted well into the thinking of the seventeenth and eighteenth centuries. It was a period of increasing optimism, of emancipation from social and legal burdens, of conviction that better times were coming, possibly a millennium. Progress was preached not only by utopians and reformers but became the theme of philosophies, particularly of the historical-idealistic schools from Herder and Schelling to Hegel and Marx (Toulmin 1982). It was strongly reenforced by the studies of the geologists and particularly by the discovery of successions of fossil faunas culminating in strata containing mammals and eventually man (Bowler 1976). It fitted well with Lamarck's theory of gradual evolutionary change, this being the first genuine theory of evolution (1809). Not all progressionism in geology led to the acceptance of evolution; in fact the majority of paleontologists from Cuvier to Louis Agassiz thought, rather, in terms of catastrophes and subsequent more progressive new creations. Fewer and fewer authors continued to insist on the constancy of the world; most of them, by contrast, saw continuous change and indeed a trend toward perfection. This can be perceived in the writings of almost all authors between 1809 and 1859, even though it was expressed in various ways by authors like Meckel, Chambers, Owen, Bronn, von Baer, and Agassiz.

The general optimism of the eighteenth century received a severe jolt through the disastrous Lisbon earthquake of 1 November 1755. It was this which induced Voltaire to satirize the Panglossian thinking of Pope and Leibniz in his *Candide*. David Hume also ridiculed claims of a harmony of nature: "Inspect a little more narrowly these living existences, the only beings worth regarding. How hostile and destructive; how insufficient all of them for their own happiness! How contemptible or odious to the spectator! The whole presents nothing but the idea of a blind nature." Kant likewise refuted the claims of natural theology; other philosophers from Bacon on denied the existence of final causes. The unhappy consequences of the French Revolution contributed to the spreading of a deep pessimism. It is reflected in the thinking of Malthus and other demographers. No longer was the growth of human populations seen as one of the benefits bestowed on man by God. Rather, it was shown that owing to limits imposed by the environment such growth inevitably leads to poverty, disaster, and death.

The more the studies of the naturalists progressed, the more phenomena were found which contradicted the excellence of design. Not every organism could have been exclusively designed for its role in nature. For, how would this account for the existence of a limited number of well defined types, such as mammals, birds, snakes, beetles, and so on. Rather, at the beginning relatively few archetypes were created and the laws of nature gave rise to the subsequent diversity. Everything, however, had been contained in the plan of creation. Thus, indirectly, even in this thinking, diversity and adaptation was due to design (Bowler 1977). But even this revision of the design argument could not silence criticism. One asked: What is so wonderful about a parasite that tortures its victims and leads to their eventual death? Even worse, how could design be perfect if it leads to such widespread extinction as documented by the fossil record? If the harmony of the living world, as described by natural theologians, is reflected by the mutual adaptation of organisms to each other and to their environment, and if these adaptations must be adjusted continuously to cope with the changes of the earth and with the restructuring of the faunas owing to extinction, what final causes could there be to govern all these *ad hoc* changes? If the environment changes, the organism has to readjust to it. But there is no necessary direction, no thought of necessary progress, and no reaching of any final goals. After evolutionary thinking had begun to spread, Schleiden (1842:61) insisted that although one can observe simple as well as complicated organisms, "it would be a totally misleading language if we would use for them the words imperfect and perfect, or lower and higher."

Natural theology, with its emphasis on design, had been virtually abandoned on the Continent by about 1800. But it continued to be strong in England, and all of Darwin's teachers and peers, particularly Sedgwick, Henslow, and Lyell, were confirmed natural theologians. Naturally this was Darwin's conceptual framework when he began to think about adaptation and the origin of species.

Darwin's Beliefs

There are many indications that when Darwin returned from the *Beagle* voyage he shared the beliefs of the natural theologians. He had wholly abandoned them, 23 years later, when he published the *Origin*. There is, however, not yet complete consensus in the Darwin literature under what influences and in what stages Darwin revised his interpretations. It was a peculiar period, since the British philosophers of science—Herschel,

Whewell, and Mill—emphasized a rigorous scientific methodology and yet all firmly believed in final causes. They believed in laws, but blind, unguided laws would lead to random disorder (Ruse 1975; 1979), hence God's guiding hand was needed.

To what did the young Darwin attribute adaptation? Prior to 1838 his ideas on this point were rather vague. He seems to have attributed adaptation to certain laws, particularly the influence of the environment on the generative system. He still thought in terms of the design of the world. This is indicated by the following statements in the early *Transmutation Notebooks.*[2]

> B5: "On the other hand generation destroys the effect of accidental injuries, which if animals lived forever would be endless (that is with our present system of body and universe. Therefore final cause of life)."

> B49: "Progressive development gives final cause for enormous periods anterior to man. Difficult for man to be unprejudiced about self."

> B252: "When we talk of higher orders, we should always say intellectually higher.—But who with the face of the earth covered with the most beautiful savannahs and forests dare to say that intellectuality is only aim in this world."

> C73: "The aberrant varieties will be formed in any kingdom of nature where scheme not filled up [Scheme means design]."

> C146: "The end [purpose] of formation of species and genera is probably to add to quantum of life possible with certain preexisting laws.—If only one kind of plants not so many—."

However, finalistic comments evidently did not dominate his thinking, for there are remarkably few teleological comments in his *Transmutation Notebooks.* The most clearly teleological statement refers to dispersal: "When I show that islands would have no plants were it not for seeds being floated about,—I must state that the mechanism by which seeds are adapted for long transportation, seems to imply knowledge of whole world—if so doubtless part of system of great harmony" (D74). Darwin's pre-1838 interpretation of evolutionary change depended on God's planning and was thus clearly a finalistic interpretation. For the Darwin of the Transmutation Notebooks (before September 1838), the seeming path of progression toward perfection was simply the result of certain laws which made such a development possible. All organic change, he thought, was an adaptive response to changes, however slight, in external conditions. These environmental influences induced the generative system to produce appropriate responses. This implied that God quite directly was involved in adaptation because only God could have made the generative system in such a way that changes in the environment would induce it to come up with an adequate response.

No other Darwin scholar has recorded as carefully Darwin's gradual emancipation from the thought of natural theology than Ospovat (1981). This relates particularly to the question how perfect all adaptations are (pp. 24–25). Yet, as Darwin's studies proceeded he discovered one phenomenon after the other that cast doubt on the perfection of adaptations. First he discovered all sorts of evidence for descent (called *propagation* or progression in Darwin's earlier notes) which served as a definite constraint on the absoluteness of adaptation. Then came the consideration of rudimentary or vestigial organs which also contradicted perfect adaptation, and so did the widespread occurrence of extinction, which was likewise incompatible with perfect adaptation. Those natural theologians, and there were others beside Darwin, who saw such inconsistencies and seeming incompatibilities with the concept of a total harmony of nature ascribed the deviations from perfect adaptation to a conflict between various laws instituted by the Creator. Organisms, said these authors, were only as perfect as is possible within the limits set by the necessity of conforming to these laws. There are, for instance, different laws required to explain the facts of structure, distribution, and succession.

Somehow such a direct reliance on eternal God-given laws for the explanation of natural phenomena must have been rather unsatisfactory to Darwin and in conflict with some of his major philosophical framework. This must be the reason why he abandoned this type of thinking so speedily after he had discovered his theory of natural selection on 28 September 1838. Natural selection gave him a purely mechanistic explanation for adaptation and for evolutionary progression. As Darwin stated it in his *Autobiography* (1958:87): "The old argument of design in nature, as given by Paley, which formerly seemed to me so conclusive, fails, now that the law of natural selection has been discovered. We can no longer argue that, for instance, the beautiful hinge of a bivalve shell must have been made by an intelligent being, like the hinge of a door by man. There seems to be no more design in the variability of organic beings and in the action of natural selection, than in the course which the wind blows."

At first (after 1838) Darwin remained enough of a natural theologian to believe that natural selection could give him perfect adaptation. But, as described by Ospovat, he abandoned this belief by the 1850s, and the *Origin* (1859) is remarkably free of any teleological language. To be sure, the word *progress* is used 10 times in this volume but almost always as a term to describe a passing of time. Only in connection with the replacement of fossil faunas of which each one seems to be "higher" than the one it has replaced does Darwin speak of a process of improvement; but he adds "I can see no way of testing this sort of progress" (p. 337). However, Darwin points out that there are differences in competitive ability even among living faunas. British faunal elements introduced to New Zealand are highly successful, while he doubts the reverse would be true. "Under this point of view," says Darwin, "the productions of Great Britain may be said to be higher than those of New Zealand" (p. 337). Yet this is not a teleological argument. The greater competitive ability of the faunal elements of Great Britain was not due to any built-in drive or final cause but simply due to the fact that the British fauna had passed through a more severe struggle for existence.

Nevertheless the concept of *perfect* and *perfection* continued to be popular with Darwin. In the *Origin* he uses the word *perfect* 77 times, *perfected* 19 times, and *perfection* 27 times. What is remarkable, however, in these uses is how carefully Darwin makes a distinction between the product of selection and the process of perfecting. We look in his explanations in vain for a drive toward perfection or a tendency toward perfection. Invariably Darwin emphasizes that it is selection which carries the evolutionary line to ever-greater perfection. This is particularly well-stated in the section of Chapter 6 with the heading *Organs of Extreme Perfection and Complication* (p. 186) which, among others, contains Darwin's well-known discussion of the evolution of eyes through natural selection (see also Salvini-Plawen and Mayr 1977). Since natural selection is not a finalistic process Darwin now sees quite clearly that "natural selection will not necessarily produce absolute perfection; nor as far as we can judge from our limited faculties, can absolute perfection be everywhere found" (p. 206). Complete perfection, of course, is not at all needed, because "natural selection tends only to make each organic being as perfect as, or slightly more perfect than, the other inhabitants of the same country with which it has to struggle for existence. And we see that this is the degree of perfection attained under nature" (p. 201). There is not even a trace of a suggestion of any final cause because perfection is simply the product of the *a posteriori* process of natural selection. With the world and its biota constantly changing, perfect creation in the beginning would have been futile. "Almost every part of every organic being is so beautifully related to its complex condition of life that it seems as improbable that any part should have been suddenly produced perfect, as that a complex machine should have been invented by man in a perfect state" (*Origin,* 6th ed., pp. 58–59).

A subsequent correspondence with Asa Gray permits us to analyze Darwin's thought even a little further. Asa Gray, even though a rather strict creationist, accepted the importance and guiding capacity of natural selection. However, "Natural selection is not the wind which propels the vessel, but the rudder which, by friction, now on this side and now on that shapes the course" (Moore 1979:316). Variation, however, for Asa Gray was guided by a divine hand. This possibility was emphatically rejected by Darwin and induced him to state his ideas, rather evidently as a direct answer to Gray, in *The Variation of Animals and Plants* (1868 II:432). Gray, however, failed to understand Darwin's argument and went even so far as to praise "Darwin's great service to natural science in bringing back to it teleology" (Gray 1876:237).

It must be admitted that Darwin in his later years, particularly in letters to his numerous correspondents, was sometimes rather careless in his language. For instance he referred to "the extreme difficulty or rather impossibility of conceiving this immense and wonderful universe, including man with his capacity of looking far backwards and far into futurity, as the result of blind chance or necessity." How could he have said this when the theory of natural selection had given him exactly the means to escape from the alternate chance *or* necessity? On another occasion he said, "The mind refuses to look at this universe, being what it is, without having been designed." It is not surprising therefore that Darwin was misclassified by a number of authors who did not understand the working of natural selection. Kölliker, for instance, accused Darwin of being "in the fullest sense of the word a teleologist." And even T. H. Huxley when defending Darwin was driven to distinguish between "the teleology of Paley and the teleology of evolution" (Moore 1979:264). We now know, as will be discussed further on, that the genetic program of organisms does not contain a blueprint for the future evolution of the phyletic line to which an organism belongs. There are severe constraints on the evolutionary potentialities of a given line, but this is, of course, something entirely different from a final cause.

The Acceptance of Darwinism

The most revolutionary aspect of Darwin's theory of natural selection was that it completely eliminated any need for a finalistic force in nature, referred to by his German opponents as a *Vervollkomnungstrieb*. Particularly those who had come to evolutionary thinking through a temporalized scala naturae felt that such a teleological force was indispensable and that such a "superior force" was necessary to account for the final emergence of man. However Darwin was not alone in his rejection of finalism. Haeckel declared emphatically that "the causes of all phenomena of nature . . . are purely mechanically acting causes, never final, the goal-directed causes" (1866 2:150). The most articulate among Darwin's supporters was Weismann, who took up the battle for natural selection again and again and refuted the theories of Darwin's opponents. He attempted to show that all the details of the coloration and pattern of butterflies could be explained by selection but that "innere Bildungsgesetze," as postulated by Eimer, were quite unthinkable. In 1868 he rejected Naegeli's hypothesis of final causes, and in 1876 he answered von Baer's long defense of teleology and also that of the philosopher E. von Hartmann (1875). To postulate teleological forces, he said, is in conflict with the basic principle of science, not to invoke occult forces, as long as one can interpret things by known forces. "For what else is science but the attempt to determine the causal mechanisms by which the phenomena of the world are caused." Finally in 1909 (p. 425) he summarizes his thoughts once more by stating "the principle of natural selection solved the puzzle how it is thinkable that adaptedness ('das Zweckmässige') can originate without the intervention of a purposive force."

The voices of Haeckel, Weismann, F. Müller, and Darwin's naturalist friends were, however, merely "cries in the wilderness," for the opposition to the mechanistic process of natural selection was almost universal. I will discuss below various of the attempts to substantiate final causes, but it might be helpful to begin the discussion by analyzing some of the reasons for the strength and virtual universality of the opposition. What one discovers when studying this literature is that none of his opponents truly understood natural selection. And this misunderstanding was to a large extent due to longstanding ideological commitments. The opposition to natural selection continued up to the evolutionary synthesis and with it an open or unspoken support of finalism. Indeed, an implicit rejection of natural selection can be found in numerous publications up to our day. An analysis of the writings of a few twentieth-century authors will help to reveal the nature of the misunderstandings that were responsible for the continuing popularity of finalism.

T. H. Morgan, who showed his lack of understanding of natural selection even in his last book on evolution in 1932 (Mayr and Provine 1980), claimed in 1910 that finalism had entered biology through natural selection because "by picking out the new variation . . . purpose enters in as a factor, for selection had an end in view," completely ignoring the randomness of variation and the statistical nature of the selection process. Some 25 years later H. F. Osborn fights natural selection with pre-Darwinian, essentialistic arguments. Certain developments in the phyletic history of elephants, he says, could not have been due to selection because "no species of elephant occupied the same geographic range as another species at any given period of geological time; thus there was no competition between species" (1934:228). Osborn consistently thought of taxa as types; and even though he refers frequently to variation, this is for him always an origin of new types. Populational variation was not appreciated by him. Consequently, it never occurred to him that there could be a struggle for existence among members of the same population or species, as conceived by Darwin. And this forced him to interpret the struggle for existence as a struggle between species types instead of a struggle for reproductive success.

In his recent influential book *Chance and Necessity* (1970), J. Monod offers as alternatives only the two processes identified in the title of his volume. Although Monod is adamantly opposed to determinism, including any invoking of final causes, he totally ignores natural selection as a creative process and ascribes all evolution to pure chance. Speaking of molecular changes during evolution he says, "We call these events accidental; we say that they are random occurrences. And since they constitute the only possible source for modifications in the genetic text, itself the *sole* repository of the organism's hereditary structures, it necessarily follows that chance alone is at the source of every innovation, of all creation in the biosphere. Pure chance, absolutely free but blind, at the very root of the stupendous edifice of evolution" (p. 112). As a reductionist, Monod looks only at the level of the gene, or more correctly, the level of the base pair, which is "mechanically and faithfully replicated and translated." Except for rare mutations, there is a total invariance of the genotype. This is why for Monod evolution is an "intensely conservative system." Chance determines mutation, favorable mutations are preserved, unfavorable ones are rejected. Since for him evolutionary change is thus entirely due to chance mutation, Monod finds the harmony of nature completely inexplicable. He says that when one is looking at the marvels produced by evolution, "One may well find oneself beginning to doubt again whether all this could conceivably be the product of an enormous lottery presided over by natural selection, blindly picking the rare winners from among the numbers drawn at utter random" (p. 138). He completely ignores that a lottery is a one-step process while natural selection is a two-step process in which the second step, the actual selection, is anything but blindly picking the winners (Mayr 1962). In his description of natural selection, which for Monod consists of replicative invariance in DNA and selection, he completely omits the process that produces the actual material upon which natural selection acts, namely the variation-producing genetic recombination. It is recombination which generates the genotypes that produce the phenotypes which are the ultimate target of selection, and recombination produces an infinite amount of variance. No two individuals in a sexual population are identical, and even though recombination itself is governed by chance, the process of selection among the results of recombination is anything but a chance process. It is not the invariance at the molecular level which counts in a population in which no two individuals are the same, but rather the selection among the countless uniquely different components of the population. By his reductionist and essentialistic approach Monod barred his own exit from the dilemma stated in the title of his book. By claiming that everything in evolution is purely a matter of chance, he virtually neutralized the effects of his otherwise convincing attack on finalism.

Finalism as an Alternative to Natural Selection

Numerous attempts were made in the years after 1859 to replace Darwin's theory of natural selection by a superior way of achieving adaptation. The best known of these theories are usually classified under the headings neo-Lamarckism (inheritance of acquired characters), orthogenesis (an intrinsic perfecting principle), and saltation. They all incorporated to a lesser or greater extent some finalistic components. It is not easy to report on these theories for a number of reasons. Not only are the descriptions of the postulated mechanism by which the changes are achieved usually quite vague, but the same author may support first one and then the other of these theories, or a mixture of them. There are some excellent recent studies on these theories (Bowler 1983) as well as the classical analysis of Kellogg (1907).

Even after the Paleyan concept of an *ad hoc* design of every, even the slightest, adaptation had lost all credibility, there remained a concept of a universal design of organic progression, an evolutionary reinterpretation of the temporalized scale of nature (Bowler 1977). Such a concept seemed, at first, to have a sound observational foundation. Considering that variation is random, as Darwin postulated, and considering that the number of environmental constellations is quite unlimited, one would expect a totally chaotic network of evolutionary phenomena. What one actually finds is the existence of a limited number of well-defined lineages and the possibility to arrange organisms into progressive series. This was described not only by paleontologists but also by students of living organisms, be they butterflies (Eimer) or birds (Whitman). Variation evidently was not random but followed well-defined pathways of change. Such evolutionary trends were ascribed to an intrinsic, direction-giving force, called ortho-genesis. It was described as a perfecting principle or (in German) *Vervollk-ommnungstrieb*. The intrinsic nature of this force seemed to be confirmed by the fact that it was possible to establish rectilinear series not only for characters that might have been advanced by natural selection, such as increasing precision of mimicry patterns or phyletic increases in body size, but also in nonutilitarian or seemingly deleterious characters. This was an argument made particularly emphatically by Eimer (Bowler 1979). Most proposals of orthogenesis were made in strict opposition to and as alter-natives to natural selection.

However, there was a group of Christian Darwinians for whom natural selection was "evidence of a directing agency and of a presiding mind" (Moore 1979). They either thought that variation as such was directive, supplying just the right material to selection, or they considered the selecting process as purposive. Clearly for them natural selection was a teleological process. Most of those who had these views were theologians, but the botanist Asa Gray had similar views. He praised "the great gain to science from [Darwin] having brought back teleology to natural his-tory." As we mentioned above, Darwin took great pains (1868) to refute Asa Gray's finalistic conception of natural selection.

Even such an ultra-mechanist as J. Sachs (1894) adopted Naegeli's perfecting principle as the agent of all major evolutionary developments, with natural selection merely being able to improve fine-grained adapta-tion. Kölliker (1886) was another adherent of an autogenetic theory ascribing all evolutionary progress to "intrinsic causes," and like Naegeli he stimulated Weismann to a reply.

Developmental Constraints

With embryology reaching its greatest flowering in the last quarter of the nineteenth and first third of the twentieth century, it is not surprising that analogies between ontogeny and phylogeny were frequently at-tempted. German embryologists were particularly susceptible to such a comparison, since the same word *Entwicklung* was used both for ontogeny and evolutionary progression. Driesch, for instance, stated that evolution is nothing but "an *Entwicklungsprozess,* like the ontogeny of an echinoid larva." There is a hint of such analogical thinking also in the writings of Eimer. Even though they were wrong in their neglect of natural selection and their belief in intrinsic perfecting forces, these writers did call attention to developmental constraints, an important evolutionary factor, emphasized again in the most recent evolutionary literature. Developmental constraints played an important role in Eimer's (1897) thinking, but even more so in that of C. O. Whitman. Every stage in ontogeny is rather rigidly determined by the preceding stage. Would it not be probable therefore that possible manifestations of variation are severely constrained by the present configuration of the genotype? Whitman formulated his thoughts as follows: "Is not every stage, from the primordial germ onward, and the whole sequence of stages, rigidly orthogenetic? If variations are deviations in the directions of the developmental processes, what wonder is there if in some directions there is less resistance to variation than in others? . . . And if we find large groups of species, all affected by like variation, moving in the same general direction, are we compelled to regard such a 'definite variation tendency' as teleological, and hence out of the pale of science? If a designer sets limits to variation in order to reach a definite end, the direction of events is teleological; but if organization and the laws of development exclude some lines of variation and favor others, there is certainly nothing supernatural in this, and nothing which is incompatible with natural selection. Natural selection may enter at any stage of orthogenetic variation, preserve and modify in various directions the results over which it may have had no previous control" (1906).

It is evident that such a constraint on the possible pathways of variation is a factor that is totally different from any built-in drive toward perfection, that is, from any teleological principle. To use the term *orthogenesis* for the mere description of rectilinear evolutionary trends was therefore quite confusing. This was fully understood by Plate (1903), who postulated that rectilinear trends were ultimately determined by natural selection and who designated their causation therefore as *orthoselection.*

No one, of course, was more familiar with rectilinear trends in evolution than the paleontologists. Every worker, whether he was a vertebrate paleontologist like Kovalevsky and Marsh (phylogeny of horses) or a student of invertebrates dealing with clams or ammonites, stressed "orthogenetic series." This literature has been well dealt with by Kellogg

(1907), Huxley (1942), Simpson (1949; 1974) and Rensch (1947). A late but very typical representative of this school of thinking was H. F. Osborn (1934). By the choice of his terminology, *aristogenesis,* he emphasized the movement toward the "best." In his descriptions, however, he very perceptively emphasizes developmental potentiality. He describes how both proboscideans and equids start in the Eocene with rather similar four-coned molars. "But differentiating this visible or phenotypic similarity is the widely divergent aristogenic potentiality of the primitive proboscidean and primitive equine molar . . . [By the end of the Tertiary] the one ends in the marvelously complicated 27-plated molar of the wooly mammoth, the other is fated to evolve into the double-columned grinder of the horse" (1934:219–220). When one reads Osborn's account, one has the impression that his aristogenic trends are due to final causes, and this is how everybody interprets Osborn. Nevertheless he takes great pains to distance his thinking from "predetermination which is the essential element in every entelechistic hypothesis." His thinking, he emphasizes, depends entirely on potentiality. Curiously, in view of his controversies with T. H. Morgan, his "potentiality" is very much the same thing as the "mutation pressure" of the mutationists. Both of them thought of taxa as types, and when speaking of variation thought of it in terms of an origin of new types. Populational variation was not appreciated by either of them. Hence a struggle for existence among members of the same population and species never occurred either to Osborn or to Morgan. That in spite of the same "potentiality" a great range of genetically different individuals would result (owing to genetic recombination) did not enter the thinking of either of them. Both of them being typologists, they were unable to integrate Darwinian natural selection into their thinking.

By the time of the evolutionary synthesis of the 1940s virtually no evolutionary biologist, in fact no competent biologist, was left who still believed in any final causation of evolution. The few biologists who still did so either had theological commitments, like Teilhard de Chardin, or were unaware of the developments of biology in the twentieth century, like Comte de Nouy.

Final causes, however, are far more plausible and pleasing to a layperson than the haphazard and opportunistic process of natural selection. For this reason, a belief in final causes has had a far greater hold outside of than within biology. Almost all philosophers, for instance, who wrote on evolutionary change in the 100 years after 1859 were confirmed finalists. All three philosophers closest to Darwin—Whewell, Herschel, and Mill—believed in final causes (Hull 1973). The German philosopher E. von

Hartmann was a strong defender of finalism, stimulating Weismann to a spirited reply. In France Bergson postulated a metaphysical force, *élan vital,* which, even though Bergson disclaimed its finalistic nature, could not have been anything else, considering its effects. There is room for a good history of finalism in the post-Darwinian philosophy, although Collingwood (*Idea of Nature*) has provided a survey. Whitehead, Polanyi, and many lesser philosophers were also finalistic. Throughout this period there have been exceptions, the most noteworthy perhaps being the German philosopher Sigwart, who as early as 1881 provided a remarkably modern treatment of the problems of teleology. Modern philosophers— that is, those who have published since the evolutionary synthesis—have, on the whole, refrained from invoking final causes when discussing evolutionary progress. Apparently they fully accept the explanation provided by the evolutionary synthesis. When they do discuss teleology, like Beckner or Nagel, they deal with adaptation and with "teleological systems." Finalism is no longer part of any respectable philosophy. One last vigorous attack on finalism was Monod's book *Chance and Necessity.* As mentioned above, Monod failed to understand the explanatory power of natural selection and opted for pure chance as having been responsible for the phenomena of nature. Such Epicureanism, however, is only rarely encountered in modern times.

What Led to the Decline, if not Demise, of Finalism?

One cannot attribute the decline of finalism to a single factor. Rather it was due to a combination of causes, the most important of which might be the following:

(1) The realization that science was unable to detect any mechanism that would ensure continuing progress and that to postulate final causes amounted therefore to the invoking of a metaphysical, if not supernatural, force. Yet such a force has no place in either science or philosophy.

(2) The realization that several rather different kinds of phenomena had been confounded under the term *final cause*. This led to a partitioning of the term *teleological* into (at least) four separate categories (Essay 3). Two of these are relevant in evolutionary discussions: (a) A goal-directed activity of an individual, effected by a learned or genetic program. Such a process being designated by Pittendrigh (1958) as teleonomic; and (b) a timeless goal-directed process, moving to a far distant goal and leading to evolutionary progress or perfection (cosmic teleology). Of these two processes

only (b) answers to the definition of final causes, while (a) is due to the efficient cause of a program.

As far back as Aristotle (Delbrück 1971) thinkers have pointed out that organisms are not like inert matter or like machines but have all sorts of properties that are found in nature only in living organisms. Those who held such views were inevitably called vitalists or finalists but since we now know that phenomena which they enumerated as evidence for the validity of their view are due to the fact that organisms have a genetic program, their argument was indeed valid. E. S. Russell (1945), who emphatically presented this viewpoint in his *The Directiveness of Organic Activities,* defended himself quite rightly against the accusation (for instance, Simpson 1949:127) that he was a finalist. The clear recognition of the purely mechanistic basis of teleonomic processes has eliminated most of the so-called evidence for the existence of final causes in organisms.

(3) The realization that certain evolutionary phenomena, claimed to be able to refute Darwinism, actually can be shown to be consistent with the theory of natural selection, as soon as one understands that natural selection is not a one-step random process, but a two-step direction-giving process. This category includes the following phenomena:

(a) The claim that perfectly rectilinear trends exist. Simpson and other paleontologists have shown that such trends do not exist. All evolutionary lines show changes of direction and often even reversals. Therefore future evolution can never be predicted, as should be possible if there were constant final causes (Monod 1970).

(b) The claim that it is impossible to explain rectilinear trends except by final causes. Simpson, Rensch, and other authors have shown that such trends can invariably be explained directly or indirectly by the action of natural selection. An author like H. F. Osborn quite rightly described how totally different the trends in tooth evolution were in the horses and the proboscideans. What he did not realize were the very severe constraints which the epigenotype (Waddington) places on evolutionary possibilities. The first author who clearly saw this was C. O. Whitman. He pointed out that evolution advances to varying degrees along the lines on which evolutionary variation is possible and that such stages can be arranged in the form of orthogenetic series. He considered this as being completely compatible with natural selection. Curiously, the observation of the existence of such developmental constraints was already well known in preevolutionary times. It was this which led Cuvier to assert that he could reconstruct an entire mammal skeleton if given merely a few teeth or bones. We now know that he was largely right because the genotype of any evolutionary lineage has extremely limited potentialities.

(c) The claim that trends in nonutilitarian characters refute natural selection. Opponents of natural selection like to point to trends in nonutilitarian characters that could "not possibly be of any selective significance." What they forget is that the phenotype as a whole is the target of selection and therefore also the genotype as a whole. As a result, many features of organisms show evolutionary trends as by-products of the whole selected genotype. As Huxley and Rensch have emphasized, allometry frequently supplies illustrations of this principle.

(d) The claim that trends in deleterious characters clearly contradict natural selection. It has frequently been asserted that only final causes could account for the evolutionary development of so-called "excessive structures." This category includes the giant antlers of the Irish Elk, the tusks of the Moluccan Babirussa, the canines of the saber-toothed tigers (*Smilodon*), and the copulatory apparatus of certain slugs, to mention only a few of the numerous similar instances listed in the anti-Darwinian literature. All these features would seem, at first sight, to be highly deleterious, and it was claimed that natural selection could not possibly have favored or even tolerated their evolution. However, the studies of Rensch, Simpson, Gould, and various other paleontologists have demonstrated that the species that had these "excessive" characters always flourished for considerable periods of time when these characters clearly were of selective advantage and that their ultimate extinction coincided with a climatic or broad faunal change which simultaneously led to the extinction of numerous other species without such "excessive" characters. It is now quite evident that natural selection is always able to prevent the development of fitness-reducing characters and that such seemingly "excessive" structures were clearly favored by natural selection at the time when they developed.

Actually, if final causes dominated evolutionary developments and controlled the attainment of perfection, one should find far more perfection in nature than one actually encounters. I have recently listed seven constraints that prevent an optimal response to a newly arising selection pressure (Essay 8). Most important among these factors are developmental interactions and, at the genetic level, the cohesion of the genotype. A failure to achieve perfection is not as serious under normal circumstances as it might seem, for, as Darwin has said so rightly,

"Natural selection tends only to make each organic being as perfect as, or slightly more perfect than, the other inhabitants of the same country with which it has to struggle for existence" (1859:201).

Evolutionary Progress without Final Causes

The *Origin* is remarkably free of any teleological language. Nevertheless the word *progress* is used 10 times and it has been suggested that this implies a teleological commitment. The evidence is opposed to this interpretation. Almost all of the uses of the word progress by Darwin refer to the passing of time. Only in connection with the replacement of fossil faunas, each of which seems better than the one it has replaced, does Darwin speak of a process of improvement, but he adds: "I can see no way of testing this sort of progress" (p. 337). As Darwin points out, however, there are differences in competitive ability even among living faunas. British faunal elements introduced into New Zealand are highly successful, while he doubts that an introduction of members of the New Zealand fauna to England would be a success. "Under this point of view," says Darwin, "the productions of Great Britain may be said to be higher than those of New Zealand" (p. 337). Yet this does not imply any final causes but simply refers to a greater competitive ability of Great Britain's fauna, the result of a preceding more severe struggle for existence.

No statement in the *Origin* sounds more teleological than one on the next to last page of this work (p. 489): "And as natural selection works solely by and for the good of each being, all corporeal and mental endowments will tend to progress toward perfection." But again there is no intimation of a teleological mechanism, for as Darwin emphasizes quite specifically such "progress toward perfection" is an *a posteriori* product of the mechanism of natural selection.

For some Darwinians the concept of evolutionary progress seems to have been somewhat of an embarrassment. How can a strictly opportunistic competitive struggle lead to progress? Darwin himself occasionally seems to have had such doubts, and they are reflected in his comment, "Never use the words higher or lower," written on the margin of his copy of the *Vestiges*. Others who also questioned progress pointed to the continued existence of the archaebacteria and other prokaryotes, to the great flourishing of the protists and lower fungi up to the present, to the parasites, and to the inhabitants of caves. None of these can be called progressive, in the sense of "higher," and yet they continue to exist. So, it was said, evolution is simply a process of specialization. And yet, who can deny that overall there is an advance from the prokaryotes that dominated the living world more than three billion years ago to the eukaryotes with their well organized nucleus and chromosomes as well as cytoplasmic organelles; from the single-celled eukaryotes to metaphytes and metazoans with a strict division of labor among their highly specialized organ systems; within the metazoans from ectotherms that are at the mercy of climate to the warm-blooded endotherms, and within the endotherms from types with a small brain and low social organization to those with a very large central nervous system, highly developed parental care, and the capacity to transmit information from generation to generation?

Attempts to define progress have been many.[3] For Lamarck, for instance, and for many nineteenth-century authors man was clearly the most perfect organism, and all forms of life were arranged in a single column on the basis of their assumed progress toward manhood. Now we know that diversification is the most characteristic attribute of evolution and that life not only very early split into five kingdoms (prokaryotes, protists, fungi, plants, and animals) but also developed literally thousands of distinct phyletic lines within each of these kingdoms, most of them not in the slightest tending toward the characteristics of man. Neither can the dominance of a group on earth be considered the criterion of progress. On that basis the vascular plants would have to be considered more dominant than man and even the insects. And man's ancestors until less than 10,000 years ago were anything but dominant on earth.

Structural complexity is sometimes mentioned as a sign of progress, but trilobites and placoderms would seem to have been more complex in structure and perhaps more specialized than modern man. Huxley (1942) considered emancipation from the environment an important index of progressiveness, and in that criterion man certainly ranks higher than any other organism. However, is independence from the environment truly an index of progressiveness?

When discussing evolutionary progress one seems to be quite unable, since one is a member of the human species, to get away from criteria that would give man supremacy. However, there are two criteria of progressiveness that would seem to have a considerable amount of objective validity. One of these is parental care (made possible by internal fertilization), which provides the potential for transferring information nongenetically from one generation to the next. And the possession of such information is of course of considerable value in the struggle for existence. This information transfer generates at the same time a selection pressure in favor of an improved storage system for such remembered information, that is, an enlarged central nervous system. And, of course, the combination of postnatal care and an enlarged central nervous system is the basis of culture, culture together with speech setting man quite aside from all other living organisms. However, even if we would designate the acquisition of these capacities as evidence for evolutionary progress, it would not strengthen the case for final causes, since these developments were clearly achieved with the help of natural selection.

Species Selection

To some students of evolution, it has seemed inconceivable that the slow process of natural selection, that is, the struggle for existence among individuals of a species, could lead to the enormous evolutionary progress observed in so many phyletic lines, whether that leading to the highest mammals and birds, to the social insects, to orchids, or to giant trees. To see all evolution simply as the result of intrapopulational competition is indeed simplistic, because superimposed on this individual selection is a process traditionally referred to as *species selection*. An individual organism competes not only with members of its own species but "struggles for existence" also against members of other species. And this process is probably the greatest source of evolutionary progress. As I stated previously, each newly formed species, if it is evolutionarily successful, must represent, in some way, evolutionary progress: "Each species is a biological experiment . . . There is no way to predict, as far as the incipient species is concerned, whether the new niche it enters is a dead end or the entrance into a large new adaptive zone . . . The evolutionary significance of species is now quite clear . . . The species are the real units of evolution, as the temporary incarnation of harmonious, well-integrated gene complexes. And speciation, the production of new gene complexes capable of ecological shifts, is the method by which evolution advances. Without speciation there would be no diversification of the organic world, no adaptive radiation, and very little evolutionary progress" (Mayr 1963:621). Any successful new species is thus, in a way, an advance over previously existing ones. Darwin explained this as follows: "But in one particular sense the more recent forms must, on my theory, be higher than the more ancient; for each new species is formed [that is, has become successful] by having had some advantage in the struggle for life over other and preceding forms" (p. 337). When the competition among individuals of different species leads to or at least contributes to the extinction of one of the competing species, it is a case of species selection (Essay 8). That competition among species could lead to the extinction of one of the competitors was of course already known to Charles Lyell and other preevolutionary authors.

A new species will be successful in the struggle for existence over a previously existing species only if it has made some, even the smallest, evolutionary invention. This might be an improvement in its digestive physiology or its nervous system or its lifestyle or any other of the countless ways by which the so-called "higher" organisms differ from the lower ones. Thus the Darwinian mechanism of variation and selection, of speciation and extinction, are fully capable of explaining all macroevolutionary developments, whether specializations, improvements, or other innovations. And none of this requires any finalistic agent.

The Incompatibility of Evolutionary Progress with Final Causes

Close study of evolutionary progress shows that its characteristics are not compatible with what one would expect from a process guided by final causes. Progressive changes in the history of life are neither predictable nor goal-directed. The observed advances are haphazard and highly diverse. It is always uncertain whether newly acquired adaptations are of permanent value. Who at the beginning of the Cretaceous would have predicted the total extinction by the end of the period of that flourishing taxon of dinosaurs? Episodes of stasis alternate with episodes of precipitous evolutionary change. Evolutionary trends are rarely rectilinear for any length of time, and when such rectilinearity occurs it can usually be shown to be due to built-in constraints.

All the evolutionary phenomena and aspects of evolutionary progress that were considered as irrefutable proof of teleology by earlier generations can now be shown to be entirely consistent with natural selection. Phenomena that are due to a chain of historical events cannot be ascribed to simple laws and can therefore not be proven in the same way as are phenomena studied in the physical sciences. However, they can be shown to be consistent with the laws of genetics and with the theory of natural selection in its modern sophisticated form. No one has refuted the finalistic thesis of evolution more convincingly than Simpson (1949; 1974). He pointed out that each evolutionary line goes its own way, and evolutionary progress can be defined only in terms of that particular lineage. Nothing seemed more progressive in the geological past than the ammonites and the dinosaurs, and yet both taxa became extinct. On the other hand many evolutionary lines have displayed no evidence of progress in hundreds or thousands of millions of years, and yet they have survived to the present day, as the archaebacteria and other prokaryotes. Progress thus is not at all a universal aspect of evolution, as it ought to be if evolution were generated by final causes.

NOTES

This essay first appeared in *Scientia* 118 (1983):97–117.

1. The literature on finalism, progress, and God's world is enormous. For further reading I recommend Lovejoy (1936), Gillispie (1951), Glacken (1967), Nisbet (1980), and Mayr (1982).

2. Darwin's notebooks were published as follows: G. de Beer, ed. 1960–1967. Darwin's notebooks on transmutation of species. *Bull. Brit. Mus. (Nat. Hist.)*, Hist. Ser. 2:25–73 (Notebook B); 2:75–118 (Notebook C); 2:119–150 (Notebook D); 2:151–183 (Notebook E). G. de Beer, M. J. Rowlands, and B. M. Skramovsky, eds. 1967. Darwin's notebooks, etc. Pages Excised by Darwin. *Bull. Brit. Mus. (Nat. Hist.)*, Hist. Ser. 3:129–176. Referred to as Notebooks B, C, D, and E with Darwin's original page numbers.

3. For biological progress see Simpson (1949; 1975), Huxley (1942), Thoday (1975), Stebbins (1969), Goudge (1961), Ayala (1974), Rensch (1947).

REFERENCES

Ayala, F. J. 1974. The concept of biological progress. In F. J. Ayala and Th. Dobzhansky, eds. *Studies in the Philosophy of Biology.* Berkeley: University of California Press.

Bowler, P. J. 1976. *Fossils and Progress.* New York: Science History Publications.

——— 1977. Darwinism and the argument from design: suggestions for a re-evaluation. *J. Hist. Biol.* 10:29–43.

——— 1979. Theodor Eimer and orthogenesis: evolution by 'definitely directed variation.' *J. Hist. Med. Allied Sci.* 34:40–73.

——— 1983. *The Eclipse of Darwinism: Anti-Darwinian Evolution Theories in the Decades around 1900.* Baltimore: Johns Hopkins University Press.

Darwin, Charles. 1859. *On the Origin of Species.* London: John Murray. Facs. rpt. Cambridge, Mass.: Harvard University Press, 1964.

——— 1868. *The Variation of Animals and Plants under Domestication.* London: John Murray. ed.

——— 1958. *The Autobiography of Charles Darwin.* Ed. Nora Barlow. London: Collins.

Darwin, F. 1887. *The Life and Letters of Charles Darwin.* 3 vols. London: John Murray.

Delbrück, M. 1971. Aristotle-totle-totle. In J. Monod and E. Borek, eds. *Of Microbes and Life.* New York: Columbia University Press.

Gillispie, C. C. 1951. *Genesis and Geology.* Cambridge, Mass.: Harvard University Press.

Glacken, C. J. 1967. *Traces on the Rhodian Shore: Nature and Culture in Western Thought from Ancient Times to the End of the Eighteenth Century.* Berkeley: University of California Press.

Goudge, T. A. 1961. *The Ascent of Life.* London: Allen and Unwin.

Gray, Asa. 1876. *Darwiniana.* New York: D. Appleton. Rpt., ed. H. Dupree. Cambridge, Mass.: Harvard University Press, 1963.

Haeckel, E. 1866. *Generelle Morphologie der Organismen.* 2 vols. Berlin: Georg Reimer.

Hull, D. L. 1973. *Darwin and His Critics.* Cambridge, Mass.: Harvard University Press.

Huxley, J. 1942. *Evolution: The Modern Synthesis.* London: Allen and Unwin.

Kant, I. 1790. *Kritik der Urtheilskraft.* Eng. trans., J. H. Bernard. *Critique of Judgment.* London, 1914.

Kellogg, V. L. 1907. Darwinism To-day. New York: Henry Holt.

Kölliker, A. 1886. Das Karyoplasma und die Vererbung, eine Kritik der Weismann' schen Theorie von der Kontinuität des Keimplasmas. In *Z. wiss. Zool.* 44:228–238.

Lennox, J. G. 1983. Robert Boyle's defense of teleological inference in experimental science. In *Isis* 74:38–52.

Lovejoy, A. O. 1936. *The Great Chain of Being.* Cambridge, Mass.: Harvard University Press.

Mayr, Ernst. 1962. Accident or design, the paradox of evolution. In G. W. Leeper, ed. *The Evolution of Living Organisms.* Melbourne: Melbourne University Press.

———— 1963. *Animal Species and Evolution.* Cambridge, Mass.: Harvard University Press.

———— 1982. *The Growth of Biological Thought.* Cambridge, Mass.: Harvard University Press.

Mayr, E., and W. Provine, eds. 1980. *The Evolutionary Synthesis.* Cambridge, Mass.: Harvard University Press.

Monod, J. 1970. *Le hasard et la nécessité.* Paris: Seuil.

Moore, J. R. 1979. *The Post-Darwinian Controversies.* Cambridge: Cambridge University Press.

Morgan, T. H. 1910. Chance or purpose in the origin and evolution of adaptation. *Science* 21:201–210.

Nisbet, R. 1980. *History of the Idea of Progress.* New York: Basic.

Osborn, H. F. 1934. Aristogenesis, the creative principle in the origin of species. *Amer. Nat.* 68:193–235.

Ospovat, Dov. 1981. *The Development of Darwin's Theory: Natural History, Natural Theology, and Natural Selection, 1838–1859.* Cambridge: Cambridge University Press.

Pittendrigh, C. 1958. Adaptation, natural selection, and behavior. In A. Roe and G. G. Simpson, eds. *Behavior and Evolution.* New Haven: Yale University Press.

Plate, L. 1903. *Ueber die Bedeutung des Darwinschen Selections—Prinzip und Probleme der Artbildung.* Leipzig: Engelmann.

Rensch, B. 1947. *Neuere Probleme der Abstammungslehre: Die transspezifische Evolution.* Stuttgart: Ferdinand Enke.

Ruse, M. 1975. *The Relationship between Science and Religion in Britain, 1830–1870. Church History.*

——— 1979. *The Darwinian Revolution.* Chicago: University of Chicago Press.

Russell, E. S. 1945. *The Directiveness of Organic Activities.* Cambridge: Cambridge University Press.

Sachs, J. 1894. Mechanomorphosen und Phylogenie. *Flora* 78:215–243.

Salvini-Plawen, L., and E. Mayr. 1977. On the evolution of photoreceptors and eyes. *Evol. Biol.* 10:207–263.

Schleiden, M. J. 1842. *Grundzüge der wissenschaftlichen Botanik.* Leipzig: Wilhelm Engelmann.

Sigwart, C. 1881. Der Kampf gegen den Zweck. *Kleine Schriften* 2:24–67.

Simpson, G. G. 1949. *The Meaning of Evolution.* New Haven: Yale University Press.

——— 1974. The concept of progress in organic evolution. *Social Research* pp. 28–51.

Stebbins, G. L. 1969. *The Basis of Progressive Evolution.* Chapel Hill: University of North Carolina Press.

Thoday, J. M. 1975. *Non-Darwinian 'evolution' and biological progress. Nature* 255:675–677.

Toulmin, S. 1982. Darwin und die Evolution der Wissenschaften. *Dialektik* 5:68–78.

Weismann, A. 1868. *Ueber die Berechtigung der Darwinschen Theorie.* Leipzig: Wilhelm Engelmann.

——— 1876. *Ueber die letzten Ursachen der Transmutationen.* Leipzig.

——— 1896. *Ueber Germinal Selektion: Eine Quelle bestimmt gerichteter Variation.* Jena: Gustav Fischer.

——— 1909. *The Selection Theory.* In A. C. Seward, ed., *Darwin and Modern Science,* pp. 18–65. Cambridge: Cambridge University Press.

Whitman, C. O. 1906. The problem of the origin of species. *Proc. Congr. Arts Sci., Universal Expos. St. Louis* 5:41–58.

THE DEATH OF DARWIN?

IN 1957 John Greene wrote *The Death of Adam*. In recent years he has seemed to want to prove the death of Darwin, as indicated by a recent essay (Greene 1986) and other recent writings. Is there any substance to Greene's endeavor to demonstrate the obsolesence of Darwin's theories and ideas? A Darwinian who reads Greene's arguments is overcome by a feeling of frustration. There we have labored for the last 50 years to make Darwin's thought better known, and have documented how brilliantly Darwin has overcome all the objections raised against his work so that we are now "closer to Darwin's basic ideas than biologists have ever been since 1859." It is discouraging to find out how completely Greene ignores this argumentation and instead arrives at conclusions that reveal quite conclusively how little he understands Darwin's thought.

Ghiselin, myself, and many of the recent Darwin scholars have emphasized what a revolutionary innovator in philosophy Darwin was. Without referring to any of that literature, Greene treats Darwin as a man hopelessly entangled in the now obsolete concepts of the Galilean-Cartesian philosophy that everything in nature is due to matter in motion. He altogether fails to recognize that Darwin has liberated us from that philosophy which is so totally unsuitable for biology. Darwin's emphasis on variation, populations, chance, and pluralism started a new era in the philosophy of nature, an insight that can no longer be ignored even though there are still some philosophers who only read each other's writings or the literature of the physical sciences. How little Greene understands Darwin's thought is well documented by the fact that he always brackets him together with Spencer, two writers whose ideas had remarkably little in common, as has been demonstrated by recent writers.

Even though Darwin, in the fashion of his period, talked a good deal about laws, it is quite evident, and even Greene recognizes this, that they were not the God-given laws of the deists but rather simple facts or what Greene calls processes. Since Darwin did not believe in fixed laws, natural selection for him was a statistical process. This went over very poorly with the deterministic physicists. To describe Darwin's entirely new way of interpreting nature as an adherence to the Cartesian matter-in-motion principle is totally misleading.

Greene seems to think that he can document Darwin's uncertainties and inconsistencies particularly well by his treatment of progress. When dealing with progress in the world of life, one must make a clear distinction between progression in the complexity of types of organisms in the long geological history of the earth, and a teleological interpretation of the causes of such progress. Through the researches of the last 25 years, we now know that life on earth after its origin about 3.5 billion years ago remained extremely simple for about 2 billion years, and that such evolutionary innovations as warm-bloodedness and highly organized central nervous systems are a product of only the last couple of hundred million years. The term *evolutionary progress* is highly inappropriate when ascribed to teleological or finalistic forces. But both Darwin (Essay 14) and Julian Huxley rejected such a usage. Yet to designate as progress the series of changes from the simplest prokaryote to a large angiosperm tree or a primate is descriptively legitimate. How else could we designate the successive, innovative acquisitions of photosynthesis, eukaryoty (development of a nucleus), multicellularity (metaphytes, metazoans), diploidy, homeothermy, central nervous systems, and parental care (Mayr 1982:532)? Greene finds it impossible to reconcile this evolutionary progression "with the mechanistic view of nature." Yet, as Darwin said so clearly, the combined forces of competition and natural selection leave no other alternative but either extinction or evolutionary progression. The analogy between evolution and industrial developments is quite legitimate. Why are modern motor cars so strikingly better than those of 75 years ago? Because all manufacturers constantly experimented with various innovations, while competition through customer demands led to enormous selection pressure. Neither in the automobile industry nor in the world of life do we find any finalistic forces at work, nor any mechanistic determinism. Hence when J. Huxley described progressive evolution in figures of speech borrowed from the progress of human technology, he did not in the least fall "into an implicit vitalism and teleology and undermined the idea that human beings are part of nature," as Greene accuses him.

How little Greene understands variational evolution is documented by his statement that there is a conflict between "competitive struggle as the sine qua non of progress in nature and history and admiring the wonderful adaptedness and interdependence of organic beings." But that is of course exactly the core concept of natural selection. Of course an essentialist— and Greene writes like an essentialist—has trouble understanding what consequences the selective survival of certain unique individuals in a large population will have on the genetic endowment of future generations. Incidentally, competition was not quite as strictly a British concept as claimed by Greene. It was well appreciated on the Continent, as documented in the writings of Herder and de Candolle. These authors, however, being essentialists, were unable to use competition as a component in a theory of natural selection.

Greene, the historian, is very much of an externalist, and seems inclined to ascribe all changes in scientific theory to ideological forces. When evolutionists (during the evolutionary synthesis) rejected orthogenesis and neo-Lamarckism, they did so, thinks Greene, because these theories were considered "metaphysical." A closer look at the literature, however, shows rather clearly that these theories were rejected for three reasons: first, numerous facts were in clear violation of these theories; second, all efforts to find biological mechanisms that would make orthogenesis or an inheritance of acquired characters possible were unsuccessful; and third, all the relevant facts could be explained quite readily by natural selection. I found no evidence in the literature that these hypotheses were rejected as being metaphysical. They simply did not stand up against a proper scientific analysis.

Greene's failure to understand Darwinian thought is well illustrated by his detailed analysis of the thought of Julian Huxley. In fact he devotes nearly 30 percent of his essay to an analysis of these views, particularly as presented in *Evolution: The Modern Synthesis* (1942). Actually, although Huxley's views upheld the main thesis of Darwin, they were not at all typical of those of the architects of the synthesis. Huxley's strong support of evolutionary progress, his definition of progress by criteria that would establish man as the highest, best-adapted organism, his claim that evolution at the species level was ultimately irrelevant as far as evolutionary trends are concerned, and several other claims he made were criticized by Simpson, Dobzhansky, Mayr, and other evolutionists long before Greene. In order to evaluate J. Huxley, one must remember that his background was in experimental biology and that he never worked in population biology or systematics. What he said on these subjects was derived from the literature and did not always reflect the best contemporary thinking. It must also be remembered that, like his grandfather, Huxley excelled in popular presentation and that he lectured far and wide to the general public. His concern for these audiences clearly colored his writings.

In many of his objections against Darwinism, Greene argues like a teleologist, with this word being used in its classical sense, in the same sense as it was by Darwin's adversaries Sedgwick and K. E. von Baer. Indeed his teleology seems to have the same ideological basis. From all of his writings it is evident that Greene is a devout Christian. Apparently he cannot adopt Darwinism because he sees God's hand in everything in nature. The evolutionary progression from the simplest prokaryotes to man is for him clear evidence of the workings of the mind of the Creator. Any attempt at a purely materialistic explanation would cause an insoluble conflict.

Evidently Greene is unable to see any difference between teleonomic and teleological. That organisms are entirely different systems from inanimate matter is something Greene evidently fails to understand. The fact that the genetic program is coded information, that it contains instructions, and that it behaves in most ways very much like a program in a computer somehow seems to make no sense to Greene. That it was convenient for the evolutionist to take over some of the computer terminology in view of the far-reaching equivalence of the two systems is for Greene simply a "highly anthropomorphic projecting onto a nonhuman process the technological aims and terminology of human engineering." Aristotle more than 2,000 years ago already understood remarkably well that a program of instruction is needed for the development of an egg, as recently shown by Aristotle scholars, and molecular genetics has discovered the nature of this program. To do away with these developments as "anthropomorphism" is a remarkably ingenuous solution.

How simplistic Greene's understanding of evolution is is well expressed by his question: "How, then, was he [J. Huxley] to avoid the scylla of finalism and vitalism without steering into the Charybdis of a mechanical determinism that reduced biology to physics and chemistry?" As if Darwin had not found in natural selection exactly the way by which the stated dilemma can be avoided.

In a curious argument which, frankly, I was quite unable to follow, Greene accuses the Darwinians of having brought telos into the world and to have placed "anthropocentrism once more in the driver's seat." He ridicules the fear expressed by J. Huxley that "the fate not only of mankind but also of life on the planet Earth . . . was now seen to depend . . . on human decisions for good or ill." Has Greene never heard of the heartrending destruction of the tropical forests and the inevitable extermination of quite literally millions of species of animals and plants, has he never heard of the thousands of sterile acid lakes, or of the dying of a large part of the European forests, or of the desertification of the Sahel and other savannah regions aggravated by over-grazing, of the greenhouse effect, not to mention the threatening nuclear winter? It is a real mystery to me how anyone in this day and age can still ignore the fatal impact of human decisions on the life occupying the planet Earth.

Philosophers of science have always emphasized that all major research traditions (Kuhn's paradigms) are a mixture of old and new, with inconsistencies and even outright contradictions as well as unopened black boxes. Darwin's paradigm was no exception, and evolutionary biologists have worked for the last 125 years to explore the black boxes and to remove inconsistencies. Would it be justified to claim that this revisionary process has led to a refutation of Darwinism, or is it rather true that it merely resulted in its clarification and purification? Greene insists on the former, we evolutionists on the second alternative. The only way to decide who is right is to look at the set of theories of which Darwin's paradigm is composed, and see whether or not they are still considered valid.

Was Darwin right about proclaiming evolution? Certainly, except that he called it a theory, while the modern biologist has such overwhelming evidence for evolution that he simply considers it a fact, as much of a fact as that the earth moves around the sun rather than the reverse.

Was Darwin right about common descent? Certainly. The last link in the chain of evidence was the demonstration by molecular biology that all organisms have the same genetic code. There is a historical unity in the entire living world which cannot help but have a deep meaning for any thinking person and for his feeling toward fellow organisms.

Was Darwin right about the gradualness of evolution? Yes, provided gradualness is properly defined. Darwin opposed typological saltation as well as any special creation by the "introduction" of new species in the form of single individuals. We now understand that evolution is a populational process, consisting of the gradual—slow or rapid—genetic reordering of populations. To be sure there is also polyploidy, a process able to produce new species instantaneously, primarily among plants. But this has not led to any macroevolutionary consequences different from populational evolution. All other speciation is populational, even in the theory of punctuated equilibria.

Finally, what about natural selection? Darwin realized that selection could not work unless it had unlimited variation at its disposal. But he

had no idea where this variation came from. So he thought that use and
disuse, and other forms of inheritance of acquired characters, might contribute to the production of variability. We now know that in this he was
wrong. He was also wrong in thinking that at least some inheritance was
"blending," that is, that it would lead to a complete fusion of paternal
and maternal characters. Otherwise, Darwin was remarkably astute. He
clearly saw (better than Wallace and most other contemporaries) that there
were two kinds of selection, one for general viability leading to the
maintenance or improvement of adaptedness, and this he called natural
selection, and another that leads to greater reproductive success, and this
he called sexual selection.

Where we differ from Darwin is almost entirely on matters of emphasis.
Darwin was fully aware of the probabilistic nature of selection, but the
modern evolutionist emphasizes this even more. Chance events play an
important role in evolution as it is seen by modern evolutionists. Nor
would we be prepared to say that "selection can do anything" (of course
Darwin never said this either). On the contrary, there are numerous
constraints on selection. And appallingly often, selection is for various
reasons unable to prevent extinction.

How different is this interpretation of the evolutionist from Greene's
ultimate conclusions? They are embodied in such sentences as "the concept
of nature presupposed by Darwin and his contemporaries was disintegrat-
ing," or "the ideas of nature and science Darwin and his contemporaries
took for granted are no longer viable"; in other words, Darwin is dead!

Actually, the basic Darwinian paradigm is as well and as alive today as
it was in Darwin's day. Some of Darwin's peripheral ideas, such as an
inheritance of acquired characters and blending inheritance, had to be
discarded. But this actually only strengthens his theory. All of Darwin's
more basic principles are far more firmly established today than they were
in Darwin's lifetime.

Greene might answer that he is not interested in technical details of
evolutionary biology, but in the more basic conceptual framework that
controlled Darwin's thinking. But here also the philosophical revolution
brought about by Darwin is more firmly established than ever. What
Darwin rebelled against were a number of dominant beliefs of his day.
One was the assumption of natural theology that the world had been
designed by the Creator and that everything in the world of life was the
result of the wise and benign thought of the Creator. How decisively
Darwin emancipated himself from this belief of his youth has been excellently described by Gillespie (1979). In connection with this, Darwin

once and for all refuted another dominant belief of his period, that there is an immanent teleology in this world that will lead to ultimate perfection of whatever telos the Creator had in mind. Darwin eliminated any reliance on supernaturalism and provided the explanatory models that made this possible. Equally important was his refutation of essentialism and its replacement by population thinking. It established a new emphasis on variation, on a potential for change, and on the uniqueness of individuals. It was this population thinking that made the theory of natural selection possible. Philosophers have not yet quite caught up with all the consequences of these revolutionary new ideas.

Contrary to Greene's assertions, these most basic ideas are not only still alive but they are infinitely better established than they were in Darwin's own day. One hundred and twenty-five years of unsuccessful refutations have resulted in an immense strengthening of Darwinism. Whatever attacks on Darwinism are made in our age are made by outsiders, jurists, journalists, and so on. The controversies within evolutionary biology about such matters as the occurrence of sympatric speciation, the existence or not of cohesive domains within the genotype, the relative frequency of complete stasis in species, the rate of speciation, and whatever other arguments there are these days all take place within the framework of Darwinism. The claims of certain outsiders that Darwinism is in the process of being refuted are entirely based on ignorance. To repeat, the basic Darwinian principles are more firmly established than ever. Paraphrasing Mark Twain, we are justified in saying that the news of "the death of Darwin" is greatly exaggerated.

NOTE

This essay first appeared in *Revue de Synthèse* 4, no. 3(1986):229–235.

REFERENCES

Gillespie, N. C. 1979. *Charles Darwin and the Problem of Creation*. Chicago: Chicago University Press.
Greene, J. C. 1959. *The Death of Adam: Evolution and Its Impact on Western Thought*. Iowa City: University of Iowa Press.
———— 1986. The history of ideas revisited. *Revue de Synthèse* 4(3):201–227.
Huxley, J. S. 1942. *Evolution: The Modern Synthesis*. London: Allen and Unwin.
Mayr, E. 1982. *The Growth of Biological Thought*. Cambridge, Mass.: Harvard University Press.

part five ❧

D I V E R S I T Y

Introduction

The production of diversity and the achievement of adaptation are the two major processes of evolution. Darwin at first was particularly concerned with the problem of the origin of diversity, that is, of new species. Later it became adaptation and its causation that concerned him most. The latter interest also dominated the attention of most evolutionary geneticists in the first third of the twentieth century. Neither Fisher nor Haldane nor Wright had any particular interest in speciation. It was also ignored in the then traditional definition of evolution as "a change of gene frequencies in populations." The exception was Dobzhansky, whose *Genetics and the Origin of Species* (1937) returned to a more balanced view of evolution. My *Systematics and the Origin of Species* (1942) was an endeavor to point out the important role of all aspects of diversity in the process of evolution, from speciation to the origin of higher taxa and evolutionary novelties. And this interest in diversity has stayed with me (see References).

The essays in Part Five address three entirely different areas of systematics. In Essay 16 I try to show what kind of theory of classification one might develop by combining the best elements of several currently feuding schools of taxonomy. At the present time we have not yet reached a true synthesis in the field of biological classification. The source of the major disagreement is whether or not recognized taxa should be reasonably homogeneous, and—correlated with this point—whether or not anagenesis (divergent evolution) should be considered in the ranking of taxa, or whether the delimitation of taxa should be determined entirely by cladogenesis, the splitting of phyletic lineages. In 1981 I was not yet aware to what extent anagensis (distance) is considered in some of the numerical methods that arose from cladistics.

Essay 17 is an address I gave in 1973 on the occasion of the opening of the laboratory wing of the Museum of Comparative Zoology at Harvard

University. When I became director of the MCZ I made it my foremost task to develop a laboratory wing and to raise the needed funds to build it. Although I completed this portion of the project, the actual building took place after my retirement, during the tenure of my successor, Professor Alfred W. Crompton. He kindly invited me to address the attending guests on the occasion of the opening ceremonies, and I used this opportunity to develop my ideas on the role of the modern natural history museum.

Systematics, the science dealing with organic diversity, has long suffered from an outdated image. Many of those only superficially acquainted with the field have failed to learn that modern systematics is something entirely different from the purely descriptive Linnaean taxonomy. Modern systematics is inseparable from evolutionary, ecological, and behavioral biology. It covers also all the research directed toward answering "Why?" questions relating to diversity. And because such researches often lead to experimental investigations which can be as much a part of systematics as description and observation are, laboratory facilities are an appropriate, indeed indispensable, necessity in modern systematics.

In Essay 18 I analyze developments in the classification of that group of organisms, birds, whose knowledge has achieved the greatest maturity. Already 50 years ago all but less than 3 percent of all species of birds had been discovered, and an endeavor had been made to assign all allopatric populations to polytypic species. How were these advances in microtaxonomy achieved, and what remains to be done? As advanced as avian taxonomy was at the species level, until recently it has been backward at the level of macrotaxonomy, the classification of the higher taxa. In 1978, when this survey was written, only the first results of Sibley's DNA hybridization work were available. No one could have foreseen at that time the degree to which this work would revolutionize avian classification (Sibley and Ahlquist 1983). I have deliberately left the wording as I wrote it in 1978, in order to highlight the change.

Anyone who studies the literature by systematics published in the last 25 years will be impressed by the vigor of the research and the controversies. The study of diversity is perhaps at the present time a more lively branch of biology than it has ever been. This is due not only to the conceptual turmoil of competing philosophies of classification but also to the opening up of a whole new realm of taxonomic characters provided by molecular biology. It would be misleading to say that the traditional morphological characters have been replaced by molecular characters, for the morphological characters are as significant as ever, particularly when combined with ecological studies. Furthermore, our understanding of the information content of different kinds of molecules is still rather rudimentary. Some molecules—for instance, hemoglobin—are rather susceptible to selection pressures and are, therefore, poor indicators of relationship. Others seem to change in evolution almost with the regularity of a clock, hence the popular term "molecular clock."

Clearly the classifier should not rely entirely on a single molecule or a single method. In particular, when morphology and a single molecular method lead to different answers, additional methods will have to be employed. In view of the strong interest that molecular biologists have developed in reconstructing phylogenetic pathways, greatly increased activity in the study of organic diversity is to be expected. This is true even for the most strictly descriptive level, considering that perhaps less than 10 percent of all species of organisms have so far been described.

REFERENCES

Dobzhansky, Th. 1937. *Genetics and the Origin of Species*. New York: Columbia University Press.

Mayr, E. 1969. *Principles of Systematic Zoology*. New York: McGraw-Hill.

———— 1976. *Evolution and the Diversity of Life*. Cambridge, Mass.: Harvard University Press.

———— 1982. *The Growth of Biological Thought*. Cambridge, Mass.: Harvard University Press.

Mayr, E., E. G. Linsley, and R. L. Usinger. 1953. *Methods and Principles of Systematic Zoology*. New York: McGraw-Hill.

Sibley, C. G., and J. E. Ahlquist. 1983. Phylogeny and classification of birds based on the data of DNA-DNA hybridization. *Current Orn.* 1:245–292.

TOWARD A SYNTHESIS IN BIOLOGICAL CLASSIFICATION

FOR nearly a century after the publication of Darwin's *Origin,*[1] no well-defined schools of classifiers were recognizable. There were no competing methodologies. Taxonomists were unanimous in their endeavor to establish classifications that would reflect "degree of relationship." What differences there were among competing classifications concerned the number and kinds of characters that were used, whether or not an author accepted the principle of recapitulation, whether he attempted to "base his classification on phylogeny," and to what extent he used the fossil record.[2] As a result of a lack of methodology, radically different classifications were sometimes proposed for the same group of organisms; also new classifications were introduced without any adequate justification except for the claim that they were "better." Dissatisfaction with such arbitrariness and seeming absence of any carefully thought out methodology led in the 1950s and 1960s to the establishment of two new schools of taxonomy, numerical phenetics and cladistics, and to a more explicit articulation of Darwin's methodology, now referred to as evolutionary classification.

The Major Schools of Taxonomy

NUMERICAL PHENETICS

From the earliest preliterary days, organisms were grouped into classes by their outward appearance, into grasses, birds, butterflies, snails, and others. Such grouping "by inspection" is the expressly stated or unspoken starting point of virtually all systems of classification. Any classification incorporating the method of grouping taxa by similarity is, to that extent, phenetic.

In the 1950s to 1960s several investigators went one step further and suggested that classifications be based exclusively on "overall similarity." They also proposed, in order to make the method more objective, that every character be given equal weight, even though this would require the use of large numbers of characters (preferably well over a hundred). In order to reduce the values of so many characters to a single measure of "overall similarity," each character is to be recorded in numerical form. Finally, the clustering of species and their taxonomic distance from each other is to be calculated by the use of algorithms that operationally manipulate characters in certain ways, usually with the help of computers. The resulting diagram of relationship is called a *phenogram*. The calculated phenetic distances can be converted directly into a classification.

The fullest statement of this methodology and its underlying conceptualization was provided by Sokal and Sneath.[3] They called their approach "numerical taxonomy," a somewhat misleading designation, since numerical methods, including numerical weighting, can be and have been applied to entirely different approaches to classification. The term *numerical phenetics* is now usually applied to this school. This has introduced some ambiguity since some authors have used the term phenetic broadly, applying it to any approach making use of the "similarity" of species and other taxa, while to the strict numerical pheneticists the term phenetic means the "theory-free" use of unweighted characters.

CLADISTICS (OR CLADISM)
This method of classification,[4] the first comprehensive statement of which was published in 1950 by Hennig,[5] bases classifications exclusively on genealogy, that is, on the branching pattern of phylogeny. For the cladist, phylogeny consists of a sequence of dichotomies,[6] each representing the splitting of a parental species into two daughter species; the ancestral species ceases to exist at the time of the dichotomy; sister groups must be given the same categorical rank; and the ancestral species together with all of its descendants must be included in a single "holophyletic" taxon.

EVOLUTIONARY CLASSIFICATION
Phenetics and cladistics were proposed as replacements for the methodology of classification that had prevailed ever since Darwin and that was variously designated as the "traditional" or the "evolutionary" school. It based its classifications on observed similarities and differences among groups of organisms, evaluated in the light of their inferred evolutionary

history.[7] The evolutionary school includes in the analysis all available attributes of these organisms, their correlations, ecological stations, and patterns of distributions and attempts to reflect both of the major evolutionary processes, branching and the subsequent diverging of the branches (clades). This school follows Darwin (and agrees in this point with the cladists) that classification must be based on genealogy and also agrees with Darwin (in contrast to the cladists) "that genealogy by itself does not give classification."[8]

The results of the evolutionary analysis are incorporated in a diagram, called a *phylogram,* which records both the branching points and the degrees of subsequent divergence. The method of inferring genealogical relationship with the help of taxonomic characters, as it was first carried out by Darwin, is an application of the hypothetico-deductive approach. Presumed relationships have to be tested again and again with the help of new characters, and the new evidence frequently leads to a revision of the inferences on relationship. This method is not circular,[9] as has sometimes been suggested.

Is There a Best Way to Classify?

Each of the three approaches to classification—phenetics, cladistics, and evolutionary classification—has virtues and weaknesses. The ideal classification would be one that would best meet as many as possible of the generally acknowledged objectives of a classification.

A biological classification, like any other, must serve as the basis of a convenient information storage and retrieval system. Since all three theories produce hierarchical systems, containing nested sets of subordinated taxa, they permit the following of information up and down the phyletic tree. But this is where the agreement among the three methods ends. Purely phenetic systems, derived from a single set of arbitrarily chosen characters, sometimes provide only low retrieval capacity as soon as other sets of characters are used. The effectiveness of the phenetic method could be improved by careful choice of selected characters. However, the method would then no longer be "automatic," because any selection of characters amounts to weighting.

Cladists use only as much information for the construction of the classification as is contained in the cladogram. They convert cladograms, quite unaltered, into classifications, only when the cladograms are strictly dichotomous. Even though cladists lose much information by this simplistic approach, the information on lines of descent can be read off their classifications directly. However, a neglect of all ancestral-descendant in-formation reduces the heuristic value of their classifications. By contrast, since evolutionary taxonomists incorporate a great deal more information in their classifications than do the cladists, they cannot express all of it directly in the names and ranking of taxa in their classifications. Therefore, they consider a classification simply to be an ordered index that refers them to the information that is stored elsewhere (in the detailed taxonomic treatments).

A far more important function of a classification, even though largely compatible with the informational one, is that it establishes groupings about which generalizations can be made. To the extent that classifications are explicitly based on the theory of common descent with modification, they postulate that members of a taxon share a common heritage and thus will have many characteristics in common. Such classifications, therefore, have great heuristic value in all comparative studies. The validity of specific observations can be generalized by testing them against other taxa in the system or against other kinds of characters.[10-12]

Pheneticists, as well as cladists, have claimed that their methods of constructing classifications are nonarbitrary, automatic, and repeatable. The criticisms of these methods over the last 15 years[13] have shown, however, that these claims cannot be substantiated. It is becoming in-creasingly evident that a one-sided methodology cannot achieve all the above-listed objectives of a good classification.

The silent assumption in the methodologies of phenetics and cladistics is that classification is essentially a single-step procedure: clustering by similarity in phenetics, and establishment of branching patterns in clad-istics. Actually a classification follows a sequence of steps, and different methods and concepts are pertinent at each of the consecutive steps. It seems to me that we might arrive at a less vulnerable methodology by developing the best method for each step consecutively. Perhaps the steps could eventually be combined in a single algorithm. In the meantime, their separate discussion contributes to the clarification of the various aspects of the classifying process.

Establishment of Similarity Classes

The first step is the grouping of species and genera by "inspection," that is, by a phenetic procedure. (I use phenetic in the broadest sense, not in the narrow one of numerical phenetics.) All of classifying consists of, or at least begins with, the establishment of similarity classes, such as a preliminary grouping of plants into trees, shrubs, herbs, and grasses. The reason why the method is so often successful is simply that—other things being equal—descendants of a common ancestor tend to be more similar to each other than they are to species that do not share immediate common descent. The method is thus excellent in principle. Numerical phenetics has nevertheless proved to be largely unsuccessful because (1) claims, such as "results objective and strictly repeatable," were not always justifiable since in practice different results are obtained when different characters are chosen or different programs of computation are used; (2) the method was inconsistent in its claim of objectivity since subjective biological criteria were used in the assigning of variants (for example, sexes, age classes, and morphs) to "operational taxonomic units" (OTUs); and, most importantly, the method insisted on the equal weighting of all characters.

It is now evident that no computing method exists that can determine "true similarity" from a set of arbitrarily chosen characters. So-called similarity is a complex phenomenon that is not necessarily closely correlated with common descent, since similarity is often due to convergence. Most major improvements in plant and animal classifications have been due to the discovery of such convergence.[14]

Different types of characters—morphological characters, chromosomal differences, enzyme genes, regulatory genes, and DNA matching—may lead to rather different grouping. Different stages in the life cycle may result in different groupings.

The ideal of phenetics has always been to discover a measure of total (overall) similarity. Since it is now evident that this cannot be achieved on the basis of a set of arbitrarily chosen characters, the question has been asked whether there is not a method to measure degrees of difference of the genotype as a whole. Improvements in the method of DNA hybridization offer hope that this method might give realistic classifications on a phenetic basis, at least up to the level of orders.[15] The larger the fraction of the nonhybridizing DNA, the less reliable this method is, because it cannot be determined whether the nonmatching DNA is only slightly or drastically different.

Testing the Naturalness of Taxa

In the first step of the classifying procedure, clusters of species were assembled that seemed to be more similar to each other than to species in other clusters. These clusters are the taxa we recognize tentatively.[16] In order to make these clusters conform to evolutionary theory, two, operationally more or less inseparable tests, must be made: (1) determine for all species of a cluster (taxon) whether they are descendants of the nearest common ancestor and (2) connect the taxa by a branching tree of common descent, that is, construct a cladogram. An indispensable preliminary of this testing is an analysis of the characters used to establish the similarity clusters.

CHARACTER ANALYSIS

A careful analysis shows almost invariably that some characters are better clues to relationship (have greater weight) than others. The fewer the number of available characters, the more carefully the weighting must be done. This weighting is one of the most controversial aspects of the classifying procedure. Investigators who come to systematics from the outside, say from mathematics, or who are beginners tend to demand objective or quantitative methods of weighting. There are such methods, principally ones based on the covariation of characters, but they are not nearly as informative as methods based on the biological evaluation of characters.[17] But such an evaluation requires an understanding of many aspects of the to-be-classified group (that is, its life history, the inferred selection pressures to which it is exposed, and its evolutionary history) that may not be available to an outsider. This creates a genuine dilemma. If strictly taxometric methods were available that would produce satisfactory weighting, everyone would surely prefer them to weighting based on experience and biological knowledge. But so far such methods are still in their infancy.

The greatest difficulty for a purely phenetic method, indeed for any method of classification, is the discordance (noncongruence) of different sets of characters. Entirely different classifications may result from the use of characters of different stages of the life cycle—for instance, larval versus adult characters. In a study of species of bees, Michener[18] obtained four different classifications when he sorted them into similarity classes on the basis of the characters of (1) larvae, (2) pupae, (3) the external morphology of the adults, and (4) male genitalic structures. Phenetic delimitation of taxa unavoidably necessitates a great deal of decision-making on the use and weighting of characters. Often, when new sets of characters become available, their use may lead to a new delimitation of taxa or to a change in ranking.

Each group (taxon) tentatively established by the phenetic method is, so to speak, a hypothesis as to common descent, the validity of which must be tested. Is the delimited taxon truly monophyletic?[19] Are the species included in this taxon nearest relatives (descendants of the nearest common ancestor)? Have all species been excluded that are only superficially or convergently similar?

Methods to answer these questions have been in use since the days of Darwin, particularly the testing of the homology of critical characteristics of the included species. However, Hennig was the first to articulate such methods explicitly, and these have been modified by some of his followers. These methods can be designated as the cladistic analysis.

Such an analysis involves first the partitioning of the joint characters of a group into ancestral ("plesiomorph" in Hennig's terminology) characters and derived ("apomorph") characters, that is, characters restricted to the descendants of the putative nearest common ancestor.[20] The joint possession of homologous derived characters proves the common ancestry of a given set of species. A character is derived in relation to the ancestral condition of the character. The end product of such a cladistic character analysis is a cladogram, that is, a diagram (dendrogram) of the branching points of the phylogeny.

Although this procedure sounds simple, numerous practical difficulties have been pointed out.[21-22] Very often the branching points are inferred by way of single or very few characters and are affected by all the weaknesses of single character classifications. More serious are two other difficulties.

(1) Polarity. A derived character is often simpler or less specialized than the ancestral condition. For this reason it can be difficult to determine polarity in a transformation series of characters, that is, to determine which end of the series is ancestral. Tattersall and Eldredge[23] stressed that "in practice it is hard, even impossible, to marshall a strong, logical argument for a given polarity for many characters in a given group." Are they primitive (ancestral) or derived? Much of the controversy concerning the phylogeny of the invertebrates, for instance, is due to differences of opinion concerning polarity. Hennig tried to elaborate methods for determining polarity but, as others have shown,[24-25] with rather indifferent success.

Since characters come and go in phyletic lines and since there is much convergence, the problem of polarity can rarely be solved unequivocally. There are three best types of evidence for polarity reconstruction. First is the fossil record. Although primitiveness and apparent ancientness are not correlated in every case, nevertheless as Simpson[26] stressed, "For any group with even a fair fossil record there is seldom any doubt that characters usual or shared by older members are almost always more primitive than those of later members." Second is sequential constraints. Consecutive chromosomal inversions (as in *Drosophila*) or sets of amino acid replacements (and presumably certain other molecular events) form definite sequences. Which end of the sequence is the beginning can usually not be read off from the sequence itself, but additional information (polarity of other character chains, geographical distribution, and the like) often permits an unequivocal determination of the polarity. Third is the reconstruction of the presumed evolutionary pathway. This can sometimes be done by studying evidence for adaptive shifts, the invasion of new competitors or the extinction of old ones, the behavior of correlated characters, and other biological evidence. Particular difficulties are posed when the polarity is reversed in the course of evolution, as documented in the fossil record.

(2) *Kinds of derived characters.*   Two taxa may resemble each other in a given character for one of three reasons: because the character existed already in the ancestry of the two groups before the evolution of the nearest common ancestor (symplesiomorphy in Hennig's terminology), because it originated in the common ancestor and is shared by all of his descendants (homologous apomorphy or synapomorphy), or because it originated independently by convergence in several descendant groups (nonhomologous or convergent apomorphy).[27] Since, according to the cladistic method, sister groups are recognized by the possession of synapomorphies, convergence poses a major problem. How are we to distinguish between homologeous and convergent apomorphies? Hennig was fully aware of the critical importance of this problem, but it has been quietly ignored by many of his followers. Both grebes and loons, two orders of diving birds, have a prominent spur on the knee and were therefore called sister groups by one cladist. However, other anatomical and biochemical differences between the two taxa indicate that the shared derived feature was acquired by convergence. The reliability of the determination of monophyly of a group depends to a large extent on the care that is taken in discriminating between these two classes of shared apomorphy.

There is a third class of derived characters, so-called autapomorphies, which are characters that were acquired by and are restricted to a phyletic line after it branched off from its sister group.

The pheneticists do not undertake a character analysis. Cladists and evolutionary taxonomists agree with each other in principle on the importance of a careful character analysis. They disagree, however, fundamentally in how to use the findings of the character analysis in the construction of classifications, particularly the ranking procedure.

The Construction of a Classification

CLADISTIC CLASSIFICATION

Cladists convert the cladogram directly into a cladistic classification. In such a classification taxa are delimited exclusively by holophyly, that is, by the possession of a common ancestor, rather than by a combination of genealogy and degree of divergence.[19] This results in such incongruous combinations as a taxon containing only crocodiles and birds, or one containing only lice and one family of Mallophaga.

Taxa based exclusively on genealogy are of limited use in most biological comparisons. Since, as Hull[28] pointed out, cladists really classify characters rather than organisms, they have to make the arbitrary assumption that new apomorph characters originate whenever a line branches from its sister line. This is unlikely in most cases. Surely the reptilian species that originated the avian lineage lacked any of the flight specializations characteristic of modern birds, except perhaps the feathers.[29]

Two principles govern the conversion of a cladogram into a cladistic classification: (1) all branchings are bifurcations that give rise to two sister groups, and (2) branchings are usually connected with a change in categorical rank. Cladistic classifications are only representations of branching patterns, with complete disregard of evolutionary divergence, ancestor-descendant relationships, and the information content of autapomorph characters. Because these aspects of evolutionary change are neglected, the cladistic method of classification "either results in lumping very similar forms (parasites and their relatives) or in recognizing a multitude of taxa (perhaps also of other categories) regardless of the extreme similarity of some of them. Such simplistic procedures do violence to most biological attributes other than the pattern of the cladistic branching system, as well as to the function of a classification for convenient information transmittal and storage," as Michener remarked.[18]

These objections show that the methodology of cladistic classification is not satisfactory. Anyone familiar with the history of taxonomy is strangely reminded of the principles of Aristotelian logical division when encountering cladistic classifications with their rigid dichotomies, the mandate that every taxon must have a sister group, and the principle of a straight-line hierarchy.

There has been much argument over the relationship between classification and phylogeny.[30] Both cladists and evolutionary taxonomists agree that all members of a taxon must have a common ancestor. A phylogenetic analysis, and in particular a clear separation of homologous apomorphies from convergences, is a necessary component of the classifying procedure. Classificatory analysis often leads to new inferences on phylogeny, and new insights on phylogeny may necessitate changes in classification. These interactions are not in the least circular.

It is quite unnecessary in most cases to know the exact species that was the common ancestor of two diverging phyletic lines. An inability to specify such an ancestral species has rarely impeded paleontological research.[31-32] For instance, it is of little importance whether *Archaeopteryx* was the first real ancestor of modern birds or some other similar species or genus. What is important to know is whether birds evolved from lizard-like, crocodile-like, or dinosaur-like ancestors. If a reasonably good fossil record is available, it is usually possible, by the backward tracing of evolutionary trends and by the backward projection of divergent phyletic lines, to reconstruct a reasonably convincing facsimile of the representative of a phyletic line at an earlier time.

Simpson has provided us with cogent arguments about why it is not permissible to reject information from the fossil record under the pretext that it fails to give their phylogenetic connections between fossil and recent taxa with absolute certainty. Hence, there is no merit in the suggestion to construct separate classifications for recent and for fossil organisms. After all, fossil species belong to the same tree of descent as living species. Indeed, enough evidence usually becomes available through a careful character analysis to permit relatively robust inferences on the most probable phylogeny. A number of recent endeavors have been made to develop a cladistic methodology that is quantitative and automatic. New methods in this area are published in rapid succession and it would seem too early to determine which is most successful and freest of possible flaws.[33]

EVOLUTIONARY CLASSIFICATION

The taxonomic task of the cladist is completed with the cladistic character analysis. The genealogy gives him the classification directly, since for him classification is nothing but genealogy. The evolutionary taxonomist carries the analysis one step further. He is interested not only in branching, but, like Darwin, also in the subsequent fate of each branch. In particular, he undertakes a comparative study of the phyletic divergence of all evolutionary lineages, since the evolutionary history of sister groups is often strikingly different. Among two related groups, derived from the same nearest common ancestor, one may hardly differ from the ancestral group, while the other may have entered a new adaptive zone and evolved into a novel type. Even though they are sister groups in the terminology of cladistics, they may deserve different categorical rank, because their biological characteristics differ to such an extent as to affect any comparative study. The importance of this consideration was stated by Darwin (*Origin,* p. 420): "I believe that the *arrangement* of the groups within each class, in due subordination and relation to the other groups, must be strictly genealogical in order to be natural, but that the *amount* of difference in the several branches or groups, though allied in the same degree in blood to their common progenitor, may differ greatly, being due to the different degrees of modification which they have undergone, and this is expressed by the forms being ranked under different genera, families, sections or orders." Darwin refers then to a diagram of three Silurian genera that have modern descendants; one has not even changed generically, but the other two have become distinct orders, one with three and the other with two families.

The question as to what extent an analysis of degrees of divergence is possible, is still debated. The cladist makes only "horizontal" comparisons, cataloging the synapomorphies of sister groups. The evolutionary taxonomist, however, also makes use of derived characters that are restricted to a single line of descent, so-called autapomorph characters (Figure 16.1), which are apomorph characters restricted to a single sister group. The importance of autapomorphy is well illustrated by a comparison of birds with their sister group.[34] Birds originated from that branch of the reptiles, the Archosauria, which also gave rise to the pterodactyls, dinosaurs, and crocodilians. The crocodilians are the sister group of the birds among living organisms; a stem group of archosaurians represents the common ancestry of birds and crocodilians. Although birds and crocodilians share a number of synapomorphies that originated after the archosaurian line had branched off from the other reptilian lines, nevertheless crocodilians are on the whole very similar to other reptiles, that is, they have developed relatively few autapomorph characters. They represent the reptilian "grade," as many morphologists call it. Birds, by contrast, have acquired a vast array of new autapomorph characters in connection with their shift

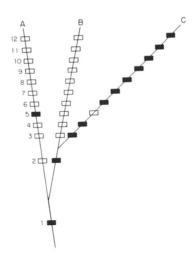

Figure 16.1   Cladogram of taxa A, B, and C. Cladists combine B and C into a single taxon because B and C are holophyletic. Evolutionary taxonomists separate C from A and B, which they combine, because C differs by nine autapomorph characters from A and B and shares only one synapomorph character (2) with B.

to aerial living. Whenever a clade (phyletic lineage) enters a new adaptive zone that leads to a drastic reorganization of the clade, greater taxonomic weight may have to be assigned to the resulting transformation than to the proximity of joint ancestry. The cladist virtually ignores this ecological component of evolution.

The main difference between cladists and evolutionary taxonomists, thus, is in the treatment of autapomorph characters. Instead of automatically giving sister groups the same rank, the evolutionary taxonomist ranks them by considering the relative weight of their autapomorphies as compared to their synapomorphies (Figure 16.1). For instance, one of the striking autapomorphies of man (in comparison with his sister group, the chimpanzee) is the possession of Broca's center in the brain, a character that is closely correlated with man's speaking ability. This single character is for most taxonomists of great weight than various synapomorphous similarities or even identities in man and the apes in certain macromolecules such as hemoglobins and cytochrome *c*. The particular importance of autapomorphies is that they reflect the occupation of new niches and new adaptive zones that may have greater biological significance than synapomorphies in some of the standard macromolecules.

I agree with Szalay[35] when he says: "The loss of biological knowledge when not using a scheme of ancestor-descendant relationship, I believe, is great. In fact,, whereas a sister group relationship may . . . tell us little, a postulated and investigated ancestor-descendant relationship may help explain a previously inexplicable character in terms of its origin and transformation, and subsequently its functional (mechanical) significance." In other words, the analysis of the ancestor-descendant relationships adds a great deal of information that cannot be supplied by the analysis of sister group relationships.

It is sometimes claimed that the analysis of ancestor-descendant relationships lacks the precision of cladistic sister group comparisons. However, as was shown above and as is also emphasized by Hull,[36] the cladistic analysis is actually full of uncertainties. The slight possible loss of precision, caused by the use of autapomorphies, is a minor disadvantage in comparison with the advantage of the large amount of additional information thus made available.

The information on autapomorphies permits the conversion of the cladogram into a phylogram. The phylogram differs from the cladogram by the placement of sister groups at different distances from the joint common ancestry (branching point) and by the expression of degree of divergence by different angles. Both of these topological devices can be translated into the respective categorical ranking of sister groups. These methods[37] generally attempt to discover the shortest possible "tree" that is compatible with the data. Yet, anyone familiar with the frequency of evolutionary reversals and of evolutionary opportunism realizes the improbability of the assumption that the tree constructed by this so-called "parsimony method" corresponds to the actual phylogenetic tree. "To regard [the shortest tree method] as parsimonious completely misconceives the intent and use of parsimony in science."[38]

It is not always immediately evident whether a tree construction algorithm is based on cladistic principles or on the methods of evolutionary classification. If the "special similarity" on which the trees are based are strictly synapomorphies, then the method is cladistic. If autapomorphies are also given strong weight, then the method falls under evolutionary classification.

The particular aspect of the method of evolutionary taxonomy found most unacceptable to cladists is the recognition of "paraphyletic" taxa. A paraphyletic taxon is a holophyletic group from which certain strikingly divergent members have been removed. For instance, the class Reptilia of the standard zoological literature is paraphyletic, because birds and mammals, two strikingly divergent descendants of the same common ancestor of all the Reptilia, are not included. Nevertheless, the traditional class Reptilia is monophyletic, because it consists exclusively of descendants from the common ancestor, even though it excludes birds and mammals owing to the high number of autapomorphies of these classes. The recognition of paraphyletic taxa is particularly useful whenever the recognition of definite grades of evolutionary change is important.

The Ranking of Taxa

Once species have been grouped into taxa, the next step in the process of biological classification is the construction of a hierarchy of these taxa, the so-called Linnaean hierarchy. The hierarchy is constructed by assigning a definite rank such as family or order to each taxon, subordinating the lower categories to the higher ones. It is a basic weakness of cladistics that it lacks a sensitive method of ranking and simply gives a new rank after each branching point. The evolutionary taxonomist, following Darwin, ranks taxa by the degree of divergence from the common ancestor, often assigning a different rank to sister groups. Rank determination is one of the most difficult and subjective decision processes in classification. One aspect of evolution that causes difficulties is mosaic evolution.[39] Rates of divergence of different characters are often drastically different. Conventionally taxa, such as those of vertebrates, are described and delimited on the basis of external morphology and of the skeleton, particularly the locomotory system. When other sets of morphological characters are used (for example, sense organs, reproductive system, central nervous system, or chromosomes), the evidence they provide is sometimes conflicting. The situation can become worse if molecular characters are also used. The anthropoid genus *Pan* (chimpanzee), for instance, is very similar to *Homo* in molecular characters, but man differs so much from the anthropoid apes in traditional characters (central nervous system and its capacities) and occupation of a highly distinct adaptive zone that Julian Huxley even proposed to raise him to the rank of a separate kingdom—Psychozoa.

It has been suggested that different classifications should be constructed for each kind of character, or at least for morphological and molecular characters. Yet there is already much evidence that the acceptance of several classifications based on different characters would lead to insurmountable complications. By taking all available data into consideration simultaneously, a classification can usually be constructed that can serve conveniently as an all-purpose classification or, as Hennig called it, "a general reference system."

It is usually possible to derive more than one classification from a phylogram, because higher taxa are usually composed of several end points of the phylogram, and different investigators differ by the degree to which they lump such terminal branches into a single higher taxon.[40] An example is the phylogram of the higher ferns on which, as Wagner[41] has shown, six different classifications have been founded (Figure 16.2) and many more are possible. The extent to which investigators "split" or "lump" higher taxa, thus, is of considerable influence on the classifications they produce.

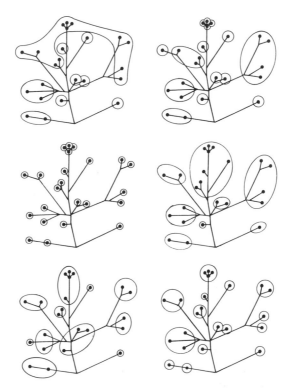

Figure 16.2. Six different possible classifications of ferns, based on the same dendrogram. Each filled circle is a genus, and each open circle is a family. The differences are due to which and how many genera are combined to make up the families. (After W. H. Wagner, "The construction of a classification," in *Systematic Biology*, Publication 1692, National Academy of Sciences, Washington, D.C., 1969, pp. 67–90.)

Comparison of the Three Major Schools

Each school believes that its classification is the "best." Pheneticists as well as cladists claim that their respective methods have also the great merit of giving automatically nonarbitrary results. These claims cannot be substantiated. To be sure, grouping by phenetic characters and determination of holophyly by cladistic analysis are valuable components of the procedure of biological classification. The great deficiency of both phenetics and cladistics is the failure to reflect adequately the past evolutionary history of taxa.

What needs to be emphasized once more is the fact that groups of organisms are the product of evolution and that no classification can hope to be satisfactory that does not take this fact fully into consideration. Both pheneticists and cladists are ambiguous in their attitude toward evolutionary theory. The pheneticists claim that their approach is completely theory-free, but they nevertheless assume that their method will produce a hierarchy of taxa that corresponds to descent with modification. On the basis of this assumption, they also claim to be "evolutionary taxonomists,"[42] but the fact that different phenetic procedures may produce very different classifications and that their procedure is not influenced by evolutionary considerations refutes this assertion. The cladists exclude most of evolutionary theory (for example, inferences on selection pressures, shifts of adaptive zones, evolutionary rates, and rates of evolutionary divergence) from their consideration[43] and tend increasingly not to classify species and taxa, but only taxonomic characters and their origin. The connection with evolutionary principles is exceedingly tenuous in many recent cladistic writings.

By contrast, the evolutionary taxonomists expressly base their classifications on evolutionary theory. They make full use of information on shifts into new adaptive zones and rates of evolutionary change and believe that the resulting classifications are a key to a far richer information content.

Although the three schools still seem rather fundamentally in disagreement, as far as the basic principles of classification are concerned, the more moderate representatives have quietly incorporated some of the criteria of the opposing schools, so that the differences among them have been partially obliterated. For instance, Farris'[44] clustering of special similarities is a phenetic method based on the weighting of characters. The evolutionary school uses phenetic criteria to establish similarity classes and to construct a classification, and cladistic criteria to test the naturalness of taxa. Comparing what McNeill[45] says in favor of phenetics (appropriately modified) and Farris against it, we find that the gap has narrowed. I have no doubt that moderates will be able to develop an eclectic methodology, one that contains a proper balance of phenetics and cladistics that will produce far more "natural classifications" than any one-sided approach that relies exclusively on a single criterion, whether it be overall similarity, parsimony of branching pattern, or what not. Evolutionary taxonomy, from Darwin on, has been characterized by the adoption of an eclectic approach that makes use of similarity, branching pattern, and degree of evolutionary divergence.

Classification and Information Retrieval

Biological classifications have two major objectives: to serve as the basis of biological generalizations in all sorts of comparative studies and to serve as the key to an information storage system. Up to this point, I have concentrated on those aspects of classifying that help to secure a sound basis for generalizations. This leaves unanswered the question of whether achievement of this first objective is, or is not, reconcilable with achievement of the second objective. Is the classification that is soundest as a basis of generalizations also most convenient for information retrieval? This, indeed, seems to have been true in most cases I have encountered. However, we can also look at this problem from another side.

It is possible at nearly each of the three major steps in the making of a classification to make a choice between several alternatives. These choices may be scientifically equivalent, but some may be more convenient in aiding information retrieval than others. If we choose one of them, it is not necessarily because the alternatives were "falsified," but rather because the chosen method is "more practical." In this respect, biological classifications are not unique. Scientific theories are nearly always judged by criteria additional to truth or falsity, for instance, by their simplicity or, in mathematics, by their "elegance." Therefore, it can be asserted that convenience in the use of a classification, including its function as key to information retrieval, is not necessarily in conflict with its more purely scientific objectives.[46-48]

NOTES

This essay first appeared in *Science* 214 (1981): 510–516, under the title "Biological classification: toward a synthesis of opposing methodologies." Copyright © 1981 by the American Association for the Advancement of Science.

1. C. Darwin, *On the Origin of Species* (Murray, London, 1859).

2. For an illuminating survey of the thinking of that period see F. A. Bather [*Proc. Geol. Soc. London* 83 LXII (1927)].

3. R. R. Sokal and P. H. A. Sneath, *Principles of Numerical Taxonomy* (Freeman, San Francisco, 1963). A drastically revised second edition was published in 1973.

4. The method was first published under the misleading name phylogenetic systematics, but since it is based on only a single one (branching) of the various processes of phylogeny, the terms cladism or cladistics have been substituted and are now widely accepted.

5. W. Hennig's original statement is *Grundzüge einer Theorie der Phylogenetischen Systematik* (Deutscher Zentralverlag, Berlin, 1950). A greatly revised second edition (reprinted in 1979) is *Phylogenetic Systematics,* D. D. Davis and R. Zangerl, Eds. (Univ. of Illinois Press, Urbana, 1966); see also W. Hennig (47). An independent phylogenetic analysis of characters was made by T. P. Maslin [*Syst. Zool.* 1, 49 (1952)]. For an overview of the more significant recent literature see D. Hull (36) and J. S. Farris [*Syst. Zool.* 28, 483 (1979)].

6. Some cladists in recent years have relaxed the requirements of strict dichotomy and have permitted tri- and polyfurcations or have quietly abandoned dichotomy by admitting empty internodes in the cladogram. Polyfurcations can be translated into several alternate bifurcations [see J. Felsenstein, *Syst. Zool.* 27, 27 (1978)], and this makes the automatic conversion of the cladogram into a classification of sister groups impossible.

7. The classical statement of this theory is to be found in C. Darwin (*1,* pp. 411–434). G. G. Simpson [*Principles of Animal Taxonomy* (Columbia Univ. Press, New York, 1961)] and E. Mayr (48) provide comprehensive modern presentations of this theory. Several critical recent analyses are: W. Bock (*11*); C. D. Michener (*18*); *Syst. Zool.* 27, 112 (1978); P. D. Ashlock (*12*).

8. F. Darwin, *Life and Letters of Charles Darwin* (Murray, London, 1887), vol. 2, p. 247.

9. D. Hull, *Evolution* 21, 174 (1967); see also W. Bock (*11*).

10. F. E. Warburton, *Syst. Zool.* 16, 241 (1967); W. Bock, *ibid.* 22, 375 (1973).

11. W. Bock, in *Major Patterns in Vertebrate Evolution,* M. K. Hecht, P. C. Goody, B. M. Hecht, Eds. (NATO Advanced Study Institute Series, Plenum, New York, 1977), vol. 14, pp. 851–895.

12. P. D. Ashlock, *Syst. Zool.* 28, 441 (1979).

13. I shall not, at this time, recount the almost interminable controversies among the three schools. For critiques of phenetics see E. Mayr (*48,* pp. 203–211), L. A. S. Johnson [*Syst. Zool.* 19, 203 (1970)], and D. Hull [*Annu. Rev. Ecol. Syst.* 1, 19 (1970)]. Some of the weaknesses pointed out by these early critics have been corrected in the 1973 edition of Sokal and Sneath (3). For critiques of cladistics see E. Mayr (*21*), R. R. Sokal (*22*), G. G. Simpson (*32*), D. Hull (*36*), P. D. Ashlock [*Annu. Rev. Ecol. Syst.* 5, 81 (1974)]; and L. van Valen *49*.

14. A particularly illuminating example is the breaking up of the plant group Amentiferae, which has been shown to consist of taxa secondarily adapted for wind pollination [R. F. Thorne, *Brittonia* 25, 395 (1973)]. Examples among animals of radical reclassifications are the Rodentia, parasitic bees, certain beetle families, and the turbellarians.

15. C. G. Sibley, in preparation.

16. There have been arguments since before the days of Linnaeus about how to determine whether or not a system, a classification, is "natural." William Whewell, at a time before Darwin had proclaimed his theory of common descent, expressed the then prevailing pragmatic consensus, "The maxim by which all systems professing to be natural must be tested is this: that the arrangement obtained from one set of characters coincides with the arrangement obtained from another set" [W. Whewell, *Philos. Inductive Sci.* 1, 521 (1840)]. Interestingly, the covariance of characters is still perhaps the best practical test of the goodness of a classification. Since Darwin, of course, that classification is considered most natural that best reflects the inferred evolutionary history of the organisms involved.

17. For a tabulation and analysis of such qualitative methods of weighting, see E. Mayr (*48*, pp. 220–228).

18. C. D. Michener, *Syst. Zool.* 26, 32 (1977).

19. I use the word monophyletic in its traditional sense, as a qualifying adjective of a taxon. Various definitions of monophyletic have been proposed but all of them for the same concept, a qualifying statement concerning a taxon. A taxon is monophyletic if all of its members are derived from the nearest common ancestor [E. Haeckel, *Natürliche Schöpfungsgeschichte* (Reimer, Berlin, 1868)]. Cladists have attempted to turn the situation upside down by placing all descendants of an ancestor into a taxon. Monophyletic thus becomes a qualifying adjective for descent, and a taxon is not recognized by its characteristics but only by its descent. The transfer of such a well-established term as monophyletic to an entirely different concept is as unscientific and unacceptable as if someone were to "redefine" mass, energy, or gravity by attaching these terms to entirely new concepts. P. D. Ashlock [*Syst. Zool.* 20, 63 (1971)] has proposed the term holophyletic for the assemblage of descendants of a common ancestor. See also P. D. Ashlock (*12*, p. 443).

20. Terms like apomorph, synapomorph, derived, ancestral, and so forth always refer to characters of taxa at all levels. A genus may have synapomorphies with another genus, and so may an order with another order. It is this applicability of the same criteria for taxa of all ranks that permits the construction of the Linnaean hierarchy.

21. E. Mayr, *Z. Zool. Syst. Evolutionsforsch.* 12, 94 (1974); reprinted in E. Mayr, *Evolution and the Diversity of Life* (Harvard Univ. Press, Cambridge, Mass., 1976), pp. 433–478.

22. R. R. Sokal, *Syst. Zool.* 24, 257 (1975).

23. I. Tattersall and N. Eldredge, *Am. Sci.* 65, 204 (1977).

24. D. S. Peters and W. Gutmann, *Z. Zool. Syst. Evolutionsforsch.* 9, 237 (1971).

25. O. Schindewolf, *Acta Biotheor,* 18, 273 (1968); H. K. Erben, *Verh. Dtsch. Zool. Ges.* 79, 116 (1979).

26. G. G. Simpson (*32*); see also L. van Valen (*49*).

27. For a diagram of these three categories of morphological resemblance see figure 1 in W. Hennig (*47*).

28. "Cladistic classifications do not represent the order of branching of sister groups, but the order of emergence of unique derived characters" [see D. Hull (*36*)].

29. G. G. Simpson [*The Major Features of Evolution* (Columbia Univ. Press, New York, 1953), p. 348] discusses the fallacy of the cladist assumption.

30. Phylogeny is equated by cladists with cladogenesis (branching), while the evolutionary taxonomist subsumes both branching and evolutionary divergence (anagenesis) under phylogeny.

31. C. W. Harper, *J. Paleontol.* 50, 180 (1976).

32. G. G. Simpson, in *Phylogeny of the Primates,* W. Pluckett and F. S. Szalay, Eds. (Plenum, New York, 1975), pp. 3–19.

33. J. H. Camin and R. R. Sokal, *Evolution* 19, 311 (1965); W. M. Fitch and E. Margoliash, *Science* 155, 279 (1967); W. M. Fitch, in *Major Patterns of Vertebrate Evolution,* M. K. Hecht, P. C. Goody, B. M. Hecht, Eds. (NATO Advanced Study Institute Series, Plenum, New York, 1977), vol. 14, pp. 169–204.

34. There are literally hundreds of cases to illustrate this situation. I use again the classical case of birds and crocodilians because even a nonbiologist will understand the situation if such familiar animals are used. The holophyletic classification of the lice (Anoplura) derived from one of the suborders of the Mallophaga is another particularly instructive example [K. C. Kim and H. W. Ludwig, *Ann. Entomol. Soc. Am.* 71, 910 (1978)].

35. F. S. Szalay, *Syst. Zool.* 26, 12 (1977).

36. D. Hull, *ibid.* 28, 416 (1979).

37. J. W. Hardin, *Brittonia* 9, 145 (1957); W. H. Wagner, in *Plant Taxonomy: Methods and Principles,* L. Benson, Ed. (Ronald, New York, 1962), pp. 415–417; A. G. Kluge and J. S. Farris, *Syst. Zool.* 18, 1 (1969); J. S. Farris, *Am. Nat.* 106, 645 (1972).

38. L. H. Throckmorton, in *Biosystematics in Agriculture,* J. A. Romberger, Ed. (Wiley, New York, 1978), p. 237. Others who have questioned the validity of the so-called parsimony principle are M. Ghiselin [*Syst. Zool* 15, 214 (1966)] and W. Bock (*11*).

39. Unequal rates of evolution for different structures or for any other components of phenotypes or genotypes are designated mosaic evolution.

40. See E. Mayr (*48*, pp. 238–241) on the differences between splitters and lumpers.

41. W. H. Wagner, "The construction of a classification," in *Systematic Biology*

(Publication 1692, National Academy of Sciences. Washington, D.C., 1969), pp. 67–90.

42. R. R. Sokal in (22), "I have yet to meet a nonevolutionary taxonomist."

43. Several leading cladists have recently published antiselectionist statements.

44. J. S. Farris, in *Major Patterns in Vertebrate Evolution,* M. K. Hecht, P. C. Goody, B. M. Hecht, Eds. (NATO Advanced Study Institute Series, Plenum, New York, 1977), vol. 14, pp. 823–850.

45. J. McNeill, *Syst. Zool.* 28, 468 (1979).

46. For criteria by which to judge the practical usefulness of biological classifications, see E. Mayr (*48,* pp. 229–242).

47. W. Hennig, *Annu. Rev. Entomol.* 10, 97 (1965).

48. E. Mayr, *Principles of Systematic Zoology* (McGraw-Hill, New York, 1969).

MUSEUMS AND
BIOLOGICAL LABORATORIES

THE opening of the Museum of Comparative Zoology's laboratory wing is a milestone in the history of the MCZ. It is an occasion to look back to the days of its founding and to look forward to its future. It is also an occasion to ask some searching questions. For instance, someone unacquainted with biology and intolerant of anything but his own hobbyhorse might ask, "Why do we still need natural history museums?" Such a question is quite legitimate, for I am a strong believer in the principle that the legitimacy and continuing value of traditional rituals and institutions should be challenged from time to time. How, then, would we answer this question?

The role of museums in science, and their image in our society, is changing from decade to decade. When natural history was revived during the Renaissance and during the seventeenth and eighteenth centuries, it expressed at first man's wonder and bewilderment at the enormous variety of life. This "diversity of nature" has been a key concept in man's world picture from the days when the Lord told Adam to give names to all the creatures in the field to the present day when species diversity is one of the central themes in the work of the ecologists.

The rich treasures brought back from exotic countries in the eighteenth and nineteenth centuries by voyages and expeditions, combined with the steady rise of a more and more scientific attitude in Western man, resulted in a changed concept of organic diversity. No longer was it merely a source of wonder, but naturalists began to raise questions concerning the reasons for the existence of so many and such strange organisms and about the meaning of their peculiar distribution in Asia, Africa, the Americas, and Australia.

I am not claiming that naturalists were always interested only in the most lofty generalizations, because there was hardly a naturalist who was not also infected by that strange virus called the collector's fever. Perhaps no one was more affected by this disease than the founder of the MCZ, Louis Agassiz, who cheerfully pawned everything he owned in order to acquire more specimens. Indeed, it is said that only a few decades ago this Museum still had unopened boxes of collections from Louis Agassiz's days.

These collections, however, were not merely the useless gatherings of pack rats. It was their study which helped bring about a conceptual revolution—the establishment by Darwin of the theory of evolution, to a considerable extent based on Darwin's own researches during the voyage of the *Beagle* and the subsequent working out of his collections. And the proposal of the theory of evolution was only one of several such conceptual revolutions in the history of natural history.

The diversity of nature has been considered, ever since Darwin, a documentation of the course of evolution. Research in the pathway of evolution indeed turned out to be an incredibly rich gold mine. And it was the museums that established and maintained leadership in this type of research. The historians of biology have clearly determined that the crucial advances in the modern interpretation of species, of the process of speciation, and of the problems of adaptation were made by systematists.

One of the greatest conceptual revolutions in biology, the replacement of essentialism by population thinking, was introduced into biology by museum systematists. From systematics it was brought into genetics by workers like Chetverikov, Timofeeff-Ressovsky, Dobzhansky, Sumner, and Edgar Anderson, all of whom had either been trained as systematists or had worked closely with systematists.

Again and again the students in special branches of biology such as biogeography have gone back to systematics for material and for novel ideas.

But why is systematics so important? This question leads right on to the further question of the position of systematics in biology as a whole. I pointed out a dozen years ago that, in spite of all of its unitary characteristics, biology really has two major divisions; indeed, one can speak of two biologies. In the first one, functional biology, "How?" questions are the important ones. This is the biology that deals with physiological mechanisms, developmental mechanisms, metabolic pathways, and with the chemical and physical basis of all aspects of life. To use modern technical language, this part of biology ultimately deals both with the translation (decoding) of genetic programs into components of the phenotype and with their subsequent functioning. This type of biology played a decisive role in disproving conclusively all vitalistic notions and in establishing firmly that nothing happens in organisms that is in conflict with the laws of chemistry and physics. This is the biology which interprets all cellular and developmental processes, both the normal ones and such abnormal ones as the origin of cancer.

The other biology is interested in the genetic programs themselves, dealing with their origin and evolutionary change. It continuously asks "Why?" questions, for instance:

Why is there such a diversity of animal and plant life?

Why are there two sexes in most species of organisms?

Why is the old faunal element of South America seemingly related to that of Africa while the new one is related to that of North America?

Why are the faunas of some areas rich in species and those of others poor?

Why are certain organisms very similar to each other, while others are utterly different?

In the last analysis, all questions in this part of biology are evolutionary questions, and museum-based collections are ultimately needed to find the facts for posing and answering all of these questions.

At this point some of the more perceptive members of this audience will think that I have painted myself into a corner. Why, they will say, do you need a laboratory wing when the method of systematic and evolutionary biology is the comparative method, based on observations? Why do you have to perform experiments?

The explanation for the seeming contradiction is that I have told only part of the story. Systematics, as it was defined by G. G. Simpson, "is the scientific study of the kinds and diversity of organisms *and of any and all relationships among them.*"

This definition has two consequences: First, it means that the systematist also must ask "How?" questions, like "How do species multiply?" or "How does an evolutionary line acquire new adaptations?" or "How did the phyletic line leading to Man emerge from the anthropoid condition?"

All these evolutionary questions deal with the history of changes, and, most importantly, with the causation of changes. Translated into Darwinian language, each of the questions I have just posed can also be stated in the following terms: "What were the selection pressures responsible for causing the stated evolutionary changes?"

Not only is it often necessary to make use of experiments to answer this type of question, but, more importantly, many of such questions cannot be answered—or at least not completely—simply by the study of preserved material.

Since the investigation of diversity includes the study of relationships, organisms must be studied alive and in the field. In the last 150 years there has hardly been an outstanding systematist who was not, at the same time, an outstanding field naturalist, and who could not have been called, with equal justification, an ecologist or a student of behavior. This is, by no means, a recent development. Re-reading recently Louis Agassiz's "Essay on Classification," published in 1857, I was astonished to find what stress he placed on the study of the "habits of animals," as he put it.

"Without a thorough knowledge of the habits of animals," he said, "it will never be possible to determine what species are and what not." Today we would call this a biological species concept. He goes on to say that we want to find out "how far animals related by their structure are similar in their habits, and how far these habits are the expression of their structure." He continues, "How interesting would be a comparative study of the mode of life of closely allied species." Indeed, Agassiz proposes a program of study which is virtually identical with that of the founders of ethology more than 50 years later: "The more I learn about the resemblances between species of the same genus and of the same family . . . the more am I struck with the similarity in the very movements, the general habits, and even in the intonation of the voices of animals belonging to the same family . . . a minute study of these habits, of these movements, of the voice of animals cannot fail, therefore, to throw additional light upon their affinities."

An interest in the behavior of animals is still a tradition in the MCZ, more than 100 years later. Half of my Ph.D. students in the last 20 years, for example, did their theses on problems of behavior. One of the outstanding characteristics of the so-called new systematics is the concern with the attributes of the living animal. Variation, adaptation, speciation, and evolutionary change cannot be fully understood unless the field work is supplemented by experimental research in population genetics, the analysis of protein and chromosomal variation in populations, the study of the relations between adaptation and functional morphology, to give merely a few examples. Laboratories for such studies are a major component of the new wing. Environmental physiology, another aspect of animal adaptation of great interest to the evolutionist, is being studied at the Countway Laboratories of the Concord Field Station.

The outside world has been largely oblivious to these developments and, I am sorry to say, unfortunately so have also many systematists. For the modern systematist, however, all this seems to be a perfectly natural development. Anyone who has read books like Huxley's *New Systematics* (1940) or my own *Systematics and the Origin of Species* (1942) knows to what an extent all these mentioned activities have been part of systematics for at least 30 years. The new wing gives us an opportunity to help correct the false image about museums which is still widely held, and replace it by the new concept, the beginnings of which were already outlined by Louis Agassiz 116 years ago.

The new wing signals to the outside world that the MCZ is not merely a repository of collections but a biological research institute that differs from the other laboratories in the Biological Laboratories only in the nature of the subject matter. While the emphasis in much of the Biological Laboratories is on cells and the molecular constituents of cells, the major emphasis in the MCZ is on the whole organism, on the diversity of organisms, and on their evolution. Since closest contact between the two groups of investigators is of the utmost mutual benefit to both of them, the organization of the Department of Biology was modified in recent years in order to integrate the staffs of the two groups. Research and teaching are the objectives of both of them.

In this day and age, science is no longer conducted merely for its own sake. Science is no longer the tenant of an ivory tower. Without wanting to minimize in any way the indispensability of basic science, we now realize that the scientist also has social obligations. When optimistically inclined, he will say that he is helping to build a better world; when pessimistically inclined, he will say he is trying to prevent a further deterioration of this world.

But he cannot do this unless he has a sound understanding of Man and of the world in which he lives. And it is precisely the study of diversity and of evolutionary history which has made a major contribution toward the development of a *new image of Man*.

In the pre-Darwinian literature, and also, in much of certain types of contemporary literature, man is conceived as a static being, created within an equally static nature that is subservient to him. Ever since Darwin this concept has increasingly been replaced by a new image, an image of an evolved and still evolving man, part of the evolutionary stream of the whole living world. And this new image, the direct product of evolutionary and natural history studies, is of critical importance, not only for our personal concept of the world in which we live but also for such quite practical issues as man's relation to the environment, to the natural resources, and indeed even to the interaction among men.

It is about time we realize that the future of mankind is not something "written in the stars," something controlled by external forces, but that it is we humans ourselves who hold the fate of our species in our hands. We now have a fairly good idea what the major ills of mankind are, and it has become quite clear that only a few of them are susceptible to purely technological solutions. Instead, most of them are of a behavioral-socio-logical nature and require a change in our value systems, a change one is not likely to accept unless one has a far better understanding of nature, of the dynamics of populations, of the biological basis of behavior, and of other components of the biology of organisms, than most of those have who are responsible for policy decisions.

It will require a deeper understanding of the mentioned problems and it will require massive education based on the findings that emerge from the type of researches that we are planning. During the planning of the wing we sometimes referred to it as a new "center for environmental and behavioral biology." Although this title was not officially adopted, it is indeed an apt description of the focus of attention of the investigators in our new facility.

There may be some who have not kept up with recent developments in biology and who might consider it far-fetched to claim that the men-tioned problems fall within the area of interest of systematics. And yet with systematics defined as the science of biological diversity and with the organism defined as something living and not merely a preserved specimen, a solid chain of links is formed from the systematics of Linnaeus through that of a Louis Agassiz to that of the modern evolutionary systematist and population biologist.

I add my vote of thanks to those who have made the creation of this new center of environmental and behavioral biology possible. I predict that it will have an impact on our knowledge and our thinking that will reach to the far corners of the earth.

NOTE

This address was presented at the opening of the laboratory wing of The Museum of Comparative Zoology on May 29, 1973. It was first published in *Breviora* 416 (1973):1–7, a publication of the Museum of Comparative Zoology.

PROBLEMS IN THE
CLASSIFICATION OF BIRDS

IT WOULD be unthinkable to hold an International Ornithological Congress in Berlin without paying tribute to the memory of that great man who for 50 years dominated the intellectual life of German and, indeed, of world ornithology. The task was assigned to me to present a progress report on the state of avian classification and more specifically, to review Stresemann's contributions to bird taxonomy. The area of avian classification in which Stresemann was most interested and in which he was a master, was the study of species. He realized that the species is the keystone of all biological research, and he considered it his major objective, as an avian taxonomist, to clarify the status of the hundreds if not thousands of doubtful nominal species which cluttered up the literature. More clearly than any of his contemporaries, he saw that three quite separate problems had to be solved in order to achieve this task:

(1) First, all sibling species-complexes had to be dissected into the component species.

(2) Second, all morphs (or "Mutationen" as Stresemann called them) had to be unmasked and assigned to the populations of which they form part.

(3) And finally, all geographical isolates had to be assigned to the polytypic species to which they belong.

Let me begin with points (1) and (2). In line with his endeavor to straighten out complexes of sibling species, Stresemann systematically revised one "difficult" genus after the other, such as *Collocalia, Accipiter, Zosterops, Cyornis, Pericrocotus,* and many others I might care to mention. These revisions were not necessarily the last word, but in each case they converted existing chaos at least into preliminary order. (For a bibliography of these revisions, see *Mitt. Zool. Mus. Berlin* 46 [1970]: 7–29, and *J. Orn.* 114 [1973]: 482–500.)

The second area to which Stresemann devoted literally scores of publications was the phenomenon he called *Mutationen,* in line with the terminology of the early Mendelians, but now called morphs. Again and again Stresemann showed that a so-called species of the ornithological literature was nothing but a color variant of another species and could be stricken from the inventory of avian species.

By far the greatest amount of effort and ingenuity, however, Stresemann devoted to the correct allocation of geographic isolates. Stresemann always considered himself a disciple of Ernst Hartert, and he followed his master in the endeavor to assign all geographic isolates to the polytypic species to which they seemed to belong. In some of his earliest papers he permitted his youthful enthusiasm to carry him too far. For instance, the forms which he combined in 1916 in *Corvus coronoides* we now assign to five good species. At that time Stresemann more or less followed the principles of Kleinschmidt, according to whom all geographic representatives had to be combined into a single *Formenkreis.* In the mid-1920s Stresemann abandoned this extremism, and we find in the *Journal of Ornithology* for 1927 a rather amusing controversy between Stresemann and Hartert on the question whether the desert horned lark of the Sahara (*Eremophila bilopha*) should or should not be included in the Holarctic *Eremophila alpestris,* with Stresemann defending the recognition of *bilopha* as a separate species.

More important than such rather technical matters was Stresemann's consistent emphasis on the primacy of biological over morphological criteria. In 1919, in his *Certhia* revision, he gave an excellent definition of the biological species. He said (1919:64) that speciation is then completed "when two speciating forms have diverged physiologically from each other to such an extent that they can come together again in nature without interbreeding." And he emphasized that "morphological divergence is independent of physiological divergence" (page 66). At a later period (1943) he was one of the first to stress the fact that speciation affects not only reproductive isolation but also ecological divergence, and that the adaptation of populations to new habitats and ecological niches is often a decisive component in the speciation process.

Again and again Stresemann was pioneering with new taxonomic ideas. He exerted a far reaching influence not only through his philosophy, which dominated what was published in the *Journal of Ornithology,* but also through his students like myself and Meise or through his associates at the Berlin Museum like Rensch.

Microtaxonomy

One can distinguish two levels at which activities of classification are performed. Classification at the species level can be referred to as *micro-taxonomy,* and classification at the level of the higher categories as *macro-taxonomy.* As far as the species level is concerned, it is generally admitted that avian taxonomy is more mature than that of any other group of organisms. Yet, even at this level, complete stability has not yet been achieved, as becomes evident when we attempt to answer the question: How many species of birds are there? We then realize that we first have to answer the question: What factors affect the recognition and delimitation of species taxa in birds? When we try to answer this question, we discover that there are three sets of such factors and that they are, in part, in conflict with each other.

> (1) The downgrading of geographical isolates from the rank of species to subspecies in the course of the revision of polytypic species;
>
> (2) The discovery of genuinely new species, including the unmasking of previously unrecognized sibling species, also the discovery of geographical overlaps of taxa previously considered subspecies;
>
> (3) The recent trend to upgrade highly distinct and geographically isolated subspecies to the rank of allospecies.

Let me now discuss these three sets of causes before coming to a final conclusion. The downgrading of thousands of geographic isolates, originally described as full species, to the rank of subspecies together with their assemblage into polytypic species was one of the major concerns of the New Systematics. Stresemann was one of the leaders in this endeavor. This phase of new systematics is now virtually completed, but a comparison of faunal lists published between the 1930s and 1970s shows how recent much of the work was.

When Peters in 1931 published the first volume of his *Checklist of Birds of the World,* he presented what he considered a definitive list of species. The advances of ornithological research, however, were so precipitous that by 1978 a revised edition of Volume 1 had become necessary. Table 18.1 documents how drastic the changes were that had occurred during 48 years.

Table 18.1.   Changes in the inventory of avian species between 1930 and 1978.

Category		Inventory
Species recognized in Volume 1 (1930) of Peters'		853
"Species" since reduced to subspecies or synonyms		101
	Subtotal	752
Valid new species described since 1930		13
Taxa listed by Peters as subspecies or synonyms but considered full species in 1978		34
Taxa considered species in 1978	Total	799

The second factor which affects the stability of species classification is the discovery of new species. The question here is: How complete is our inventory of avian species? There is no doubt that an appreciably large number of species was still to be discovered in the 1920s and 1930s. At that time within less than 10 years I myself was able to describe more than 20 new species, all of them from the circumscribed island region east of New Guinea. The law of diminishing returns, however, has asserted itself in the meantime, and most young bird taxonomists will never have the opportunity to describe even a single new species. But when will the last bird be discovered?

I'm afraid I have consistently underestimated the number of still undiscovered species. In 1935 (page 22) I predicted, "I believe that the number of still undescribed species of birds is below 100." About 140 good new species have already been described since that date. Surely, this is an impressive figure, even though it is only 1.6% of the total of species known in 1935.

More interesting is the question *where* the new species are being discovered. Prior to 1960 more than half of the new species came from the Indo-Australian island region, from Southeast Asia, or from more or less isolated mountain areas in Africa, in other words from the Old World. But the number of such islands and mountains is limited, while the unexplored portions of South America, particularly the slopes of the Andes, seem inexhaustible. As a result, the percentage of American species among the new species has been steadily increasing (Table 18.2). Yet, it

Table 18.2. New species of birds, 1938–1980.

Category	1938–1941	1941–1955	1956–1965	1966–1975	1976–1980
Presumably valid new species described	25	36	33	34	12
Number per year	6.2	2.5	3.3	3.4	2.4
Percent New World	44%	47%	52%	56%	83%

would seem improbable that an annual rate of more than 3 new species can be continued much longer.

The final number of species to be recognized by the ornithologists will far less depend on future new discoveries than on the third factor I have listed above, future changes in the species concept.

The Species Problem Today

It is now evident that the combining of geographical isolates into polytypic species was not the final solution to the species problem in ornithology. As a matter of fact, as in other groups of animals there are two kinds of species problems, the first being the delimitation of taxa and the second being the decision concerning the appropriate rank (species or subspecies) of these taxa. All the cases where we deal with complexes of sibling species—for example, the tree creepers, the gray titmice, and *Phylloscopus* in Europe, or elsewhere in the world the genus *Collocalia,* the *Meliphaga analoga* group of honey eaters in New Guinea, and certain tyrant flycatchers (*Empidonax, Elainia, Myiarchus,* and so forth)—pose a problem of the recognition of species taxa. In these cases field work (the study of vocalization, and so on) may be necessary for a final answer. A typical case is the African genus *Vidua,* whose exceedingly similar sympatric brood parasites are considered good species by some authors but not by others. In a case such as *Vidua,* where the males seem to learn the song of the host species, the final decision may be very difficult, and it may be necessary to test specific distinctness through protein analysis or other methods. Nevertheless, when we add up the totality of such difficult sibling species complexes in all orders of birds in all parts of the world, we arrive at an astonishingly small number. In other words, the task of the correct discrimination of species taxa in birds is virtually completed.

The situation is, however, drastically different when it comes to ranking these taxa. Let me make this clear by a rapid historical survey. Ever since geographic variation in birds was recognized, and this recognition goes as far back as Pallas at the end of the eighteenth century, there has been great uncertainty as to how far to go in the inclusion of geographically representative forms in a single species. The American school of Baird, Coues, and Ridgway adopted the principle to include only those forms in polytypic species that intergrade with each other. The German ornithologist Kleinschmidt went to the other extreme by proposing to combine all related species into a single taxon as long as they were more or less representing each other geographically. He called such assemblages *Formenkreise,* realizing himself that such an aggregate of allopatric populations was far more inclusive than biological species. Nevertheless, he found a number of followers, and for a while even Stresemann combined allopatric species with an impetuous enthusiasm, until he realized that such assemblages of allopatric forms are highly heterogeneous. In view of this situation, Rensch (1928) proposed to recognize two kinds of Formenkreise, those that are composed of genuine geographic races, which he called *Rassenkreise,* and those which also included allopatric species, which Rensch called *Artenkreise.* In order to make Rensch's concepts more accessible internationally, I translated the term Rassenkreis into *polytypic species* and the term Artenkreis into *superspecies* (1931). This terminological distinction was important in order to make clear that geographical replacement is not sufficient to prove conspecificity.

At first, this distinction had little influence on taxonomic practice in ornithology, and only a few authors adopted the superspecies concept. When in doubt, geographically representative forms continued to be called subspecies since this ranking seemed to convey more information. As a consequence, the polytypic species of the period 1920 to about 1955 was very broadly defined. Eventually, however, it was recognized that ranking all geographical isolates as subspecies concealed important evolutionary differences, for these subspecies included forms for which morphological and other criteria clearly indicated that they had already reached species level. The result has been an ever greater use of the category superspecies in the recent ornithological literature.

The use of the superspecies concept raised many geographical isolates, which had previously been called subspecies, to the rank of allospecies, particularly in island regions. However, treating allospecies as equivalent to species that are *not* members of superspecies leads to a sharp increase in the number of recognized species, independent of any discovery of actual new species. Let me give you some examples (based on Vuilleumier 1976). In the genus *Diglossa,* for instance, Hellmayr (1935) and all subsequent authors recognized 11 species, while Vuilleumier raised the total to 17 by recognizing 6 additional allospecies. Similarly, in the Genus *Ortalis* some recent authors recognize only 6 species, others 11; in *Penelope* some authors recognize 6 species, others 15 for the same populations. Many birds from the Polynesian Islands, the Solomon Islands, and New Guinea, which I treated as subspecies in the 1930s and 1940s, I now consider to be allospecies in more widespread superspecies. In the 1940s, prior to this development, counts of the total number of species of birds in the world fluctuated around 8,600. At that time, even though new species were added at the rate of 3 to 5 per year, many isolated forms that had been previously listed as species were reduced at about the same rate to the rank of subspecies and thus, the grand total did not change materially over many years. However, in the most recent count of all bird species (by Bock), the total has risen to 9039 and it may rise to 9,500 when the new criteria for species recognition are consistently applied to all bird faunas of the world.

The recognition of allospecies raises various problems, particularly for zoogeography, as was seen by Rensch at an early stage. When one compares different faunas, it would be quite misleading to give the same weight to allospecies as to genuine species that are not members of superspecies. Many superspecies of New Guinea birds, for instance, have a different allospecies on almost every mountain range (for example, *Astrapia, Parotia*). Superspecies are particularly widespread in archipelagos, like the West Indies or the Solomon Islands, and when the fauna of such island regions are compared with that of the nearest mainland, the unit of comparison must be the superspecies, or more correctly the new category *zoogeographic species,* composed of superspecies and all those isolated species that are not members of superspecies. The importance of this distinction for zoogeographic comparisons can be shown by a few examples. The total number of species breeding in Northern Melanesia is 237. By allowing only one species per superspecies, the figure reduces to 187 zoogeographic species, a reduction of 21%. The total number of breeding North American birds is 518, but the number of zoogeographic species is only 471, a reduction of 9.1%. For the world as a whole, I estimate the number of zoogeographic species to be about 7,000 ± 200.

The question has been raised (Vuilleumier 1976) whether this new development in the ranking of isolated populations has weakened the biological species concept. In my opinion it has not. The biological species concept has validity only in what I have called the "non-dimensional situation," that is, where populations are actually in contact with each other. The allospecies is one of the forms of semi-species, that is, of forms that are in the transition from incipient to full species. Since they are geographically isolated from each other, their species status cannot be tested directly but can only be inferred. The biological species concept itself is not weakened by the fact that there are situations in nature where it cannot be tested.

Let me close this part of my discussion by emphasizing two points. The first is that in ornithology the recognition of species taxa and their delimitation has reached great maturity. The addition of a few unrecognized sibling species or undiscovered species will change the total of known bird species by less than one percent. The situation is different with respect to geographical isolates, where we have experienced a quiet revolution during the last 25 years owing to the application of the superspecies concept. This has necessitated the recognition of an additional category, the zoogeographic species, the use of which facilitates the comparison of different bird faunas.

Let me add a few comments on the categories below and above the species. The subspecies has by the 1970s lost the importance it had when I was a budding ornithologist. I am sure that at the present time more subspecies names are placed in synonymy than new ones proposed.

As far as the genus is concerned, we have left the 1920s far behind, when Mathews (Australia), Austin Roberts (South Africa), and Oberholser (North America) placed just about every good species in a different genus. Professor Bock very kindly has made the latest count of genera available to me. According to the standards of the latest revisions 1905 genera are now recognized for 9039 species and allospecies. This makes for an overall average of 4.74 species per genus. With a consistent adoption of super-species, it might seem advisable to broaden the genus concept and to reduce the total number of genera to about 1,750. At the same time this would be in conflict with the function of a stable binomen as an information retrieval device. Time will tell how this conflict will be resolved.

Macrotaxonomy

Let me now turn to macrotaxonomy, the science of classification. It deals with the determination of relationship among families and orders and with converting it into a useful classification. For a biologist the primary objective of a classification is to provide the foundation for comparative studies, and one might even go so far, as was done by Bock (1976:178), to say that the best classification is the one that permits the most useful comparative investigations.

Those who are not personally active in taxonomic research do not appreciate the immense amount of work that is required for the improvement of a classification. Let us say there are 2,000 genera of birds. For each of these genera one must answer numerous questions. First, what species should be included in this genus; second, is the respective genus justified or could it just as well be combined with another genus; third, what other genus is nearest to it, or more broadly stated, to what family does the genus belong? And finally, into which suborders and orders should these families be combined?

Taxonomic research above the level of the species and particularly above that of the genus was badly neglected during the period of the new systematics, let us say up to the 1950s. Stresemann, for instance, was not particularly interested in this area of taxonomy. Essentially, he simply adopted the classification of Fürbringer. He, who at the level of species and genera was such a lumper, was definitely a splitter at the level of the higher categories. He recognized 51 orders of birds, as against the 27 or 28 orders recognized by most other ornithologists. His attitude about the relationships of these orders was quite agnostic as expressed in his well-known statement: "In view of the continuing absence of trustworthy information on the relationship of the higher categories of birds to each other, it becomes strictly a matter of convention how to group them into orders. Science ends where comparative morphology, comparative physiology, comparative ethology have failed us after nearly 200 years of efforts. The rest is silence" (1959).

Such pessimism, however, was an extreme and rather isolated attitude among ornithologists. Nevertheless, the backward state of avian classification right up to the 1960s cannot be denied. For instance, there is not a single bird fauna in the world that does not have its share of genera that raise questions. In North America, for instance, the dickcissel (*Spiza*), the pinon jay (*Gymnorhinus*), the verdin (*Auriparus*), the wren tit (*Chamaea*), the dipper (*Cinclus*), the solitaire (*Myadestes*), the Phainopepla, the olive warbler (*Peucedramus*), etc.

As far as the New Guinea fauna is concerned (in part shared with Australia), one might mention the following genera: *Daphoenositta, Drymodes, Eulacestoma, Eupetes, Ifrita, Machaerirhynchus, Melampitta, Oreocharis, Orthonyx, Pachycare, Paramythia, Peltops,* and *Timeliopsis.*

Among Australian genera, one might mention the following: *Artamus, Ashbyia, Atrichornis, Cincloramphus, Cinclosoma, Climacteris, Corcorax, Dasyornis, Ephthianura, Eremiornis, Falcunculus, Grallina, Menura, Neositta, Psophodes,* and *Struthidea.*

In all these cases either the assignment to the family is in doubt or else the placement within a family. All of the 37 genera I have listed are passerine birds. Even more serious is the problem with such isolated genera as those of the whale-headed stork (*Balaeniceps*) or the hoatzin (*Opisthocomus*) or of isolated orders like those of the flamingos and tinamous, the placement of which is still quite controversial. We will never attain a perfect classification of birds until all these genera and isolated families and orders throughout the world have been properly assigned to their right places. Since there are numerous such genera in the tropics of Asia, Africa, and South American, there is still a very large task ahead of us.

The uncertainty in classification of birds is perhaps best illustrated by pointing out that for most of the 28 orders of birds now usually recognized, there is no general agreement as to which other order is a given order's nearest relative. Let us take the Passeriformes, for example: some authors postulate that they were derived from the Piciformes, others from the Apodiformes, and still others from the Coraciiformes.

As I said, during the flowering of the new systematics, let us say from the 1920s to the 1950s, classification studies were, for a number of reasons, badly neglected. But the situation has now changed. I entirely agree with Bock (1976:176), who said at Canberra: "The past 25 years have been, without doubt, the most active period in the history of avian classification." This is even more true in 1978, as documented by the symposia and discussion groups at this Congress.

What are the reasons for this renewed activity? There is first of all the realization how sadly the classification of birds had been neglected during the era of the new systematics and in what deplorably bad condition it is. This new interest led to the search for, and the discovery of, whole sets of new characters which made the task of finding an improved classification far more hopeful. Equally important was the fact that a new interest in the concepts of classification developed in the period after 1950 and that the heated controversies on some of the new theories helped to revive an interest in macrotaxonomic research. Both preconditions were obviously highly necessary because in order to establish a sound classification one must first have a sufficient number of characters to work with, and one must secondly have a sound conceptual framework within which to develop one's conclusions.

Taxonomic Characters

Let us begin with the problem of taxonomic character. In the early stages of taxonomy, for instance in ornithology in the eighteenth and early nineteenth centuries, classifications served simultaneously also as identification keys. This necessitated the basing of classifications on single characters: feet with webs or not; bill hooked or not. Perceptive authors, from the very beginning, recognized that the tyranny of single characters leads to the establishment of artificial groups based on convergence. John Ray, as well as Buffon and his followers, proposed instead, in contrast to the Linnaeans, to base classifications on an ensemble of characters.

As much as ornithologists pay lip service to the fallibility of single characters, nevertheless whenever a new taxonomically useful character is found, it tends to be made the basis of rather sweeping taxonomic proposals. This has been true for the structure of the syrinx, certain muscle patterns, the scutellation of the tarsus, the number of notches on the sternum, or the form of the stapes. All of these characters have given us valuable information, but I rather suspect that not a single one of them is infallible and that all of them occasionally break down.

The foundation of all of our classifications is still morphological characters. Since Bock (1976) has reviewed at Canberra the contributions to avian taxonomy made by morphologists, I shall not cover the same ground again. He gave a characterization of what are "good taxonomic characters" and has emphasized correctly that different characters must be used at different levels of the taxonomic hierarchy. Morphological characters are particularly valuable because some of them permit a connection with the fossil record. Yet, much of the morphological analysis concerns characters that do not fossilize, like the syrinx, the various muscle systems, the intestinal tract, and the sense organs. Exactly in this area some excellent studies have been made in recent years, such as those of Ziswiler on the intestinal tract, and some unexpected discoveries have been made. Olson (1973) has shown us how much the classification of a family of birds can be improved simply by an intelligent evaluation of morphological characters, the rails being the group he studied.

The boat-billed heron (*Cochlearius*) is a typical case where a conspicuous morphological specialization has not yet been fully elucidated. Owing to its peculiar bill, Wetmore (1930) considered this bird the representative of a separate family, while Heinroth and, following him, Bock (1958) thought that this heron, except for its broad bill, was a typical night heron and should not be given more than generic rank; Payne and Risley

(1976) removed the bird again from the night herons and gave it the rank of a tribe. It was generally assumed that the broad bill had developed as a special feeding adaptation, but the latest reports seem to indicate that the feeding habits of the boat-bill are identical with those of the night herons. What then is the meaning of this unusual bill? Is it a courtship adaptation (connected with bill clapping), and if so how highly should it be regarded? I am using this case simply to illustrate the kinds of difficulties one encounters in the evaluation of morphological characters.

Before concluding my remarks on morphological characters, let me emphasize how great a contribution to avian classification I still expect from morphology. There are numerous aspects of avian morphology that have not yet been investigated at all, and the analysis of the characters that have been used in the past could be considerably expanded in breadth and depth.

Nonmorphological Characters

Disappointed by the seeming failure of morphological characters and of the fossil evidence to give us definitive answers in our search for the relationship of the major taxa of birds, some ornithologists in recent years have increasingly turned to a search for new kinds of characters. The large repertory of behavioral characters seemed at first sight to be able to provide a most useful set of such characters. This included vocalizations, courtship displays, predator thwarting activities, comfort movements, activities connected with nest building and the raising of young and many others. This expectation was not altogether disappointed, and after the pioneering efforts of Whitman and Heinroth numerous ornithologists employed behavioral characters in their endeavors to improve classification. I have summarized much of this endeavor at an earlier occasion (Mayr 1958). Eventually, it turned out, unfortunately, that behavioral characters are the less useful, the higher up we go in the taxonomic hierarchy. They are most useful at the level of the species and the genus, less so at the level of the family, and virtually useless at the ordinal level, precisely where we are in the greatest need of finding additional characters. There are a few exceptions, like the behavioral similarities between Galliformes and Anseriformes, but otherwise this generalization is valid. Behavorial characters are particularly subject to multiple, convergent origins, as demonstrated by "scratching over or under the wing" or the mode of drinking.

The convergence in behavioral characteristics, correlated with habitat or food niche, deserves far more attention than it has so far received. The hawking behavior of a true flycatcher, let us say *Muscicapa striata,* is extraordinarily similar to that of certain tyrant flycatchers, let us say, the phoebe (*Sayornis*). The Australian flycatchers (*Microeca* group) are, as Sibley has shown, a third convergent development. To give another example, we find in virtually every continent a genus of birds, resembling the European wagtail (*Motacilla*), and adapted to life on torrents or brooks. Even though these genera are quite unrelated, they all have similar behavior specializations. Let me mention one other case of behavior convergence which led me to wrong conclusions. It concerns the Australian scrub robin, *Drymodes,* which some authors have placed with the thrushes, others with the babblers. When I was able to observe the bird in Australia in 1959, I discovered that it seemed to agree with thrushes like *Hylocichla* in its hopping, wing and tail movements, grasping of food, and so on. Clearly, I said, *Drymodes* must be a thrush and not a babbler. Now Sibley's DNA matching has revealed that *Drymodes* is neither a thrush nor a babbler, but the derivative of an autochthonous Australian song bird assemblage. And this poses the interesting question: What is the selective advantage of having movements like a thrush if one is a ground feeder like a thrush? Furthermore, it once more opens up the whole problem of the taxonomic information content of behavioral adaptations. Meise (1963) has shown very convincingly in how many behavioral characteristics the various orders of ratites resemble each other, but there is still the remote possibility that some of these similarities are adaptive, acquired secondarily after the loss of flight and after the acquisition of a terrestrial, running mode of life.

As with all characters, one must exercise great caution with behavioral characters. Even in the comparison of vocalization, there is always the possibility that certain sibling species do not differ noticeably in their songs. There is a group of sibling species of honey eaters in New Guinea, the *Meliphaga analoga* group, in which several field workers up to now have been unable to find differences in vocalization. Frankly, I would be greatly surprised if differences were not eventually found, but it is evident that they cannot be very conspicuous.

More dangerous for the taxonomist is another problem. Owing to the fact that good sympatric species nearly always differ in their songs, the conclusion is sometimes wrongly made that geographic isolates must be raised to species rank if they differ in their songs. However, this is not true. We now know that in many species song varies geographically in a rather drastic manner. Since the deviating populations are connected, in most cases, by intermediates, we know that they are conspecific. Hence, it is not admissible to treat an allopatric population as a different species merely because it has a different song.

Macromolecules as Taxonomic Characters

A major new complex of taxonomic characters has been discovered in the last 25 years, the macromolecules. Only the specialist can appreciate the extraordinary complexity of the proteins, nucleic acids and other macromolecules, as well as the enormous amount of information they contain. One approach to the study of proteins, serological comparisons, is actually more than 75 years old, but, as applied to avian taxonomy, it was very inaccurate and produced no useful results. The first ornithologist who applied the methods of macromolecular analysis consistently for systematic purposes was Charles Sibley. With great determination he employed one method after the other and even though, as he would be the first to admit, some of his earlier methods were not reliable, he discovered the misclassification of a number of avian genera, as for instance *Zeledonia*. I shall not attempt to point out the strengths and weaknesses of the various methods beyond saying that the method of electrophoresis is more ueful for the comparison of populations and closely related species than for the purposes of macrotaxonomy. The method of amino acid sequencing has produced some extremely interesting results, but is very costly in terms of time and equipment and cannot be done routinely.

For a number of reasons various methods of DNA analysis are actually to be preferred, and Drs. Sibley and Ahlquist have shown at this Congress what exciting results they were able to obtain with the relatively simple technique of DNA hybridizing. I share Dr. Sibley's conviction that molecular methods, together with the morphological evidence, will give us in due time what Stresemann had considered as impossible, that is, clearcut evidence as to the relationship of the higher taxa of birds.

Some of the results produced by the new methods were quite unexpected. The American Turkey, for instance, which looks so strikingly different from any kind of gallinaceous bird found in the Old World, turns out, on the basis of several independent tests, to go right in with the pheasants, and to be actually closer to *Phasianus* than is the chicken.

Where the new methods are particularly useful is in revealing cases of convergence. It had of course long been known that the tyrant flycatchers and American woodwarblers are not at all related to their Old World counterparts, the true flycatchers (Muscicapidac) and the true warblers (Sylviidae). What we did not appreciate until this was revealed through

Dr. Sibley's researches was that the Australian flycatchers (Monarchinae) and robins (*Eopsaltria* and relatives) as well as the Australian warblers (Acanthizinae and Malurinae) and the shrike-like whistlers (Pachycephalinae) have nothing to do with their north-temperate counterparts but are an autochthonous adaptive radiation in Australia, all these types being more closely related to each other than to their Eurasian-African counterparts. Since these Australian flycatchers and warblers are species-rich groups, the new assignment has a major impact on the classification of the song birds (Oscines).

Let us tentatively assume that newly discovered characters will eventually give us all the information needed to determine the nearest relatives of each higher taxon of birds. But this will not be the end of our quest, for determining the relationshop of individual taxa is only the first step in making a classification. One also has to determine in what sequence one wants to list these taxa, since they are arranged in a linear sequence in faunas, books, and museum collections. And to arrive at the best possible linear sequence is difficult at best.

In the days when one still believed that evolution was a progression from lower to higher, one thought that the method of classification was quite simple. One started the linear sequence with the most primitive taxon, and moved up to the highest, to the most advanced one. But the facts do not confirm this ideal. Among birds there was apparently a very early radiation leading to the development of a multi-branched phylogenetic bush, with many, more or less equivalent, branches. The same broad radiation seems to be true for the families of songbirds as confirmed by Professor Sibley's researches.

Each of these branches may specialize in an evolutionary elaboration of some special structure or organ system, be it the wing, or the central nervous system, and there are no independent criteria that would tell us which ones of these specialized groups should be considered higher and which other ones lower.

But this absence of linearity in avian evolution is not the only obstacle in our path toward the perfect classification. We also have to find a solution for the problem of highly unequal rates of divergence, (1) of different structures and organ systems (mosaic evolution), and (2) of the different branches of the tree. The case of the relationship of gallinaceous birds and Anseriformes illustrates the kind of difficulties I have in mind, and so does the relationship of man and chimpanzee.

Could these difficulties perhaps be overcome by making use of the so-called "molecular clock," to determine the relative age of the various avian taxa? If rate of evolutionary change at the molecular level were uniform, one could—and this has actually been proposed by A. C. Wilson—list all taxa according to the degree of molecular difference from a common ancestor. With the crocodiles being the nearest living relatives of the extinct proavis, one could use them as evolutionary base line and list all avian taxa according to their degree of difference from the crocodiles. However, it is doubtful that this ingenious solution would work, since there are more advanced and less advanced genera and families on each of the branches of the avian bush.

For all these reasons, I am afraid, there is no simple solution for the problem of finding the perfect classification of birds. No instructions exist that would tell us how to convert a phylogenetic bush into a linear sequence of the higher taxa of birds. I believe that the only way this problem can ever be solved is by international agreement. And that is the current status of the problem of avian classification.

NOTE

This essay is an abridged version of one which first appeared in *Acta XVII Congressus Internationalis Ornithologicus,* pp. 95–112, under the title "Problems of the classification of birds: a progress report." Berlin: Deutsche Ornithologen Gesellschaft, 1980.

REFERENCES

Bock, W. J. 1956. A generic review of the family Ardeidae (Aves). *Amer. Mus. Novit.* 1779:1–49.
——— 1976. Recent advances and the future of avian classification. In H. J. Frith and J. H. Calaby, eds., *Proc. XVI Intern. Orn. Congr., Canberra, Australia, 12–17 August 1974,* pp. 176–184. Canberra City: Australian Academy of Sciences.
Cracraft, J. 1969. Systematics and evolution of the Gruiformes (Class, Aves): the Eocene family Geranoididae and the early history of the Gruiformes. *Amer. Mus. Novit.* 2388:1–41.
——— 1976. The species of Moas (Aves): Dinornithidae. *Smith Contrib. Paleobiol.* 27:189–205.
Feduccia, A. 1973. A new Eocene Zygodactyl bird. *J. Paleont.* 47:501–503.
——— 1975. *Morphology of the Bony Stapes (Columella) in the Passeriformes and*

Related Groups: Evolutionary Implications. Lawrence: University of Kansas Museum of Natural History, Miscellaneous Publications, No. 63.

———— 1977. A model for the evolution of perching birds. *Syst. Zool.* 26:19–31.

Mayr, E. 1935. How many birds are known? *Proc. Linn. Soc. New York* 45/46:19–23.

———— 1957. New species of birds described from 1941 to 1955. *J. Orn.* 98:21–35.

———— 1958. Behavior and systematics. In A. Roe, and G. G. Simpson, eds., *Behavior and Evolution,* pp. 341–362. New Haven: Yale University Press.

———— 1959. Trends in avian systematics. *Ibis* 101:293–302.

———— 1963. The role of ornithological research in biology. *Proc. XIII Intern. Orn. Congr. Ithaca,* 1962:27–38.

———— 1971. New species of birds described from 1956 to 1965. *J. Orn.* 112:302–316.

Mayr, E., and J. T. Zimmer. 1943. New species of birds described from 1938 to 1941. *Auk* 60:249–262.

Meise, W. 1963. Verhalten der Straussartigen Vögel und Monophylie der Ratitae. *Proc. XIII Intern. Orn. Congr., Ithaca,* 1962:115–125.

Olson, S. L. 1973. A classification of the Rallidae. *Wilson Bull.* 85(4):381–416.

Payne, R. B., and C. J. Risley. 1976. Systematics and evolutionary relationships among the herons (Ardeidae). *Misc. Publ., Mus. Zool. Michigan* 150:1–115.

Peters, J. L. 1931. *Check-List of Birds of the World.* Vol. 1. Cambridge, Mass.: Harvard University Press. Rev. ed., 1979. Ed. E. Mayr and G. W. Cottrell.

Rensch, B. 1928. Grenzfälle von Art und Rasse. *J. Orn.* 76:222–231.

Stresemann, E. 1919. Ueber die europäischen Baumläufer. *Verh. Orn. Ges. Bayern* 14:39–74.

———— 1927. Grenzfälle des Artbegriffs. *J. Orn.* 75:436–440.

———— 1943. Ökologische Sippen-, Rassen-, und Artunterschiede bei Vögeln. *J. Orn.* 91:305–324.

———— 1959. The status of avian systematics and its unsolved problems. *Auk* 76(3):269–280.

Vuilleumier, F. 1976. La notion d'espèce en ornithologie. In C. Bocquet, J. Génermont, and M. Lamotte, eds., *Les problèmes de l'espèce dans le règne animal,* pp. 29–65. *Memoire Soc. Zool. France* 38 (1976); 39 (1977); 40 (1980).

Wetmore, A. 1951. Additional forms of birds from Colombia and Panama. *Smithsonian Misc. Coll.* 117(2):1–22.

part six 🙞

S P E C I E S

Introduction

Despite decades of literature on the species concept, including many essays and chapters written by me (see References), the last word has not yet been spoken. Ever new problems seem to arise, to be studied with the help of new material and new arguments. Perhaps it is a vain hope to expect an ultimate resolution of the species problem, because the species plays such different roles in the thinking of various kinds of biologists. Anyone trying to understand the turmoil that surrounds this issue must be aware of these differences in the interests and approaches of systematists, evolutionary biologists, ecologists, behavioral biologists, biogeographers, and many other kinds of biological scientists.

The classical species concept, rightly or wrongly often referred to as the Linnaean or typological species, conceived the species as a distinctive class of objects. This concept fitted equally well with essentialism ("that which has a separate essence") and with Christian creationism ("that which was separately created by God"). The naturalists, however, well aware of the populational aspects of species, conceived of species in an entirely different manner, as the units of behavior and ecology in local biota; this is most clearly expressed in the nondimensional situation, that is, without consideration of changes over time and space. The thinking of the naturalists is best reflected in the biological species concept. Finally, paleontologists and other students of macroevolution, together with certain philosophers, see in the species primarily an evolutionary entity.

There is a certain legitimacy to all three of these ways of looking at species. Which of the three one adopts may depend on when and how one deals with species in one's research. The museum taxonomist, as well as the stratigrapher, may find the typological species concept most useful, never mind how clearly it is refuted by the existence of sibling species and strikingly different phena. But anyone working with living populations, restricted to one place and one time, finds any species concept other than the biological one to be unsatisfactory. For this researcher, the species status of a population can be determined only by its natural interaction with other populations, and this is possible only in the nondimensional situation. Not so for the paleontologist, part of whose job it is to delimit fossil species taxa in the vertical sequence of strata; he cannot help but pay attention to the time dimension. These fundamental differences in approach must be borne in mind if one is to understand why the species controversy generates so much heat.

Essay 19 deals with the meaning of the concept *species*. The species category is the class in which all taxa are included that qualify under the species definition. In this essay I take a position on several recent controversies, such as: to what extent niche occupation should be considered in the recognition of species; whether an evolutionary (vertical) concept is superior to the standard (horizontal) biological species, and whether there are "asexual species," that is, whether uniparental lineages qualify as species.

Essay 20 deals with the ontology of species taxa. Are species taxa "classes" or "individuals," to use the technical language of the philosophers, or is this traditional dichotomy insufficient to reflect the unique properties of biological species? Special attention is paid in this discussion to the historical developments in the treatment of this problem. In the past, logicians have tended to deal with species as though they were inanimate objects. Recognizing that biological species are populations united by a joint gene pool would seem to require a revision of the conceptual framework of logic. I hope that my discussions will contribute toward a clarification of the ontological status of species.

REFERENCES

Mayr, E. 1942. *Systematics and the Origin of Species*. New York: Columbia University Press.
——— 1963. *Animal Species and Evolution*. Cambridge, Mass.: Harvard University Press.
——— 1969. *Principles of Systematic Zoology*. New York: McGraw-Hill.
——— 1970. *Populations, Species, and Evolution*. Cambridge, Mass.: Harvard University Press.
——— 1976. *Evolution and the Diversity of Life*. Cambridge, Mass.: Harvard University Press.
——— 1982. *The Growth of Biological Thought*. Cambridge, Mass.: Harvard University Press.

THE SPECIES CATEGORY

IT IS often said that there is no other problem in biology that is as refractory to solution as is the species problem. And the solutions that have been proposed, from Aristotle, the scholastics, and Linnaeus to the present time, are highly diverse and often incompatible with one another. To an outsider, the picture that emerges may seem utterly bewildering. It almost seems as if every author in the history of biology who has ever written on the species problem has his own personal concept, and that we are as far from consensus as ever.

Fortunately, the situation is not quite so discouraging. There is actually a very limited number of different species concepts, and the species problem can be brought rather close to solution by a careful philosophical analysis of the terms and concepts used in this field and by an analysis of the process of speciation.

In order to refute erroneous opinions of some philosophers, it must be emphasized that the species is not an invention of taxonomists or philosophers, but that it has reality in nature. The existence of species is known to the most primitive human tribes, who recognize species that exist among the animals and plants of their environment and designate them by name.

Species Concepts

When one is dealing with evolving biological populations—and that is what species of organisms are—one cannot expect the simplicity and unambiguousness that one encounters among parameters in the physical sciences. Also, the manifestations of species status may differ quite strikingly in different groups of organisms. Yet I am prepared to suggest that nearly all the species concepts and species definitions that have ever been advanced can be grouped under four headings.

(1) THE TYPOLOGICAL SPECIES CONCEPT

The word species in its simplest conception simply means "kind," such as when a mineralogist speaks of species of crystals and a physicist of nuclear species. A typological species is an entity that differs from other species by constant diagnostic characteristics. This was the species concept of Linnaeus and Lyell and was supported by those philosophers from Plato to modern times who consider species to be "natural kinds" or "classes." Members of such a class are characterized by sharing the same species essence.

Philosophical objections to this concept will be discussed below, but it turned out that this species concept lacks even practical utility, because it forces its adherents to consider as species even different phena within a population, and to lump groups of sibling species in a single species (Mayr 1969). As recently as 1944 the geneticist Sturtevant insisted on combining *Drosophila pseudoobscura* and *D. persimilis* in a single species because this would permit "identifying wild specimens without breeding from them or examining their chromosomes" (1944:476).

The discovery of the high frequency of morphologically indistinguishable species (sibling species) has demonstrated the invalidity of the morphological species better than anything else.

Numerical pheneticists adopted the designation *Operational Taxonomic Unit* (OTU) for taxa at various hierarchical levels. One of the putative advantages of this designation at the species level was that it bypassed the need for assembling phena into biological species. In practice, however, it leads to absurdities to treat phena as OTUs, and the OTUs that are actually recognized in the phenetic analysis are the result of careful, but by necessity subjective, analysis, because otherwise morphs in polymorphic species, age stages, sex differences, and other individual variants, would have to be treated as different OTUs.

Rejection of the morphological species concept, however, does not invalidate the usefulness of morphological criteria for the drawing of inferences on species status. Furthermore, as a first approach in a preliminary analysis of the diversity of a fauna and flora, it is sometimes necessary to recognize provisional species based exclusively on morphological criteria. As will be shown below, such tentative arrangements will be confirmed or rejected or at least modified as soon as additional information becomes available. "It must be emphasized that there is a complete difference between basing one's species concept on morphology and using morphological evidence as inference for the application of a biological species concept" (Mayr 1969:25).

The typological species concept has retained its usefulness in the classification of inanimate objects. Such species are classes of objects characterized by the same defining criteria.

(2) THE NOMINALIST SPECIES CONCEPT

According to this concept, only individual objects exist in nature. Such objects or organisms are bracketed together by a name, and it is by the subjective action of the classifier that it is decided which objects are combined into one species. Species, therefore, are merely arbitrary mental constructs. Species have no reality in nature, according to the nominalist. This has been the claim of the nominalists since the Middle Ages and of numerous authors in the eighteenth century, but also of such recent authors as Gilmour (1940), Burma (1949), and Ehrlich and Holm (1962). Some recent philosophers also have intimated that species are only conventions but have no reality in nature. "The concept of species was introduced as answer to certain theoretical desiderata" (Kitcher 1984). This claim is refuted by the fact that the concept was developed by the best of the naturalists, beginning with the herbalists of the sixteenth century and continued by John Ray, Linnaeus, and virtually all naturalists up to the present day. I have always thought that there is no more devastating refutation of the nominalistic claims than the fact that primitive natives in New Guinea, with a Stone Age culture, recognize as species exactly the same entities of nature as western taxonomists. If species were something purely arbitrary, it would be totally improbable for representatives of two drastically different cultures to arrive at the identical species delimitations. Although a few nominalists still survive, it is now almost unanimously agreed that there are real discontinuities in nature, delimiting different species. Each of these species has different biological characteristics, and the analysis and comparison of these differences is a prerequisite for all other research in biology. Whether he realizes it or not, every biologist works with species.

A Shift to New Species Concepts

As the naturalists, beginning with the seventeenth century, began to make increasingly careful studies of species of organisms in nature, evidence began to accumulate that these species were something different from so-called species of inanimate objects. These naturalists showed quite conclusively that biological species not only had reality in nature, but also that in many, if not most cases, they were sharply distinguishable from each other by a natural discontinuity. However, the creation dogma, as well as the prevailing essentialistic philosophy, also postulated such discontinuities, and this favored a continued adherence to the typological concept. This was reenforced by the daily practice of classifying specimens, where the most convenient strategy was to consider as a different species "that which is different," in other words, that which could be distinguished by morphological criteria.

Not until the Darwinian revolution were attempts made to introduce into biology a species concept that reflects the unique properties of biological populations. The criterion "degree of difference" was replaced by the new criterion "absence of interbreeding." Furthermore, species were now seen as evolving populations rather than as being eternally constant.

(3) THE BIOLOGICAL SPECIES CONCEPT

This species concept is based on the observation of local naturalists that at a given locality different populations coexist that do not interbreed with each other. This is articulated in the definition, "Species are groups of interbreeding natural populations that are reproductively isolated from other such groups."

Certain aspects and implications of this definition will now be discussed. For instance, since hybridization may occur occasionally between individuals of two different good species, without a breakdown of the isolation between these species, it is important to stress that the term reproductive isolation refers to the integrity of populations, even though an occasional individual may go astray.

If one believes in evolution, then one must accept the conclusion that every species is the product of evolution, or more precisely the product of speciation, and that it must have certain qualities that are the consequence of such a history. It permits the making of certain predictions concerning properties to be assigned to a natural entity that one considers to be a species. These properties are as follows: a new species must have acquired reproductive isolation as a result of the process of speciation; it must also have acquired a new, stabilized, well-integrated genotype, and it will, in most cases, have acquired a species-specific niche.

It is this potential for making predictions that gives the biological species its great strength. It permits at all times the testing of clusters of similar (or not so similar) populations for their species status. It is the biological species that is the material with which population geneticists work as well as ecologists and students of behavior. When one concentrates on the mentioned biological attributes of species, one almost inevitably arrives at the same species delimitation, whether one is a primitive native of the New Guinea mountains or an ethologist or a student of ecosystems. The clear-cut discontinuities found in nature, when populations are studied at a given locality, are a real phenomenon. It is the existence of these discontinuities that makes the work of the student of behavior and of the ecologist possible, because every animal species has species-specific behavior patterns as well as a species-specific niche utilization.

Nothing is more misleading than the claim that the biological species concept is not practical. Actually, the biological species concept serves as a yardstick with the help of which controversial or simply obscure situations can be clarified. In an analysis of about 80 controversial cases concerning the status of avian species in North America, Mayr and Short (1970) were able to clarify all but a single one by applying the yardstick of the biological species definition. In an analysis of a local North American flora, Mayr and Wood (unpublished) were likewise able to show that by far the majority of the controversial issues (and there were literally hundreds of them) could be resolved quite unequivocally by applying the biological species concept. Shapiro (1982) has shown how helpful the biological species was in clarifying confused taxonomic situations among butterflies. But more than that, he stresses correctly the great heuristic value of this concept. "The prime virtue of the biological species concept, and the reason for its survival despite so many withering polemics, has been its ability to generate interesting questions of evolutionary, biogeographic, and systematic interest about real organisms in the real world" (p. 217).

The Protection of the Species-Specific Genotype

The question is sometimes asked, Why are there species? Why do we not find in nature simply an unbroken continuum of similar or more and more widely diverging individuals? (Mayr 1957a). It is now clear that the isolating mechanisms of a species are a protective device for well-integrated genotypes. Any interbreeding between different species would lead to a breakdown of well-balanced, harmonious genotypes, and would quickly be counteracted by natural selection. Such counterselection against hybridization has been demonstrated in nature in literally thousands of cases, even though cases of successful hybridization, particularly in plants, have also been demonstrated.

The high selective value of mechanisms that help to maintain the integrity of species-specific genotypes is documented by the diversity of isolating mechanisms (Mayr 1963, chap. 5) and by the fact that the isolation is usually effected not by a single mechanism but by a set of them. In numerous cases, particularly in plants but also in certain groups of insects (morabine grasshoppers), sterility is the principal or sometimes even exclusive isolating mechanism. In most animals, however, behavioral barriers are all-important. Males approach females and test them for their receptivity. There is an interchange of stimuli or signals between male and female, and if the female is receptive and the signals sent out by the males appropriate (that is, conspecific), the courtship will proceed to copulation.

The specific reaction of males and females toward each other is often referred to loosely as "species recognition." "This term is somewhat misleading, since it implies consciousness, a higher level of brain function than is found in lower animals (see also Spurway 1955)" (Mayr 1963:95). In recent years Paterson (1976; 1985) has again revived the recognition definition of the behavioral isolating mechanisms, without referring to the arguments why this definition had been rejected by earlier authors. It must be emphasized that there is no argument about the biological facts but only about the terminology. As I said in 1963: "Species recognition, then, is simply the exchange of appropriate stimuli between male and female, to insure the mating of conspecific individuals and to prevent hybridization of individuals belonging to different species" (1963:95).

In the process of the rejection of nonconspecific mates and the acceptance of conspecific mates, one can focus attention on either one or the other of these two processes. Rejection means maintaining isolation, and acceptance means recognition. These are simply two sides of the same coin. When I prefer the term isolation to the term recognition, which I had been fully aware of but rejected in 1963, it is for two reasons: first because the term recognition, as I said, postulates a degree of conscious cognitive activity that is not to be expected in "lower" animals, and secondly because isolation among species is effected in numerous species or organisms by isolating mechanisms other than behavioral ones. There is, for instance, no mate behavior recognition among morabine grasshoppers and many other organisms with post-zygotic isolating mechanisms.

Where Paterson is right is in pointing out the probable primacy in the acquisition of behavioral isolation as compared with niche acquisition (see Essay 22). On the other hand, acceptance of the term isolating mechanisms does not necessitate acceptance of either the Darwin-Muller-Mayr or Wallace-Dobzhansky theories of the origin of reproductive isolation. Likewise (contra Vrba 1984) it does not matter whether one accepts the isolation or recognition terminology when it comes to the evaluation of the species status of isolated populations.

Evolutionary Theory and the Biological Species Concept

Species are the product of evolution, and, owing to the genetic turnover in populations, all species are evolving all the time. This fact has been used both as evidence in favor of the biological species concept and as evidence against it. Every deme evolves and so does every subspecies as well as every higher taxon (aggregate of species). Therefore the fact of evolving is not in the slightest diagnostic for the species. Alfred Emerson (ms. and 1945) had "evolving" in his species definition, but in 1942 I rejected this inclusion as obviously not being a defining characteristic of species.

The internal cohesion of the species is continuously reinforced by interbreeding. However, such interbreeding is also stressed in the preevolutionary species definition of Cuvier, and (by implication) even in that of John Ray. But such a recombination of the genetic characteristics of populations could simply be a reshuffling of the features of the originally created ancestors. This is evidently the way it was seen by authors like Ray and Cuvier. Being spatio-temporally restricted and internally cohesive they had all the characteristics of "individuals" (Ghiselin 1966; Hull 1976) and yet they were not evolving (Essay 20).

From the beginning, various authors have rejected the biological species concept more or less emphatically (Sokal and Crovello 1970; Sokal 1973; Ehrlich 1961; 1980; 1982; and Raven 1977a; 1977b). The objections are virtually never raised against the biological species concept as such, that is, the occurrence in nature at a given locality of coexisting, reproductively isolated populations, but rather against the expansion of this nondimensional concept in space and time. Almost invariably there is a confusion between the species as a category (as expressed in the species concept) and the species as a taxon, a more or less well delimitable aggregate of populations. Nowhere are these two aspects of the species (category vs. taxon) more completely confounded than in the discussion by Sokal and Crovello (1970). Those who come from mathematics into biology would like species to be typologically fixed in space and time, and as invariable as possible. When they find that the population structure of species as well as the processes of evolution make such an ideal typological species impossible, they consider this a refutation of the biological species concept. Actually, such variation and evolutionary change can be used as an objection to any species concept so far proposed (see also below). Species taxa, as we shall see, are based on inference from the species concept. That there are operational difficulties connected with the making of such inferences does not refute the concept as such, as Hull, Wiley, and others have pointed out. It is rather ironic that some authors (such as Raven) who have criticized the biological species concept in theoretical papers seem to adopt it completely in their own taxonomic monographs and in those of their students. Numerous authors in recent years have refuted the putative objections to the biological species concept (Hull 1970; White 1978; Willmann 1985:52–54; De Jong 1982:233. Ghiselin 1987), papers in which the various objections to the biological species concept are carefully considered.

Niche Occupancy and Species Definition

Coexisting species are not only reproductively isolated, but occupy niches that are sufficiently different as to preclude competitive exclusion (Gause). Numerous distributions of closely related species are known in which these species are separated by a parapatric border. Various kinds of evidence lead in these cases to the inference that the parapatric species compete for the same niche but that one is superior on one side of the borderline, the other species on the other side. Sympatry can be achieved in such cases only when these species acquire niche differences. I incorporated this conclusion in a recently suggested species definition: "A species is a reproductive community of populations . . . that occupies a specific niche in nature" (1982:273). This reference to the ecological specificity of a species was criticized by Ghiselin (1987) as either superfluous or as confounding "professions with organizations." There is some justification for this criticism, but it fails to come to grips with the important fact that absence of reproductive isolation is not the only factor that prevents the harmonious coexistence of species. Somehow I have the feeling that a neospecies that has not yet acquired the necessary ecological autonomy to coexist with a sister species has not yet acquired full species status.

Attempts have been made to make ecological differentiation the primary species criterion rather than reproductive isolation (Van Valen 1976). This, however, leads to insuperable difficulties. In most species with geographic variation, particularly in polytypic species with numerous isolates, it is found that many of the local populations differ in their niche utilization. One would have to make each of these variant subspecies a different species. Furthermore, there is now good evidence for polymorphism in niche utilization even within a single population. All this makes it impossible to make ecological differentiation the primary species criterion.

(4) THE EVOLUTIONARY SPECIES CONCEPT

The application of the biological species concept to multidimensional assemblages of populations always encounters difficulties. Paleontologists who study species distributed in the time dimension have been searching for some time for a species concept that would be particularly suitable for the discrimination of fossil species. After Meglitsch (1954) it was particularly Simpson (1961), Wiley (1980; 1981), and Willmann (1985) who have attempted to develop a species concept particularly suitable for paleontology and phylogenetic researches. Simpson's definition (1961:153) was: "an evolutionary species is a lineage (an ancestral-descendant sequence of populations) evolving separately from others and with its own unitary evolutionary role and tendencies."

Wiley (1981:25) has slightly modified it to read: "an evolutionary species is a single lineage of ancestor-descendant populations, which maintains its identity from other such lineages and which has its own evolutionary tendencies and historical fate."

It would seem that such definitions abandon the clear-cut criterion of the biological species definition (reproductive isolation) and replace it by such undefined vague terms as "maintains its identity" (does this include geographical barriers?), "evolutionary tendencies," and "historical fate." What population in nature can we ever classify by its "historical fate," when this is *entirely in the future*? In addition, the evolutionary species concept encounters three major difficulties.

(a) It is applicable only to monotypic species. It is unable to cope with polytypic species that contain geographical isolates, because each of these isolates answers the evolutionary species definition, being a single lineage which maintains its identity. Thus the definition provides no yardstick for the placement of isolated populations. When very similar morphospecies are encountered at different exposures, the evolutionary concept provides no criterion that would permit a decision whether or not they are conspecific. Rather, as in the biological concept, one has to infer from the amount of morphological difference, exactly as one does with geographical isolates in the living fauna.

(b) The qualification "own evolutionary tendency and historical fate" does not permit discrimination between good species and isolates. There are no empirical criteria by which either evolutionary tendency or historical fate could be observed in a given fossil sample. Simpson himself (1961:154–160) realized this, and admitted arbitrariness in application.

(c) The hoped for capacity of an evolutionary concept to help in the delimitation of chronospecies did not materialize. This is well documented by the inability of the evolutionary species concept to arbitrate in the controversy between Gingerich and his followers, who believe in a frequent occurrence of phyletic speciation, and Stanley, Eldredge, and followers, who believe in a complete stasis of all neospecies. Indeed, the three most articulate proponents of the evolutionary species concept (Simpson, Wiley, Willmann) agree that it does not provide a nonarbitrary method for the delimitation of species in the time dimension. This is most curious, since the main reason why the evolutionary species concept was introduced was in order to deal with the time dimension, which is not considered in the nondimensional biological species concept.

Since time is one of the factors involved in the problem of speciation, it is necessary to discuss the problem of the modes of speciation in order to have a basis for developing criteria for the delimitation of chronospecies. A chronospecies is delimited by its birth (origin) and its death (extinction). With respect to the delimitation of species taxa, the following tabulation of modes of origins of new species might be helpful:

Potential Origins of New Species

    (A) Through a speciation event
        (a) Instantaneous (for example, polyploidy, stabilized hybrid)
        (b) Very rapid (peripatric speciation, conceivably sympatric speciation)
    (B) Without a speciation event (parental species transformed)
        (c) Dichopatric[1] speciation (split by a geographic barrier with gradual divergence)
        (d) Gradual phyletic transformation of a single lineage

The question to be answered is: In which of the four potential modes of speciation is it possible to determine the point of origin of a new species? This is clearly possible in the two cases in which the new daughter species originates either instantaneously (Aa) or so rapidly (Ab) as to be instantaneous as far as paleontology is concerned. However, the parental species remains in both cases (Aa and Ab) unchanged or at least does not change its species status.

Neither Hennig (1950) nor Willmann (1985) allows for the process of peripatric speciation, that is, for the budding-off of a new daughter species, and thus their claim that every speciation event inevitably terminates the life of the parent species is clearly wrong. This is fully understood by Wiley (1981:34–35). For instance, when the New Guinea kingfisher *Tanysiptera galatea* produced daughter species on Koffiao, Biak, Numfor, Tagula, and the Aru islands, this did not affect the parent population on the mainland of New Guinea (Mayr 1954). It would have been quite absurd to have ranked the mainland population as a new species every time it sent off a founder population to one of the adjacent islands.

This same species (*Tanysiptera galatea*) also shows the inapplicability of another of Hennig's principles for the delimitation of populations and closely related allopatric species. The daughter populations on the adjacent islands are not derived from the mainland species as a whole, but it is always a single local population of the widespread continental species that produced a given founder population. Surely, one cannot divide the mainland species into a number of separate species because five local populations have become the "stem species" of the five island species.

In all cases where several isolated populations of a polytypic species have completed speciation and the polytypic species has become a superspecies with a set of allospecies, it is impossible to arrange these in the form of an unambiguous cladogram, or at least a dichotomous cladogram. It is made impossible by the fact, as explained above, that each of the derived species is most closely related to one localized population of the mainland species, and by two additional considerations. Evolution in the isolates may represent mosaic evolution, and aspects of both reproductive isolation and morphological characters may be acquired unequally in the different isolates. Furthermore, with the derivatives of the parental species all having the same or similar potential, it is likely if not probable that several derived species will acquire the same character independently. This is well illustrated by the allospecies of the *Xiphophorus helleri* superspecies of Central America (Rosen 1979) and the species of *Lapidochromis* of Lake Malawi (Lewis 1982).

In categories Bc and Bd, the parental species is being transformed. Even in the cases where a species splits into two (Bc), the two daughter species are virtually identical at the moment of the split, and if any species differences evolve in the two separated lines, it is by gradual transformation. This makes it impossible to designate a precise point of origin of the new daughter species. This would be equally true if the process of gradual phyletic transformation (Bd) occurs, as is supported by many paleontologists. They believe that a temporal sequence of species can indeed be produced within a single lineage by gradual phyletic evolution and that in such a case the line between two sequential species must be arbitrarily placed between one generation of the parental species and its daughter generation, which now belongs to the new species. That a "subdivision of unbroken successions of species . . . must be arbitrary" was frankly admitted by Simpson (1961). These findings can be summarized by saying that the origin of a new species can be reasonably well pinpointed in the two cases Aa and Ab, but that placement of the point of origin of the new species originating in categories Bc and Bd is largely arbitrary.

The *termination* of chronospecies encounters equivalent difficulties. There are three potential modes of a termination of species.

(a) By traditional extinction, like that of the dodo or the passenger pigeon, without any connection with speciation. What percentage of species terminate in this manner is still controversial. According to the more extreme punctuationists (Stanley, Eldredge/Gould), this is the normal fate of all species. Once they have originated and have become populous and successful they enter a period of stasis that is terminated by extinction. Other authors, although not denying the frequent occurrence of extinction, believe to have also evidence for a great deal of phyletic transformation.

(b) Termination of a species in a single lineage through its transformation (by phyletic gradualism) into one descendant species. As stated above, the point of termination of the parental species is strictly arbitrary in such a case.

(c) "Extinction" of a species when it subdivides by dichopatric speciation into two or more daughter species. The change that occurs in both of the daughter lineages is a slow transformational change. There is actually no genuine extinction of the parental species, only its gradual transformation into daughter species.

It is evident from this analysis that an unequivocal determination of the end point of a chronospecies can be made only when it becomes extinct in the traditional sense of the word (category a).

(5) "OPERATIONAL" SPECIES DEFINITIONS

Some recent authors have rejected the establishment of any species concept. Instead they propose to adopt an operational principle that would serve like a recipe telling the taxonomist in each case whether a given population is or is not a different species. Among such operational principles the following are employed most widely.

(*a*) *The difference principle.*   According to this principle, species are "the smallest groups that are consistently and persistently distinct, and distinguishable by ordinary means" (Cronquist 1978:15). This means that every isolated subspecies would have to be called a species, while sibling species could not be recognized. Furthermore, since the word "groups" is not defined, males and females of sexually dimorphic species and any phena in polymorphic species would have to be treated as different species. Such a procedure is, of course, nothing but the consistent application of the old typological species concept. It would not only lead to an enormous inflation in the number of recognized species, but these so-called species would be biologically completely meaningless. As we shall see, a somewhat similar approach is sometimes temporarily adopted by the working taxonomist when dealing with a group of species that have not yet been previously revised; but the subsequent testing with the biological species concept eventually permits the sorting of the tentatively recognized morphotypes into biological species.

(*b*) *Recognizing all distinguishable geographical isolates as species.*   This principle was widely adopted in ornithology in the 1880s and 90s but was rejected when the polytypic species concept was adopted (Stresemann 1975).

(*c*) *Recognizing species by cladistic principles.*   Hennig (1950), as explained above, recommended that any species be terminated when a sister or daughter species split off from it. This recommendation dealt with the delimitation of species in the time dimension. Rosen (1978; 1979) extended the principles of cladistics to the pattern of the geographic variation of species, and since he considered absence of reproductive isolation as a plesiomorphic character, the unerring logic of cladistics forbade him to use it as a character, and this forced him to consider every geographic isolate that could be distinguished as a separate species. He then defined the species as "the smallest natural aggregation of individuals with a specifiable geographic integrity that can be defined by any current set of analytical techniques" (1979:277). Implicitly this is, of course, also a return to the pre-Darwinian concept of the typologically defined morphospecies.

Species delimited by such purely operational procedures are devoid of all biological significance. Their recognition may simplify the work of the museum curator, but they are of no use to a student of behavior, of ecosystems, or of biogeography. The weaknesses of operationism have been stressed sufficiently by Hull (1968) to make it unnecessary for me to add anything further.

From the Species Definition to the Delimitation of Species Taxa

Both Ghiselin and Hull have rightly emphasized that one cannot define individuals (or populations), one can only describe them. By contrast, a class is characterized by its defining properties. The category species, being a class, therefore can be defined. A species taxon, on the other hand, can only be described or delimited. To understand this difference is of the utmost importance for theory and practice of taxonomy at the species level. It makes clear why it is so important to distinguish between taxon (individual) and category (class). One only needs to read the paper by Sokal and Crovello (1970) to see what utter confusion results from neglecting to make that distinction.

There can be little doubt that the concept of the biological species is what most modern biologists have in mind when they talk about species. They have had too much experience with all forms of polymorphism and with sibling species to find a typological or morphological species concept useful.

This poses a dilemma. How can one use the species definition when one's task is the delimitation of species taxa? The species definition is based on the nondimensional situation of the noninterbreeding of coexisting populations, where reproductive isolation can be observed directly. Species taxa, however, are multidimensional in area and in time. The question thus becomes: How can the "ideal" nondimensional species be expanded into a concrete multidimensional species taxon?

As long as there is a continuity of populations, there is no problem. The gradual merging of such populations is de facto proof of interbreeding. A problem arises, however, whenever there are discontinuities either in nature or in our records. How can one combine into a species populations that are isolated geographically or in the time column (as encountered by the paleontologists)? In such cases conspecificity can be made probable only through inference; and such an inference can be made only by applying the yardstick of the biological species concept.

HOW TO DEVELOP INFERENCES?

In the case of populations separated by a gap, species status is determined by inference. In the case of geographical isolates as well as in phyletic series, one assumes conspecificity when no phenetic change exists.

On the other hand, one assumes specific difference when the morphological difference is of the same order of magnitude as is found among sympatric species of the same higher taxon. Here one must be aware of what is the premise and what is the conclusion. As Simpson has stated so beautifully (1961:68-69), two individuals are not identical twins because they are so similar, but they are so similar because they are identical twins. We do not combine two populations into one species because they are similar; rather, we conclude that they are so similar because they belong to the same species.

Morphological similarity is only a relative yardstick. It provides no certainty. We may find in the same genus (such as Drosophila) sibling species which by molecular analysis can be shown to be only rather distantly related, and morphologically distinct species that are very closely related. Not only may the rate of divergence in different isolates be different, but so may also be the rate of change in different characters. There are of course other clues concerning probable conspecificity, such as distribution, role in the ecosystem, life history attributes, and so on. Much of that information is not available for fossil samples.

The great similarity of isolates to the parental species has often in the past been attributed to gene flow (Mayr 1963). This factor is indeed in part responsible for such uniformity. However, as Ehrlich and Raven (1969) have rightly pointed out, almost perfect similarity may be maintained in the virtual absence of gene flow. It is highly probable that such stasis is maintained by an internal cohesion of the genotype reinforced by stabilizing selection (Mayr 1970). This cohesion of the genotype affects in most cases not only the morphology but also the isolating mechanisms. It is one of the major problems of evolutionary biology to determine under what conditions these stability-maintaining mechanisms are broken down.

Difficulties in the Application of the Biological Species Concept

Species taxa consist of populations extending in space and time. The application of the biological species concept during the process of delimiting species taxa often encounters difficulties (Mayr 1963:21-29). Very often a lack of pertinent information necessitates a provisional assignment of names to taxa. The morphological criteria generally used for a first approach can often be tested against behavioral, cytological, or molecular evidence.

It is sometimes claimed that different modes of speciation produce different kinds of species. On the whole this does not seem to be true.

Species produced by dichopatric, peripatric, and sympatric (where it oc-curs) speciation are of the same kind. Changes in the chromosome number may cause difficulties, but polyploidy and other changes in chromosome number are rare in animals. They occur commonly in plants, and the status of some of the biotypes in plants that differ in chromosome number is still controversial.

Hybridization, when resulting in fertile offspring, could give rise to reticulate evolution and thus be a special process of speciation. Again, this seems to be far more common in plants than in animals. In animals, as White (1978) has shown, successful hybridization almost invariably results in a switch to uniparental reproduction. In certain fishes and amphibians, stabilized hybrid species occur in which during meiosis—or already during fertilization—the paternal genome is eliminated, and mat-ing with a male of the paternal species is required to produce zygotes of the next generation (Dubois and Günther 1982). These stabilized hybrids behave in nature like good species. It is likely that similar cases will be found in other groups of organisms.

It is now quite evident that the evolutionary intermediacy of certain populations is the factor that poses the greatest difficulties in the delim-itation of species taxa. All populations of a species that are geographically isolated follow their own evolutionary path. They may diverge more and more from the parental population and become incipient new species. If all their characteristics would evolve at the same rate, one might be able to accept a standard that would indicate completion of speciation. This, unfortunately, is not the case, and some populations may acquire repro-ductive isolation but only minimal morphological difference (resulting in sibling species), while others may acquire conspicuously different mor-phologies but no isolating mechanisms (Mayr 1963). Even closely related species usually differ in their enzymes, as revealed by electrophoresis, but in other cases electrophoresis has so far been unable to reveal such differ-ences even though the investigated populations are genuine species by other criteria. Finally, different populations may acquire different niche utilizations but no differences in their isolating mechanisms. Such mosaic evolution forces the investigator to make subjective decisions concerning species status. Alas, evolution, being as capricious and opportunistic as it is, will inevitably produce situations that do not allow for a totally satisfactory resolution.

I have discussed such cases of evolutionary intermediacy in great detail in two previous publications (1963; 1969:181-197). Actually, neither sibling species nor drastic polymorphism nor occasional hybridization pose many practical problems when properly analyzed. What is controversial most frequently is the question whether certain geographical isolates should or should not be considered full species (allospecies), and in a mature taxonomy, as that of birds, this uncertainty causes the only major disagreement among different workers. A similar situation seems to prevail in freshwater fishes, where certain workers (such as Rosen 1979) have recently raised all subspecies to species level. Whether to call such isolates subspecies or species is in most cases quite irrelevant, because the ecologist, and in most cases also the student of behavior, deals with the interaction of sympatric populations, and the potential status of an allopatric population is usually biologically rather uninteresting.

The Significance of Species in Biology

Modern biologists are almost unanimously agreed that there are real discontinuities in organic nature, which delimit natural entities that are designated as species. Therefore the species is one of the basic foundations of almost all biological disciplines. Each species has different biological characteristics, and the analysis and comparison of these differences is a prerequisite for all other research in ecology, behavioral biology, comparative morphology and physiology, molecular biology, and indeed all branches of biology. Whether he realizes it or not, every biologist works with species.

The diversity of organic life, consisting of species and groups of species (higher taxa), is the product of evolution. This necessitates the study of the origin and evolutionary history of every species and higher taxon. The study of species is thus shown to be one of the fundamental preoccupations of biology. And no such study will be constructive unless based on a sound species concept. This is the reason why a deep understanding of the nature of species is of such utmost importance.

NOTES

This essay was first presented at a symposium held in May 1985 at the Fondation Singer-Polignac, Paris. It was published in J. Roger and J. L. Fischer, eds., *Histoire du concept d'espèce dans les sciences de la vie*, pp. 294–311, under the title "The species as category, taxon and population." Paris: Fondation Singer-Polignac, 1986.

1. *Dichopatric* (a term proposed by J. Cracraft) *speciation* occurs when the contiguous range of a species is secondarily divided by a geographical or vegetational barrier into two isolates. Dichopatric and peripatric are two forms of allopatric speciation.

REFERENCES

Ax, P. 1984. *Das Phylogenetische System,* pp. 22–30. Stuttgart: Gustav Fischer.

Bocquet, C., J. Génermont, and M. Lamotte, eds. 1976–80. *Les problèmes de l'espèce dans le règne animal. Memoire Soc. Zool. France* 38(1976); 39(1977); 40(1980).

Burma, B. 1949. The species concept: a semantic review. *Evolution* 3:369–370.

Cronquist, A. 1978. Once again, what is a species? In *Biosystematics in Agriculture.* Beltsville Symposia in Agricultural Research, 2. New York: Wiley.

Darwin, C. 1859. *On the Origin of Species.* London: John Murray. Facs. ed., Cambridge, Mass.: Harvard University Press, 1964.

DeJong, R. 1982. The biological species concept and the aims of taxonomy. *J. Res. Lepidoptera* 21:226–237.

Dubois, A., and R. Günther. 1982. Klepton and synklepton: two new evolutionary systematics categories. *Zool. Jahrb. Syst.* 109:290–305.

Ehrlich, P. R. 1961. Has the biological species concept outlived its usefulness? *Syst. Zool.* 10:167–176.

———— 1980. Colorado checkerspot butterflies: isolation, neutrality, and the biospecies. *Amer. Nat.* 115:328–341.

Ehrlich, P. R., and R. W. Holm. 1962. Patterns and populations. *Science* 137:652–657.

Ehrlich, P. R., and P. Raven. 1969. Differentiation of populations. *Science* 165:1228–1232.

Ehrlich, P. R., and D. D. Murphy. 1982. Butterflies and superspecies. *J. Res. Lepidoptera* 21:219–225.

Emerson, A. E. 1945. Taxonomic categories and population genetics. *Ent. News.* 56:14–19.

Ghiselin, M. T. 1966. On psychologism in the logic of taxonomic controversies. *Syst. Zool.* 15:207–215.

———— 1987. Species concepts, individuality, and objectivity. *Biol. and Phil.* 2:127–143.

Gilmour, J. S. L. 1940. Taxonomy and philosophy. In J. Huxley, ed., *The New Systematics,* pp. 461–474. London: Clarendon Press.

Haffer, J. 1986. Superspecies and species limits in vertebrates. *Z. f. zool. Systematik u. Evolutionsforschung* 24:169–190.

Hennig, W. 1950. *Grundzüge einer Theorie der phylogenetischen Systematik.* Berlin: Deutscher Zentralverlag.

Hull, D. L. 1968. The operational imperative—sense and nonsense in operationism. *Syst. Zool.* 17:438–457.

———— 1971. Contemporary systematic philosophies. *Ann. Rev. Ecol. Syst.* 1:19–54.

———— 1976. Are species really individuals? *Syst. Zool.* 25:174–191.

Kitcher, P. 1984. Species. *Phil. Sci.* 51:308–333.

Lewis, D. S. C. 1982. A revision of the genus *Lapidochromis* from Lake Malawi. *Zool. J. Linn. Soc.* 75:189–265.

Mayr, E. 1954. Change of genetic environment and evolution. In J. Huxley, A. C. Hardy, and E. B. Ford, eds. *Evolution as a Process.* London: Allen and Unwin. Rptd. Mayr 1976.

——— 1957a. Species concepts and definition. In *The Species Problem.* Washington, D.C.: Amer. Assoc. Adv. Sci., Publ. no. 50.

——— 1957b. Difficulties and importance of the biological species concept. In *The Species Problem,* pp. 371–388. Washington, D.C.: Amer. Assoc. Adv. Sci., Publ. no. 50.

——— 1963. *Animal Species and Evolution.* Cambridge, Mass: Harvard University Press.

——— 1969. *Principles of Systematic Zoology.* New York: McGraw-Hill.

——— 1970. *Populations, Species, and Evolution.* Cambridge, Mass: Harvard University Press.

——— 1976. *Evolution and the Diversity of Life.* Cambridge, Mass.: Harvard University Press.

——— 1982. *The Growth of Biological Thought.* Cambridge, Mass.: Harvard University Press.

Mayr, E., and L. Short. 1970. *Species Taxa of North American Birds: A Contribution to Comparative Systematics.* Cambridge, Mass.: Nuttall Ornithological Club Publications, No. 9.

Mayr, E., and C. Wood. (Unpublished.) Species delimitation in the plants of Concord township.

Meglitsch, P. A. 1954. On the nature of the species. *Syst. Zool.* 3:49–65.

Osche, G., ed. 1984. Phylogenetisches Symposion über den Artbegriff und Artbildung. *Zeitschr. zool. Syst. Evol. forschung* 22(3):161–288 (with contributions by G. Osche, D. Sperlich, W. Sudhaus, H. Zwölfer, G. L. Bush, F. Ehrendorfer, and W. E. Reif).

Paterson, H. E. 1976. Symposium address. 15th International Congress of Entomology. Washington, D.C.

——— 1985. The recognition concept of species. In E. S. Vrba, ed., *Species and Speciation,* pp. 21–29. Pretoria: Transvaal Museum Monograph No. 4.

Paterson, H. E., and M. Macnamara. 1984. The recognition concept of species. *South African Journal of Science* 80:312–318.

Raubenheimer, D., and T. M. Crowe. 1987. The recognition species concept: is it really an alternative? *South Africal J. Sci.*

Raven, P. 1977a. The systematics and evolution of higher plants. In C. E. Goulden, ed., *Changing Scenes in Natural Sciences,* 1776–1976. Philadelphia: Academy of Natural Sciences.

——— 1977b. Systematics and plant population biology. *Syst. Bot.* 1:284–316.

Rosen, D. 1978. Vicariant patterns and historical explanation in biogeography. *Syst. Zool.* 27:159–188.

————— 1979. Fishes from the upland intermontane basins of Guatemala. *Bull. Amer. Mus. Nat. Hist.* 162:269–375.

Shapiro, A. M. 1982. Taxonomic uncertainty, the biological species concept, and the Nearctic butterflies: a reappraisal after twenty years. *J. Res. Lepidoptera* 21:212–218.

Simpson, G. G. 1961. *Principles of Animal Taxonomy.* New York: Columbia University Press.

Sokal, R. R. 1973. The species problem reconsidered. *Syst. Zool.* 22:360–374.

Sokal, R. R., and T. J. Crovello. 1970. The biological species concept: a critical evaluation. *Amer. Nat.* 104:127–153.

Spurway, H. 1955. The sub-human capacities for species recognition and their correlation with reproductive isolation. *Acta XI Congr. Int. Orn.,* pp. 340–349. Basel, 1954.

Stresemann, E. 1975. *Ornithology: From Aristotle to the Present.* Cambridge, Mass.: Harvard University Press.

Sturtevant, A. H. 1944. Book review: *Drosophila pseudoobscura. Ecology* 25:476.

Van Valen, L. 1976. Ecological species, multispecies, and oaks. *Taxon* 25:233–239.

Vrba, E. S. 1984. Patterns in the fossil record and evolutionary processes. In M. W. Ho and P. T. Saunders, eds. *Beyond Neo-Darwinism: An Introduction to the New Evolutionary Paradigm.* New York: Academic.

White, M. J. D. 1978. *Modes of Speciation.* San Francisco: Freeman.

Wiley, E. O. 1980. Is the evolutionary species fiction? A consideration of classes, individuals, and historical entities. *Syst. Zool.* 29:76–80.

————— 1981. *Phylogenetics: The Theory and Practice of Phylogenetic Systematics.* New York: Wiley.

Willmann, R. 1985. *Die Art in Raum und Zeit.* Berlin und Hamburg: Parey.

THE ONTOLOGY OF THE SPECIES TAXON

CERTAIN problems of ontology seem to be exceedingly refractory to solution. Opposing viewpoints continue to be firm, neither side being able to produce the kind of arguments that would convert their opponents. Such a situation seems to pertain at the present time to the ontological status of species. Are species classes, are they individuals, or, if neither, what are they?

My major objective in the following essay is not so much to produce a final solution to this problem as to discuss the reasons for the stalemate. I shall try to show that a purely philosophical solution is impossible until the factual basis is clearly established. In other words, clarity must first be achieved on the *biological* nature of species before this can be expressed appropriately in philosophical terminology. A second reason that a controversy such as this may be prolonged is because the ontological vocabulary of both camps is inadequate. Each of these factors clearly enters into the ontological species problem. More than 50 papers on this question have been published in the last several decades, and we still seem to be far from any consensus.

Two matters require immediate clarification. First, there is the question of whether observed species have reality in nature. This question can be answered only for *nondimensional species* at a given place and a given time. No naturalist would question the reality of the species he may find in his garden, whether it is a catbird, chickadee, robin, or starling. And the same is true for trees or flowering plants. Species at a given locality are almost invariably separated from each other by a distinct gap. Actually, the question of the "realism" of species has nothing to do with the current ontological controversy, since both individuals and classes can be real.

Second, a major advance in conceptualization and terminology among taxonomists is curiously unknown to many philosophers, and this has created misunderstandings. I am referring to the difference between the species as a *taxon* (a distinct object *in nature* that the taxonomist recognizes and describes—a process called *species delimitation*) and the species as a *category* (a rank given to the species taxon by the taxonomist—the definition of this category being called *species definition*). That the species category is a class is not disputed by anyone. What is at issue is the ontological status of the species taxon. It is a serious shortcoming in the writings of most philosophers that they confound taxon and category so often in their analyses, although in principle most of them are fully aware of the distinction. In order to determine the species status of a population or taxon, one must attempt to apply to it the species definition and see whether the situation is consistent with the definition.

What is involved has been excellently illustrated by Simpson (1961) in the case of monozygotic twins. Two persons are monozygotic twins not because they are so similar, but they are so similar because they are monozygotic twins. Two populations belong to two different species not because they can tolerate coexistence without interbreeding, but rather they can tolerate such sympatry because they are species. Probabilistic inference is the major method in systematics for determining the species status of a population. It can be observed directly only in the nondimensional situation and not always even there. In all dimensional situations (that is, when longitude, latitude, or time is added), species status can be determined only by inference. No working naturalist can escape this simple fact of nature.

The objective of the present analysis is to remove, as much as possible, all semantic ambiguities and factual misunderstandings. In my treatment I shall concentrate on the crucial issues, and largely ignore a discussion of certain side issues that have been introduced into some of the arguments, such as the role of exemplars (Mayr 1983) and the applicability of laws to individuals and classes.

The Species Concept of Classical Taxonomy

In classical taxonomy, species were simply defined as groups of similar individuals that are different from individuals belonging to other species. Thus, a species is a group of animals or plants having in common one or several characteristics. Each species represents a different type of organism. The diversity of nature is seen as the reflection of a limited number of unchanging universals. This concept ultimately goes back to Plato's concept of the *eidos* or, as it was referred to by some later authors, to the "essence" or "nature" of some object or organism. The similarity of the members of a species was due to the joint possession of this eidos or essence. Variation was interpreted as due to an imperfect manifestation of the eidos which resulted in "accidental" attributes.

SPECIES AS CLASSES

This essentialistic species concept developed by the early taxonomists was based on exactly the same criteria as those adopted by the philosophers for their own concept of species. The time-honored species concept held by the philosophers was that of a class. Membership in a class is determined strictly on the basis of similarity, that is, on the possession of certain characteristics shared by all and only members of that class. In order to be included in a given class, items must share certain features which are the criteria of membership or, as they are usually called, the "defining properties." Members of a class can have more in common than the defining properties, but they need not. These other properties may be variable—an important point in connection with the problem of whether or not classes may have a history.

The class concept is widely applicable to inanimate objects. For instance, one can recognize a class of chairs as consisting of pieces of furniture built in such a way that a human can sit on them. By far the most important defining property of a class is its constancy, a necessary correlate of its being based on an essence. At the same time, class membership is not spatiotemporally restricted. If two-legged hominoids on Mars were to construct pieces of furniture with the defining properties of chairs, these would belong to the class of chairs. Nor is there any special relationship among members of a class, such as one finds among parts of an individual. Clearly there is no relation among the members of the class "chair." In other classes, there is sometimes an indication of relationship, but this relationship is not part of the definition.

The class concept of species adopted by the philosophers, which "treats species as random aggregates of individuals that have in common the essential properties of the type of the species," has been referred to as the *typological species concept* (Mayr 1963:20–21). For the traditional philosopher, the word "species" simply meant "kind of" and designated a degree of similarity (Quine 1969). There was no special relationship among the members of a species other than their similarity. The species was merely a lower ranking class than the genus; as stated by Jevons (1877:701), "A genus means any class whatever which is regarded as composed of minor classes or species."

The typological species concept, endorsed by Linnaeus and his contemporaries and widely adopted in the nineteenth century, was one of the major impediments not only to the acceptance of the general concept of evolution but also to the acceptance of particular theories of evolutionary change. (For instance, the typological species concept required speciation to be saltational, that is, to produce each new species esence by a single mutation.) Most modern taxonomists have fully understood and rejected the typological species concept because it does not correspond to the situation found in nature among biological species. Nevertheless, a few schools, such as the pheneticists, certain pattern cladists (Nelson, Patterson), and a few botanists (Cronquist 1978), still uphold it.

The nonphilosopher has considerable difficulty in reading the analyses of philosophers because class is not the only term used for aggregates of items. I have encountered also the terms natural kinds, clusters, and in recent years most frequently, sets. Not fully understanding the differences among these terminologies and rather bewildered by opposing statements by philosophers themselves, I shall not attempt to discuss this heterogeneity of terminology in full detail.

Let me say a few words, however. Natural kinds have been discussed by philosophers from Mill to Quine (1969). Most recently this designation has been supported by Schwartz (1981:301) and by Kitts and Kitts (1979), but its usefulness has been questioned by Ghiselin (1981:271) and by Sober (1984:335). Since the defenders of the natural-kinds terminology apparently make no distinction between inanimate objects and living populations, I agree with those who find it a sterile terminology. The same would seem to be true for so-called cluster concepts, favored at least by some modern philosophers (Caplan 1981:285).

What puzzles me particularly is why certain philosophers want to apply the term *set* to species (Kitcher 1984a,b). In contrast with a class, a set (if I understand Kitcher correctly) does not require any defining properties. According to his definition, any aggregate consisting of more than a single item is a set, no matter how heterogeneous. To illustrate his concept of a set without a defining property Kitcher lists "Queen Victoria, the manuscript of *Finnegan's Wake,* and the number 7." By abandoning the traditional defining properties of a class, Kitcher arrives at a definition of set which by necessity must include species, for every species taxon is composed of more than one item (organism). This concept of set is so different from the traditional definition of class that none of the arguments against class in the classical controversy on the ontological status of species is any longer applicable.

Frankly, a biologist is utterly bewildered. For more than 100 years biologists have worked very hard to discover and describe how biological species differ from other phenomena of nature and whether there is only one kind of species or several different ones. All this is obliterated if all variable phenomena of nature are dumped into a single highly heterogeneous receptacle, the set. Nor does Kitcher seem to make any distinction between species category and species taxon. Items in a set and members of a species are consistently equated. This omits any consideration of variation and population thinking. I have not found a criterion in Kitcher's discussion indicating how one could distinguish definitionally a biological species from a set characterized by some arbitrary property. For instance, hairy objects including mammals, hairy caterpillars, hairy seeds of certain plants, and other hairy objects, would make a legitimate set for Kitcher. Likewise, not only species taxa but also higher taxa, the whole animal kingdom, the whole living world, not to mention the contents of the top drawer of my desk or my wastepaper basket, would be sets. At the same time I sense in his discussions seeming inconsistencies such as his disclaimer that his analysis would imply "a general way of replacing talk of objects with talk of sets" (1984b:617), and yet this is what he seems to suggest in the rest of his discussion. Occasionally he seems to notice where his axioms lead him, and thus he rejects the idea "that organisms are just sets of cells." And yet this is precisely what they are on the basis of his definitions. I found no criterion in his writings that would make such a definition of organisms inadmissible. There is no criterion by which one can decide which multiples of objects are sets and which others are not. Nor is there any way to know how a set that is a species differs from a set of hairy objects. Indeed, Kitcher seems to consider it a virtue to have such a sweepingly defined and heterogeneous category like set. This permits him to consider any set of individuals a species which any biologist or taxonomist had ever called a species (on the basis of no matter what criterion). To me it would seem that a species concept based on such a broad and vague set of theoretical criteria loses all scientific usefulness. As both Hull and Ghiselin have pointed out, the great virtue of the newer ways of defining species is that they permit all sorts of evolutionary predictions.

FROM THE TYPOLOGICAL SPECIES TO THE REPRODUCTIVE COMMUNITY

Almost up to modern times it was similarity which revealed whether two items belonged to the same species. In the world of inanimate objects this criterion worked very satisfactorily. In the world of life, however, as the knowledge of kinds of animals and plants increased, similarity failed occasionally. In spite of their great difference, caterpillar and butterfly clearly belong to the same species. In other cases, male and female of a species were more different from each other than one of these sexes in a series of related species. How could one cope with such variability (Mayr 1982:256)? Toward the end of the seventeenth century John Ray proposed an entirely novel solution. Regardless of degrees of variation, all those variants should be considered members of a species that had sprung "from the seed of one and the same plant" or, in the case of animals, been produced by the same parents. Reproduction was here for the first time introduced into the species definition. In the case of Ray, it was problems of practical taxonomy that led to this shift. In the case of Buffon, a similar shift occurred, but on an entirely different basis (Sloan 1986). Buffon had a great interest in individual development, and it greatly puzzled him why the offspring on the whole was always so similar to the parent. This led to his famous proposal of the internal mold, derived from Plato's developmental eidos, which controlled the constancy throughout time of the species-specific characteristics. "This power of producing its likeness, this chain of successive existences of individuals . . . constitutes the real existence of the species" (Buffon 1954:233,238). Buffon explicitly rejected the class concept of species because then each organism would be "a creature by itself, isolated and detached, which has nothing in common with other beings, except in that which it resembles or differs from them" (p. 355). As Sloan points out quite rightly, this is not an intrinsically biological argument. Even though there is reproductive continuity from parent to offspring and its offspring, one finds in Buffon no trace of population thinking or of the concept of a reproductive community. Buffon does not think and talk like a naturalist (much less like an evolutionist), but rather like a student of embryology. What Buffon had actually done, and this clearly distinguishes him from Linnaeus and other taxonomists of the period, was to replace Plato's eidos (which is a typological essence) by Aristotle's eidos (which is a genetic program). In spite of the forward step, Buffon's *moule intérieure* is as constant and eternal as Plato's essence. Yet it was a step in the right direction, and it was much easier to go from Buffon's species (as constant as it was) to the reproductive community of the biological species than it would have been to take this step from the Platonian species of Linnaeus. What helped the subsequent developments even more was Buffon's idea that if two parents had the same *moule intérieure* they would be able to produce fertile offspring. And this, indeed, would permit a reproductive community.

Buffon's species concept, sometimes combined with that of Ray, rapidly displaced the strictly typological species definition of Linnaeus. Happily, it was also consistent with creationist dogma. The members of a species were the descendants of the first pair created by God at the beginning. This is what the antievolutionist von Baer (1828) had in mind when he defined the species as "the sum of the individuals that are united by common descent." Cuvier and many naturalists in the first half of the nineteenth century supported similar definitions. Increasingly often not only generation ("descent") but also common reproduction was made part of the definition. As stated by Gloger (1856), "A species is what belongs together either by descent or for the sake of reproduction." Voigt (1817), Oken (1830), and others of their contemporaries adopted similar definitions (Mayr 1957:9). In one species definition after the other during the ensuing 125 years the emphasis was on the reproductive community. This is well stated in the species definitions of Poulton (1903) and Jordan (1905). Plate (1914) stated it in these words: "The members of a species are tied together by the fact that they recognize each other as belonging together and reproduce only with each other." It was the species concept of the naturalists who participated in the evolutionary synthesis. There are numerous unequivocal statements on the nonclass nature of species in the writings of Dobzhansky and of myself. Dobzhansky stated it particularly clearly: "A Mendelian population is a reproductive community of sexual and cross-fertilizing individuals which share in a common gene pool . . . The biological species is the largest and most inclusive Mendelian population" (1951:577). And in another place he called the species "a supraindividual biological system . . . more than a group concept. A species is composed of individuals as an individual is composed of cells . . . A biological species is an inclusive Mendelian population; it is integrated by the bonds of sexual reproduction and parentage" (1970:353–354). None of these characteristics are those of a class.

My own species concept (1942–1982) has clearly always been that of a concrete entity. Eldredge, in a recent historical analysis, stated: "If it is true that Mayr did not invent the notion that species are real entities, it is nonetheless indubitable that the underlying basis for all the various recent approaches to looking at species as entities or individuals rather than as classes of individuals stems from Mayr's species concept" (Eldredge 1985:49). In 1963 I enumerated three properties of species which "raised the species above the typological interpretation of a class of objects. The nonarbitrariness of the biological species is the result of this internal cohesion of the gene pool" (1963:21). In 1969 I characterized the species as follows: "The members of a species form a reproductive community. The individuals of a species of animals recognize each other as potential mates and seek each other for the purpose of reproduction . . . The species, finally, is a genetic unit consisting of a large, intercommunicating gene pool . . . These . . . properties raise the species above the typological interpretation of a 'class of objects.'" (Mayr 1969:26).

These developments were not altogether ignored by philosophers, for J. Gregg reports that two different taxonomists had advanced the notion "that species are composed of organisms just as organisms are composed of cells: according to this argument, a species is just as much a concrete, spatio-temporal thing, as is an individual organism, though it is of a less integrated, more spatio-temporally scattered sort" (Gregg 1950:424–425). I know that I was one of these taxonomists, and either A. J. Cain or Dobzhansky was the other. Gregg rejected this notion, and all other philosophers up to the 1970s ignored it. They ignored it so much that as recently as 1984 Kitcher refers to the "traditional thesis that species are sets." Hull (1976), Caplan, Rosenberg, and Sober have made similar statements. Owing to their ignorance of the taxonomic literature, they all credit a shift in thinking on the subject to Hull's discovery (1976; 1978) of Ghiselin's papers (1966; 1974a,b) in which the species-as-class concept is rejected. It is remarkable that among philosophers Hull made only few converts for the nonclass concept of species (Rosenberg, Sober, and M. Williams). Most other philosophers who joined the controversy in recent years have continued to defend the species-as-class concept, failing to see a difference between "species" of inanimate objects and biological species.

Anyone reviewing this literature is puzzled why the virtually universal rejection of the concept of the species as a class by evolutionary biologists was so completely ignored by the philosophers. Undoubtedly there were several reasons. One of them is that philosophers of former periods almost completely ignored the scientific literature, so that it was necessary for a biologist (Ghiselin) to bring the shift in thinking to their attention. Another reason, as I shall discuss below, is the insufficiency of the terminological and conceptual repertory of the philosophy of science. A few examples of the writings of philosophers will show how little they understood the biological nature of species. For instance: "The concept of species was introduced to answer certain theoretical desiderata" (Kitcher

1984b:636). Or, species are primarily "an answer to a practical need" (Rosenberg 1985:203).

THE SPECIES AS INDIVIDUAL

In 1966 Ghiselin clearly stated, "Biological species are, in the logical sense, individuals . . . a species name is a proper noun" (1966:208–209). But this went unnoticed, and Hull in 1967 still treated the species as a class. Not until Ghiselin had amplified his thesis (1974a,b) and had pointed out that it was by no means a new proposition, but went back at least to Buffon, did Hull (1976) take notice and become converted. Considering how unequivocally the nonclass status of species had previously been maintained by one biologist after the other, it is quite possible that Ghiselin's claim would have remained equally ignored if Hull had not taken up the issue and given it a broad foundation couched in philosophical terms.

I am deliberately using the terminology of those who subscribe to the nonclass status of species because, as the ensuing controversy reveals, two different issues were involved. The first is indicated by the question, Is it correct to consider the biological species a class? and the second by the question, If it is decided that the biological species is not a class, is the term "individual" the appropriate term for it, or should a different terminology be chosen?

THE NONCLASS PROPERTIES OF BIOLOGICAL SPECIES

Ghiselin (1974a; 1981), Hull (1976; 1978; 1981; 1984), Beatty (1982), Holsinger (1984), and Sober (1984) have specified a lengthy list of properties of species that are not properties of classes as traditionally defined. Some of these putative nonclass properties of species raise problems which I shall now attempt to discuss.

Species are spatiotemporally localized. They occur at specific places and at specific times. A given species taxon cannot occur on earth and also in the Andromeda nebula. Within their spatiotemporal localization, species are on the whole continuous, and it is this continuity which, in cases of doubt, often permits the inference as to what belongs to a single individual. For instance, that caterpillar and butterfly are the same individual is inferred not from any similarity in their appearance but from this continuity. The continuity of different organisms within a species is provided by their historical (common descent) connection. Ghiselin (1974b:536) correctly stresses that the spatiotemporal continuity does not need to be physically continuous. The fact that Alaska and Hawaii are not physically contiguous with the other 48 states does not exclude them from being parts of the "individual" called the United States.

Physical discontinuity of parts of an "individual" is one of the major causes of difficulty for the student of polytypic species. How can one determine whether or not an isolated "individual" is part of a larger individual? That Alaska is part of the United States can be documented because she adheres to the Constitution of the United States. In the case of animal populations, the relation of isolated populations to other populations must be inferred on the basis of the biological species concept. The Ipswich sparrow, although completely isolated on Sable Island off Nova Scotia, is part of the savannah sparrow because it is inferred that it shares with the savannah sparrow the same isolating mechanisms. Species status or not is determined on the basis of inference on the reproductive relation to other presumably related "species." Ghiselin missed this need for testing species status when he said "if a species is an individual it hardly matters whether it is interbreeding at any given time" (1974b:538). This had things upside down, because, in the case of every isolated population, it is precisely the inference we make about its reproductive isolation which tells us whether it is a separate individual or a part of a larger individual. Until the inference about reproductive status is made, we do not know whether or not the population is a separate individual.

Discreteness. Individuals are reasonably discrete in space and time, they are bounded rather than being potentially unlimited, as classes are. However, because species speciate and occasionally merge, their borders are sometimes "fuzzy" and the point at which one species leaves off and another begins is often arbitrary, as is also true for mountains and many other objects of nature that are individuals (Rosenberg 1985:208). Owing to their discreteness, species are particular things with proper names.

Internal organization. By far the most important *definens* of an individual is its internal cohesiveness. Organisms that together form a species have intimate connections with each other not found among the members of a class of objects, as classically defined. This is due to the fact that they are derived from the joint gene pool of the species, and that they jointly contribute their genotypes to form the gene pool of the next generation. There is thus a continuity and interconnection among the organisms of which a species is composed that is altogether absent from the members of a class. Organisms that belong to a species are part of the species, not its members. This statement to its full extent is true only for the sexually reproducing biological species in which genetic recombination among the various genotypes occurs normally in every generation. The compatibility of the genotypes of conspecific mates, as documented by the production of viable new genotypes in their offspring, indicates that the species population has the kind of internal harmony one would expect to find in parts of a single system. The numerous, often fatal, incompatibilities one usually encounters in hybrids produced in species crosses reveal that the parental species are not part of the same system, that they are not parts of a single individual. The parts of an "individual" (in the broad sense of the terminology of logic) are "integrated in one way or another—joined as by physical or social forces or common descent" (Ghiselin 1981:271). Also there must be some mechanisms by which the unity of the biological species is maintained, through the joint possession of isolating mechanisms (for example, reciprocal fertility and a behavioral recognition capacity).

Does a Species Have an Essence?

No one questions the fact that organisms which belong to a species are united by the joint possession of a set of isolating mechanisms, perhaps also of a niche occupation potential, structural characteristics, and other joint properties of the genotype. Proponents of the species-as-class concept have suggested that this joint genotype should be considered the essence of the species and that this proves that species are classes. Indeed, I have heard a philosopher ask: Is not any reference to the joint properties of the members of a species "a flirtation with essentialism?" This suggestion reveals a regrettable confusion between essence and properties in common. Of course the parts of an individual have certain properties in common. No one will question "that tigers have an underlying trait that makes them tigers and not giraffes or turtles, just as there is an underlying trait that makes gold gold, or water water" (Schwartz 1981:302). Furthermore, all monophyletic groupings of organisms, from the population to the highest taxon, have of course something in common. How else would we otherwise determine whether a certain organism is a butterfly or a vertebrate? But a property in common and an essence are two entirely different things. To be sure, every essence is characterized by properties in common, but a group sharing properties in common does not need to have an essence. The outstanding characteristic of an essence is its permanence, its immutability. By contrast, the properties that a biological group have in common may be variable and have the propensity for evolutionary change. What is typical for a taxon may change through evolution at any time and then no longer be typical.

It was precisely the variability of species populations that led to population thinking, a dramatic departure from essentialism. The shift from the concept of species-as-a-class to species-as-individuals was an inevitable byproduct of the shift from essentialistic to population thinking. Unfortunately the word "essential" with reference to species has often been employed ambiguously. If someone refers to the essential characters of a species, he might mean (and usually does mean) certain indispensable properties, such as the isolating mechanisms which maintain the integrity of the species. Some philosophers, however, have interpreted this as a reference to Platonic species essences. When Kitts (1984) argues that species have an essence, it seems to me that he has fallen victim to this equivocation. There is nothing in a biological species that corresponds to the Platonic concept of a fixed and transcendental essence. If species had such an essence, gradual evolution would be impossible. The fact of their evolution shows that they have no essence. And since they do not have an essence, they do not form classes. Some philosophers recognize classes without an essence, but such a shift in terminology is too vague to be of any practical use.

Parts of a species display a three-fold variability. First there is the variability of organisms within a population. As a result, no two individuals in a sexually reproducing species have the same genotype. Such individuals may agree with each other in 85 percent or 90 percent or 95 percent or even 98 percent of their genotype, but even this identical component of the genotype of conspecific individuals has the potential not only to vary but also to evolve. The Platonic essence has neither capacity. This is the crucial distinction between a genotype and a Platonic essence. Furthermore, there is variation in space, that is, the geographic variation of populations of a species. Finally, there is evolution, that is, variation in time. Individual variability, variation in space, and variation in time are compatible with the species concept if organisms and populations within a species are considered parts of an individual; but these phenomena are incompatible with the species concept if the species is considered a class with the defining properties of an essence.

The Individual and Evolution

The properties of a particular object, an individual, may change over time. Applied to the species-as-individual, it means that the evolutionary capacity of species confirms their status as individuals. Indeed, all biological species seen today, except a few that originated by an instantaneous process (for example, polyploidy), are the result of evolution. The process of geographic speciation illustrates the difference between class and species particularly well. Dividing a class never results in a change of the properties of the new subdivisions. But if a species is split by a geographic barrier, the two separate populations inevitably begin to differ and have even the capacity to evolve into two separate species.

To include the capacity to evolve among the evidence in support of the claim that the species is an individual has been frequently criticized. Is evolutionary change a diagnostic difference between class and individual? Is it not also possible for a class to evolve? Has there not been a historical change concerning the class of chairs in the last several hundred years? Indeed, if we compare three sets of chairs, those in use in 1780, those in use in 1880, and those in use in 1980, we may discover considerable change. If it had been part of the defining criterion of the 1780 chair that the supporting structure is of wood, this definition would no longer be appropriate in an age of steel and plastic. But "being made of wood" was never a truly defining criterion of chair. What had changed over the centuries was not the essence of chairness, but rather some accidental properties of chairs that are not part of the defining essence. A class, having a constant essence, cannot evolve.

Species have the capacity to evolve. Yet, the provision "develops continuously through time" is presumably not a necessary aspect of the species-as-individual. The validity of this assumption is indicated by the fact that many punctuationists believe that most species do not evolve at all after the original speciation event but may go into complete stasis until they become extinct. It may be well to remember at this time the preevolutionary species concept of Buffon (Sloan 1986), Cuvier, and von Baer. The aspect of the species they emphasized most was the internal continuity and coherence of the species, resulting from the fact that all members of the species were derived from the original pair created by God. In that respect their species clearly had more the characteristics of an individual than those of a class, even though the criterion of constancy made them a class. Owing to mutation, genetic recombination, and selection, every species is changing over time. But such change does not need to qualify as evolution. For instance, as a thought experiment, one could conceive of a very widespread, very populous species existing in our time that lives in such a stable environment that it has reached a steady state stability without evidence of any seeming evolution.

Species have some other nonclass properties: they can speciate (split into two), they can hybridize (fuse), and they can become extinct. No class can ever become extinct. Generalizations made concerning particular species are not laws, but simply facts about particular spatio-temporally localized objects.

There is no conflict between individual and class; they are simply different aspects of nature. Even though organisms are parts of a species, one may also recognize classes of organisms such as sexes, age stages, sessile, hermaphroditic, migratory, and other assemblages based on defining properties. Even though cells are part of an individual and genes are part of the genotype, one can recognize classes of cells such as red blood cells, connective tissue cells, epidermal cells, or classes of genes such as lethal genes, enzyme-producing genes, and so on. None of this is in conflict with the fact that species, organisms, cells, and genes are individuals. Both individuals and classes can be hierarchically organized. Hierarchical organization thus is not a feature which distinguishes individuals from classes.

Objections to the Species-as-Individual Terminology

When attempting to call the attention of the philosophers to the fact that the biologists consider the species a nonclass, Ghiselin (1966; 1974a) chose for these composite wholes the term *individual,* so frequently used by philosophers as the opposite of class. Ghiselin realized the seeming inappropriateness of this term and commented, "Some readers will find it easier to call composite wholes 'collections,' 'aggregations,' or the like, but 'individuals' is the traditional term in logic" (1981:270). Actually Ghiselin was not the first to apply this terminology, since exactly 100 years earlier Haeckel had called the species *ein Individuum* (Haeckel 1866 II:30). It would require a rather tedious search through the literature to determine whether this terminology had also been used by others. Be that as it may, the choice of this terminology raises two kinds of objections. Even several of those authors who fully agreed that species are not classes were unhappy over the selection of this particular term. After all, the word individual stresses singularity more than anything else, and is used in the daily language of nonphilosophers almost exclusively to express such singularity. It is quite counterintuitive to apply it to an assemblage of individuals. Indeed, as Kitts and Kitts (1979:617) have remarked correctly, it is "flying in the face of convention." The human species consists of more than 4 billion individuals, so how can one call it *an* individual? Even Hull, a vigorous defender of the nonclass status of species, was seemingly uncomfortable with the term individual because he ended his major discussion of the problem with the statement, "I think I have adduced ample reasons in this paper for concluding that, at the very least, species are not classes" (1976:190).

Some of the objections, however, have gone much further. The word individual, derived from the Latin, basically means that which is indivisible. This is indeed essentially true for most real individuals, whose integrity is ordinarily seriously compromised by removing any parts of it. A human without his head, his heart, his parathyroid, his liver, and various other organs is unable to live. Even the removal of less vital organs, like the eyes, an arm, a leg, the stomach, or many other parts, seriously alter the individual, converting it into something it was not before. By contrast, a species of a million organisms is not seriously affected if 10,000 or even 100,000 of them should be removed by sudden death. This sort of thing happens in nature periodically as a result of drought, disease, or other catastrophes. The damage is quickly repaired in the ensuing seasons. It is only in the lower invertebrates and in many kinds of plants that a seriously mutilated individual can be restored as quickly as a decimated species.

The nature of organization (cohesiveness) is also very different. In a genuine individual all parts interact with each other and do so directly. By contrast, the interaction of parts of a species is for most of its members quite loose and indirect, consisting only of the propensity for gene exchange. But such gene exchange is virtually nonexistent in those many species which consist of highly isolated colonies or subspecies. The "organization" of such species is less cohesive by several orders of magnitude than that of a true individual.

Furthermore the species-as-individual terminology invites misunderstandings in discussions of natural selection. There is now rather broad agreement that the individual is normally the target of natural selection. Calling the species an individual, does this imply that the species as a whole is one of the targets of natural selection, as has indeed been suggested by some authors? Since most biological groups answer the logician's definition of individual, does this terminology legitimize the concept of group selection? These are not frivolous questions, because these questions have actually been raised in the literature.

Those who transfer the term individual to composite wholes are forced to find another term for what everybody up to now has called an individual. In biology they use the word organism as replacement term; but organism is actually a term that is used at various hierarchical levels. There is no implication as to level in such usages, as in "turtles are long-lived organisms." When using the word organism, one may have individuals in mind, or species, or higher taxa. It avoids a great deal of misunderstanding to return the word individual to its traditional usage.

The Problem of Terminology

Neither the term class nor the term individual expresses the ontology of biological species satisfactorily. It would seem necessary to introduce a new term, one not yet in use in the terminological repertory of philosophy. Here is not the only place in recent years where a deficiency in vocabulary seems to have been the major cause for controversy in philosophy. What happens particularly often is that philosophers use a single technical term for several objects or processes that are factually quite different from each other. Scientific analysis often leads to a partitioning of a phenomenon which up to that time had been referred to by a single term. A philosophical analysis of such problems will be incomplete or even misleading if it does not keep up with these scientific developments. For examples, see Essay 3 on the four uses of the term *teleology*; Essay 12 on Darwin's five theories of evolution; Essay 7 on the four definitions of *group* in *group selection*. Other equivocal terms are *category* (including taxon), *selection* (including elimination), *evolution* (both transformational and variational; Lewontin 1983), and *development* (ontogeny and phylogeny), to mention merely a few (see also Ghiselin 1984). It is quite impossible to carry out a constructive ontological analysis unless it has been preceded by a careful factual analysis. All heterogeneity must first be eliminated. Not surprisingly, it strikes me as an unfortunate backward step when Kitcher lumps the biological species together with all sorts of other unrelated phenomena into his category "set."

As Sober has stated quite correctly "that species are populations (i.e. physical objects—individuals—of a certain kind) was the idea that filled the vacuum left by the demise of the view that species are natural kinds" (1984:336). Bunge (1981:284) and Salthe (1981:301) also endorse population as the proper designation of species: "The preferable term for this entity is population because that appears to be the simplest or smallest unit of which most of these things can be predicated."

For a biologist, the individual-like properties of biopopulations are obvious. Unfortunately, a few philosophers coming from mathematics or the physical sciences have equated populations with classes. This became particularly clear to me when I read Caplan's discussion in which he claimed that Stebbins had ascribed all sorts of biological and evolutionary phenomena to "sets of organisms" (1981:132). Such a usage by a biologist seemed to me so improbable that I consulted the publication to which Caplan referred (Stebbins 1977), where I discovered that Stebbins had not used the word set (or for that matter, class) even a single time. What he referred to on page after page was "population." Evidently Caplan was not trying to misrepresent Stebbins, but for him populations were sets of organisms. He thereby revealed the deep difference between an essentialist, for whom populations are classes or sets, and a populational biologist, for whom populations are of the nature of individuals, as correctly recognized by Sober (1984:336). How much difficulty some philosophers have in understanding what a biological population is, is indicated by Kitcher's assertion that "there seems to be no implication that if x and y belong to the same population, then they are conspecific" (1984:622). The concept of *population* has up to now been totally missing from the conceptual framework of nearly all philosophers of science. As a result they were forced to translate the term population into something with which they were familiar, usually the term class or set. It is Ghiselin's outstanding achievement to have called attention to this error.

There is no real conflict between the terms individual and population, for a biopopulation has the spatiotemporal properties, internal cohesion, and potential for change of an individual. If a single organism is considered a "singular individual," a population can perhaps be considered a "multiple individual." The crucial properties of the biopopulation that make it an individual are maintained by the two-way street—generation after generation—between conspecific organisms and the gene pool of the species.

The term population for a species is preferable to the term individual for three reasons:

(1) because it has been traditionally used by biologists for species ever since the typological concept of natural kinds was given up,

(2) because the term conveys the impression of the multiplicity and composite nature of species, whereas the term individual implies a nonexisting singularity, and

(3) because it provides a far more suitable basis for the discussion of evolutionary phenomena, particularly speciation.

Most of the objections to the species-as-individual terminology listed above are resolved by the species-as-population terminology.

Difficulties with the Term Population

Like almost all terms in science and philosophy the term *population* does not have a single well-defined connotation. Its original application in the

English language was to the subjects of a ruler, the population of a politically defined area. Eventually it was used for any component of the human species, and later of animal and plant species. Unfortunately, the meaning of the term was eventually still further expanded, and mathematicians sometimes spoke of populations when referring to sets. Several population geneticists used the term more in this mathematical sense than in the sense of actual populations of organisms. Finally, the ecologists expanded it even further, beyond the limits of a single species, when speaking of the plankton population of a lake or of the population of savannah animals. Populations in such a broadened sense lack the most characteristic features of biological populations, their internal cohesion and variability. This was clearly seen by Bunge (1981:284), who referred to the interbreeding, conspecific population of the biologist as the *biopopulation*. It would be most helpful if the use of the word *population* or at least biopopulation could be restricted to this focal meaning of the word population, and in particular, if the misleading use of the word population in the sense of set were abandoned. The heterogeneity of usages of the word population has been evident for some time, and I have referred to it on a previous occasion (Mayr 1963:136–138).

One can recognize a hierarchy of biopopulations ranging from the local population (deme) up to the species ("the largest Mendelian population" of Dobzhansky). This is, of course, no more an argument against the adoption of the population terminology than it would be in the case of class and individual, which can also be organized hierarchically.

Are There Different Kinds of Species?

Systematists, presumably all the way back to Darwin, have been aware of the heterogeneity of species-like phenomena in nature. I considered this heterogeneity of such importance that I devoted an entire chapter (pp. 400–423) to kinds of species in my 1963 book. In the most elaborate attempt to discriminate among kinds of species, Camp and Gilly (1943) recognized 12 categories of plant species. One other well-known enumeration of kinds of species is that of Cain (1954), who recognizes four kinds.

The clarification of the ontological status of species makes it mandatory to examine these so-called kinds of species and determine whether they all have the qualifying criteria of biological species, particularly the internal cohesion and spatiotemporal definiteness of individuals-populations. Even a quick survey of the literature makes it quite clear that there are entities in nature (like asexual clones or strains of prokaryotes) which do not qualify under the biological species concept. The frequency of these exceptions is very different in different groups of organisms. In birds there are none, so far as I know. In prokaryotes they might amount to about 100 percent. In the vascular plants of a local flora, they may comprise about 15 percent. Actually, an improved understanding of the biological species concept has led to a reduction in the percentage of recognized exceptions. In 1957 Grant wrote, "There are a few botanists who adopt the so-called biological species concept" (p. 43). Twenty-five years later Stebbins implied that a majority of plants can be classified as biological species (1982).

To the species that unequivocally qualify as populations belong all regularly sexually reproducing species, all those allopolyploids in which sexuality has been fully restored, and all evolutionary species (sensu Simpson-Wiley).

Uniparentally reproducing organisms. There are many lineages of organisms which perpetuate themselves without sexual reproduction. This includes, for instance, stabilized species hybrids that have reverted to uniparental (asexual) reproductions. It includes all other forms of permanent uniparental reproduction, particularly obligatory parthenogenesis and obligatory hermaphroditism, as well as various forms of vegetative reproduction. All these lineages are clones that are independent of each other, that is, they do not share in a common gene pool. Owing to their mode of origin, such uniparental entities could theoretically originate several times independently, and yet be indistinguishable for all practical purposes. Such a case was actually discovered by M. J. D. White in Australia. The parthenogenetic grasshopper *Warramaba virgo* consists of two allopatric populations that were derived from different strains of the two parental species (White 1980).

All these asexual entities fail to quality as biological species, and the question has been much discussed in the literature as to what ontological status they should be given. Many different solutions have been proposed, which are best presented as three different options.

(1) To accept a concept of species that would apply across all the kingdoms of living organisms. Kitcher's proposal to consider species as sets would be such a solution. It would mean watering down the species concept to the extent that it would apply to any kinds of organisms, even the most aberrant ones. There would seem to be little that recommends this solution. It has two very serious flaws. By confounding various fundamentally different natural entities under one name, it would do again what has already caused so much trouble in the history of philosophy

(as in the definition of teleological, selection, group, adaptation, and so on). To apply a single term to a heterogeneous assemblage is bound to cause trouble sooner or later. Inevitably in such a broadened concept of species, all those criteria will have to be excluded that are particularly characteristic for the majority of species, that is, the characteristics of biological species.

The second flaw is that this solution can be adopted only by going back to a strictly nominalistic or typological species concept such as that suggested by the botanist Cronquist, for whom "species are the smallest groups that are consistently and persistently distinct, and distinguishable by ordinary means" (1978:3). A group of taxonomists, the so-called pattern cladists (such as Rosen 1979), have also returned to the typological species of essentialism. In one of their papers they defined the species as "the smallest detected samples of self-perpetuating organisms that have unique sets of characters" (Nelson and Platnick 1981:12). Since asexual organisms are the only self-perpetuating organisms in nature, this species definition curiously (presumably contrary to the intentions of the authors) applies only to asexual organisms. Also each clone that differs by even a single detectable mutation would be a separate species. In conclusion, it would seem rather obvious that it is an unrealistic objective to try to find a useful species concept that applies equally to all forms of life on earth.

(2) To recognize several different kinds of species. Such a pluralism has been the traditional way of dealing with this problem. It consisted of recognizing biological species, agamospecies, chronospecies, and so on (Cain 1954). Although this solution seemingly gets rid of the problem, it actually does nothing of the sort. For instance, it does not come to grips at all with the problem as to whether these several kinds of species are individuals, populations, or classes. It is by no means a scientifically or philosophically adequate solution.

(3) To restrict the term species to the biological species, the largest cohesive population. This solution has recently been proposed by Ghiselin (1986) and is clearly the most honest solution. This is the species which the modern biologist, whether evolutionist, geneticist, or ethologist, means when he uses the word species. Only that which has the propensity for speciation, says Ghiselin, deserves to be called a species, and this does not include asexual entities.

Unfortunately, even this is not a perfect solution. The fact remains that there are entities in nature that do not qualify as biological species, but which fill the same place in the ecosystem as do biological species. They cannot be ignored in any study of niche occupation, utilization of resources, and competition. An ecologist will say that some of them occupy the same place in nature as do genuine species, and that some way of dealing with them must be found. Although by descent they have continuity in time within the clone, they lack the internal cohesion characterizing a gene pool. Therefore they are classes and not individuals-populations. One possible compromise solution would be to recognize them under a special name, let us say as paraspecies.

Higher Taxa

That higher taxa are not ordinary classes is agreed upon by all recent authors who have given serious thought to their ontological status (Ghiselin 1981; Hull 1976; Fink 1981; Wiley 1981a). They share many characteristics with individuals, particularly their spatiotemporal restriction and their capacity to change (evolve). On the other hand, they differ from individuals (populations), because they lack the most characteristic property of the individual, the internal cohesion. This lack of cohesion is well demonstrated by every striking evolutionary departure, such as that of the birds from the archosaurians or the mammals from the therapsids. Wiley (1980; 1981a) has proposed the felicitous term *historical groups* for higher taxa. This is not the place to discuss to what kind of groups the term historical groups should be delimited, but personally I can see no merit in restricting it to holophyletic groups.

Grades are characterized by certain joint properties but not by any internal cohesion. Clearly they are therefore not individuals. When monophyletic they might qualify for the status of historical groups; when polyphyletic they are clearly classes.

NOTE

This essay originally appeared in *Biology and Philosophy* 2 (1987):145–166, under the title "The ontological status of species."

REFERENCES

Baer, K. E. von. 1828. *Entwickelungsgeschichte der Thiere.* Königsberg.
Beatty, J. 1982. Classes and cladists. *Syst. Zool.* 31:25–34.
Buffon, G. L. 1954, 1753 *Oeuvres Philosophiques,* ed. J. Piveteau. Paris: Presses Universaires de France.
Bunge, M. 1981. Biopopulations, not biospecies, are individuals and evolve. *Behav. and Brain Sci.* 4:284.

Cain, A. J. 1954. *Animal Species and Their Evolution.* London: Hutchinson's University Library.

Camp, W. H., and C. L. Gilly. 1943. The structure and origin of species. *Brittonia* 4:323-385.

Caplan, A. L. 1980. Have species become déclassé? *PSA* 1:71–82.

———— 1981. Back to class: a note on the ontology of species. *Phil. Sci.* 48:130–140.

Cronquist, A. 1978. Once again, what is a species? *Biosystematics in Agriculture,* pp. 3–20. Beltsville Symposia in Agricultural Research, No. 2. New York: Wiley.

Dobzhansky, T. 1951. Mendelian populations and their evolution. In L. C. Dunn, ed., *Genetics in the 20th Century,* pp. 573–589. New York: Macmillan.

———— 1970. *Genetics of the Evolutionary Process.* New York: Columbia University Press.

Eldredge, N. 1985. *Unfinished synthesis: Biological Hierarchies and Modern Evolutionary Thought,* p. 49. New York: Oxford University Press.

Fink, W. L. 1981. Individuality and comparative biology. *Behav. and Brain Sci.* 4:288.

Ghiselin, M. T. 1966. On psychologism in the logic of taxonomic controversies. *Syst. Zool.* 15:207–215.

———— 1974a. A radical solution to the species problem. *Syst. Zool.* 23:536–544.

———— 1974b. *The Economy of Nature and the Evolution of Sex.* Berkeley: University of California Press.

———— 1980. Natural kinds and literary accomplishments. *Mich. Quart. Rev.* 19:73–88.

———— 1981. Categories, life and thinking. *Behav. and Brain Sci.* 4:269–283.

———— 1984. "Definition", "character", and other equivocal terms. *Syst. Zool.* 33:104–110.

———— 1986. Species concepts, individuality, and objectivity. *Biol. and Phil.* 2:127–143.

Grant, V. 1957. The plant species in theory and practice. In E. Mayr, ed., *The Species Problem.* Washington, D.C.: AAAS, publication 50.

Gregg, J. R. 1950. Taxonomy, language, and reality. *Amer. Nat.* 84:421–433.

———— 1954. *The Language of Taxonomy.* New York: Columbia University Press.

Haeckel, E. 1866. *Generelle Morphologie,* vol. 2, p. 30. Berlin: Georg Reimer.

Holsinger, K. E. 1984. The nature of biological species. *Phil. Sci.* 51:293–307.

Hull, D. L. 1967. Metaphysics of evolution. *Brit. J. Hist. Sci.* 3:309–337.

———— 1976. Are species really individuals? *Syst. Zool.* 25:174–191.

———— 1978. A matter of individuality. *Phil. Sci.* 45:335–360.

———— 1980. Individuality and selection. *Ann. Rev. Ecol. and Syst.* 11:311-332.

———— 1981. Metaphysics and common usage. *Behav. and Brain Sci.* 4:290–291.

————— 1984. Can Kripke alone save essentialism? A reply to Kitts. *Syst. Zool.* 33:110–112.

Jevons, W. S. 1877. *Principles of Science*. London: Macmillan.

Jordan, K. 1905. Der Gegensatz zwischen geographischer und nichtgeographischer Variation. *Z. wiss. Zool.* 83:151–210.

Kitcher, P. 1984a. Species. *Phil. Sci.* 51:308–333.

————— 1984b. Against the monism of the moment: a reply to Elliott Sober. *Phil. Sci.* 51:616–630.

Kitts, D. B. 1983. Can baptism alone save a species? *Syst. Zool.* 32:27–33.

————— 1984. The names of species: a reply to Hull. *Syst. Zool.* 33:112–115.

Kitts, D. B., and D. J. Kitts. 1979. Biological species as natural kinds. *Phil. Sci.* 46:613–622.

Lewontin, R. 1983. The organism as the subject and object of evolution. *Scientia* 118:63–82.

Mayr, E. 1957. Species concepts and definitions. In E. Mayr, ed., *The Species Problem*, pp. 1–22. Washington, D.C.: AAAS, publication 50.

————— 1963. *Animal Species and Evolution*. Cambridge, Mass.: Harvard University Press.

————— 1976a. *Evolution and the Diversity of Life*. Cambridge, Mass.: Harvard University Press.

————— 1976b. Is the species a class or an individual? *Syst. Zool.* 25:192.

————— 1982. *The Growth of Biological Thought*. Cambridge, Mass.: Harvard University Press.

————— 1983. Comments on David Hull's paper on exemplars and type-specimens. *PSA 1982* 2, 504–511.

Nelson, G., and N. Platnick. 1981. *Systematics and Biogeography*. New York: Columbia University Press.

Plate, L. 1914. Prinzipien der Systematik mit besonderer Berücksichtigung des Systems der Tiere. In *Die Kultur der Gegenwart*, III (IV, 4), pp. 92–164.

Poulton, E. B. 1903. What is a species? *Proc. Entomol. Soc. London*, pp. lxxvi–cxvi.

Quine, W. V. 1969. Natural kinds. In *Ontological Relativity and Other Essays*. New York: Columbia University Press.

Rosen, D. 1979. Fishes from the upland intermontane basins of Guatemala: revisionary studies and comparative geography. *Bull. Amer. Mus. Nat. Hist.* 162:269–375.

Rosenberg, A. 1985. *The Structure of Biological Science*. Cambridge: Cambridge University Press.

Salthe, S. N. 1981. The world represented as a hierarchy of nature may not require "Species." *Behav. and Brain Sci.* 4:300–301.

Schwartz, S. P. 1981. Natural kinds. *Behav. and Brain Sci.* 4:301–302.

Simpson, G. G. 1961. *Principles of Animal Taxonomy*. New York: Columbia University Press.

Sloan, P. R. 1986. From logical universals to historical individuals: Buffon's idea of biological species. In J. Roger and J.-L. Fischer, eds., *Histoire du concept d'espèce dans les sciences de la vie.* Paris: Fondation Singer-Polignac.

Sober, E. 1984. Sets, species, and evolution: comments on Philip Kitcher's "species." *Phil. Sci.* 51:334–341.

Stebbins, G. L. 1977. *Processes of Organic Evolution.* Englewood Cliffs, N.J.: Prentice-Hall.

———— 1982. Plant speciation. In C. Barigozzi, ed., *Mechanisms of Speciation,* pp. 21–39. New York: Alan R. Liss.

White, M. J. D. 1980. The genetic system of the parthenogenetic grasshopper *Warramaba* (formerly *Moraba*) *virgo* and its sexual relatives. In R. L. Blackman, G. M. Hewitt, and M. Ashburner, eds., *Insect Cytogenetics,* p. 119. Symposium 10, Royal Entomological Society of London. London: Blackwell.

Wiley, E. O. 1980. Is the evolutionary species fiction? A consideration of classes, individuals, and historical entities. *Syst. Zool.* 29:76–80.

———— 1981a. *Phylogenetics.* New York: Wiley.

———— 1981b. The metaphysics of individuality and its consequences for systematic biology. *Behav. and Brain Sci.* 4:302–303.

part seven 㲂

S P E C I A T I O N

Introduction

One might presume that everything about the origin of new species is
understood, a century and a quarter after the publication of the *Origin of
Species,* but this is not the case. New papers on speciation are published
in almost every issue of the evolutionary-taxonomic journals, and symposia
on the subject are still being organized (Atchley and Woodruff 1981;
Barigozzi 1982). The remaining controversies are best discussed under
three headings:

(1) *What is the geographical relation of an incipient or new species to its parental
population?* This relationship is irrelevant in the case of instantaneous
speciation, where a single individual becomes the founder of a new species.
However, such saltational speciation is limited to cases of polyploidy, the
shift to uniparental reproduction in hybrids (White 1978), and a few
other rather rare phenomena. In higher animals and in most groups of
plants, speciation is populational, that is, a more or less isolated population becomes a new species through genetic restructuring.

Even though it is now recognized that most speciation is populational,
this still permits a number of options. A belief in sympatric speciation—
the splitting of a population in two through ecological divergence and
disruptive selection—is still reasonably widespread. I have listed in Essay
21 the titles of a number of recent analyses which refute the claim that
such speciation has been demonstrated. Virtually all those who have made
such claims have ignored the criteria spelled out by Maynard Smith (1966)
and Mayr (1947; 1976) that must be met to substantiate these claims.
Kondrashov and Mina (1986) have provided an unfortunately rather un-
critical review of this problem. Bush and Howard (1986) have summarized
the most recent claims in favor of sympatric speciation. See also Diehl
and Bush (1984).

A second theory is White's theory of stasipatric speciation (White

1978). It claims that individuals with chromosomal mutations can establish local populations within a species and succeed in spreading out if the new chromosomal homozygotes are of superior fitness. I have shown in Essay 21 that the predictions that follow from this theory are not encountered. All recent authors who have carefully studied White's theory have likewise found it to be either invalid or at least highly improbable (John 1981; Barton and Hewitt 1981; Bickham and Baker 1980; and Thompson and Sites 1986).

White was led to his theory by posing the unfortunate and misleading alternative, Is speciation chromosomal or geographical? As I had shown in 1970, these are two entirely independent parameters. It must never be forgotten during the analysis of processes of speciation that a number of different processes occur simultaneously and that it leads to misunderstanding to pose the question of either/or.

According to the theory of parapatric speciation, a species may break apart along an ecological "escarpment," and two closely related neospecies meet, as a result of this process, along a parapatric contact zone, either still hybridizing in this belt or meeting abruptly without hybridization. Actually, no good case of such speciation has ever been demonstrated, while it has been possible to show that existing parapatric distribution patterns can be explained as secondary contact zones of species that had acquired their reproductive isolation during a preceding period of isolation. An attempt by Endler (1982) to refute the explanation of parapatry by allopatric speciation was shown to be based on a misinterpretation of the data (Mayr and O'Hara 1986).

Much recent ecological research has demonstrated that the isolation of populations is closely correlated with the dispersal capacities of the members of these populations. This is particularly well illustrated by marine organisms as summarized by Jablonski and Lutz (1983). Species in which the larvae are dispersed in the plancton have a different speciation pattern from those with direct development without such a larval stage. A correlation between dispersal facility and pattern of speciation has also been demonstrated for island birds (Mayr and Diamond ms.). No general rules can be established for the amount of isolation needed for successful speciation, because it depends on numerous accessory factors. In the African lake cichlids with very strong sexual selection, the distances between incipient species can be very small (Essay 22). In most species, however, a temporary isolation of demes is not sufficient to achieve speciation, since the two gene pools again merge rather rapidly on secondary contact (Lande 1979).

One of the most important insights of recent times is that one must distinguish between two very different kinds of geographic or allopatric speciation. In one of these the previously continuous species range is secondarily interrupted by a barrier to gene flow (White's "dumbbell model" of speciation). The other represents a very different process, the establishment of a founder population beyond the periphery of the species range through a process of primary isolation. This process of peripatric speciation (Essay 21) is apparently far more frequent than that of secondary isolation and may also be of greater importance in macroevolution (see Essays 23 and 25).

All these processes of speciation are populational, hence gradual.

(2) How do new isolating mechanisms evolve? Two theories have been opposed to each other since the days of Darwin and A. R. Wallace. According to the Darwin–Muller–Mayr theory, isolating mechanisms between two potential new species cannot be produced by *ad hoc* natural selection but must have been established prior to the contact of such neospecies. According to the Wallace–Dobzhansky theory, rudiments of isolating mechanisms may have been acquired during prior isolation but are greatly added to and perfected after a contact zone has been established.

Most modern authors agree that there is no evidence that would support the Wallace–Dobzhansky theory. If the isolating mechanisms have not yet been essentially completed, a secondary hybrid zone develops and such a hybrid zone may continue to exist for many thousands of years, as has been demonstrated for the hybrid zones between subspecies and allospecies that had evolved in Pleistocene refuges (Meise 1928 and many other authors). A theoretical analysis of this problem by Spencer et al. (1986) shows that various highly unrealistic assumptions must be made in order to make the Wallace reinforcement theory work. Indeed, it is adopted by hardly any modern author.

It is quite possible that through stabilizing selection there is a slight improvement of the isolating mechanisms of two neospecies when they first meet. However, as I stated at one time (Mayr 1967:85), this would involve only the last 3 percent or less of the to-be-completed isolating mechanisms.

The main objection that was raised against the Darwin theory is reflected by the question, How can isolating mechanisms between two widely separated populations evolve? After all, it was said, isolating mechanisms are *ad hoc* devices that have no other function but to protect the integrity of gene pools; how could natural selection favor such devices? The traditional answer was that they were the incidental byproduct of the reorganization of the genotype of the isolated population necessitated by its adaptation to new ecological conditions characterizing the niche of the new species. Since there is no way of testing this hypothesis, owing to our ignorance of the genetics of speciation, there remained misgivings as to the validity of such an ad hoc hypothesis.

An alternate theory recently proposed ascribes the origin of isolating mechanisms, at least behavioral ones, not to ecological factors but to sexual selection. This is discussed in Essay 22. This, of course, does not explain the origin of postzygotic mechanisms.

(3) What is the genetic basis of speciation? It is quite obvious now that no simple answer to this question is possible. The answer is different for a chromosomal restructuring than for the acquisition of female preferences in connection with sexual selection. In the latter case, changes in a few genes may effect speciation. The observation that the comparison of enzyme genes of closely related species sometimes fails to reveal any difference whatsoever confirms the possibility that relatively few genes may be involved in certain speciation events. On the other hand, peripatric speciation in a small founder population provides the opportunity for a drastic and rather rapid genetic restructuring. Therefore, whenever speciation was accompanied by a relatively drastic change, one can be reasonably certain that it happened during a bout of peripatric speciation. The role which peripatric speciation may play in macroevolution is discussed in Essay 25.

Most of the recent discussions of speciation are flawed by the failure of authors to consider the possibility of such pluralism. Not only is the genetics of different kinds of isolating mechanisms very different but also the geographic situation. Population size, dispersal capacity, and many other factors play a role in controlling or at least affecting the genetics of speciation. This pluralism cannot be stressed strongly enough.

REFERENCES

Atchley, W. R., and D. Woodruff. 1981. *Evolution and Speciation.* Cambridge: Cambridge University Press.

Barigozzi, C., ed. 1982. *Mechanisms of Speciation.* New York: Alan R. Liss.

Barton N. H., and G. M. Hewitt. 1981. Hybrid zones and speciation. In Atchley and Woodruff 1981:109–145.

Bickham, J. W., and R. J. Baker. 1980. Reassessment of the nature of chromosomal evolution in *Mus musculus. Syst. Zool.* 29:159–162.

Blair, F., ed. 1967. *Vertebrate Speciation*. Austin: University of Texas Press.

Bush, G. L., and D. J. Howard. 1986. Allopatric and nonallopatric speciation, assumptions and evidence. In S. Karlin and E. Nevo, eds., *Evolutionary Processes and Theory*, pp. 411–438. New York: Academic Press.

Diehl, S. R., and G. L. Bush. 1984. An evolutionary and applied perspective of insect biotypes. *Ann. Rev. Entomol.* 29:471–504.

Endler, J. A. 1982. Pleistocene forest refuges: fact or fancy? In G. T. Prance, ed., *Biological Diversification in the Tropics*, pp. 641–657. New York: Columbia University Press.

Jablonski, D., and R. A. Lutz. 1983. Larval ecology of marine benthic invertebrates: Paleobiological implications. *Biol. Rev.* 58:21–89.

John, B. 1981. Chromosome change and evolutionary change: a critique. In Atcheley and Woodruff 1981:23–51.

Kondrashov, A. S., and M. V. Mina. 1986. Sympatric speciation: when is it possible? *Biol. J. Linn. Soc.* London 27:201–223.

Lande, R. 1979. Effective deme sizes during long term evolution estimated from rates of chromosomal rearrangements. *Evolution* 33:234–251.

Maynard Smith, J. 1966. Sympatric speciation. *Amer. Nat.* 100:637–650.

Mayr, E. 1947. Ecological factors in speciation. *Evolution* 1:263–288.

—— 1967. In Blair 1967:85.

—— 1970. *Populations, Species and Evolution*. Cambridge, Mass.: Harvard University Press.

—— 1976. *Evolution and the Diversity of Life*. Cambridge, Mass.: Harvard University Press.

Mayr, E., and R. O'Hara. 1986. The biogeographic evidence supporting the Pleistocene forest refuge hypothesis. *Evolution* 40:55–67.

Meise, W. 1928. Die Verbreitung der Aaskrähe (Formenkreis Corvus corone L.). *J. Orn.* 76:1–203.

Spencer, H. G., B. H. McArdle, and D. M. Lambert. 1986. A theoretical investigation of speciation by reinforcement. *Amer. Nat.* 128:241–262.

Thompson, P., and J. W. Sites. 1986. Comparison of population structure in chromosomally polytypic and monotypic species of *Sceloporus* in relation to chromosomally-mediated speciation. *Evolution* 40:303–314.

White, M. J. D. 1978. *Modes of Speciation*. San Francisco: Freeman.

PROCESSES OF SPECIATION
IN ANIMALS

SPECIATION, as M. J. D. White has recently stated quite rightly, now appears to be the key problem of evolution. It is remarkable how many problems of evolution cannot be fully understood until speciation is understood. In view of this, it is somewhat puzzling why genetics has so long neglected a serious study of this process. This has now changed, and White's "Modes of Speciation" (1978)[1] provides a comprehensive summary of the current state of speciation research. The availability of this volume greatly facilitates my task. For many aspects of the speciation problem, I shall simply refer to the relevant chapters of White's book. There have been attempts by authors such as Bateson, de Vries, Morgan, and Goldschmidt to explain speciation by macromutations, or of others to describe it simplistically merely as a change of gene frequencies in populations, but these endeavors failed to do justice to the real situation in nature. To state it factually, speciation research did not become part of the tradition of genetics until very recently.

Some of the reasons for this neglect are obvious. Speciation, except for polyploidy and some other chromosomal processes, is too slow to be observed directly. Therefore, the method of speciation research must consist of an attempt to reconstruct the historical precedents, derive from this reconstruction certain deductive generalizations, and test their validity by proper comparative methods. Since the conditions prevailing at the time of the division of an ancient gene pool into two cannot be observed directly, they must be inferred. In order to arrive at the most probable course of events, one must have a great deal of experience with cases of incipient speciation, and with the genetic, chromosomal, ecological, and geographical circumstances of recently speciated populations and of those that seem just now to be undergoing the process of speciation. There are nearly always several possible scenarios, and it is not surprising that different authors may differ in the choice of their explanations. Owing to the slowness of the speciation process, it is not possible to study the same individual or population "just before" and "just after" speciation. By necessity there is some arbitrariness in the sequence of events one postulates to have occurred, and authors with different professional backgrounds— let us say a geneticist, a cytologist, a paleontologist, a zoologist, and a botanist—may select different scenarios as most probable.

I shall merely mention a second difficulty: the pluralism of speciation processes. Plants offer particularly striking illustrations of this; but even among animals, different species groups within the same genus may have rather different speciation patterns, as shown by the speciation of *Thomomys talpoides* compared with other pocket gophers. This pluralism of speciation processes shows that it is not permissible to apply an answer that is correct in one case automatically to the other situations. Speciation patterns in groups with high dispersal facility are different from those with low dispersal; speciation in groups in which premating isolating mechanisms are acquired first and postmating isolation only subsequently is different from that in groups for which the reverse is true.

A third difficulty, now largely overcome, was that many authors had only vague ideas as to what the term *speciation* really meant. It is now generally agreed that it signifies a multiplication of species—that is, the production of new, reproductively isolated individuals (allopolyploids) or populations.

Perhaps even more damaging and causing greater confusion was the failure to recognize that certain phenomena or processes occur simulta- neously, rather than being "either-or" alternatives. Let me begin with *the chromosomal vs geographical alternative.* White, Bush, and other recent pro- ponents of stasipatric speciation have repeatedly made statements approx- imately of this form: "Chromosomal rearrangements and not geographical isolation must have been responsible for speciation" in a given case. Such a statement implies that one must make a choice between either one or the other factor. By contrast I have long insisted that they are not mutually exclusive and therefore that the two processes are independent of each other and can, but need not necessarily, coincide with each other.[2] Failure to recognize this means that an author will fail to make a distinction between necessary and sufficient conditions. Even if we were to make the extreme assumption that speciation would always require chromosomal reorganization, it could well be possible that such a reorganization could be completed successfully only in an isolated population of minimal size. I shall come back to this question in the discussion of stasipatric speciation.

A second erroneous alternative is that between the *reproductive* and the *ecological* aspects of speciation. Speciation is *not* either the acquisition of reproductive isolation *or* of the ability to compete successfully with other species. It is always primarily the acquisition of reproductive isolation, but sympatry cannot be achieved until sufficient niche segregation has evolved to prevent fatal competition. To this problem I shall also return later in my discussion.

Geographic or Allopatric Speciation

The first authors in the nineteenth century to speculate about the origin of species were local naturalists who studied the variation of species in their home country and attempted to find such deviations from the type as could be considered incipient species. This nondimensional approach was particularly favored by the early botanists, but it has remained the approach of certain zoologists and botanists right up to the present time.

The application of multidimensional thinking to the speciation problem by L. von Buch (1825), Darwin (March 1837), and M. Wagner (1841) was a conceptual breakthrough of the first order.[3] By adopting a new sequence of events, namely by first splitting off a population from the parental species and then accumulating the genetic differences that permit speciation, most of the difficulties were avoided that had previously bedeviled the typological and nondimensional (single-population) approach.

THE DUMBBELL MODEL

The classical concept of the process of geographical speciation was that of a widespread species, the range of which was divided into two halves by a geographical barrier. It is illustrated in the so-called dumbbell diagram. It was this concept most authors had in mind when talking about allopatric speciation, and particularly geneticists and cytologists without personal acquaintance with actual speciation in the field. Actually, there exist many forms of geographic speciation, depending primarily on the size of the isolated populations. One extreme is represented by the dumbbell model, in which two large, subequal portions of the species are separated from each other. No special name exists for this extreme, even though it is what most people had in mind when referring to geographic speciation. More interesting, and apparently far more important for speciation, is the other extreme, where a strong disparity in size exists between the isolated populations.

PERIPATRIC SPECIATION

Zoologists, and some botanists, who analyzed actual speciation patterns among closely related species, however, have long been aware of the disparity situation. Already Darwin, when comparing South American with Galapagos species, knew that speciation is far more active among small island populations than among large continental species. That there is a roughly inverse relation between population size and rate of speciation had long been intuitively appreciated by many students of mammals, birds, fishes, and certain groups of insects, as is evident from the taxonomic literature. Ever since the publication in 1942 of my *Systematics and the Origin of Species,* I puzzled over this difference of rates, but it was not until 1954 that I proposed an entirely new theory of allopatric speciation.[4] How drastically different from traditional geographic speciation this new theory was is missed by all those who, like Michael White, lump the two models and still speak of *"the* allopatric model of speciation." Actually, the two allopatric models are worlds apart. To make this even clearer than it has been in the past, and in order to preclude the continuous confounding of my new model of speciation with traditional geographic speciation of large populations, I propose that my 1954 model be designated as *peripatric speciation.*

Here, the gene pool of a small either founder or relict population is rapidly, and more or less drastically, reorganized, resulting in the quick acquisition of isolating mechanisms and usually also in drastic morphological modifications and ecological shifts. It involves populations that pass through a bottleneck in population size.

I illustrated this process by the distribution pattern of the *Tanysiptera galathea* species group of birds which shows hardly any geographic variation on the mainland of New Guinea over a distance of more than 1,000 kilometers and with a distribution over several climatic zones and across several geographical barriers, whereas all populations on islands off New Guinea (for example, Koffiao, Biak, Numfor) are so strikingly different that they were considered to be separate species. Each of these islands, none of which could have been in recent continental connection with New Guinea, was almost certainly colonized by a founding pair of birds, giving rise to a highly distinct population. My analysis of distribution patterns in the Indo-Pacific archipelagoes revealed an even more spectacular phenomenon. In many cases, when I traced a series of closely related allopatric species, I found that the most distant, the most peripheral species, was so distinct that ornithologists had either described it as a separate genus (*Serresius* derived from *Ducula pacifica, Dicranostephes* derived from *Dicrurus hottentottus*) or had, at least, failed to recognize its true relationship (*Dicaeum tristrami* as an allospecies of *Dicaeum erythrothorax*). In genus after genus I found the most peripheral species to be the most distinct.

It is on this *strictly observational* basis that I proposed in 1954 my theory of drastic speciation in peripheral founder populations—that is, my theory of peripatric speciation. Nothing has been discovered in the more than 25 years since my publication that would have weakened my case. On the contrary, it has been enormously strengthened by the brilliant researches of Carson on speciation of *Drosophila* in the Hawaiian Islands, which have put flesh on the skeleton of my theory.

When it came to provide a genetic interpretation for my observations, all I could offer were hypotheses, because no adequate genetic analysis of any peripherally isolated founder population was as yet available. Still, I saw a number of things rather clearly and emphasized them in my original paper: (1) that there is always a considerable if not drastic loss of genetic variability in a founder population; (2) that there is greatly increased homozygosity in the new population and that this will affect the selective value of many genes as well as the total internal balance of the genotype; (3) that conditions are provided for the occasional occurrence of a veritable genetic revolution; (4) that such a genetic revolution occurs only sometimes, but by no means in all cases of speciation in founder populations; (5) that such founder populations can pass quickly through a condition of heterozygosity in those cases in which the heterozygotes are of lowered fitness as is, for instance, often the case with chromosomal rearrangements; (6) that such genetically unbalanced populations may be ideally situated to shift into new niches; (7) that the genetic reorganization might sufficiently loosen up the genetic homeostasis to facilitate the acquisition of morphological innovations; and (8) that the genetic revolution is a population phenomenon, not like Goldschmidt's hopeful monsters, one of individuals, and that therefore the change, no matter how drastic and rapid it may be, is gradual and continuous; it is not a saltation (or transilience as Galton had called it).[5]

In 1954 geneticists thought that all DNA of the nucleus consisted of a single kind of traditional genes, and, not being a geneticist myself, I naturally based my own tentative explanation on this assumption. It is a source of great satisfaction to me that my interpretation is equally applicable to the newer genetic explanations that have been suggested in recent years. Indeed, any more drastic disturbance of the genotype, particularly one that produces a deleterious heterozygous condition, is coped with much more easily in a small, inbred, founder population than in any other kind of population.

A few years later the botanist Harlan Lewis proposed independently a very similar theory of rapid speciation in peripherally isolated populations of plants and, as I had, emphasized strongly the importance of selection in such genetic revolutions.[6]

As Key has quite rightly remarked, one can recognize an "internal" allopatry in species that have vacant areas inside the overall species range, owing to vegetational or physiographic barriers.[7] Species with low dispersal power may be able to establish founder populations in these vacant areas if they find suitable spots. This is consistent with all requirements of peripatric speciation. The origin of new karyotypic "species" within the range of *Mus musculus* in isolated human habitations in remote peripheral Alpine valleys may be an illustration of "internal" peripatric speciation. It is not parapatric speciation which, according to its defenders, is a rupture of population continuity through selection. Nor is it sympatric, because in the isolated area of the founder population there is no other population of the species. In questions of speciation one must think in terms of populations, and not in terms of gross geography.

One major addition must be made to the theory of peripatric speciation. Haffer has clearly demonstrated relatively rapid speciation in Amazonian rain forest birds which had been isolated in forest refugia during Pleistocene drought periods.[8] Even though these populations were subjected to a severe reduction of population size, they did not go through nearly as drastic a bottleneck as do founder populations. Nevertheless, there has been active speciation in these forest refugia, as confirmed by Vanzolini and Williams for reptiles and by Turner for butterflies. None of these neospecies are so strikingly different as those that I found among Pacific birds, produced by genetic revolutions in peripheral founder populations. Since these refuge populations consisted even at their smallest size presumably of hundreds if not thousands of individuals, their rapid change is probably due more to greatly increased selection pressure (particularly owing to the drastically altered biotic environment) than to the genetic consequences of inbreeding. Relatively rapid speciation occurred apparently also in some Pleistocene refuges in the Holarctic.

Speciation on Continents

The two theories of allopatric speciation demand the existence of natural barriers that can isolate a population sufficiently long and effectively that it can pass through the bottleneck of deleterious heterozygosity and reorganize itself genetically. From the beginning some naturalists have had difficulties in conceiving the existence of such barriers on continents, even though these same authors freely adopted allopatric speciation for islands. Darwin himself accepted various schemes of sympatric speciation to explain speciation on continents, and so have other authors right up to the present. Mr. White's major reason for proposing stasipatric speciation on continents was that "there is no plausible geographic or paleographic reason for believing that the two populations were ever geographically isolated" (p. 108). The known facts thoroughly refute White's opinion. Vegetational, climatic, and other physiographic barriers abound on continents, particularly in the tropics, as demonstrated by literally scores of zoologists. These areas are rich in allopatric species of greater or lesser recency, separated from each other by physiographic barriers. During the Pleistocene and post-Pleistocene climatic fluctuations many, perhaps most, species ranges were fractured and the remnants were pushed into refugia. A series of splendid monographs has appeared in recent decades documenting the causative role of vegetational-climatic barriers for the speciation of Australian birds (Keast), African birds (Moreau-Hall), South American birds (Haffer), South American reptiles (Vanzolini and Williams), and South American butterflies (Turner), to mention only a few papers of a rich literature. Ken Key, Michael White's collaborator in the work on the Australian grasshoppers,[9] has adopted the same interpretation, and the allopatric distribution pattern of nearly all closely related species of morabine grasshoppers makes this interpretation, of course, almost inevitable. Ironically, one of the best listing of cases of allopatric speciation on continents is provided by White himself, labeled, however, as cases of stasipatric speciation.

Fate of Neospecies and Incipient Neospecies

The distribution pattern of neospecies raises the question, What happens during the isolation? Here we must clearly distinguish two very different topics: (1) phenotypic manifestations of species status, and (2) the nature of the genetic mechanisms responsible for these manifestations. Let me begin with the phenotypic manifestations. When such an isolated population survives and expands its range after the obliteration of the barrier, it will sooner or later encounter the range either of its parental species or that of a sister neospecies. The fate of such an encounter depends on two sets of factors: (1) the nature and extent of the isolating mechanisms acquired during the spatial isolation, and (2) the degree of ecological compatibility acquired. If complete reproductive isolation as well as ecological compatibility were acquired, the two neospecies can widely overlap but usually do so only partly, owing to a previous acquisition of different habitat preferences.

If complete postmating reproductive isolation was acquired, but not premating isolation, either no hybrids at all or sterile hybrids will be produced in the zone of contact. Single invaders into the range of the alien species will therefore leave no offspring, as is documented by most parapatric species pairs of morabine grasshoppers. A parapatric pattern, however, can also develop through competitive exclusion when reproductive isolation but no ecological compatibility has developed during the isolation. This seems to be the major cause for parapatric distribution in birds.

Finally, if neither premating nor effective postmating isolating mechanisms had developed during the isolation, a wide or narrow hybrid belt will develop in the zone of contact, the width of the belt depending both on ecological and genetic factors.

In this analysis I have repeatedly used the term *parapatric pattern*. Let me explain this in more detail.

Parapatric Pattern of Distribution

A pattern of distribution in which the borders of closely related species touch, often for considerable distances, without major overlap or massive hybridization is called parapatric distribution. Some cases of this pattern had long been known, but only the careful mapping of tropical and Australian species has revealed how frequent parapatry is. The problem we have is to explain the reasons for the existence of this pattern. We are facing here an explanatory problem that applies to all sciences and which deals with historical narratives. Our task is, as I pointed out at the beginning, to infer from the present pattern through what past events and processes it had originated.

There are two possible interpretations of the causation of parapatry. According to the traditional explanation, which I have just presented, the two parapatric species speciated in isolation, expanded their ranges subsequently, and finally met in the zone of parapatric contact. Mutual overlap of their ranges is prevented either for ecological reasons (competition) or owing to the production of sterile hybrids resulting from a lack of premating isolating mechanisms.

White and other authors have advanced an alternative theory according to which parapatric patterns are due to in situ speciation by which the previously continuous range of an ancestral species is disrupted along a step in an environmental gradient. I find this explanation singularly unconvincing. [10]

The fact that for many species pairs the parapatric zones coincide, in the absence of any physiographic or vegetational line, is particularly convincing evidence for the secondary origin of these zones. [11] I must confess that in all the cases of postulated parapatric speciation listed by White in chapter 5 of his book, there is not a single one in which the nonallopatric mode of origin would seem more probable to me than the allopatric one. Parapatric patterns of distribution, thus, as I see it, are always caused by the fact that the development of the isolating mechanisms between the two neospecies had not yet reached perfection.

A sterility barrier, caused by chromosomal restructuring, is only one among the known kinds of isolating mechanisms between species. The true nature of isolating mechanisms was appreciated by most evolutionary biologists remarkably late. Dobzhansky in 1941 and H. J. Muller in 1942 still included geographical barriers among the isolating mechanisms. The true intrinsic isolating mechanisms can be divided into premating (mostly behavioral or ecological) mechanisms, and postmating mechanisms. Because it was found in birds and some other groups of animals that often there were behavioral (premating) but no sterility barriers between closely related species, it was long assumed that in animals the evolution of premating barriers precedes that of postmating (sterility) barriers, whereas in plants speciation would always begin with the development of sterility barriers. However, we now also know several groups of animals in which the evolution of sterility barriers preceded that of behavioral barriers. This is, for instance, true for the Australian grasshoppers on which White and Key are working, and apparently also for some groups of amphibians. The literature reveals that authors who work with groups of animals or plants in which sterility is the primary isolating mechanism find it very difficult to accept speciation primarily based on the origin of behavioral barriers. Yet, such a process is now abundantly established for numerous animal groups. Unfortunately, I do not have the time to discuss various aspects of behavior that affect patterns of speciation in animals.

The Genetics of Allopatric Kinds of Speciation

The traditional presentation of speciation in genetics textbooks was that of a continuous piling up of gene differences, leading to a gradual divergence of populations separated by a barrier. The discovery of the method of enzyme electrophoresis raised the hope that this method would reveal the nature of the genetic changes occurring during speciation. This hope was not fulfilled. Nevertheless, this method supplied the extremely important evidence that in speciation there is no major involvement, if any at all, of the classical enzyme genes. There are multiple proofs for this conclusion, perhaps the most convincing being that the rate of isozyme replacement is no more rapid in actively speciating phyletic lines than in conservative ones. With the enzyme genes removed from consideration, we must ask what other kinds of genetic factors are responsible for speciation.

Let me begin with a consideration of the chromosome as a whole. Since in the eukaryotes virtually all the genetic material is located on the chromosomes, no one will question that the chromosomes are important in speciation—the only question being, in what way? To be sure, the occurrence of homosequential speciation in many Hawaiian species groups of *Drosophila* and in other genera of dipterans demonstrates that speciation is possible without gross chromosomal changes. Nevertheless, the frequency with which closely related species of most groups of organisms differ in chromosome structure testifies to the frequency at which chromosomal reorganization seems to accompany speciation.

The explanatory value of this observation, however, is quite limited. First of all, chromosomal rearrangements have at least two very different effects. The first is that any breaks may affect the neighborhood of genes, hence various kinds of position effects in the broadest sense of the word. The second effect is that such rearrangements often inhibit or prevent crossing over and thus protect coadapted linkage groups against being broken up. Also rearrangements often cause difficulties during meiosis. There is another difficulty: The discovery of transpositional DNA and of the various kind of repetitive DNA have somewhat blurred the sharp demarcation between genes and gross chromosomal changes. How does this affect the question, Is the genetic basis of speciation genic or chromosomal?

There is one point, however, that needs to be emphasized. Michael White and other cytogeneticists have shown that, broadly speaking, chromosomal rearrangements fall into two classes. In one class heterozygotes are of normal viability or even superior—that is, heterotic—and this situation leads to chromosomal polymorphism. In the other class, heterozygotes are inferior or even semilethal. This is precisely the class to which so often the chromosomal differences between closely related species belong. As I have emphasized for more than 10 years, there is no surer way to get quickly through this deleterious heterozygous condition to a new superior homozygosity than in a highly inbred founder population. Such a rapid chromosomal replacement may accompany or even cause a genetic revolution. The new homozygote may be established within two generations, but the chance of two heterozygotes finding each other in a large population is infinitely smaller.

A third possible major genetic cause of speciation is some as yet unidentified fraction of DNA—a possibility I mentioned more than 15 years ago.[12] Numerous cases of incipient or complete speciation, evidently caused by genetic factors other than enzyme genes, have been described in the literature. For instance, in several North American species of *Lepidoptera* there is no discernible geographic variation of enzyme genes, yet there is a considerable amount of hybrid inviability in crosses between different populations.[13] To what class of DNA these inviability (or in other cases sterility) factors belong is unknown. It could be repetitive DNA, but it could also be a category of single-coding genes, genes outside the class of enzyme genes revealed by electrophoresis.

It has become fashionable to say that regulatory genes are the genetic agents of speciation. Hedrick and McDonald rightly remark that "changes in genetic regulation would be the favored genetic strategy for a population adapting to a sudden and substantial environmental change."[14] For no other kind of population would this be so true as for a founder population undergoing a genetic revolution.

When finally identified, I am quite certain that the genetic speciation factors will consist of several classes. For instance, they might well be different in species in which the premating isolating mechanisms developed first, from species in which postmating mechanisms (such as sterility) developed first.

That a breaking up of the internal balance of the genotype may be an important component of peripatric speciation is indicated by the frequency with which the resulting neospecies give rise to further new rapidly speciating founder populations. As each neospecies becomes adapted to the new environment beyond the ancestral species range, it becomes especially suited to serve as source of colonists of new environments. The distribution of the neospecies of *Spalax* and *Proechimys,* as well as the successive speciational steps in Hawaiian *Drosophila,* illustrates this process particularly well.

Nongeographical Kinds of Speciation

Although no one can any longer question the importance of the various categories of allopatric speciation, and although many zoologists think that they are the prevailing modes of speciation among animals, it must not be forgotten that other forms of speciation have been proposed, and I would like to comment on them now.

A saltational origin of new species was postulated long before there was any theory of evolution and was popular among many of Darwin's contemporaries, such as T. H. Huxley, Galton, and Koelliker. Bateson, de Vries, and most early Mendelians favored it, and so did several later paleontologists, including Schindewolf and the geneticist Goldschmidt, with his "hopeful monsters." The crucial aspect of this postulated type of speciation is the typological concept of the production of a single individual, which then becomes the progenitor of a new species.

It took a long time before the virtual impossibility of this happening in a sexually reproducing species was understood. The demonstration that de Vries's new Oenothera species were not species greatly helped to clear the air. There are, however, three potential processes by which such instantaneous speciation can indeed occur. Two of these are autopolyploidy in a sexual species, and any kind of macromutations in uniparentally reproducing organisms. There is still much uncertainty as to what role these two types of speciation play. Autopolyploidy presumably is quite unimportant.

There is, however, a third and reasonably common type of instantaneous speciation, and that is the production of stabilized species hybrids. This was first discovered for plants in which the doubling of the chromosome number of more or less sterile species hybrids may produce *allopolyploids,* capable of sexual reproduction. Such polyploidy is very common in plants, but, as White has once more clearly demonstrated in Chapter 8 of his *Modes of Speciation,* it is exceedingly rare in animals. What happens, however, not infrequently in animals is that a product of hybridization shifts to uniparental reproduction and forms a new parthenogenetic or thelytokic species (see Chap. 9 in White). There are some special cases of persistent hybridity, typified by *Rana esculenta* and some American freshwater fishes, in which the paternal chromosome set is invariably lost in meiosis, and restored by mating with the respective paternal species. As successful as some of these stabilized species hybrids may be in the short run, they appear to have little long-range evolutionary success.

Sympatric Speciation

The form of nongeographic speciation that is considered most probable and most frequent by many students of speciation is *sympatric speciation* by host specialization. Sympatric speciation is favored particularly by authors who have little experience with geographical variation. I must have in my files some 15 or 20 reprints proposing sympatric speciation in which the authors have ignored even the most elementary potential objections to their schemes. I tabulated in 1947[15] the difficulties encountered by hypotheses of sympatric speciation, but most of these difficulties are not at all taken into consideration by these modern authors. The difficulties encountered by the proposals of sympatric speciation have recently also been emphasized by others.[16]

I am sometimes accused of having always adamantly denied any possibility of sympatric speciation. A glance at my previous publications shows that this accusation is not justified.[17] I have always taken the possibility of sympatric speciation seriously, but what I objected to was the superficial manner by which the claims of sympatric speciation were advanced. In nearly all cases listed in the literature, allopatric speciation is at least as probable an explanation for the particular example cited as is sympatric speciation. Let me mention only one case. The existence of sympatric groups of closely related species of beetles on small oceanic islands (St. Helena, Rapa) would seem, at first sight, conclusive proof of sympatric speciation. But when one looks at these cases more carefully, one is struck by three facts: (1) most of these species are not at all host-specific; (2) they are flightless; and (3) these are volcanic islands, crisscrossed by old lava flows. Carson quite rightly points to the parallel situation with speciation of Drosophila on the island of Hawaii. As he says: "Oceanic volcanic shields invariably have a long period of violent growth wherein lava flows from the rifts periodically destroy the forests in haphazard mosaic patterns (Kipuka formation). These continuing cycles of building, destruction, isolation, and recolonization would be expected to impose a pattern of spatial isolation and founder effects even over very short distances. These influences would appear to leave ample opportunity for allopatric speciation."[18] I fully agree with this interpretation. I do not object to the proposals of sympatric speciation, but I deplore it when authors completely ignore alternate explanations.

Among the numerous claims of sympatric speciation, none has been promoted with more vigor than the derivation of an "apple species" of *Rhagoletis pomonella* fruit flies from a hawthorn population.[19] However, not even this case is securely established, and serious questions have been raised about it. Even the nearest relative of *pomonella,* the *Rhagoletis suavis* group, clearly exhibits a pattern of allopatric speciation. Furthermore, as I pointed out in 1947, the existence of species-rich genera of host specialists by no means proves sympatric speciation. It merely means that they have more niches available than generalists.

The conclusion we can draw from the literature is that sympatric speciation by a shift of host preference is a possibility but that it has not yet been conclusively substantiated. After analyzing the documentation, Paterson concludes: "I am yet to be convinced that sympatric speciation has ever occurred."[20] I likewise doubt that any of the putative cases of sympatric speciation reported in the recent literature will survive a truly critical analysis. This includes the numerous cases of postulated sympatric speciation in freshwater.

Bush et al have suggested that in the Equidae (horse family) the social-reproductive structure of the small herds was inducive to sympatric-stasipatric speciation.[21] This ignores the fact that the young individuals of one sex always join other herds, or establish their own herd, and that this results in very active gene flow. The distribution pattern of related species of horses, asses, and zebras documents allopatric speciation quite graphically. The rather drastic chromosomal repatterning that seems to have occurred during some of the speciation events in this group was surely much more easily achieved in a peripherally isolated founder population than inside the population continuity of the parent species.

Stasipatric Speciation

The fact of chromosomal speciation poses a problem. If a chromosomal rearrangement produces superior heterozygotes, its origin will lead to polymorphism but not to speciation. If it produces deleterious heterozygotes, these will be eliminated by natural selection in a large panmictic population before they have any chance to attain the relatively high frequency required before the production of the new type of homozygotes becomes probable. Yet, heterozygously deleterious chromosomal rearrangements must sometimes be able to pass through the bottleneck of heterozygosity: How else could we explain the frequent presence of chromosomal differences between closely related species, which are deleterious in the heterozygous condition? I have previously discussed this problem in detail and have proposed that these chromosomal reconstructions are best achieved peripatrically in small inbred founder populations, where homozygosity of the new chromosomal type can be attained in two generations.

White, by contrast, has proposed a different process, which he calls *stasipatric speciation*.[23] According to this model, a new gene arrangement may turn up anywhere within the range of a species, gradually become more and more common, and spread out from the place of origin, until the number of heterozygotes has reached such a high frequency that homozygotes will finally occur. White states clearly the two reasons that induced him to reject the model of chromosomal speciation in peripheral isolates. First, White could not conceive of any isolating barriers on continents, and secondly, he believed that genetic revolutions in founder populations are restricted to genic changes. Both of these assumptions are misconceptions, as I have pointed out previously.

What is involved is the question of primacy. White has stated: "More and more it appears as if chromosomal rearrangements . . . have played the primary role in the majority of speciation events,"[24] and he makes it quite clear that by "primary" he means preceding in time any kind of spatial segregation of the speciating individuals. For White, the chromosomal change comes first, and the new chromosomal type creates its own deme. For me, the new chromosomal type has no chance to achieve homozygosity unless it occurs in an isolated deme; hence the isolation of the deme comes first. In other words, peripatric speciation is involved.

Some of White's own followers are confused about the problem of chronological primacy. When Bush et al. describe chromosomal speciation as follows: "A karyotypic mutation that has become fixed in a given deme can act as a sterility barrier,"[25] they imply that the existence of a (an isolated) deme precedes the fixation of the karyotypic mutation, but such a process would clearly be peripatric and not stasipatric speciation.

The causal relation between chromosomal rearrangements and genetic revolutions in peripheral isolates is asymmetrical: The homosequential *Drosophila* species of Hawaii prove that not every speciation in a peripheral isolate is correlated with a chromosomal rearrangement; on the other hand, there is much evidence to suggest that chromosomal restructuring occurs preferentially, if not always, in peripheral isolates. I have been unable to find any evidence to substantiate White's claim of a temporal precedence of chromosomal over geographical factors. The proof he offers, that the distribution maps of currently allopatric species indicates that the most primitive species are peripheral, is inconclusive since both Ken Key and I, looking at these very same maps, have concluded exactly the reverse.

The model of stasipatric speciation has one great virtue: It can be tested readily. According to White, a new center of stasipatry (a new chromosomal heterozygote) may turn up anywhere in the species range. Consequently, in any widespread species subject to chromosomal speciation, one should find numerous enclaves of new chromosomal species with smaller or more extended ranges, like oases in a desert. Actually, such a distribution pattern is totally unknown. New chromosomal rearrangements that can serve as isolating mechanisms invariably occur peripherally parapatric, as is consistent with the founder model, but not with the stasi-patric one. I am sorry to have to say that the stasipatric model has been completely falsified by this test. If we look at cases of chromosomally speciating mammals, like *Spalax*, *Proechimys*, or *Thomomys*, to mention only three genera, the patterns of the new chromosomal species are invariably allopatric. The same is, of course, true even for the species groups of Australian morabine grasshoppers. Their ranges are invariably allopatric or parapatric, as one would predict from the peripatric explanation, and no oases or enclaves are ever found, as the stasipatric model would predict. As a consequence, White's own co-worker in the grasshopper work, Ken Key, rejects the stasipatric hypothesis and has accepted allopatric specia-tion for the origin of species in these animals. The adherents of stasipatric speciation may object to my claim that no cases of stasipatric enclaves are known anywhere and may cite the rodent *Ellobius talpinus*. This species has indeed 54 chromosomes from the Crimea to Mongolia but also a 31 chromosome population seemingly in the middle of that area. However, this aberrant population occurs in a single isolated valley in the Pamir Mountains, which is peripheral in the total species range.

Key has rightly pointed out that in species with low dispersal facilities (and/or rather specific habitat requirements), one can also find "internal" founder populations—that is, populations colonizing previously unoccu-pied spots within the species range.[26] These new founder populations may be quite isolated for some time and undergo peripatric speciation during this isolation exactly like peripherally isolated populations. The chromosomal races of *Mus musculus* in the Alpine valleys may be due to such a process. Mouse populations in these valleys are furthermore confined (as commensals) to isolated human habitations. In these small founder populations the occurrence of chromosomal fusion homozygotes is made possible by stochastic processes, and if the new karyotype is adaptively superior, the newly formed population will be able to spread even against other karyotypes. The same interpretation is applicable to the origin of new mouse karyotypes at other localities.

The Ecology of Speciation

Since speciation is as much a population phenomenon as a phenomenon of genetic mechanisms, any aspect of the structure of populations is also of potential relevance to speciation. The botanists were the first to under-stand this, perhaps because plants have a so much richer repertory of modes of reproduction. Population size, as I mentioned already, is inversely related to rate of speciation. When J. Diamond and I (ms in preparation) assigned the species of Northern Melanesian birds to five classes, based on their propensity for dispersal, we discovered that each of these five classes had a class-specific speciation pattern. High rates of dispersal mean high rates of gene flow, and such gene flow leads to the formation of widespread, relatively uniform populations and species.

As stated by Patton and Young, "Gene flow is a key element in the population biology of *T(homomys) bottae* and perhaps other gophers as well."[27] The claims of the unimportance of gene flow, based on highly arbitrary and unrealistic assumptions made by some recent authors, have been rejected with sound reasons by Jackson and Pounds.[28] The uniformity of species is, of course, not caused solely by gene flow, but also by the regulatory system of species, but gene flow is an important factor in determining species structure and pattern of speciation. A comparison of the speciation patterns of marine species with high and those with low larval dispersal confirms this most impressively. Great care must be exercised in all species of animals with a social population structure in which, for instance, all the young of only one sex may be expelled from the family group. Gene flow in such species is largely controlled by the expelled individuals. Factors such as this have, unfortunately, been entirely ignored in some recent speculations.

NOTES

This essay first appeared in C. Barigozzi, ed., *Mechanisms of Speciation,* pp. 1–19. New York: Alan R. Liss, 1982.

1. White, M. J. D. 1978. *Modes of Speciation.* San Francisco: Freeman.

2. Mayr, E. *Populations, Species, and Evolution.* Cambridge, Mass.: Harvard University Press, 1970.

3. Mayr, E. 1963. *Animal Species and Evolution,* pp. 483–486. Cambridge, Mass.: Harvard University Press.

4. Mayr, E. 1954. Change of genetic environment and evolution. In J. Huxley, A. C. Hardy, and E. B. Ford, eds. *Evolution as a Process,* pp. 157–180. London: Allen and Unwin. Also in *Evolution and the Diversity of Life,* pp. 188–210. Cambridge, Mass.: Harvard University Press, 1976.

5. Templeton has recently proposed reviving Galton's term *transilience* (1894) for the process I have referred to as *genetic revolution.* Galton, however, believed in blending inheritance and, therefore, postulated that evolutionary change de-

pended on "sports" or "transiliences"—that is, on single deviant individuals that could not "regress backwards towards the typical centre" (p. 368). The process suggested by Galton is so completely different from the rapid reorganization of populations during genetic revolutions that it would be altogether misleading to apply Galton's term to this gradual populational phenomenon.

6. Lewis, H. 1962. Catastrophic selection as a factor in speciation. *Evolution* 16:257–271.

7. Key, K. H. L. 1968. The concept of stasipatric speciation. *Syst. Zool.* 17:14–22.

8. Haffer, J. 1969. Speciation in Amazonian forestbirds. *Science* 165:131–137.

9. Key, K. 1981. Species, parapatry and the morabine grasshoppers. *Syst. Zool.* 30: 425–458.

10. Parapatric speciation is particularly favored by Clarke (1968) and Murray (1972): B. Clark. Balanced polymorphism and regional differentiation in land snails. In E. T. Drake, ed., *Evolution and Environment*, pp. 351–368. New Haven: Yale University Press. J. J. Murray. *Genetic Diversity and Natural Selection*. Edinburgh: Oliver and Boyd. Mosaic distributions where rather different populations meet discontinuously (as in cases of area effects) are frequent in land snails. This does not lead to speciation, as shown by *Cepaea nemoralis*, and numerous other examples.

11. Meise, W. 1928. Rassenkreuzungen an den Arealgrenzen. *Verh. Dtsch. Zool. Ges.,* pp. 96–105. Many of the suture lines go for hundred or thousands of kilometers through regions that had been uninhabitable during the preceding isolation thus documenting their secondary origin.

12. Mayr, E. 1965. Selektion und die gerichtete Evolution. *Naturwissenschaften* 52:173–180.

13. Oliver, C. G. 1979. Genetic differentiation and hybrid viability within and between lepidopterous species. *Amer. Nat.* 114:681–694.

14. Hedrick, P. W., and J. F. McDonald. 1980. Regulatory gene adaptation: an evolutionary model. *Heredity* 45(1):83–97.

15. Mayr, E. 1947. Ecological factors in speciation. *Evolution* 1:263–288. (Rptd. in *Evolution and the Diversity of Life*, pp. 144–175.)

16. Futuyma, D. J., and G. C. Mayer. 1980. Non-allopatric speciation in animals. *Syst. Zool.* 29:254–271. Jaenike, J. 1981. Criteria for ascertaining the existence of host races. *Amer. Nat.* 117:830–834. Paterson, E. H. 1981. The continuing search for the unknown and unknowable: a critique of contemporary ideas on speciation. *South Afr. J. Sci.* 77:113–119. Sympatric speciation was scheduled to be treated at the Barigozzi symposium by Professor G. Bush, and thus I limited myself to just a few comments on this mode of speciation. At the last minute Professor Bush was prevented from attending, and thus the topic was inadequately covered at the conference. For a recent treatment see G. L. Bush and D. J. Howard, "Allopatric and nonallopatric speciation; assumptions

and evidence" in S. Karlin and E. Nevo, eds., *Evolutionary Processes and Theory* (New York: Academic, 1986), pp. 411–438.

17. Mayr, E. 1942. *Systematics and the Origin of Species*, p. 215. New York: Columbia University Press. *Animal Species and Evolution*, pp. 449–480. *Populations, Species, and Evolution*, pp. 256–275.

18. Carson, H. L. 1979. Chromosomes and species formation. *Evolution* 32:925–927.

19. Bush, G. L. 1975. Modes of animal speciation. *Ann. Rev. Ecol. Syst.* 6:339–364.

20. See note 16, above.

21. Bush, G. L., S. M. Case, A. C. Wilson, and J. F. Patton. 1977. Rapid speciation and chromosomal evolution in mammals. *Proc. Natl. Acad. Sci. USA* 74:3942–3946.

22. *Populations, Species, and Evolution*, pp. 310–319.

23. White, M. J. D. 1968. Models of speciation. *Science* 159:1065–1070.

24. *Modes of Speciation*, p. 336.

25. See note 21, p. 3945.

26. See note 9, above.

27. Patton, J. L., and S. Y. Yang. 1977. Genetic variation in *Thomomys bottae* pocket gophers: macrogeographic patterns. *Evolution* 31:697–720.

28. Jackson, J. F., and J. A. Pounds. 1981. Comments on assessing the dedifferentiating effect of gene flow. *Syst. Zool.* 28:78–85. Also 1981. Riverine barriers to gene flow and the differentiation of fence lizard populations. *Evolution* 35:516–528.

essay twenty-two ❧

EVOLUTION OF
FISH SPECIES FLOCKS

❧ According to the Darwin-Muller-Mayr theory of allopatric speciation, isolating mechanisms are acquired by an incipient species population while isolated from the parental species. However, there remained considerable uncertainty concerning the mechanism or the selection force responsible for this acquisition. The most persuasive proposal was that such isolating mechanisms were an incidental byproduct of the general genetic restructuring of the genotype during isolation, particularly under the influence of selection forces exerted by the novel physical and biotic environment. But there was no real evidence to support this hypothesis, and it was only accepted in the absence of a better solution. In this essay I discuss the recently proposed theory that behavioral isolating mechanisms may be acquired through sexual selection. This theory simultaneously explains that under certain circumstances such a genetic change can take place in neighboring populations.

THAT freshwater lakes, particularly old ones, sometimes have very species-rich endemic faunas of fishes, snails, crustaceans, and other invertebrates has long been known. Seemingly monophyletic groups of closely related species coexisting in the same area are often referred to as species flocks (Greenwood 1984b). There is no need to define this term of convenience too rigidly. Common descent and coexistence in the same lake are the earmarks of lacustrine species flocks.

The existence of species flocks has raised a number of interesting questions, two of which have generated controversies that have not yet been muted: (1) Can the presence of scores, if not hundreds, of closely related species in a single lake be explained by allopatric speciation? (2) Do these species flocks refute the principles of competitive exclusion? Or, stated differently, how can so many sympatric species partition the available resources in such a way as to avoid too intense competition?

Mode of Speciation

Being unable to discover any major geographic barriers within lakes supporting species flocks, most early authors postulated some process of sympatric speciation. This is not the place to describe the manifold difficulties of the sympatric model, which ignores the genetic difficulties (Mayr 1976) and is not substantiated by any incipient cases (Greenwood 1984a). Such a model is difficult to visualize, since speciation is a populational process and therefore "niche space," always having a spatial dimension, is at least microallopatric.

Some sympatric sibling species of fishes have only slight differences in their allozymes (Echelle and Kornfield 1984). However, this cannot be considered evidence for sympatric speciation. It is now quite evident that the allozymes, isozymes, or electromorphs have little or nothing to do with the speciation event (consensus in Barigozzi 1983). The amount of difference in allozymes between two species is a rough indication of the time that has passed since their phyletic lines split apart from each other. In rapidly speciating groups, perfectly good species may have adequate isolating mechanisms and yet show no electrophoretic differences at all. Allozyme differences among sibling species may have evolved subsequent to the speciation event, and even subsequent to the reestablishment of sympatry. What is now needed is a careful comparison of closely related sympatric and allopatric species to determine the average amount of allozyme difference between these two groups of species. The genus *Labidochromis* in which many geographically isolated species occur would seem particularly well suited for such a comparison.

The discovery of two markedly different trophic morphs in *Cichlasoma minckleyi* has revived the theory of speciation by disruptive selection. If assortative mating should develop among the members of two morphs, the papilliform and the molariform, it would represent a case of sympatric speciation. How probable is such a development? The hypothesis of disruptive speciation in polymorphic species must be regarded from the viewpoint of natural selection. We must ask which of two individuals would in the long run leave more offspring: individual *A* which can produce several morphs, each partially removed from competing with its brothers and sisters; or individual *B* which has only one kind of offspring restricted to a single niche? My guess is that individual *A* would leave more offspring, because it would be far better able to cope with resource fluctuations and changes in the environment. A narrowing down of the food niche through increased specialization would be favored by selection only when the range of the polymorphic population was invaded by a related and competing species. If the latter were equally polymorphic, each species would presumably be the superior competitor in a different subniche. If the new invader were already specializing in one of the subniches, it would almost surely displace the resident polymorphic species from that subniche. The occupation of several subniches by an ecologically polymorphic species thus is like hedging a bet: it permits an instantaneous retreat to a subniche not occupied by an invading competitor. Certain papers in this volume propose speciation by disruptive selection owing to the sympatric disruption of a polymorphic population (for example, Humphries 1984), but I did not find the arguments particularly convincing.

Rensch (1933) was apparently the first to attempt to explain the rich species flocks in freshwater lakes by various modes of allopatric speciation. I adopted this interpretation in 1942, and during the 25 years after 1949 most authors endeavored to show that the species flocks in freshwater lakes could be explained by allopatric speciation (Brooks 1950; Fryer and Iles 1972; Greenwood 1974). However in the 1970s theories of sympatric speciation again found favor owing to the discovery of the distinct trophic morphs in Cuatro Cienegas (Sage and Selander 1975) and the unexpectedly high number of sibling species of cichlids in East African lakes. The question was raised again: Can such rich species flocks be explained by the process of allopatric speciation? While investigating this question, it is particularly important to remember two aspects of these species flocks: other higher taxa in the same lakes have not evolved species flocks, and the sister groups of lacustrine species flocks in nearby river systems likewise have not evolved species flocks. It is thus evident that there must be some aspect in the life history and perhaps the genome of certain taxa which permit them to speciate rapidly inside of lakes.

The solution to the problem of explosive speciation in freshwater lakes must be looked for in the behavior and ecology of the speciating taxa. Alas, we will probably never learn the answer for the 18 endemic species of cyprinids of Lake Lanao on Mindanao because 15 or 16 of these species are now extinct (Kornfield and Carpenter 1984), and nothing can be said about their niche occupation and behavior. Nor can one speculate on the contribution of behavior or ecology to the speciation of the 25 or so endemic species of the cyprinodont *Orestias* of Lake Titicaca in the Andes because there are little or no life history data. Many of the 43 species of this genus occur in Andean stream systems of Peru, Chile, and Bolivia, and several colonizations of the Lake Titicaca Basin are evident (Parenti 1984). Nevertheless, there are monophyletic species assemblages inside the lake which evidently originated through intralacustrine speciation.

Brooks (1950) reports on the early history of the problem of speciation in freshwater lakes. Rensch (1933:38), in an attempt to refute Woltereck's (1931) hypothesis of sympatric speciation, proposed three mechanisms to account for the richness of the lake faunas: (1) Repeated colonizations from rivers leading into and out of the lakes; (2) allopatric speciation in different portions of the lake; and (3) an accumulation of species through a fusion of previously existing bodies of fresh water.

Truly intralacustrine speciation was minimized in this scheme and in my own (1942) presentation, which was largely based on Rensch's. Our knowledge of species flocks in freshwater lakes has increased enormously in the past 30 years, and the need for a new analysis of the problem of species flocks in freshwater lakes is apparent. Does the new evidence substantiate the occurrence of sympatric speciation and, if not, what is the relative importance of various allopatric speciation processes?

The reason why no definitive answers to these questions have yet been reached in spite of the excellent research carried out in recent years are the following:

(1) There is still great uncertainty concerning the age of the species flocks and of the lakes in which they occur. The age of Lake Victoria, for instance, is given as 750,000 years by some, but as only 100,000 years by others, and still others suggest that the lake basin had almost entirely dried up during one phase of the Pleistocene. Or to take another example, the nearest relatives of *Orestias* of Lake Titicaca are still uncertain, and the origin of the genus would have to be placed in the early Mesozoic if the unlikely suggestion were true that the Anatolian cyprinodonts are the sister group of *Orestias* (Parenti 1984). Similar uncertainties exist for nearly all the lakes and the taxa they contain. For instance, the age of the Lake Lanao species flock evidently is far greater than the 10,000 years postulated in the earlier literature (Kornfield and Carpenter 1984). Even if the maximum age is accepted for Lake Victoria, this is much shorter than the time intervals normally available to the student of fossils, and it is a time span seemingly totally inadequate to explain the origin of a flock of 250 haplochromine species.

(2) Taxonomic analysis is still incomplete. Sibling species abound in most species flocks and many earlier type series consist of two or three sibling species. Yet, as Greenwood (1984a) points out, these are unquestionably good biological species. There is no intergradation among them and where their breeding habits were studied, they have been found to be completely isolated reproductively. Genetically they are good species, no matter how similar they may be morphologically. The problems escalate in the case of fossil species.

(3) Limitations to the cladistic technique. Groups of closely related species have exceedingly similar genotypes. There is a great potential

among such species to develop independently the same synapomorph characters. Not surprisingly Lewis (1982), during a cladistic analysis of the species of *Labidochromis* from Lake Malawi, came to the conclusion that these species do not form sister groups. That such pluripotentiality is a major obstacle to the application of the cladistic method to close relatives has long been known. Other problems with applying cladistic methods to species flocks are discussed in Barbour and Chernoff 1984; Echelle and Echelle 1984.

These various classes of difficulties must be carefully kept in mind during the analysis of potential patterns of speciation.

It is now rather generally agreed that there are three possible major causes for species flocks in lakes: (1) multiple colonization; (2) amalgamation of smaller lakes owing to a rise in water level; (3) intralacustrine speciation by one or another of several possible mechanisms (Smith and Todd 1984). Although Rensch and I thought that the first-mentioned two causes make the major contribution to the size of the species flocks, it is now evident, particularly for the immense species flocks of African cichlids, that intralacustrine speciation was largely responsible. Even though multiple colonization is not entirely excluded from having contributed to the species flocks in the African lakes, this process was apparently more important for temperate zone lakes, for Lake Titicaca, and for atherinids in lakes on the Mesa Central of Mexico (Echelle and Echelle 1984). A fusion of smaller lakes might have contributed to the size of the species flock in Lake Victoria (Greenwood 1951). On the other hand the Great Basin lakes in North America, in spite of a series of Pleistocene lake level fluctuations, have not produced species flocks (Smith and Todd 1984).

As far as intralacustrine allopatric speciation is concerned, three possibilities have been suggested.

Lake basins. A giant lake like Lake Baikal (636 km length) has several partly separated basins and would permit some extent of allopatric speciation within the lake, particularly if there were major fluctuations of lake level. Good basins are also accepted for Lake Tanganyika (Fryer and Iles 1972). However, major basins are not present in most other lakes, although the five Great Lakes in North America are actually five basins of one great lake. There is indeed some indication of incipient speciation in the ciscoes (*Coregonus*) of the Great Lakes of North America (Smith and Todd 1984).

Allopatric speciation within a single lake may be possible even in lakes not divided into separate lake basins. Most species in the larger lakes occur only in a portion of the lake, and such spotty distribution favors intralacustrine speciation. Only about a quarter of the haplochromines of Lake Victoria seem to have a lakewide distribution. The cichlids of Lake Malawi seem to be even more localized.

Substrate localization.   The frequently made assumption that a lake is a rather homogeneous habitat is quite erroneous. In nearly all lakes, and this is particularly true for the large East African lakes, there is an alternation of rocky, sandy, and muddy stretches of sublittoral substrates. Contrary to earlier claims, this is also true for Lake Victoria (Witte 1984). Many species of cichlids are completely restricted to one kind of habitat. Such substrate dependence is greatest in algal grazers, less so in piscivores and omnivores. Many species of the rock-dwelling genus *Labidochromis* in Lake Malawi are known only from a single rock outcrop (Lewis 1982). Nearly all haplochromines are mouth-brooders, and have always been assumed to be, on the whole, poor dispersers. As a result, many closely related species are still allopatric (Lewis 1982). It appears from available descriptions that habitats in a lake that are suitable for a given species are more or less scattered and isolated like islands in an archipelago. Intralacustrine speciation has therefore been sometimes compared to archipelago speciation in terrestrial organisms (Brooks 1950; Mayr 1963:507). Rock-inhabiting species found on offshore islands illustrate the process of peripatric speciation particularly well. Of 150 mbuna species on the Malawi coast of Lake Malawi, only three are recorded for the entire area.

Of particular interest are the experiments of McKaye and Gray (1984) of building artificial "reefs" in sandy areas. These reefs were colonized by more than 20 species, most of them plankton feeders and not found elsewhere in surveys of sandy habitats. There is a seeming and not yet resolved contradiction between this capacity for dispersal and the fact that these plankton feeders belong to the most speciose of all the haplochromine cichlids.

Spawning site isolation.   Discontinuities between spawning sites, even in the absence of visible barriers, might serve as the means by which the isolation is achieved which is the prerequisite for the acquisition of isolating mechanisms. It must be stressed that there are two classical theories concerning this process. According to one theory, that of Darwin, H. J. Muller, and myself, the isolating mechanisms are acquired during spatial isolation as a byproduct of the genetic changes of the isolated population. According to the other theory, that of A. R. Wallace and Dobzhansky, the isolating mechanisms are formed during isolation only very incompletely, and are perfected through natural selection (owing to the inferiority of the hybrids) only after the secondary encounter of the neospecies. The almost complete absence of hybrids in the lake cichlids is convincing evidence for the validity of the Darwin theory as far as the species flocks in lakes are concerned.

Two sets of factors, then, must be demonstrated in order to explain any instance of allopatric speciation. These are (1) the isolating factors which safeguard the integrity of the incipient species during the period of its genetic reconstruction, and (2) the factors responsible for the building up of isolating mechanisms before there is any contact with the species against which the isolating mechanisms eventually will be employed.

As far as spatial isolation is concerned, there are many indications that such coarse substrate criteria as rock, sand, or mud are not sufficient to explain the exuberant speciation of cichlids in the East African lakes. There is, however, one aspect of intralacustrine speciation that has not been sufficiently emphasized. While speciation in terrestrial organisms is constrained by their restriction to a two-dimensional space, lake inhabitants have a third available dimension. There are good indications that this additional dimension facilitates speciation. But there are still other factors. Foremost among these is the fact that the water mass of a lake is not homogeneous, and this is particularly true for feeding conditions. As is well known, mortality in fishes is highest among the freshly hatched fry. If they hatch in an area of abundant food supply there will be high survival. This fact sets up a selection force for the appropriate site finding (homing) and timing behavior of the spawning parents (Smith and Todd 1984). Individuals not spawning at the right time and place face the risk of lower reproductive success. This will lead to a local clustering of spawning populations and to a continuing increase of philopatry.

But how will such spawning clusters acquire isolating mechanisms against each other? This has always been a weakness of the Darwin theory as interpreted by recent evolutionists. Along with others, I had (1963:551) thought that the isolating mechanisms arose as a byproduct of the ecological divergence of the incipient species, but this interpretation was not very convincing, owing to the slightness of the niche differences in many incipient species, illustrated by the spawning clusters of lake fishes.

The probable causal factors in the origin of isolating mechanisms has recently been discovered in sexual selection. I find Dominey's (1984) explanation, largely based on Thornhill and Alcock (1983) and West-Eberhard (1983), completely convincing. Local fashions in mate preference can apparently change rather rapidly in small isolated populations, and these can develop into behavioral isolating mechanisms. Furthermore, they serve as recognition features of the individuals that belong to such a spawning cluster, and thus reinforce the spatial isolation. Dominey rightly stresses the important role played by the mating system of cichlids in this process of speciation. Although the distances among the spawning clusters may be small, the isolation nevertheless is spatial, and the process can be justly designated as micro-allopatric.

One of the consequences of this model of intralacustrine speciation (Smith and Todd 1984) is that the first characters to diverge should relate to courtship, while characters relating to competition for shared resources will evolve later, particularly after sympatry is established. The frequency of male color differences in morphological sibling species is consistent with this prediction. By contrast, the morphological consequences of sympatric speciation would be that characters related to resource partitioning (food utilization) should be the first to diverge. Eventually there will be character displacement wherever such newly evolved species come into competition with each other: morphological differences which reinforce resource partitioning will evolve.

Much of intralacustrine speciation, according to this model, is thus due to an interaction of two sets of factors, one having to do with the spawning site (appropriate territoriality and philopatry), while the other concerns the development of genetic mechanisms for localized recognition characters which isolate the spawners from straying individuals of other populations. Such a recognition pattern (behavioral isolation) could presumably be acquired by very few genetic changes. Once acquired, its strength seems to be considerable, as documented by the virtual absence of hybrids in lake cichlids. The complex courtship presumably prevents cases of mismating.

The Smith-Todd-Dominey model is at this time not much more than a working hypothesis, even though it fits the known facts extremely well. It will require a considerable amount of field work to test the validity of its various assumptions. How permanent are the spawning areas, and how reliable is the food supply for the hatched fry? How much dispersal is there among the offspring of a given female? How selective are females when faced with a choice of two kinds of male? Life history studies will presumably shed far more light on speciation in lakes than attempts at a genetic analysis. It is unfortunately true that we still know virtually nothing about the actual genetic factors responsible for speciation even in *Drosophila* and other genetically well-known terrestrial species (Barigozzi 1983).

The above postulated process of intralacustrine speciation is applicable primarily to the cichlid species flocks in the lakes of tropical East Africa. By contrast, intralacustrine speciation in temperate zone lakes is dominated by other factors. Here, seasonal phenomena are apparently utilized to produce isolation. Time of spawning as well as spawning migrations seem to play a far greater role than in tropical lakes, and some of the spawning migrations may have been a byproduct of multiple colonization (Smith and Todd 1984).

Fish Flocks and Evolutionary Theory

The study of fish species flocks in freshwater lakes is making considerable contributions to evolutionary theory, as is evident from the papers in this volume. For instance, the two morphotypes in the Cuatro Cienegas cichlid remind us that the term *gradual evolution* has two independent meanings that are almost invariably confounded. Populational gradualism was most important for Darwin in his fight against essentialism. However, this is not the same as phenotypic gradualism, since a single population can be polymorphic for sharply distinct phenotypes. Phenotypic discontinuities are not necessarily incompatible with populational gradualism.

There is an interesting correlation in the East African lakes between the age of the lake and the average morphological distance among the species. In the youngest of these lakes, Lake Victoria, which is at most 750,000 years old, there are no extreme morphotypes, and related species are connected by intermediate species forming a morphocline (Greenwood 1984a). By contrast, some rather diverse types have evolved in Lake Tanganyika, which is at least two million years old, with a much higher average morphological distance between species. Lake Malawi, apparently younger, is intermediate in the morphological distance between species. The evidence would seem to indicate that natural selection continues to push species toward a more extreme divergence and that intermediate types are more vulnerable to extinction, presumably by competition.

The specializations in the East African lakes illustrate the opportunism of natural selection. Parallel utilization of the same resources (trophic types) has evolved in the various lakes, and correlated with it parallel morphological specializations. Apparently, all the major potential resources are utilized in all the lakes. Different kinds of organisms may evolve along rather different pathways, even under similar conditions. Such evolutionary pluralism, as rightly stressed by several recent authors, is abundantly illustrated by the fish flocks. The cichlids and coregonids, for instance, exhibit rather different evolutionary processes from other fish with which they share their habitat as well as from freshwater inhabiting invertebrates. To mention another instance of pluralism, founder populations may sometimes acquire rather different phenotypes or else evolve as sibling species with hardly any morphological change. Even though most of the species must have originated through peripatric speciation, they nevertheless have retained an average level of heterozygosity, and the same polymorphism may turn up in rather distantly related species (Sage et al. 1984). It must be remembered that heterozygosity does not necessarily have to be largely eliminated during peripatric speciation provided the period of the bottleneck is very short (Nei et al. 1975).

Cichlid species provide many illustrations of mosaic evolution. Trophic specialization, for instance, and reproductive isolation would seem to evolve independently. In *Cichlasoma minckleyi* trophic polymorphism evolved without reproductive isolation. In the numerous sibling species in the East African lakes, reproductive isolation evolved without trophic specialization.

Cichlid fishes have long provided by far the best illustration for rapid speciation (Mayr 1967). There is much evidence for speciation within three to four thousand years, as documented for instance by the cichlids of Lake Nabugabo (Greenwood 1965). In view of this rapidity, the slightness of the morphological change is not surprising. What is surprising, perhaps, is the rapidity by which the presumed isolating mechanisms are acquired (Dominey 1984).

It has been suggested that Lake Victoria might have completely dried up during one of the arid periods of the Pleistocene, about 14,000 years ago. The production of over 200 endemic species of haplochromines since that recent date would seem utterly improbable. However, if we were to assume that each bout of speciation would require about 2,800 years, and that the founder species of the new Lake Victoria gave rise by peripatric speciation to three daughter species in different parts of the new lake, and that this process was repeated every 2,800 years, there could be 243 species after 14,000 years.

I do not know of any other organism for which such a rapid rate of speciation could even be considered. Actually, Lake Victoria presumably never dried up completely, but was merely reduced to a series of ponds in which at least some of the previous fish fauna survived. Under such an assumption, the length of individual speciation bouts could be increased considerably beyond 2,800 years.

What all this teaches us, and this is in line with the admonitions of several leading evolutionists, is that there is a great diversity of evolutionary processes, and the parameters that are valid for one group of organisms may not necessarily be at all applicable to another group. There are few, if any, universal laws in evolutionary biology (Mayr 1967).

Adaptive Radiation and Competitive Exclusion

The coexistence of hundreds of closely related species in the same lake poses some fundamental questions concerning competition and resource utilization. To what extent, if any, is the existence of fish flocks in freshwater lakes in conflict with the concept of competitive exclusion? Greenwood's analysis of the problem (1984a) is superb. The fact that competition must exist is self-evident. That it sets up selection forces is indicated by the fact that there is more morphological divergence among species in Lake Tanganyika, the oldest of the African lakes, than in the relatively recent Lake Victoria. The severity of the competition is also indicated by the fact that the haplochromines utilize every potentially available resource, and that every truly lacustrine habitat is occupied by one or several haplochromine species. Nevertheless, there are far more species than trophic types; for instance, of the 11 trophic types among 250+ species of haplochromines in Lake Victoria, 30 to 40 percent are piscivorous.

What factors permit the coexistence of so many competitors and prevent them from exterminating each other? The adaptive radiation was evidently facilitated by the seeming flexibility of the exploratory tendencies in haplochromines. This agrees with the observation that most organisms have a rather open genetic program concerning habitat and food utilization (Mayr 1974). Not surprisingly, in different lakes where similar diversities of resources are available, parallel adaptations have evolved. In such cases, I would prefer not to speak of "the deterministic action of selection" (Greenwood 1984a) since I cannot see anything deterministic in the opportunism of natural selection. If those individuals of a certain population that can better crush snails leave more offspring than other members of their population, one would expect that natural selection would favor those aspects of behavior or structural modifications that facilitate use of snails as a food resource.

What, then, are the factors that mitigate competition? First, many species are localized, and even though there may be 250 haplochromines in Lake Victoria, they will not all coexist at the same locality. Further, syntopy is often reduced by exclusion in time or space. A preference for different water depth of potential competitors often reduces competition

(March et al. 1981). Witte (1984) has made a particularly careful analysis of niche differences among similar species. For instance, in the zooplanktivores of Lake Victoria, each case of asserted "total interspecific overlap" that was studied in detail revealed niche segregation. A thorough study of the ecology of these species is beginning only now, and little is so far known about components of the life cycle, seasonal differences in food utilization, and other aspects that play a role in reducing competition.

One of the surprising findings is that some structurally rather specialized trophic types are nevertheless facultative feeders, utilizing a variety of different resources. Greenwood calls attention to the seeming paradox that the specialists among the cichlids usually can also exploit the more generalized food niches, while the generalists cannot do the reverse. Hence, widespread assumptions notwithstanding, it is the specialists who have the potential for the exploitation of a wider range of food niches. Indeed, as Liem and Kaufman (1984) point out, the prey-capture apparatus in cichlids is built in such a way that it can meet multiple problems. Profound functional shifts can be made without any major structural alteration of jaws and associated muscles. And a seemingly rather drastic difference, such as between the molariform and the papilliform morphs in *Cichlasoma minckleyi,* can be effected by a very small shift in regulatory genes.

The recent works, particularly on the haplochromines in East Africa, are shedding considerable light on the relation between behavior, structure, speciation, and food niche. The fact that most closely related species show no morphological differences at all or only very minor ones in trophic adaptations indicates that morphological differentiation is neither a prerequisite nor even a necessary corollary of speciation. The motto that behavior is the pacemaker of evolution is massively supported by the evolutionary changes in the cichlid species flocks. Also interesting is the degree to which evolutionary change is restricted to trophic structures. "There is remarkably little diversity in body form" (Greenwood 1984a). The cichlids thus have experienced a major ecological evolution without a corresponding morphological one. This parallels the situation in the songbirds (Oscines), where one also finds remarkably little divergence except for coloration and features of bill and feet.

There are, of course, many unanswered questions, and additional puzzling aspects of evolution will surely be discovered during the further study of the ecology and behavior of the cichlid species flocks. Greenwood (1984a) echoes a question often raised in the history of Darwinism when he states, "It is difficult to imagine how selection can be invoked to explain the evolution of highly derived features (dental specialization) when functionally and morphologically intermediate stages are still extant and present in species occurring syntopically with the more derived forms." In a way this is merely a new version of the question raised by anti-Darwinians in the last century: "Why would higher organisms evolve when simple ones like bacteria and protists are so eminently successful?" A single component, of the phenotype, like dental specialization, is of course not the whole of the adaptive value of an organism. Each newly evolving species presumably has some character, by no means necessarily a structural one, which gives it the capacity to compete successfully, and thus to coexist, with other species that may have acquired a structural specialization.

Many important biological problems can be illuminated by the study of even so specialized a subject as that of fish species flocks in freshwater lakes. And yet, the analysis of this intriguing evolutionary phenomenon has only begun. There is every reason to believe that further studies will result in raising tantalizing new questions but also contribute to the solution of still open problems. What is most important now is to push these studies as hard as possible as long as these diversified faunas still exist.[1] Lake Lanao and other lakes where the endemic faunas have been largely exterminated are an ominous warning of what the future might have in store.

NOTES

This essay is an abridged version of one which first appeared in A. A. Echelle and I. Kornfield, eds., *Evolution of Fish Species Flocks,* pp. 3–11, under the title "Evolution of fish species flocks: a commentary." Orono, Maine: University of Maine Press, 1984.

1. The rich cichlid fauna of Lake Victoria has been nearly exterminated since this was written, owing to the introduction of a predatory fish, the Niles Perch (Barel, C. D. N., ed. 1986. *The Decline of Lake Victoria's Cichlid Species Flock.* Leiden, The Netherlands: Research-Group Ecological Morphology, Zoologisch Laboratorium).

REFERENCES

Barbour, C., and D. B. Chernoff. 1984. Comparative morphology and morpho-
    metrics of the pescados blancos (Genus *Chirostoma*) from Lake Chapala, Mexico.
    In Echelle and Kornfield (1984:111–128).

Barigozzi, C. 1983. *Mechanisms in Speciation.* New York: Alan R. Liss.

Brooks, J. L. 1950. Speciation in ancient lakes. *Quart Rev. Biol.* 25:30–176.

Dominey, W. J. 1984. Effects of sexual selection and life history on speciation: species flocks in African cichlids and Hawaiian *Drosophila.* In Echelle and Kornfield (1984:231–250).

Echelle, A. A., and A. F. Echelle. 1984. Evolutionary genetics of a "species flock": atherinid fishes on the Mesa Central of Mexico. In Echelle and Kornfield (1984:93–110).

Echelle, A. A., and I. Kornfield, eds. 1984. *Evolution of Fish Species Flocks.* Orono, Maine: University of Maine at Orono Press.

Fryer, G., and T. D. Iles. 1972. *The Cichlid Fishes of the Great Lakes of Africa: Their Biology and Evolution.* Edinburgh: Oliver and Boyd.

Greenwood, P. H. 1951. Evolution of the African cichlid fishes: the haplochromine species flock in Lake Victoria. *Nature* 167:19–20.

——— 1965. The cichlid fishes of Lake Nabugabo, Uganda. *Bull. Brit. Mus. Nat. Hist. (Zool.)* 12:315–357.

——— 1974. The cichlid fishes of Lake Victoria, East Africa: the biology and evolution of a species flock. *Bull. Brit. Mus. Nat. Hist. (Zool.)* suppl. 6:1–134.

——— 1984a. African cichlids and evolutionary theories. In Echelle and Kornfield (1984:141–154).

——— 1984b. What is a species flock? In Echelle and Kornfield (1984:13–20).

Humphries, J. M. 1984. Genetics of speciation in pupfish from Laguna Chichancanab, Mexico. In Echelle and Kornfield (1984:129–140).

Kornfield, I., and K. Carpenter. 1984. Cyprinids of Lake Lanao, Philippines: Taxonomic validity, evolutionary rates and speciation scenarios. In Echelle and Kornfield (1984:69–84).

Lewis, D. S. C. 1982. A revision of the genus *Labidochromis* from Lake Malawi. *Zool. J. Linn. Soc.* 75:189–265.

Liem, J. F., and L. Kaufman. 1984. Intraspecific macroevolution: functional biology of the polymorphic cichlid species *Cichlasoma minckleyi.* In Echelle and Kornfield, (1984:203–216).

Marsh, A. C., A. J. Ribbink, and B. A. Marsh. 1981. Sibling species complexes in sympatric populations of *Petrotilapia* Trewavas (Cichlidae, Lake Malawi). *Zool. J. Linn. Soc.* 71:253–264.

Mayr, E. 1942. *Systematics and the Origin of Species.* New York: Columbia University Press.

——— 1963. *Animal Species and Evolution.* Cambridge, Mass.: Harvard University Press.

——— 1967. Evolutionary challenges to the mathematical interpretation of evolution. In P. S. Moorhead and M. M. Kaplan, eds., *Mathematical Challenges to the Neo-Darwinian Interpretation of Evolution.* Wistar Institute Symposium, monograph 5. Philadelphia: Wistar Institute Press. (Rptd. in Mayr 1976:53–63.)

——— 1974. Behavior programs and evolutionary strategies. *Amer. Sci.* 62:650–659. (Rptd. in Mayr 1976:694–711.)

——— 1976. *Evolution and the Diversity of Life.* Cambridge, Mass.: Harvard University Press.

McKaye, K. R., and W. N. Gray. 1984. Extrinsic barriers to gene flow in rock-dwelling cichlids of Lake Malawi: macrohabitat heterogeneity and reef colonization. In Echelle and Kornfield (1984:169–184).

McKaye, K. R., T. Kocher, P. Reimtal, and I. Kornfield. 1982. A sympatric sibling species complex of *Petrotilapia* Trewavas from Lake Malawi analyzed by enzyme electrophoresis (Pisces, Cichlidae). *Zool. J. Linn Soc.* 76:91–96.

Nei, M., T. Mayurama, and R. Chakraborty. 1975. The bottleneck effect and the genetic variability in populations. *Evolution* 29:1–10.

Parenti, L. R. 1984. The species flock concept as it relates to the phylogeny and biogeography of the Andean killifish *Orestias*. In Echelle and Kornfield (1984:85–92).

Rensch, B. 1933. Zoologische Systematik und Artbildungsproblem. *Verh. Deutsch Zool. Ges.,* pp. 19–83.

Sage, R. D., P. Loiselle, P. Basasibwaki, and A. C. Wilson. 1984. Molecular versus morphological change among cichlid fishes of Lake Victoria. In Echelle and Kornfield (1984:185–202).

Sage, R. D., and R. K. Selander. 1975. Trophic radiation through polymorphism in cichlid fishes. *Proc. Nat. Acad. Sci.* 72:4669–4673.

Smith, G. R., and T. N. Todd. 1984. Evolution of species flocks of fishes in north-temperate lakes. In Echelle and Kornfield (1984:47–68).

Thornhill, R., and J. Alcock. 1983. *The Evolution of Insect Mating Systems.* Cambridge, Mass.: Harvard University Press.

West-Eberhard, M. J. 1983. Sexual selection, social competition, and speciation. *Quart. Rev. Biol.* 58:155–183.

Witte, F. 1984. Ecological differentiation in Lake Victoria haplochromines: comparison of cichlid species flocks in African lakes. In Echelle and Kornfield (1984:155–168).

Woltereck, R. 1931. Beobachtungen und Versuche zum Fragenkomplex der Artbildung 1. *Biol. Centralblatt* 51:231–253.

MACROEVOLUTION

Introduction

Ever since Darwin there has been a widespread impression among evolutionists that the evolutionary processes occurring in populations and species might be different from the evolutionary phenomena characterizing higher categories and occurring during geological time. Eventually, it became customary to designate evolution at and below the species level as microevolution, and evolution above the species level as macroevolution. Almost automatically this definition generated a controversy, because for some evolutionists it implied a fundamental difference between the two kinds of evolution. Others, particularly most geneticists, asserted that the genetic mechanisms known to operate in populations were also able to explain such macroevolutionary phenomena as the origin of new higher taxa and the origin of new structural innovations and of functional shifts, an assertion also supported by all the architects of the evolutionary synthesis. In reality, all the geneticists were able to document was that macroevolutionary phenomena are consistent with the known genetic processes. There was, however, a singular absence of any proof that macroevolutionary phenomena are indeed caused by the same genetic mechanisms that characterize intrapopulational variation.

A study of the writings published during the evolutionary synthesis reveals that the problems of macroevolution were rather neglected. This had primarily two reasons. The first was that, prior to the development of molecular biology, macroevolutionary phenomena were quite inaccessible through the classical genetic methods, because one cannot cross representatives of higher taxa. The second reason was that the interests of paleontologists and comparative anatomists were remarkably limited to the vertical (transformational) component of evolution. This is quite evident from the writings of Simpson, Rensch, and Huxley. A thorough study of the actual origin of organic diversity, and particularly of the connection between speciation and the origin of higher taxa, was rather conspicuously missing from the literature. Finally, macroevolutionary processes, more than almost any other processes in evolutionary biology, can be determined only by inference, and there was considerable reluctance to make such inferences during an excessively experimental-reductionist period. It was overlooked that some of the most interesting aspects of evolution would remain forever obscure unless one made an effort to reconstruct past processes.

Every event in the evolutionary past raises intriguing questions. For example, one might ask why mammals were so comparatively unsuccessful during a 150 million year period from the Triassic to the Paleocene and then suddenly became so super-successful? To the anatomists' suggestion that a secondary palate (and the new functions that this permits) might have been the key, Gould (1982:385) quite rightly replies that it may not have been the palate at all but rather the small size and nocturnal habits of these early mammals which allowed them to coexist with the dinosaurs and eventually to outlive them. Although we may never be able to reconstruct the correct scenario, even merely to ask challenging questions greatly enriches our understanding of past periods.

The major bone of contention in the controversy about macroevolution was the claim made by Darwin and his followers that macroevolution is nothing but a magnified extension of evolution at the level of populations and species. As I stated it in 1942 (p. 298): "All the processes and phenomena of macroevolution and the origin of higher categories can be traced back to intraspecific variation, even though the first steps of such processes are usually very minute." This statement is a necessary corollary of the fact that an individual has only a single genotype, whether we study him as a member of a population, a species, or a higher taxon. During the evolutionary synthesis, when my statement was made, all genes were considered to be equivalent. Since then the great heterogeneity of the components of the genotype, as discovered by molecular biologists, has raised numerous questions that so far have not been answered.

In Essay 24 I review some of the evidence that, up to a point, certain domains in the genotype behave as units in macroevolution. (For further evidence in support of a peculiar cohesion of components of the genotype see Chapter 10 in Mayr 1963 and 1970.)

In Essay 25 I speculate on the possible role of founder populations in the restructuring of such domains in the genotype. I suggest that relatively small genetic changes might have rather far-reaching effects on the niche occupation and adaptedness of such populations. And such "revolutionary" events might be important in macroevolution. The paleontologist, owing to the incompleteness of the fossil record, would interpret such events as saltations, while in reality they are perfectly normal Darwinian processes—gradual, despite the rapidity with which they occur.

A detailed factual and historical analysis of the theory of punctuated equilibria (Eldredge and Gould) is the subject of Essay 26. I attempt to show that a moderate version of this theory, which is based on my theory of speciational evolution, is not in conflict with Darwinian views and agrees better with the observed phenomena of evolution than theories which ignore the role of speciation in evolution.

DOES MICROEVOLUTION
EXPLAIN MACROEVOLUTION?

AMONG all the claims made during the evolutionary synthesis, perhaps the one that found least acceptance was the assertion that all phenomena of macroevolution can be "reduced to," that is, explained by, microevolutionary genetic processes. Not surprisingly, this claim was usually supported by geneticists but was widely rejected by the very biologists who dealt with macroevolution, the morphologists and paleontologists. Many of them insisted that there is a more or less complete discontinuity between the processes at the two levels—that what happens at the species level is entirely different from what happens at the level of the higher categories. Now, 50 years later the controversy still seems undecided.

Why are these two viewpoints seemingly so incompatible? It has now become rather evident that we are dealing with two separate levels in a hierarchy of levels, each level having its own subject matter and methodology. Genetics explains microevolution through the study of the behavior of genes in populations—that is, it concentrates on changes in the genotype. This method cannot be directly applied at the level of higher categories. Macroevolution, by contrast, is based on the study of phenotypes. Such macroevolutionary phenomena as preadaptation, change of function, and the origin of evolutionary novelties or of higher taxa must be studied through a comparison of phenotypes. Until recently, no one knew of a way by which to get at the underlying genetic mechanisms of the macroevolutionary changes of the phenotype. For this reason, the geneticists Stebbins and Ayala admit quite freely that "the decision as to which among alternative [macroevolutionary] hypotheses is correct cannot be reached by recourse to microevolutionary principles . . . thus, macroevolution is an autonomous field of evolutionary study and, in this epistemologically very important sense, macroevolution is decoupled from microevolution" (1981:971).

Is the study of macroevolution an autonomous branch of evolutionary biology, as Stebbins and Ayala suggest, or does microevolution grade imperceptibly into macroevolution, so that the findings of intraspecific genetics also explains macroevolution? The literature on this question leaves one with the impression of an irreconcilable conflict, but the opposing viewpoints in this controversy can be narrowed down considerably if we recognize that two rather different issues are involved:

(1) Genetic changes, even those that produce orders, classes, or phyla, occur in the genotypes of individuals and obey the same laws as any intraspecific variation. Any evolutionary change of macroevolutionary significance is simultaneously a change in a local population. In this respect there is no decoupling of macro- and microevolution.

(2) Yet the traditional definition of evolution adopted by the geneticists ("changes in gene frequencies") is highly misleading. Evolution is a change in adaptation and organic diversity (Mayr 1977). All macroevolutionary phenomena and processes such as adaptation, convergence, rate of evolution, and shift of adaptive zones relate to phenotypes and can be studied without reference to their genetic basis. Indeed, up to the present time, everything we know about macroevolution has been learned by the study of phenotypes; evolutionary genetics has made virtually no direct contribution so far. In this respect, indeed, macroevolution as a field of study is completely decoupled from microevolution.

When we study macroevolution—say the origin of birds from reptiles or hominid evolution from ape to man—we have traditionally studied phenotypes, particularly changes in morphology. And, as has been remarked so often, the differences among the phenotypes of the higher taxa are usually of quite a different order of magnitude from the differences among individuals of a population or a species. Thus, the phenotypic study of microevolution ordinarily involves a rather different set of phenomena from the study of macroevolution. (Admittedly this belief is not unopposed.) Stebbins and Ayala are quite correct in stating that "macroevolution is an autonomous field of study, that must develop and test its own theories"—theories that "are not reducible (at least at the present state of knowledge, and probably in principle) to microevolutionary theories" (1981:970). Numerous monographs and symposium volumes have been published in recent decades demonstrating how much a purely phenotypic approach, combined with ecological and behavioral *Fragestellungen,* can contribute to an understanding of macroevolution.

And yet the fact that the individual is the target of selection, and the population the locus of evolutionary change, automatically connects all macroevolutionary processes with the microevolutionary level. How then can a bridge be built across the gap between intrapopulational genetic analysis and the macroevolutionary study of phenotypes? To be sure, mutation, stochastic processes, and selection are as much involved in macroevolution as in microevolution. But the claim that these three factors fully explain macroevolution falls considerably short of reality.

The gap appears so unbridgeable because most geneticists and traditional students of macroevolution failed to appreciate the multidimensional nature of evolution. Both presented macroevolution as a strictly vertical process, that is, as the change of a phyletic lineage improving its adaptation through a change in gene frequencies. Curiously, neither side has availed itself of the rich evidence on macroevolution presented by neontology. The study of polytypic species and of superspecies reveals what rich macroevolutionary potential is provided by peripherally isolated populations and allospecies. Paleontologists have failed to see that a series of geographic races or allospecies often represent a horizontal evolutionary sequence corresponding to a vertical sequence that has been transferred to the geographical dimension. It is this recognition which permitted Bock (1970) to reconstruct the pattern of speciation among the Hawaiian finches (Drepanididae) that led to morphological changes in bill structure almost as extensive as those in the bills of all families of song birds. Even more convincing and far more detailed is the analysis of the speciational and macroevolutionary history of Hawaiian *Drosophila* flies, which has the additional immense advantage that the timing of the various events can be inferred rather precisely, and even more importantly, can be analyzed genetically (Carson and Yoon 1982). Carson has been spectacularly successful in resolving macroevolutionary into populational phenomena.

In the case of the Hawaiian fauna, it has thus been possible to trace macroevolutionary phenomena back to their beginnings in local populations, largely at the phenotypic level, but in *Drosophila* also through genetic analysis.. This research fully substantiates the claim that all macroevolutionary phenomena can be traced back to intraspecific variation (Mayr 1942). And yet the traditional formulation of population genetics, in which "mutation" is credited with the nature of the change, is quite unsatisfactory, because it implies that mutation is a unitary process and that all mutational changes of the genotype are of equivalent macroevolutionary significance. However, as wrong as Goldschmidt (1940) and Schindewolf (1950) have been in their saltational explanation of evolution, they may well have been right in suggesting that there is a fundamental difference between the "housekeeping" kinds of genetic adjustments in local populations and the truly significant genetic changes during major macroevolutionary shifts. Several molecular biologists (for instance, Shapiro 1983) have called attention to the many recently discovered molecular mechanisms that might have the propensity to produce changes of the genotype that are far more drastic than the reshuffling of allozymes that has been studied so intensely by electrophoresis.

Numerous papers have been published in recent years dealing with such molecular changes and their possible evolutionary impact (for instance Arnheim 1983; Doolittle 1986; Campbell 1985; Milkman 1982; Nei and Koehn 1983). There is still too much uncertainty about this subject for it to be discussed profitably by a nonmolecular biologist. One weakness of much of this literature is that many authors consider the gene, rather than the whole organism, to be the target of selection. Its strength is that some of these authors focus on the study of that part of the genotype that is not composed of structural genes; this is probably the most promising object of study for evolutionary genetics.

Population geneticists in making the silent assumption that all genes are equivalent, have so far failed to make a contribution to the resolution of this problem. By cleverly employing mathematics and making numerous arbitrary assumptions, one can develop macroevolutionary models based on beanbag genetics. However, there is no way to test these models for their validity. At the present moment, unfortunately, the genetics of microevolutionary processes has been unable to provide a full explanation of macroevolution, nor has the analysis of macroevolutionary phenomena provided any answers as to the nature of the genetic processes characterizing macroevolutionary events.

Some of the particularly puzzling macroevolutionary phenomena and processes that have not yet been satisfactorily interpreted in terms of known genetics are:

(1) What happens to the genotype during speciation?

(2) What happens in the genotype during drastic ("saltational") evolutionary innovations of the phenotype?

(3) What structures of the genotype are responsible for long-time stasis, including the preservation in ontogeny of ancestral developmental stages (such as gill arches in tetrapods) and the extraordinary stability of the *Bauplan* of the major types of organisms?

The Stability of the Bauplan

One of the great new insights acquired by the animal systematists early in the nineteenth century was that animals cannot be seriated in a smooth, continuous chain from the simplest to the most perfect, as the proponents of the *scala naturae* had presumed. Instead, a limited number of discrete types can be recognized, such as vertebrates, insects, and mollusks. This new insight is connected with the names of Cuvier, Oken, Owen, Agassiz, and their followers, the so-called "idealistic morphologists" (Desmond 1982). Comparative morphologists recognized in the animal kingdom from this point on a limited number of types, archetypes, or *Baupläne*. Unfortunately the word "type" in the ensuing period was used in two rather different senses, and this created major misunderstandings. Those who attacked the typology of the idealistic morphologists did not question the propriety of recognizing purely descriptively the major archetypes of organisms reflected in the taxonomically recognized 27 or so phyla of animals. Rather they criticized what was later called essentialism, the concept that a limited number of constant, invariant types existed in nature which were separated from each other by bridgeless gaps. Several recent discussions of types and typology, even though otherwise valuable and informative, suffer from the failure to recognize these two very different meanings of the word type (Schindewolf 1969; van der Hammen 1981).

The archetypes of the idealistic morphologists became the major higher taxa of the post-1859 classifications. In spite of their inferred common descent, they remained discrete entities, each with its own *Bauplan*. The origin of these *Baupläne* is a separate topic (Lange 1985), which I shall not discuss at this time. Atomistic genetics has been quite incapable of coming up with an explanation for the stability of these *Baupläne*. Perhaps it can be functionally explained why all tetrapods have basically one pair of anterior and one pair of posterior extremities. Yet why all insects, nearly one million species, should have three pairs of extremities, and all spiders four pairs, can be explained only by the conservatism of the developmental system built into their genotype. The same is true for the five-rayed foot of the terrestrial vertebrates. Even where it is reduced, as in horses, birds, and some amphibians, or supplemented by additional rays, as in certain marine vertebrates, the structure is always laid down in ontogeny with five rays. One could cite one example after the other of such conservative aspects of the *Bauplan,* either extended into the adult phenotype or visible only during ontogeny (recapitulation), but such a list adds nothing to the solution of this problem. There are evidently internal structures of the genotype which account for the conservation of the basic structure.

In a critical discussion of recent controversies concerning macroevolution, Maynard Smith (1983) states that, according to new proposals, "the fundamental structures or bauplans, of major taxa represent a set of possibilities constrained by the laws of development; in contrast, Darwin argued that the structures common to a taxon exist because they were present in the ancestors of that taxon, and that they arose in those ancestors as naturally-selected adaptations to particular ways of life."

By presenting these two opposing views as mutually exclusive hypotheses, Maynard Smith misrepresents the current controversy. He fails to realize that those who, for instance, believe that the retention of the chorda and the embryonic gill arches are due to developmental constraints would not question for a moment that these structures had originally arisen in their "ancestors as naturally-selected adaptations to particular ways of life." In this case Maynard Smith has failed to discriminate between proximate and evolutionary causes.

What the analysis of the ancestral heritage does not reveal is why so many vestiges of the past, including the metamerism of birds and mammals, are still retained long after they have lost their function "as naturally-selected adaptations." So far no one has come up with a better explanation than that a tightly cohesive genotype and developmental constraints have prevented the elimination of these features.

One further aspect of these *Baupläne* is important. They seem to become more and more inflexible in the course of evolution. When the eukaryotes originated during the Precambrian, some 60 or more different archetypes evolved. While more than half of them became extinct in the ensuing geological periods, not a single new archetype has originated since that era. And even the evolutionary potential of the existing archetypes seems to have been narrowed down drastically in the last 400 million years.

If speciation is the moment at which evolutionary innovation is most likely to occur (as will be discussed below), one might expect to encounter the origin of incipient new archetypes among highly speciose taxa, as for instance the teleost fishes, song birds, and rodents. However, all of these radiations are comparatively recent and have nowhere succeeded in breaking out of the constraints of their archetypes. Different adaptive types in the cichlids of the East African lakes, perhaps the group of animals now speciating most rapidly, are all very similar to each other and still connected by intermediates (Greenwood 1979). The stability of the structural types thus demands an explanation as well as the processes by which such developmental stability can be broken down.

The Origin of Evolutionary Novelties

Darwin, in his argumentation against creationism, always stressed the gradualness of evolutionary change. The cases of the ancon sheep and the turn-spit dog, mentioned by him, indicate that he was well aware of the possibility of rather drastic sudden phenotypic changes, but these played no role in his populational concept of evolutionary change. This gradualistic interpretation of evolution was quite unacceptable to most paleontologists and morphologists. Bateson (1894) in particular, and the Mendelians, postulated instead that evolutionary innovations were introduced by sudden saltations. Being typologists, unused to populational thinking, they visualized the process of evolutionary innovation as the production of an individual with the new feature, this individual starting a new taxon with the new character. An individual with an innovative new mutation was "preadapted," according to Cuénot (1901) and Davenport (1903), if it could find a suitable niche. For Goldschmidt (the most recent representative of the saltational school), a systemic mutation might result in the origin of a new, drastically changed individual ("hopeful monster") which would become the first member of an entirely new higher taxon, perhaps an order, a class, or even a phylum (Goldschmidt 1952:91–92).

According to the opposing Darwinian viewpoint, all evolution, including the origin of evolutionary novelties, is "gradual." But what is gradual? In recent years it has become quite evident that the term gradual is ambiguous. For the morphologist it always means a gliding transition among phenotypes; for the Darwinian it means a change consisting of the slow restructuring of populations. Within the latter process, phenotypic polymorphism is possible and complete phenotypic gradualness not a necessity.

In animals, almost invariably, a change in behavior is the crucial factor initiating evolutionary innovation. As has been stated so often, behavior is the pacemaker of evolution. Primarily there are two pathways by which this can be effected. When an arboreal bird becomes more terrestrial, as did the mockingbird-like ancestor of the thrashers (*Toxostoma*), this shift set up a selection pressure on strengthening and elongating the legs and strengthening the bill used for digging in the leaf-mold and soil (Miller 1949). In birds the form of the bill is particularly plastic and apt to respond to shifts in behavior. This has been shown by Lack (1947) and Grant (1986) for the bills of the Galapagos finches and by Bock (1970) for the bills of the Hawaiian finches. Minor adjustments to new sources of food or changes in the physical environment are simply called adaptations rather than evolutionary novelties. The end stage of such an adaptive development, however, can amount to a veritable innovation, particularly when the intermediate stages become extinct. Severtsov (1931) has spoken in such cases of an *intensification of function*. The most striking case, of course, is that of the evolution of eyes. As Darwin already suggested, nothing is needed for their evolution but the existence of light-sensitive cells at the surface of an organism to initiate a chain of structural and functional events, culminating in a more or less highly developed photosensitive organ. There is no need to postulate rare or unique events, since eyes evolved in the animal kingdom at least 40 times independently (Salvini-Plawen and Mayr 1977).

A very different class of evolutionary innovations is involved whenever a structure or other attribute of an organism is able to assume a new function. Darwin was the first to recognize the importance of this principle (1859:454), now called *change of function principle*. It was later elaborated by Dohrn (1875), Severtsov (1931), and Mayr (1960). For instance, it is now generally believed that feathers originated in the ancestors of birds in connection with temperature regulation (Regal 1975). It was this function which provided the selection pressure for the modification of scales into feathers. When an entirely different selection pressure favored the evolution of flight, selection made use of the existing structure, elaborating those particular feathers (wing and tail feathers) that are essential for efficient flight. Thus, birds were preadapted for flight because of their possession of feathers. Numerous other cases are cited in the literature where such preadaptations permitted a change of function.

What is so often forgotten in the discussion of evolutionary discontinuities is that they represent not only a structural discontinuity but also an ecological one. A new niche and, particularly, a new adaptive zone is often separated from the ancestral one by a pronounced gap. There is no well-adapted condition for the area between the two adaptive zones. Hence, when a new adaptive zone is first invaded, a zone of maladaptedness has to be crossed. Simpson (1944; 1953) emphasized this point correctly. However, he failed to suggest a plausible solution to this problem. The question is, under what conditions can the gap be crossed most easily. His suggestion of quantum evolution had so little factual or theoretical support that Simpson himself more or less abandoned it later. Indeed, contrary to Simpson, no large widespread, populous species, even if consisting of numerous semi-isolated demes, is likely to achieve such a transition. On the other hand, a founder population is ideally situated for such a rapid ecological shift.

No organism can invade a new adaptive zone unless it has a minimum of structural, physiological, and behavioral attributes that preadapt it to succeed in this shift. Preadaptations are another illustration of the propensity of natural selection for tinkering. Natural selection makes use of whatever genic or phenotypic material is available in order to answer a newly arising need. This is difficult to understand for those who see in natural selection a deterministic, quasiteleological force. Bock (1959) has perceptively called attention to the multiple pathways by which newly arising needs can be satisfied. All one has to think of are the many different methods by which nature has made pelagic floating possible (Rensch 1947), or how many different components of the plumage have been made use of in the birds of paradise to generate display organs. When only a limited number of answers to a certain ecological challenge are available, genuine convergence results.

Particularly frequent is the parallel acquisition or loss of adaptations among related taxa. This often makes it difficult to determine what is a derived and what an ancestral character. Phylogenetic analysis has revealed how often even rather complex evolutionary novelties have been acquired independently. I mentioned already eyes, with at least 40 independent origins. This is matched by bioluminescence, which apparently evolved some 30 times independently but using rather similar mechanisms (Hastings 1983). Even more frequent was the shift from planktonic to yolk-rich nonplanktonic larvae among marine invertebrates (Jablonski 1986).

How Gradual Is Evolution?

Among recent controversies concerning the nature of macroevolution, none has been as heated and prolonged as the question whether or not evolution is always gradual. This argument originates with Darwin's claim (1859:71): "As natural selection acts solely by accumulating slight, successive, favorable variations, it can produce no great or sudden modifications; it can act only by very short and slow steps." Darwin was so convinced of the validity of this principle that he was willing to assert: "If it could be demonstrated that any complex organ existed which could not possibly have been formed by numerous, successive, slight modifications, my theory would absolutely break down" (1859:189). He was at once challenged by T. H. Huxley and others of his friends who thought that they found evidence for saltations throughout evolutionary history.

Those historians who have been puzzled by Darwin's stubbornness in insisting on the gradualness of evolution forget that a new theory is nearly always proposed in opposition to an existing one. When Darwin advanced his theory of gradual evolution, it was in opposition to Lyell's instantaneous "introduction of new species," and other creationist theories of sudden origin. When one proposes a new theory, it is good strategy to make it as drastically different from the existing theory as possible. This is the best way to draw attention to it and to emphasize its novelty.

What Darwin opposed in his theory of gradualness was the saltationism implicit in two ideologies dominant in his time, creationism and essentialism. To insist on gradualness, thus, meant for him to reject both creationism and essentialism. This is forgotten by some of those who object to even the slightest deviation from perfect gradualism. The architects of the evolutionary synthesis still had to defend gradualism against the saltationism of a Goldschmidt, Schindewolf, and Willis. And even today some evolutionists have attacked the theory of punctuated equilibria as a departure from Darwin's gradualism. A number of confusions are the reason why this controversy has been so vehement and prolonged. For instance, it was overlooked by some opponents that one can have a pluralistic attitude toward the problem, admitting that some evolutionary advance is achieved by completely gradual phyletic evolution while in other cases evolution proceeds by spurts and stops. More serious was the failure, as will be pointed out below, to recognize that the word "gradual" meant different things to different authors.

To what extent Darwin himself departed from his principle is somewhat controversial (Rhodes 1983) and shall not occupy us here. But as Grant (1983:151) has rightly pointed out, alternatives to complete gradualism have been part of the evolutionary synthesis all along.

Some of those who entered the controversy considered the existence of different rates of evolution as evidence against gradualness. These two aspects of evolution are, of course, not necessarily correlated at all. Darwin himself was fully aware that certain evolutionary lines had evolved much more slowly than others. And, more importantly, he had learned from the paleontologist Falconer that a single phyletic lineage could have successive phases of rapid change and of stasis (Darwin 1872). The same idea has been one of the cornerstones of Simpson's quantum evolution and of his recognition of bradytelic and tachytelic modes of evolution (Laporte 1983). There is no contradiction whatsoever between a belief in evolutionary gradualism and the belief in an existence of varying rates of evolution, including very rapid evolution.

More serious is the confusion created by the failure of most modern authors to distinguish between taxic gradualness and phenotypic gradualness. What Darwin mostly argued against was the thesis that evolutionary novelties could originate through the production of a single individual representing a new type, a new taxon. Instead, he proposed that all evolutionary innovation is effected through the gradual transformation of populations. Such a populational approach was quite alien to the essentialists with whom he argued. Gradualism, made possible by the transformation of populations, was primarily defended by the naturalists, while saltationism was popular in all those biological disciplines such as embryology, paleontology, and early Mendelism that were dominated by essentialistic thinking. Goldschmidt (1940) accepted gradual changes of an adaptive nature within species but insisted in several of his writings that a new taxon, even a higher taxon, is established by a "hopeful monster," the product of a drastic systemic mutation. Accordingly he cited with approval Schindewolf's suggestion that the first bird hatched out of a reptilian egg. Referring to the Linnaean hierarchy, he said:

> We see the picture represented as a pedigree or a tree of descent. This means that a phylum consists of a number of classes all of which are basically recognizable as belonging to the phylum, but, in addition, are different from each other. The same principle is repeated at each taxonomic level. All the genera of a family have in common the traits which characterize the family; e.g., all genera of penguins are penguins. But among themselves they differ from genus to genus. So it goes on down to the level of the species. Can this mean anything but that the type of the phylum was evolved first and later separated into the types of the classes, then into orders, and so on down the line? This natural, naive interpretation of the existing hierarchy of forms actually agrees with the historical facts furnished by paleontology. The phyla existing to-day can be followed farthest back into remote geological time. Classes are a little younger, still younger are the orders, and so on . . . Thus logic as well as historical fact tell us that the big categories existed first, and that in time they split in the form of the genealogical tree into lower and still lower categories (Goldschmidt 1952:91–92).

Goldschmidt clearly thought in terms of an origin of new types rather than a gradual transformation.

One must sharply distinguish populational gradualness from phenotypic gradualness. Darwin failed to do this, and so have many of those who participated in the recent controversies. R. A. Fisher (1930), in reaction to the saltationism of De Vries and other Mendelians, promoted the idea of extreme phenotypic gradualism. All evolutionary change, according to him, was due to mutations with very small phenotypic effects. Allowing for mutations with large phenotypic effects was falsely assumed to be a concession to saltationism. Turner (1983), in the meantime, showed that in most cases of mimicry one must postulate an initial mutation with a rather large phenotypic effect. Indeed, cases of polymorphisms have been found involving considerable phenotypic differences among the morphs. Gould (1980:127) concluded from this the possibility of a saltational origin of the essential features of key adaptations. "Why may we not imagine that gill arch bones of an ancestral agnathan moved forward in one step to surround the mouth and form proto-jaws?" Indeed this is conceivable, but how much evidence is there that would favor such an explanation? Goldschmidt (1940:6–7) considered the gradual evolution of the following features by accumulation and selection of small mutations impossible: "Hair in mammals, feathers in birds, segmentation of arthropods, the transformation of the gill arches in phylogeny including the aortic arches, muscles, nerves, etc.; further, teeth, shells of molluscs, ectoskeletons, compound eyes, blood circulation, alternation of generations, statocysts, ambulacral systems of echinoderms . . . poison apparatus of snakes, whale bone, and, finally, primary chemical differences like hemoglobin versus hemocyanin." It is quite probable that in some of the listed evolutionary innovations major phenotypic changes were involved. However, this does not in the slightest necessitate adopting a non-Darwinian interpretation. As long as new genotypes can coexist successfully with the parental genotype in a polymorphic population, a gradual shift from one to the other genotype is possible.

Recent genetic analysis has revealed, however, that even when close relatives differ drastically in some aspect of their genotype, this does not need to be the result of a single mutation. For instance, at least 8 loci are involved in the genetic basis of the projecting eyes of the Hawaiian *Drosophila heteroneura* as compared with its very close relative *D. sylvestris*. There is, however, also at least some evidence in favor of the occurrence of mutations with major phenotypic effects. Iltis (1983), for instance, has postulated that the drastic shift in flower structure from teosinte to cultivated maize was due to a single major mutation. Hilu (1983) has reported on a large number of successful mutations in plants with drastic phenotypic effects.

For animals, it has been suggested by various authors that a drastic phenotypic modification would be most easily achieved by the mutation of a regulatory gene affecting a relatively early stage in ontogeny. At the same time it has been pointed out by Waddington that deviations early in ontogeny tend to be subsequently corrected owing to a preexisting system of canalizations. That development is involved in all phenotypic changes is self-evident, but how often a more drastic mutation can lead to a selectively superior phenotype is still uncertain. In the nearly one million known currently living species of animals there are very few with drastic phenotypic polymorphisms.

Even though this survey indicates that beneficial phenotypic saltations are possible, there is likewise evidence that successful ones are far too rare to account for the numerous evolutionary innovations one finds throughout the animal and plant kingdoms. This raises the question whether such innovations could be explained by a steady accumulation of minimal mutations. The answer is that this is possible in principle. Both of the two major methods of the acquisition of evolutionary novelties—intensification of function and change of function—could be effective under such a system of gradualism. A detailed analysis would be necessary to determine whether Goldschmidt's list of saltations could be resolved into gradual series. Several paleontologists have demonstrated the occurrence of long-continued phyletic gradualism (Bretsky 1976; Kellogg 1983). Particularly impressive is the change in a series of Eurasian microtine rodents extending over 1.5 million years and comprising four consecutive species. The continuous trend involves tooth hypsodonty and the progressive and continuous development of lateral enamel tracks. Since the continuous area from Spain to Siberia participates in this evolution, the changes could hardly be due to local founder populations (Chaline and Laurin 1986). What is involved in this case, however, is a trend, an intensification of existing characters rather than a genuine evolutionary novelty.

Recent authors are virtually unanimous in the opinion that saltations as envisioned by Bateson, De Vries, Schindewolf, and Goldschmidt do not occur. There is no evidence for their existence, and the suggested genetic mechanisms are not workable. Does this mean that now everyone agrees that the origin of higher taxa and the acquisition of evolutionary novelties proceeds by such extreme gradualness as proclaimed by Darwin: "only by very short and slow steps", "by the preservation and accumulation of infinitesimally small inherited modifications"? Unfortunately, there is no such consensus because a number of evolutionary processes and phenomena suggest that not all evolution conforms to such a characterization of gradual phyletic change. For this reason a new evolutionary theory has been proposed: peripatric speciation with genetic revolutions (Mayr 1954) and, based on it, punctuated equilibria (Eldredge and Gould 1972).

The evidence in support of a theory in which macroevolutionary advances are due to relatively short pulses connected with speciation events, followed by stasis or ordinary phyletic evolution, is threefold: (1) the fossil record, (2) the phenomenon of peripatric speciation, and (3) evidence for a peculiar cohesion of the genotype. The following essays are devoted to peripatric speciation (Essay 25), the cohesion of genotype (Essay 24), and punctuated equilibria (Essay 26). I shall limit myself in the present essay to a few words about the fossil record.

Gaps in the Fossil Record

Almost every careful analysis of fossil sequences has revealed that a multiplication of species does not take place through a gradual splitting of single lineages into two and their subsequent divergence but rather through the sudden appearance of a new species. Early paleontologists interpreted this as evidence for instantaneous sympatric speciation, but it is now rather generally recognized that the new species had originated somewhere in a peripheral isolate and had subsequently spread to the area where it is suddenly encountered in the fossil record. The parental species which had budded off the neospecies showed virtually no change during this period. The punctuation is thus caused by a localized event in an isolated founder population, while the main species displays no significant change.

The number of such occurrences in the fossil record is legion. Stanley (1979) has listed numerous such cases, and additional ones have been recorded in every recent volume of paleontological journals. Such punctuationism may well have played a role in our own ancestry. No transition from *Australopithecus africanus* (including *afarensis*) to *Homo habilis* has so far been found, nor from *Homo habilis* to *H. erectus*. The more recent species in these cases suddenly appeared in the range of ancestral species. There is every reason to believe that it originated in a peripheral isolate of the ancestral species.

What characterizes most of the well-documented cases (which does *not* include the hominid ancestry) is the relative stability of the parental species and also of the neospecies once it has completed the speciation process. One has the impression that there is some kind of a force—Lerner has referred to it as genetic homeostasis—which normally prevents any major departure from the well-established genotype. This is a phenomenon that does not fit at all well into the reductionist, atomistic conceptual framework of classical population genetics and has, therefore, been unduly neglected (but see Mayr 1963; 1970; and Essay 24). The interaction of genes is a subject about which few geneticists venture to speculate. A commendable exception is Zuckerkandl (1983), whose major conclusion is that the topology (circuitry) of gene interactions seems to be very conservative and that even changes that appear to be quite drastic may simply be due to quantitative modifications.

As everybody realizes, most seeming "saltations" in the fossil record are due to extinction. And the gaps produced by extinction prevent regrettably often the smooth connection of micro- with macroevolution. It would seem important therefore to say a few words about extinction.

Extinction

Lamarck still resisted the concept of extinction, since this was in conflict with his belief in plenitude and in a benevolent and thoughtful creator (Mayr 1972). By the 1830s the phenomenon was widely acknowledged, and Darwin's mentor, Charles Lyell, was therefore forced to search anxiously but in vain for a law that would explain "the introduction of new species." The importance of extinction as an integral component of the overall process of evolutionary change has received special attention in recent years, and several volumes have been devoted to this subject (Stanley 1979; Nitecky 1984; Valentine 1985).

At least two kinds of extinction must be recognized: a steady slow extinction of species, resulting in a regular faunal turnover; and occasional phases of cataclysmic extinction, as at the end of the Permian and the Cretaceous, when a massive extermination of the biota took place. And, of course, man has caused a massive extinction of the macrofauna during the last 12,000 years, on islands as well as on the continents, and is continuing to do so (Ehrlich 1981). Although less conspicuous, gradual extinction is actually more informative for the student of the evolutionary process than cataclysmic extinction. For instance, there is a straight line (logarithmic) inverse correlation between island size and rate of extinction (Mayr 1965; Diamond 1984). To what extent this extinction is due to interspecific competition, predation, new pathogens, or lack of climatic tolerance is still controversial, but recent researches indicate that—overall—any (or several) of these factors may be involved in a given case, each extinction being a special situation. However, it is evident that the larger a population is, the less vulnerable it is on the average.

Darwin gave much thought to extinction, and the evident frequency of extinction contributed to his rejection of natural selection as a process resulting in perfection (Darwin 1859:201). Each case of extinction vacates, so to speak, a niche or at least makes some resources available. Hence, extinction has also an innovative influence. The flowering of the mammals in the Paleocene and Eocene, for instance, would surely not have been possible except for the prior extinction of the dinosaurs.

Evolution, a Hierarchical Phenomenon

For the extreme reductionists among the geneticists, who looked at evolution entirely in terms of changing frequencies of genes, there is a complete continuity among all phenomena of evolution. For those who thought of evolution as change of species and higher taxa, and this includes Darwin, evolution has always been considered as hierarchically structured. No other component of Darwin's thinking was as readily and widely adopted as his theory of common descent, a strictly hierarchical theory. And most of the paleontological literature, largely devoted to an elucidation of common descent, was strongly hierarchical in its approach.

In recent years there has been a new enthusiasm for a hierarchical approach to evolution (Eldredge and Salthe 1984; Eldredge 1985; Salthe 1985). Although some of the authors of this literature have claimed that by neglecting hierarchical levels "the modern synthesis has failed to provide a workable (testable) evolutionary theory," they do not demonstrate how. Perhaps this is not surprising, since as I have said, most evolutionists have been thinking hierarchically "ever since Darwin," even when not using the term "hierarchy" in their discussions. Of course, organic evolution is a hierarchical process, and nothing demonstrates this better than the difference between the gene frequency variation phenomena within populations and the phenotypic changes in macroevolution. I am prepared to assert that all those evolutionists who objected to strict reductionism were thinking in terms of hierarchies, at least as one significant aspect of evolution.

To be sure, there is a need for stressing hierarchical levels. This need is particularly well documented by the long-time neglect of the study of diversity. In the first third of the century only one of the two great aspects of evolution was emphasized, the continuing improvement of adaptation in the time dimension. As rightly stressed by Gould, such a strictly vertical approach characterized the writings not only of the mathematical geneticists but even of the paleontologists, including G. G. Simpson. I was greatly disturbed by this in the 1930s, and it was the major theme of my *Systematics and the Origin of Species* (1942) that the origin of organic diversity, effected by speciation, was an equally important aspect of evolution. It was this publication which, more than any other, introduced into modern evolutionary biology what Eldredge (1979) has called the "taxic" approach. And in the taxic approach one deals quite automatically with hierarchical levels. Curiously, my various papers and books (1954; 1963; 1970) in which I stressed this importance of the study of diversity for the understanding of macroevolution were almost entirely ignored by the paleontologists and students of macroevolution. I say "almost," because Bock (1970) and a few others did analyze the connections between speciation and macroevolutionary processes. The real revival of an interest in the hierarchical aspects of evolution did not come until the proposal of the theory of punctuated equilibria in 1972 (see Essay 26).

NOTE

This essay is previously unpublished.

REFERENCES

Anon. 1981. How true is the theory of evolution? *Nature* 290:75–76.

Arnheim, N. 1983. Concerted evolution of multigene families. In Nei and Koehn, eds., *Evolution of Genes and Proteins,* pp. 38–61. Sunderland, Mass.: Sinauer.

Ayala, F. J. 1983. Beyond Darwinism? The challenge of macroevolution to the synthetic theory of evolution. *PSA 1982* 2:275–291. East Lansing: Phil. Sci. Assoc.

———— 1985. Reduction in biology: a recent challenge. In Depew and Weber 1985:65–79.

Bateson, W. 1894. *Materials for the Study of Variation.* London: Macmillan.

Bock, W. 1959. Preadaptation and multiple evolutionary pathways. *Evolution* 13:194–211.

———— 1970. Microevolutionary sequences as a fundamental concept in macroevolutionary models. *Evolution* 24:704–722.

———— 1979. The synthetic explanation of macroevolutionary change—a reductionist approach. *Bull. Carnegie Mus. Nat. Hist.* 13:20–69.

Bretsky, S. S. 1976. Evolution and classification of the Lucinidae (Mollusca, Bivalvia) *Palaeont. Amer.* 50:217–337.

Campbell, J. H. 1982. Autonomy in evolution. In R. Milkman, ed., *Perspectives on Evolution,* pp. 190–201. Sunderland, Mass.: Sinauer.

Campbell, J. H. 1985. An organizational interpretation of evolution. In D. J. Depew and B. H. Weber 1985:133–167.

Carson, H. L., and K. Y. Kaneshiro. 1976. Drosophila of Hawaii: systematics and ecological genetics. *Ann. Rev. Ecol. Syst.* 7:311–345.

Carson, H. L., and J. S. Yoon. 1982. Genetics and evolution of Hawaiian Drosophila. In *The Genetics and Biology of Drosophila,* Vol. 3b, pp. 298–344. New York: Academic.

Chaline, J., and B. Laurin. 1986. Phyletic gradualism in a European Plio-Pleistocene *Mimomyx* lineage (Arvicolidae, Rodentia). *Paleobiology* 12:203–216.

Cuénot, L. 1901. L'évolution des théories transformistes. *Rev. gen. sci. pur. applic.* 12:264–269. [The concept of preadaptation was developed in this essay, but the term did not appear until the 1909 essay.]

———— 1909. Le peuplement des places vides dans la nature et l'origine des adaptations. *Rev. gen. sci. pur. applic.* 20:8–14.

Darwin, C. 1859. *On the Origin of Species by Means of Natural Selection or the Preservation of Favored Races in the Struggle for Life.* London: Murray.

———— 1872. *The Origin of Species.* 6th ed. London: Murray.

Davenport, C. B. 1903. The animal ecology of the Cold Spring Harbor sandspit, with remarks on the theory of adaptation. *Univ. Chicago Decennial Pubs.* (1)10:155–176.

Depew, D. J., and B. H. Weber, eds. 1985. *Evolution at a Crossroads: The New Biology and the New Philosophy of Science.* Cambridge, Mass.: MIT Press.

Desmond, A. 1982. *Archetypes and Ancestors.* London: Blond and Briggs.

Diamond, J. M. 1984. "Normal" extinctions of isolated populations. In Nitecki 1984:191–246.

Dohrn, A. 1875. *Der Ursprung der Wirbelthiere und das Princip des Functionswechsels.* Leipzig: Engelmann.

Doolittle, W. F. 1986. Some broader evolutionary issues which emerge from contemporary molecular biological data. *PSA 1984* 2:129–144. East Lansing: Phil. Sci. Assoc.

Dover, G. 1982. Molecular drive: a cohesive mode of species evolution. *Nature* 299:111–117.

Ehrlich, P., and A. Ehrlich. 1981. *Extinction: The Causes and Consequences of the Disappearance of Species.* New York: Random House.

Eldredge, N. 1979. Alternative approaches to evolutionary theory. *Bull. Carnegie Mus. Nat. Hist.* 13:7–19.

———— 1985. *Unfinished Synthesis: Biological Hierarchies and Modern Evolutionary Thought.* New York: Oxford University Press.

Eldredge, N., and S. J. Gould. 1972. Punctuated equilibria: an alternative to phyletic gradualism. In T. J. M. Schopf, ed., *Models in Paleobiology,* pp. 82–115. San Francisco: Freeman and Cooper.

Eldredge, N., and S. N. Salthe. 1984. Hierarchy and evolution. *Oxford Surveys in Evol. Biol.* 1:184–208.

Ferguson, A. 1976. Can evolutionary theory predict? *Amer. Nat.* 110:1101–1104.

Fisher, R. A. 1930. *The Genetical Theory of Natural Selection.* Oxford: Clarendon Press.

Goldschmidt, R. B. 1940. *The Material Basis of Evolution.* New Haven: Yale University Press.

———— 1952. Evolution, as viewed by one geneticist. *Amer. Sci.* 40:84–98.

Goodman, M., M. L. Weiss, and J. Czelusniak. 1982. Molecular evolution above the species level. *Syst. Zool.* 31:376–399.

Gould, S. J. 1980. Is a new and general theory of evolution emerging? *Paleobiology* 6:119–130.

———— 1982. Darwinism and the expansion of evolutionary theory. *Science* 216:380–387.

Grant, P. R. 1986. *Ecology and Evolution of Darwin's Finches*. Princeton: Princeton University Press.

Grant, V. 1983. The synthetic theory strikes back. *Biol. Zentralbl.* 102:149–158.

———— 1985. *The Evolutionary Process: A Critical Review of Evolutionary Theory.* New York: Columbia University Press.

Greenwood, P. H. 1979. Macroevolution—myth or reality. *Biol. J. Linn. Soc.* 12:293–304.

Grene, M. 1983. *Dimensions of Darwinism*. Cambridge: Cambridge University Press.

Hammen, L. v. d. 1981. Typeconcept, higher classification, and evolution. *Acta biotheoretica* 30:3–48.

Hastings, J. W. 1983. Biological diversity, chemical mechanisms, and the evolutionary origin of bioluminescent systems. *J. Molec. Evol.* 19:309–321.

Hilu, K. W. 1983. The role of single-gene mutations in the evolution of flowering plants. *Evol. Biol.* 16:97–128.

Huxley, A. 1982. Anniversary Address. *Proc. Roy. Soc. London* A 379:IX–XVII.

Iltis, H. H. 1983. From teosinte to maize: the catastrophic sexual transmutation. *Science* 222:886–894.

Jablonski, D. 1986. Larval ecology and macroevolution in marine invertebrates. *Bull. Marine Sci.* 39:565–587.

Jablonski, D., and R. A. Lutz. 1983. Larval ecology of marine benthic invertebrates: Paleobiological implications. *Biol. Rev.* 58:21–89.

Kellogg, D. 1975. The role of phyletic change in the evolution of *Eudocubus vema* (Radiolaria). *Paleobiology* 1:359–370.

———— 1983. Phenology of morphologic change in radiolarian lineages from deep-sea cores: implications for macroevolution. *Paleobiology* 9:355–362.

Lack, D. 1947. *Darwin's Finches*. Cambridge: Cambridge University Press.

Lange, E. 1985. Ursachen und Entstehen der Gliederung des Organismenreichs. *Biol. Zentralbl.* 104:497–510.

Laporte, L. F. 1983. Simpson's *Tempo and Mode in Evolution* revisited. *Proc. Amer. Phil. Soc.* 127:365–417.

Levinton, J. S. 1983. Stasis in progress: the empirical basis of macroevolution. *Ann. Rev. Ecol. Syst.* 14:103–37.

Lewontin, R. 1983a. [Optimization and perfectionist natural selection.] *N.Y. Rev. Bks.,* June 16, 1983, pp. 21–27.

———— 1983b. The organism as the subject and object of evolution. *Scientia* 118:63–82.

Maynard Smith, J. 1983. Current controversies in evolutionary biology. In Grene 1983:273–286.

Mayr, E. 1942. *Systematics and the Origin of Species.* New York: Columbia University Press.

———— 1954. Change of genetic environment and evolution. In J. Huxley, A. C. Hardy, and E. B. Ford, eds. *Evolution as a Process,* pp. 157–180. London: Allen and Unwin.

———— 1960. The emergence of evolutionary novelties. In S. Tax 1960:349–380. [See also Mayr 1976:88–113.]

———— 1963. *Animal Species and Evolution.* Cambridge, Mass.: Harvard University Press.

———— 1965. Avifauna: turnover on islands. *Science* 150:1587–1588.

———— 1967. Evolutionary challenges to the mathematical interpretation of evolution. *Wistar Symposium Monograph* 5:47–58 [See also Mayr 1976:53–63.]

———— 1970. *Populations, Species, and Evolution.* Cambridge, Mass.: Harvard University Press.

———— 1972. Lamarck revisited. *J. Hist. Biol.* 5:55–94 [See also Mayr 1976:222–250.]

———— 1976. *Evolution and the Diversity of Life.* Cambridge, Mass.: Harvard University Press.

———— 1977. The study of evolution historically viewed. In C. E. Goulden, ed., *The Changing Scenes in Natural Sciences 1776–1976,* pp. 39–58. Philadelphia: Acad. Nat. Sci., Special Pub. 12.

Miller, A. H. 1949. Some ecologic and morphologic considerations in the evolution of higher taxonomic categories. In E. Mayr and E. Schüz, eds., *Ornithologie biol. Wiss.,* pp. 84–88. Heidelberg: Carl Winter.

Nei, M., and R. K. Koehn, eds. 1983. *Evolution of Genes and Proteins.* Sunderland, Mass.: Sinauer Associates.

Nitecki, M. H., ed. 1984. *Extinctions.* Chicago: Chicago University Press.

Regal, P. 1975. The evolutionary origin of feathers. *Quart. Rev. Biol.* 50:35–66.

Reif, W.-E. 1983. Evolutionary theory in German paleontology. In Grene 1983:173–203.

Rensch, B. 1947. *Neuere Probleme der Abstammungslehre.* Stuttgart: Enke.

Rhodes, F. H.T. 1983. Gradualism, punctuated equilibrium and the *Origin of Species. Nature* 305:269–272.

Rose, M. R., and W. F. Doolittle. 1983. Molecular biological mechanism of speciation. *Science* 220:157–162.

Salthe, S. N. 1985. *Evolving Hierarchical Systems.* New York: Columbia University Press.

Salvini-Plawen, L. v., and E. Mayr. 1977. On the evolution of photoreceptors and eyes. *Evol. Biol.* 10:207–263.

Schindewolf, O. H. 1950. *Grundfragen der Palaeontologie.* Stuttgart: Schweizerbart.

——— 1969. Über den 'Typus' in der morphologischen und phylogenetischen Biologie. *Abh. Akad. Wiss. Lit., Mainz, Math.-Nat. Kl.* 4:58–131.

Severtzoff, A. N. 1931. *Morphologische Gesetzmässigkeiten der Evolution*. Jena: Gustav Fischer.

Shapiro, J. A. 1983. [Letter.] *Nature* 303:196.

Simpson, G. G. 1944. *Tempo and Mode in Evolution*. New York: Columbia University Press.

——— 1953. *The Major Features of Evolution*. New York: Columbia University Press.

Spieth, H. T. 1982. Behavioral biology and evolution of the Hawaiian picture-winged species group of Drosophila. *Evol. Biol.* 14:351–437.

Stanley, S. M. 1979. *Macroevolution: Pattern and Process*. San Francisco: Freeman.

Stebbins, G. L., and F. J. Ayala. 1981. Is a new evolutionary synthesis necessary? *Science* 213:967–971.

Tax, S. 1960. *The Evolution of Life*. Chicago: University of Chicago Press.

Turner, J. R. G. 1983. Mimetic butterflies and punctuated equilibria: some old light on a new paradigm. *Biol. J. Linn. Soc.* 20:277–300.

——— 1984. Darwin's coffin and Doctor Pangloss: Do adaptationist models explain mimicry? In B. Shorrocks, ed., *Evolutionary Ecology,* pp. 313–361. Oxford: Blackwell Scientific Publications.

Valentine, J. W., ed. 1985. *Phanerozoic Diversity Patterns: Profiles in Macroevolution*. Princeton, N.J.: Princeton University Press.

Watt, W. B. 1985. Allelic isozymes and the mechanistic study of evolution. *Isozymes: Current Topics in Biological and Medical Research* 12:89–132.

Wilson, A. C., V. M. Sarich, and L. R. Maxson. 1974. The importance of gene arrangement in evolution: evidence from studies on rates of chromosomal, protein, and anatomical evolution. *Proc. Nat. Acad. Sci.* 71:3028–3030.

Zuckerkandl, E. 1983. Topological and quantitative relationships in evolving genomes. In C. Hélène, ed., *Structure, Dynamics, Interactions and Evolution of Biological Macromolecules,* pp. 395–412. Dordrecht: Reidel.

THE UNITY OF THE GENOTYPE

THE historian of science often finds that a particular area of research alternates, in its interests and interpretations, between two extremes. This is certainly true for the attitude of geneticists toward the genotype. The pre-Mendelian breeders and hybridizers had discovered many striking manifestations of segregation, including 3 to 1 ratios, without coming even near to a Mendelian interpretation. One of the reasons for their blindness is now obvious. These forerunners considered the essence of the species an indivisible whole. Questions about individual genetic factors simply made no sense to an essentialist.

After the rediscovery of the Mendelian rules, the pendulum swung to the other extreme. The approach now became entirely atomistic and, for the sake of convenience, each gene was treated as if it were quite independent of all others. In due time all sorts of phenomena were discovered which contradicted *this* interpretation, such as the linkage of genes, epistasis, pleiotropy, and polygeny, and yet in evolutionary discussions only lip service was paid to these complications. Differences among populations and species continued to be simply described in terms of gene frequencies. Evolution, as recently as the 1950s, was defined as a change in gene frequencies, the replacement of one allele by another; thus is was treated as a purely additive phenomenon.

As useful as this approach was, and as magnificent as the results which it produced, it did not provide the whole answer. Almost as far back as the rediscovery of Mendel's work there was a minority of authors who stressed interaction among genes, the integration of the genotype. The purely analytical school thought that such an integrative attitude was incompatible with a meaningful analysis and dangerously close to such a stultifying concept as holism. Chetverikov (1926), with his concept of the genetic milieu, was perhaps the first author to stress constructive aspects of the study of the interaction of genes, and through him and his school the study of the cohesion of the genotype has become an increasingly important branch of evolutionary genetics. Progress, however, was slow. If we want to single out a definite year which signalizes the new interest in the interactions of the genotype, it is the year 1954 in which Lerner's *Genetic Homeostasis* was published as well as my own paper on the importance of the genetic environment in evolution (Mayr 1954). The newer views can be summarized as follows: Free variability is found only in a limited portion of the genotype. Most genes are tied together into balanced complexes that resist change. The fitness of genes tied up in these complexes is determined far more by the fitness of the complex as a whole than by any functional qualities of individual genes.

Evidence for Unity of the Genotype

Even though the unity of the genotype was grossly neglected for nearly 100 years, naturalists and breeders had long been aware of it. It is what morphologists and systematists had in mind when they spoke of the mammalian or chordate type. Darwin and others spoke of "the mysterious laws of correlation." When referring to *correlation of growth,* Darwin said: "I mean by this expression that the whole organization is so tied together during its growth and development, that when slight variations in any part occur, and are accumulated through natural selection, other parts become modified" (1859:143). E. Geoffroy St. Hilaire's *loi de balancement* is an indication of similar ideas (1818).

Other aspects of the unity of the genotype became apparent already rather early in the history of genetics. The phenomenon of pleiotropy, the capacity of a gene to affect several different aspects of the phenotype, led Chetverikov (1926) "to the idea of the genetic milieu which acts from the inside on the manifestation of every gene in its character. An individual is indivisible not only in its soma but also in the manifestation of every gene it has."

Evidence for an interaction among genes is manifold. Most convincing are the consequences of *artificial selection.* Darwin already had stated: "If man goes on selecting, and thus augmenting any peculiarity, he will almost certainly modify unconsciously other parts of the structure, owing to the mysterious laws of the correlation of growth." Dog fanciers and the breeders of race horses are sadly aware of the undesirable side effects of a breeder's concentration on a particular character. In almost all recent experiments on maximizing a particular character in *Drosophila,* such as bristle number, some selected lines died out owing to sterility. I do not know of a single intensive selection experiment during the past 50 years during which some such undesirable side effects have not appeared.

The phenomenon which Lerner has designated as *genetic homeostasis* is additional evidence for the organization of genes into co-adapted systems. When selection directed toward the maximization of one particular character is terminated after a series of generations and the population is permitted to reach its own genetic equilibrium, the strongly selected character very often loses part, if not most, of the phenotypic advance it had achieved during the preceding period of intense selection. There is, as it were, an internal balance of selection pressures, restoring the co-adapted system to its previous balance. One can infer from this phenomenon of genetic homeostasis that genes, which are tied up in a co-adapted complex, participate in a number of independent metabolic operations (or ontogenetic processes), and that a rather specific frequency of the various interacting genes is needed for the optimal performance of these physiological processes. If this frequency has been distorted by one-sided selection, internal selection pressures will tend to restore it to the original optimal frequency. The original fitness will be restored by this return to the original balance of genes.

I will refrain, at this point, from becoming specific on the nature of the interaction among genes because we are skating here on thin ice. We have become dimly aware that structural genes, the material mainly studied by the geneticist, may well be only a small minority of all the genes, let us say 50,000 among 5 million. What percentage of the other genes are regulator genes of all sorts and what else the remaining DNA does is currently the subject of intense research. When one studies long-term evolution one has the uncomfortable feeling that it is no longer sufficient to interpret it entirely in terms of the traditional genes. Britten and Davidson (1969:356) might well be right in saying: "At higher grades of organization, evolution might indeed be considered in terms of changes in the regulatory systems." Much that is now explained as "epistatic interactions between different loci" might well be due to the activities of regulatory genes. Fortunately, this possibility does not affect very greatly what I am now attempting to present.

Among the many manifestations of the cohesion of the genotype encountered by the naturalist, I will discuss here only a single one, the narrowness of hybrid zones. When the geographic barrier between two incipient species breaks down after they had been isolated for a lengthy period of time, they will either have completed the acquisition of isolating mechanisms or else they will hybridize. One would expect that such hybridization would lead to a free introgression of genes of either population into the other one. This, in turn, should lead to a rapid widening of the hybrid zone until the introgressing genes have reached the opposite species border. But this is not at all what happens in most cases of hybrid zones. The hooded and the carrion crows met in Central Europe at the end of the Ice Age some 8,000 years ago and hybridized freely. Yet, as Meise (1928) pointed out, this hybrid belt is still only about 100 miles wide. This was correctly interpreted by Dobzhansky as early as 1941 to mean that the introgressing genes of the hooded crow are selected against in the genotype of the carrion crow and vice versa.

A far more precise analysis of the same phenomenon was recently provided by Hunt and Selander (1973) for the hybrid zone between *Mus m. musculus* and *M. m. domesticus* in Denmark. Allozymic variation in 13 proteins controlled by 41 loci was studied and 7 polymorphic enzymes were selected for detailed analysis. As in the case of the crows, the width of the hybrid zone is still relatively narrow, even though the two kinds of mice must have first met many thousands of years ago. The rate of introgression is distinctive for each locus. Some alleles have penetrated into the alien genotype for a much larger distance than others. Although several additional factors, such as a slow dispersal rate of the mice and local climatic fluctuations, affect the rate of introgression, there can be little doubt that selection against the alien genes is an important if not the major factor for the slowness of the introgression. The fact that the width of the hybrid belt is asymmetrical, that is, that introgression into *musculus* is more extensive than into *domesticus,* also shows that the dispersal of genes is not simply a "mechanical" phenomenon.

A particularly elegant demonstration for the unity of the genotype is provided for by the so-called "lethal chromosomes," chromosomes that are lethal when homozygous. As summarized in his *Topics in Population Genetics* (1968), Wallace has shown how often such chromosomes are normal or superior in heterozygous condition when placed in the gene pool of their own population. Furthermore the "lethality" of a newly arisen lethal chromosome becomes steadily less through natural selection when maintained in its population. Anderson (1969) showed in an analysis of data provided by Dobzhansky and Spassky that among 45 lethal chromosomes tested on the genetic background of their own populations none were significantly deleterious in heterozygous conditions, but 5 were significantly heterotic. However, when tested on foreign genetic backgrounds, 10 of these chromosomes were significantly deleterious and none were significantly heterotic.

The Mechanisms of Cohesion

Nothing is explained, however, by simply saying such and such a phenomenon is due to the unity of the genotype. We also want to know by what mechanisms this unity of the genotype is accomplished. Why is it that not all genes are entirely independent during recombination? What holds them together in the face of the centrifugal forces of recombination? To be frank, our answers to these questions are still incomplete. Until rather recently, genetic analysis was able to focus on only one locus or at best one gene arrangement at a time. But this has been changed by technological advances. The method of gel electrophoresis and the availability of large computers for linkage studies have greatly enlarged the power of the analysis. The literature is expanding rapidly, and I will give only a short résumé, since I want to concentrate on the evolutionary consequences of this cohesion.

To summarize it in a single sentence, the interaction of macromolecules at the cellular level produces individual phenotypes of different fitness, and natural selection will therefore tend to hold together those alleles at different loci that produce individuals of the greatest selective value.

Let me now present some evidence for the individual links of this causal chain.

Interaction of macromolecules. I am not a molecular biologist and am not at all qualified to talk about this subject. But let me remind you that the "morphology" of a macromolecule, that is, its structural contour, caused by the folding of its protein chain (or other constituents) must match the structure of the membranes to which it is attached or of other macromolecules with which it interacts. Furthermore, a single enzyme is often involved in several reasonably different functions (O'Brien et al. 1972) and must have the optimal constitution for each of these separate demands. The literature of molecular biology reports numerous cases where even a seemingly very slight change had a drastic effect on the efficient interaction of such a molecule.

Also, let us remember that the products of one locus may serve as repressors or inducers at other loci. Finally, the activity of regulator genes provides almost unlimited scope for the interaction of genes. These are merely reminders of the importance of macromolecules for cohesion.

Selection. How does selection cope with this situation? Let us say allele 1 at locus A gives superior fitness when in epistatic balance with allele 3 at locus B, but that there is simultaneously a selective premium for heterozygosity at each of these loci. Free recombination in such a system would, obviously, produce a tremendous genetic load, through inferior recombinants. If I would ask you how many such gene pairs a population could maintain, without accumulating an intolerable genetic load, you would probably give some figure between 5 and 10. Actually the figure is much larger than one might think at first sight (Mayr 1963: 261). On the average only 2 offspring need to be successful in an animal pair that produces 1,000 or 100,000 zygotes. This means that well over 99 percent of these zygotes are expendable and the genetic load can, so to speak, be charged against the surplus, provided the few survivors carry the superior gene combinations.

The production of a large number of offspring so that an inevitable genetic load of inferior gene combinations can be charged to this reproductive surplus is one possible evolutionary strategy.

Linkage. An alternate strategy is to prevent free recombination. Cytogeneticists, from the earliest times on, have emphasized that the organization of the genetic material into chromosomes is a powerful mechanism for keeping superior gene combinations together. Yet, chromosomes are not permanently inviolable structures because genes are separated from each other in every generation owing to the process of crossing-over during meiosis. Linkage, for this reason, was for a long time considered a rather inefficient mechanism for a long-term tying together of specific alleles at different loci of the same chromosome, unless reinforced by special crossover inhibitors. It would indeed be highly advantageous if favorable gene combinations could be tied together into unbreakable linkage groups, as was stressed by R. A. Fisher and others since the 1920s. The great advantage of such "supergenes" is that their existence drastically reduces the frequency of deleterious recombinants, and with it the genetic load.

There are two lines of evidence to indicate that chromosomes and chromosome arms are far more permanent structures than had been envisioned by the earlier students of linkage and of crossing-over. One of these are detailed studies of chromosome structure in insects, amphibians, mammals, and other animals. They show that the chromosomes contain balanced gene combinations and that speciation, in a high proportion of cases, involves a more or less extensive repatterning of the chromosome (White 1973). We shall come back to these speciational aspects later.

The other evidence is provided by Franklin and Lewontin's linkage simulation studies (1970). These show that linkage is far more powerful than had been previously imagined and that superior chromosomes go through meiosis from generation to generation virtually untouched.

Although this conclusion was at first largely conjectural, concrete evidence for it has now been found in several cases, again through the gel-electrophoresis method. Clegg and Allard (1972) for a number of cases in *Avena* and Webster (1973) for one case in the salamander *Plethodon cinereus* have demonstrated that certain alleles at different loci go always (or preferably) together. Cytogeneticists refer to this occurrence by the quaint term "linkage disequilibrium."

A tight linkage of alleles at two different loci, such as found for *Avena* and *Plethodon,* has so far been found in *Drosophila* only for inversions or near breakage points. How widespread and important linkage disequilibrium in other organisms is still remains to be determined.

There are, of course, various chromosomal mechanisms which can induce an increase or decrease in the level of recombination and in chiasma localization.

I am afraid I have barely scratched the surface, because there are surely numerous other mechanisms either to tighten or to loosen the cohesion of the genotype.

Consequences of the Cohesion of the Genotype

As long as genes were considered to be independent of each other, it was difficult to explain certain evolutionary phenomena. Indeed, some of the anti-Darwinian arguments arose from such an inability to explain the "mysterious laws of correlation" and other phenomena that we shall presently mention. However, as soon as it is realized that the genotype consists of a number of co-adapted gene complexes and that even the gene pool of a population as a whole is well integrated and co-adapted, such arguments lose their validity.

Let us now look in more detail at what kind of consequences one may derive from the concept of the unity of the genotype:

(1) Since the fitness of a gene depends in part on the success of its interaction with its genetic background, it is no longer possible to assign an absolute selective value to a gene. A gene has potentially as many selective values as it has possible genetic backgrounds.

(2) The target of selection does not consist of single genes but rather of such components of the phenotype as the eye, the legs, the flower, the thermo-regulatory or photo-synthetic apparatus, etc. As a result any given selection pressure affects simultaneously whole packages of genes which may or may not be tied together by special devices, such as linkage, epistatic balances, etc.

(3) At no stage in the life of an individual is the interaction of genes more obvious than during ontogeny. The adult individual is the end-product of the entire epigenetic process. The endeavor to dissect this into the effects of individual genes can only rarely be successful.

(4) Most species are remarkably uniform over vast areas. The climatic adaptations of species are conspicuous and have been described through the study of climatic rules, clines, ecotypes, and the many other man-ifestations of local or regional selection pressures. Much of geographic variation is conspicuously adaptive and has supplied some of the most convincing evidence for Darwinian selection. Yet, as I pointed out some 20 years ago, pronounced geographic variation of the phenotype is far less universal, or at least far less conspicuous, than is usually assumed. When I first encountered the phenomenon of a seeming uniformity of a species over wide areas in spite of drastic environmental differences, I attempted to explain it as the result of gene flow. Gene flow, indeed, is an important factor, more so than admitted by some contemporary critics, but when the phenotypic uniformity of a species extends across all of North America from the Pacific to the Atlantic or across the entire Eurasian continent, it becomes obvious that additional factors must be involved.

At one time it was suggested that the seeming phenotypic uniformity of such species is only apparent, being of the type of similarity found in sibling species while concealing great cryptic variation of the genotype under this superficial cloak. The method of gel electrophoresis has refuted this assumption. Prakash and Lewontin (1968) for *D. pseudoobscura* were the first to show, and this was later confirmed by Ayala et al. (1972) for *D. willistoni,* that rare allozymes are often found throughout the range of these species at very similar low frequencies. If these genes were either neutral or serving strictly local adaptation, one would expect stochastic processes or selection pressures to produce strong geographic variation in their frequencies. Nor can gene flow be responsible for this universally even frequency, as Ayala et al. demonstrated convincingly. Admittedly the universal distribution of certain alleles at the same low frequencies is not explained easily. The interpretations that make the most sense are (1) that these alleles are normally inferior, but sufficiently valuable in certain epistatic interactions or in certain subniches to be retained in the gene pool of the population as a necessary component of the co-adapted gene complex, or (2) that frequency-dependent selection is involved.

If we make the assumption that the frequency of these alleles is the result of neither stochastic processes nor ecotypic selection but is determined by the unity of the genotype then we must postulate that a different set of such alleles will be favored in every species, because speciation entails a thorough reorganization of the genotype and new balances of interacting genes will result from such a genetic revolution. Ayala and Anderson (1973) have shown that this is indeed the case for a certain enzyme in the three closely related sibling species *Drosophila willistoni, D. equinoxialis,* and *D. tropicalis.* Experimental evidence indicates that selection favors that allele which is most common ("wild type") in the particular species. Since these three species are largely sympatric, it is clearly not the absolute activity of the allele which is selected but rather how the particular allozyme fits with the physiological and genetic background of the particular species in which it finds itself.

The uniformity of the frequency of rare alleles and many of the other phenomena just discussed demonstrate what a conservative influence the unity of the genotype is. As soon as one has fully grasped this point, the whole problem of speciation appears in a new light.

Speciation

The classical theory of geographic speciation postulates the following sequence of events. Some newly arisen geographical barrier divides the range of a species in two, the two subdivisions of the species acquire isolating mechanisms during their period of isolation, and are able to coexist after the break-down of the barrier. Massive evidence has accumulated in recent decades to indicate that this is probably the exceptional rather than the standard case of geographic speciation. What apparently happens far more frequently is that a founder population becomes peripherally isolated and undergoes a genetic revolution (Mayr 1954) during which it rather rapidly acquires isolating mechanisms and species status. The evidence for this thesis is far too massive to be presented here in detail and all I can do is to outline it and give a few striking examples.

It has long been known that in addition to the conventional species with wide-ranging, continental distribution patterns there is a second category of species with an "insular" or "colonial" population structure (Kinsey 1937; Mayr 1942). Wingless grasshoppers and subterranean mammals provide particularly instructive examples. The pocket gophers (*Thomomys*), the mole rats (*Spalax*) of eastern Europe and the Near East, and the tuco-tucos (*Ctenomys*) of Argentina all have a distribution pattern consisting of numerous isolated populations. We used to make fun of the pocket gopher specialists for describing hundreds of new subspecies, but—up to a point—they now have the last laugh. The work of Patton and others has shown, at least for some groups of pocket gophers, that there is strong variation in chromosome number and chromosome repatterning from colony to colony. In spite of minimal morphological differences, some of the populations seem to have reached species level.

In the case of species with an insular distribution pattern, new outposts are established by founders and this provides an opportunity either for genic or for chromosomal genetic revolutions. If the colonization of a *terra nova* is successful owing to the greatly increased opportunity for the incorporation of new adaptive gene combinations (Lewis 1973), this will place a selective premium on any and all genic or chromosomal mechanisms that facilitate genetic revolutions.

At this point I would like to call attention to the history of our thinking about speciation. The early Mendelians thought that a single mutation could create a new species. This was replaced in the early days of population genetics by the postulate that a relatively small number of genes was involved in speciation, geographic or otherwise. I recall a report in an early issue of *Evolution* in which the author postulated a number of less than 10 gene fixations for speciation in beetles. In recent decades it has become quite apparent that a purely additive approach to this problem is misleading. Perhaps one can go to the opposite extreme and postulate that successful speciation always requires the acquisition of a new balance or, to describe it more appropriately, it requires the breaking up, the dissolution, of the previous cohesion and its replacement by a new cohesion.

Carson (1970) has demonstrated through his discovery of homosequential species of Hawaiian Drosophilae that such a reorganization can be accomplished purely on the gene level. The same has been demonstrated by the genetic revolution in the founder population of *Drosophila pseudoobscura* at Bogota. This population is still remarkably similar to the Central and North American populations of the species and no behavioral isolation of either males or females of the Bogota population from North American flies was found. Yet, F_1 males obtained from the cross of *D. pseudoobscura* females from Bogota with males of this species from the North American mainland (from Guatemala north) are sterile. This sterility is due to about two genes located on the X chromosome and one each on two of the other chromosomes (Prakash 1972). The most likely explanation is that the reproductive isolation is due to the incorporation of new genes on the four stated loci to improve the fertility of the males in the highly inbred Bogota founder population. Additional fixation at other loci is, of course, also possible since only 24 of approximately 10,000 enzyme loci were tested.

An increasing amount of evidence is now accumulating to indicate that purely genic speciation, such as found by Carson, not involving chromosome repatterning, is comparatively rare. There has been a gradual return in recent years to earlier views of a great importance of karyotypic events in speciation. In contrast to the earlier view, however, the concepts of chromosomal reorganization are now integrated into the biological species concept and combined with population thinking. Harlan Lewis (1966, 1973) and his school for plants, and M. J. D. White in a series of papers for animals (summarized in White 1973), have demonstrated two facts:

(1) That closely related species differ in the majority of cases by chromosomal rearrangements while such differences are rarely found among contiguous populations of the same species.

(2) That in cases of incipient speciation, for instance peripherally isolated populations, and in cases of recently separated parapatric species, one can usually demonstrate a lowered fitness of genotypes heterozygous for the chromosomal rearrangements. It is highly probable that the passing through the bottleneck of deleterious heterozygosity cannot occur in a populous widespread population but only during a genetic revolution in a founder population.

Macroevolutionary Consequences

The recognition of the fact that there is a strong cohesion of the genotype has greatly helped in the understanding of certain previously puzzling macroevolutionary phenomena. These had caused the early Darwinians a good deal of trouble and had induced certain evolutionists, particularly many paleontologists between 1880 and 1930, to reject the Darwinian interpretation of macroevolution.

(1) Why, for example, do the embryos of terrestrial vertebrates still go through the gill-arch stage? Or more broadly, why are all those ancestral conditions maintained in ontogeny which at one time were interpreted as evidence for recapitulation?

(2) What is the explanation for the extreme evolutionary inertia (stagnation) of certain evolutionary lines, as indicated by such types as *Gingko, Equisetum, Limulus, Lingula, Triops,* and *Nautilus* (Mayr 1970:367)?

(3) What is the explanation for the sudden flowering ("explosive evolution") of certain previously long-stagnant evolutionary lines?

(4) What is the explanation for the conservative nature of the *Baupläne* of the major animal types? The basic mammalian *Bauplan,* for instance, is maintained in elephants, bats, man, and whales. Why do all terrestrial vertebrates develop as tetrapods, all insects as hexapods? Obviously it would be absurd to claim that the type of locomotion of insects requires three pairs of legs and that of terrestrial vertebrates two pairs of extremities.

In terms of the cohesion of the genotype the situation is no longer a puzzle. A restructuring of the major morphogenetic pathways (resulting in a different number of pairs of extremities) would require such a drastic interference with the cohesion of the genotype—as-a-whole—that it would be selected against quite vigorously. That it is not the characters themselves but their integration into the total genotype that is important is obvious when we compare the stability of certain wing veins and bristle patterns in insects from one genus or family to another.

The most difficult feat of evolution is to break out of the straight-jacket of this cohesion. This is the reason why only so relatively few new structural types have arisen in the last 500 million years, and this may well also be the reason why 99.999% of all evolutionary lines have become extinct. They did so because the cohesion prevented them from responding quickly to sudden new demands by the environment.

We recognize about 24 distinct phyla among the recent animals. There is much evidence to indicate that the basic features of the structural plan of these phyla is derived from that of the founders of these types. If the founder crawled on the surface, there was a premium on the development of metamerism; if he tunnelled through the soft substrate, the development of a compressible coelom was favored. The same basic adaptation could evolve independently in different founders. Once established and properly consolidated in the genotype, such a *Bauplan* remained remarkably stable regardless of subsequent developments. The metamerism of the annelids is still apparent in the tube-living sessile polychaetes, and coeloms are retained by most of the coelomates that have long since given up tunnelling.

The same phenomenon is illustrated by the gill arches that still dominate the ontogeny of land-living vertebrates. It is obvious in all these cases that development is controlled by such a large number of interacting genes that the selection pressure to eliminate vestigial structures is less effective than the selection to maintain the efficiency of well established developmental pathways. This is, of course, not a new explanation, but it has been remarkably little explored by the students of developmental physiology.

Two other sets of phenomena confirm this conclusion.

Evolutionary trends. In many, if not most, phyletic lines there is an indication of trends. Many evolutionists prior to 1940 considered such "orthogenetic" trends as a refutation of "Darwinian evolution by chance," as they phrased it. It is now quite obvious that such trends are the necessary consequence of the unity of the genotype which greatly constrains evolutionary potential. Each family or order of animals and plants has its special potential such as that of the ungulates for horns, or of the drongos (Dicruridae) for elaborations of the tail feathers.

Evolutionary rates. Evolutionary rates are highly dissimilar at the species level. Large, populous species with wide distribution show great evolutionary inertia, not so much, as was once said, because it takes so long for an allele to disperse through the entire species range, but rather because the co-adapted gene complex is highly resistant to the incorporation of new genes. On the other hand, founder populations may undergo a genetic evolution within an extremely short time and may possibly replace alleles at 30 to 50 percent of the loci within 5,000 to 10,000 years. Such cases as the Bogota population of *D. pseudoobscura* or the five new species of *Haplochromis* fishes in Lake Nabugabo which is less than 4,000 years old, and the rapid speciation of subterranean mammals all indicate exceedingly rapid rates of speciation.

Considering these drastic differences at the species level it is difficult to advance generalizations as far as phyletic evolution is concerned. There is little doubt, however, that those who enter biology from the physical sciences make rather unrealistic assumptions on the evolutionary rate of changes in macromolecules. From the fact that some molecules such as the histones evolve very slowly, while others have intermediate, or very rapid evolution, some authors have drawn the conclusion that such molecules have a built-in, so to speak orthogenetic, rate of evolution. Depending on the particular macromolecule they assume that there will be an amino-acid replacement every 2 million, 5 million, or 10 million years. All the known facts contradict this naive assumption.

It seems to me that it leads to a far more realistic interpretation of evolutionary rates if we assume that the evolutionary change of a given type of macromolecule is not due to a built-in rate but rather due to the need for co-adaptation with other molecules with which it has to interact. As the other molecules change in the course of time in response to *ad hoc* selection pressures, every molecule occasionally has to adjust to its changed molecular environment; it must adjust to the cohesion of the genotype. This, undoubtedly, is the reason for the continuous evolutionary change of molecules, even of those in which the active site has not changed since the days of the most primitive eukaryotes or even prokaryotes.

Let me now summarize:

The genes are not the units of evolution nor are they, as such, the targets of natural selection. Rather, genes are tied together into balanced adaptive complexes, the integrity of which is favored by natural selection.

The study of the mechanisms by which the cohesion of the genotype is achieved is a promising area of evolutionary research.

It is important to understand this cohesion of the genotype, because it permits the explanation of many previously puzzling phenomena of speciation and macroevolution.

NOTE

This essay was originally presented as a lecture to the First International Congress of Systematics and Evolutionary Biology (ICSEB), Boulder Colorado, on August 9, 1973. It was first published in *Biol. Zentralblatt* 94 (1975):377–388.

REFERENCES

Anderson, W. W. 1969. The selection coefficients of heterozygotes for lethal chromosomes in Drosophila on different genetic backgrounds. *Genetics* 62:827–836.

Ayala, F. J., J. R. Powell, M. L. Tracey, C. A. Mourao, and S. Perez-Salas. 1972. Genic variation in natural populations of *Drosophila willistoni*. *Genetics* 70:113–139.

Ayala, F. J., and W. W. Anderson. 1973. Evidence of natural selection in molecular evolution. *Nature* (New Biology) 241:274–276.

Britten, R. J., and E. H. Davidson. 1969. Gene regulation for higher cells: a theory. *Science* 165:349–357.

Carson, H. L. 1970. Chromosome tracers of the origin of species. *Science* 168:1414–1418.

Chetverikov, S. S. 1926. On certain aspects of the evolutionary process from the standpoint of modern genetics. *J. Expltl. Biol.* (Russian) A2. Eng. Trans. 1961. *Proc. Amer. Phil. Soc.* 105:167–195.

Clegg, M. T., and R. W. Allard. 1972. Patterns for genetic differentiation in the slender wild oat species *Avena barbata*. *Proc. Nat. Acad. Sci.* 69:1820–1824.

Darwin, C. 1859. *On the Origin of Species by Means of Natural Selection*. London: John Murray.

Fisher, R. A. 1930. *The Genetical Theory of Natural Selection*. Oxford: Clarendon Press.

Dobzhansky, Th. 1941. *Genetics and the Origin of Species*. 2nd ed. New York: Columbia University Press.

Franklin, J., and R. C. Lewontin, 1970. Is the gene the unit of selection? *Genetics* 65:707–734.

Geoffroy St. Hilaire, E. 1818. *Philosophie anatomique*. Paris.

Hunt, W. G., and R. K. Selander. 1973. Biochemical genetics of hybridization in European house mice. *Heredity* 31:11–33.

Kinsey, A. C. 1937. An evolutionary analysis of insular and continental species. *Proc. Nat. Acad. Sci.* 23:5–11.

Lerner, I. M. 1954. *Genetic Homeostasis*. Edinburgh: Oliver and Boyd.

Lewis, H. 1966. Speciation in flowering plants. *Science* 152:167–172.

——— 1973. The origin of diploid neospecies in *Clarkia. Amer. Nat.* 107:161–170.

Mayr, E. 1942. *Systematics and the Origin of Species*. New York: Columbia University Press.

——— 1954. Change of genetic environment and evolution. In J. Huxley, A. C. Hardy, and E. B. Ford, eds., *Evolution as a process*, pp. 157–180. London: Allen and Unwin.

——— 1963. *Animal Species and Evolution*. Cambridge, Mass.: Harvard University Press.

——— 1970. *Populations, Species, and Evolution*. Cambridge, Mass.: Harvard University Press.

Meise, W. 1928. Die Verbreitung der Aaskrähe (Formenkreis *Corvus corone* L.). *J. Orn.* 76:1–203.

Nevo, E., J. K. Yung, C. R. Shaw, and C. S. Thaeler, Jr. 1974. Genetic variation, selection and speciation in *Thomomys talpoides* pocket gophers. *Evolution* 28:1–23.

O'Brien, S. J., B. Wallace, and R. J. Macintyre. 1972. The x-Glycerophosphate cycle in *Drosophila melanogaster* III. *Amer. Nat.* 106:767–771.

Patton, J. L. 1973. Patterns of geographic variation in karyotype in the pocket gopher, *Thomomys bottae. Evolution* 26:574–586.

Prakash, S. 1972. Origin of reproductive isolation in the absence of apparent genic differentiation in a geographic isolate of *Drosophila pseudoobscura. Genetics* 72:143–155.

Prakash, S., and R. C. Lewontin. 1968. Direct evidence of co-adaptation in gene arrangements of Drosophila. *Proc. Nat. Acad. Sci.* 59:398–405.

Wallace, B. 1968. *Topics in Population Genetics.* New York: Norton.

Webster, T. P. 1973. Adaptive linkage disequilibrium between two esterase loci of a salamander. *Proc. Nat. Acad. Sci.* 70:1156–1160.

White, J. M. D. 1973. *Animal Cytology and Evolution.* 3rd ed. London: Cambridge University Press.

SPECIATION AND
MACROEVOLUTION

SPECIES and higher taxa represent, phenomenologically, two very different levels in the hierarchical organization of the living world. Ever since there has been a concept of evolution, there has been the problem how one can get from the species level to that of the higher categories. Darwin, the champion of gradualism, declared that it was a purely quantitative problem. If one would simply pile enough small differences on top of each other, one would eventually get something that is qualitatively different, that is, a higher taxon or an evolutionary novelty. Why Darwin was so intent on defending gradualism is something I do not want to take up at this time (Gruber 1974; Stanley 1979; Ospovat 1982). But Darwin, at that time, was vitually alone in this insistence on gradualism (as far as I know the literature). Virtually all other evolutionists of his period were so impressed by the gaps between genera and by the even greater gaps among the higher taxa that they felt they could not do without saltations. T. H. Huxley was characteristic of this thinking. Saltationism became even more popular after the publications of Bateson (1894) and de Vries (1901–1903), even though there were occasional voices (such as Scott 1894) championing gradual evolution. Despite a last flare-up of saltationism in the 1940s and 1950s (represented by Goldschmidt 1940, Willis 1940, and Schindewolf 1950), gradualism was triumphant in the evolutionary synthesis. I want to point out, however, that it was, on the whole, a gradualism in the vertical (Lamarckian) tradition. It was simply a thinking in terms of phyletic lines that gradually and inexorably moved upward to ever better adaptations or ever greater specializations.

The Darwinian "horizontal" tradition of an origin of diversity, that is, of a multiplication of species, and the role of this diversification in macroevolution was totally ignored. When one studies the writings of "Darwinian" paleontologists, one discovers that in their argumentation they proceed directly from the genetic variation–mutation level to that of macroevolutionary processes (new higher taxa, evolutionary novelties). The same was true for the geneticists who (except for Dobzhansky and a few others with a natural history background) moved straight from the gene level to that of macroevolution.

It is only in the writings of the naturalists, particularly the zoologists, that the complete transition from the gene level to that of macroevolutionary processes is considered by inserting and analyzing the role of the species, in the transition from population to species to higher taxon.

The difference between the thinking of the geneticists-paleontologists on one hand, and Darwin and the naturalists on the other hand, is far more fundamental than is appreciated by most evolutionists. If we define evolution as changes in adaptation and diversity, then the students of adaptation deal with what we might call the vertical dimension of evolution, while the students of diversity deal with the horizontal dimension, that is, with the changes of populations in latitude and longitude. Actually, of course, both processes take place simultaneously, but most workers in the different subdisciplines of evolutionary biology have paid attention only to one of the two dimensions. Paleontologists, when studying macroevolution, traditionally never come to grips with the problem of the origin of the taxa or types that evolved "higher" or experienced adaptive radiations. Simpson (1944), for example, makes no reference to species or speciation in his *Tempo and Mode in Evolution.* The inconvenient fact that the study of phyletic lines through time seems to reveal only minimal changes and no evidence of a change from one genus into a new one or of the gradual origin of an evolutionary novelty, was quietly ignored. Most often it was blamed on the incompleteness of the fossil record.

The students of speciation, and more broadly the neontologists as a whole, accepted as an article of faith not only that all macroevolutionary phenomena were consistent with the laws of genetics but also that they could be explained in terms of the phenomena of geographic variation and speciation. I concluded my 1942 book with the claim that "all the available evidence indicates that the origin of the higher categories is a process which is nothing but an extrapolation of speciation." (Mayr 1942:298).

Simpson was not alone in his neglect of the problem of speciation. Most paleontologists, during the 100 years after the publication of the *Origin,* entirely ignored the problem of the origin of organic diversity. They did so in part because their "data just aren't sensitive enough to analyze evolutionary kinetics" (Carson 1981), and in part because neither the existing reductionist theory of genetics nor the classical speciation theories of the systematists provided appropriate scenarios that would have connected with the findings of paleontology. Their attitude, thus, was not unreasonable.

Even though both Simpson (1944) and I (1942) vigorously opposed essentialistic speciation by saltation, this model of macroevolution was adopted by those few paleontologists who did make the attempt to explain the origin of diversity. Essentialism still had a strong hold, particularly in non-English-speaking countries, and even some of those who accepted common descent went back to pre-Darwinian theories of origins by sudden saltations, thereby consciously rejecting Darwin's gradualism.

The observed facts, as was at that time emphasized by Goldschmidt (1940), Willis (1940), and Schindewolf (1950), seemed to confirm the claims of the saltationists. Nowhere in nature, so they said, does one find any transitions between genera, families, orders, and still higher taxa. Even the fossil record fails to substantiate any continuity, and all novelties appear in the fossil record quite suddenly. Therefore, concluded the opponents of gradualism, the claim of the adherents of the evolutionary synthesis that the genetics of populations and the facts of geographic speciation can explain macroevolution is a hypothesis without factual substantiation. The very same objection to gradualism has again been raised during the last couple of years. If, as I have always claimed, speciation is the key to the solution of the problem of macroevolution, it is necessary to review recent developments in the theory of speciation.

Kinds of Speciation

The term *speciation* has been used ambiguously throughout much of the history of evolutionary biology. For evolutionists in the vertical tradition it meant phyletic speciation, that is, the transformation of one species into another one. For those in the horizontal tradition it meant the multiplication of species, that is, the establishment of separate populations that are incipient species. Much of the current conflict about the validity of punctuated equilibria is actually the subconscious perpetuation of the old ambiguity as to what speciation really is. Some of those who support phyletic gradualism are still thinking in terms of phyletic speciation.

Complications remain, however, even if one clearly defines speciation as the production of new daughter species. Three or four legitimate theories exist as to how such new species might originate, but fortunately the validity or not of these various forms of potential speciation is of only very limited relevance to the punctuationist argument.

I have recently presented a detailed analysis of the problem of speciation (Essay 21), and I will here only summarize my final conclusions. By necessity they will sound somewhat categorical.

There is now little doubt that, at least as far as animals are concerned, the prevailing mode of speciation is allopatric. I defined this in 1942 as follows: "A new species develops if a population which has become geographically isolated from its parental species acquires during this period of isolation characters which promote or guarantee reproductive isolation when the external barriers break down." In addition to allopatric speciation, several forms of nonallopatric speciation have also been postulated. The most important ones among those discussed by White (1978) are the following:

Sympatric speciation.   This is defined as the origin of isolating mechanisms within the dispersal area of the offspring of a single deme. When I pointed out in 1942, and even more decisively in 1947 (reprinted in 1976), what genetic difficulties the sympatric splitting of a deme poses in a diploid sexually reproducing species and, furthermore, that all the cases cited as demonstrating sympatric speciation could just as easily or even more easily be explained as the result of geographic speciation, the assumption of a prevalence of sympatric speciation lost in favor. However, in the 1960s and 1970s it was again vigorously promoted, particularly by Guy Bush and M. J. D. White. In the last couple of years, however, several authors have independently questioned the prevalence if not the occurrence of sympatric speciation. Even though it is a conceivable mechanism of speciation, I doubt that it is of major importance.

One special form of sympatric speciation is speciation by disruptive selection. According to this hypothesis, new species might originate through the acquisition of reproductive isolation by morphs in a polymorphic population. There is no doubt that one can set up a set of conditions that would make this possible, but it would seem to me that selection would forcefully resist in natural populations such a drastic reduction in the reproductive potential of an individual. The ideal situation for such speciation to occur would be freshwater fishes, where Selander has discovered trophic "species," that is, morphs so discontinuously specialized that they have the phenotypic earmarks of several good morphological species, yet are members of a single population. Indeed, until far better evidence for the prevalence of sympatric speciation is provided than is so far available, it can be omitted from a consideration of the role of speciation in macroevolution. As a matter of fact, it has never been shown to have any particular relevance to macroevolution.

Stasipatric speciation.   As I have shown previously (Essay 21), there is no evidence whatsoever in existence to support the occurrence of this form of speciation (as postulated by M. J. D. White).

Parapatric speciation.   This model of speciation, supported by B. Clarke, J. Endler, and others, likewise fails to produce convincing evidence. All the cases listed are more easily explained as secondary contact zones of previously isolated populations. It is only among partially autogamous plants that one finds situations that might be considered as cases of incipient parapatric speciation. Lack of space precludes a discussion of gene flow, the proper analysis of which is essential in the consideration of the possibility of parapatric speciation.

It must be emphasized once more that for the argument of phyletic versus punctuational speciation it is quite irrelevant whether new species originate peripatrically (see below), sympatrically, or by disruption; they all would result in a punctuational situation.

Kinds of Allopatric Speciation

After the evolutionary synthesis, allopatric speciation was generally accepted as the prevailing mode of speciation, at least in animals. Those who had no personal acquaintance with speciation problems usually pictured speciation as the separation of a species into two halves through a newly arising geographical barrier and the gradual genetic divergence of the two isolated halves. White (1978) has designated this version of allopatric speciation as the "dumbbell" model of allopatric speciation.

When I studied actual cases of incipient speciation in various groups of animals and plants, I dicovered that such cases rarely conformed to this textbook model. There were two aspects of allopatric speciation, in particular, which remained puzzling. One was a frequent failure of a divergence of widely discontinuous portions of a species. Such discontinuities had been known since before the age of Darwin, and had in fact been the major reason for Agassiz's theory of multiple creations of the same species and for the correspondence between Darwin and Asa Gray.

More interesting was the fact, as was first emphatically pointed out by me (1942:281–285), that highly isolated portions of species are sometimes so drastically different from the main body of the species that taxonomists had not only made them different species but had even sometimes raised them to the rank of separate genera. Indeed, more and more evidence accumulated, suggesting drastically different rates of speciation under different circumstances. I puzzled over this difference for many years after the publication of my book *Systematics and the Origin of Species* and finally concluded that one must distinguish in addition to the "dumbbell" model of allopatric speciation an additional process of speciation in peripherally isolated populations (Mayr 1954), a process which I now designate as *peripatric speciation*.

There have been so many misrepresentations of my theory in the recent literature that it is important to emphasize the actual contents of my original papers (Mayr 1954; 1963; 1970). The fundamental fact on which my theory was based is that when a highly divergent population or taxon is found in a superspecies or species group, it is invariably found in a peripherally isolated location. In many cases, when I traced a series of closely related allopatric species, I found that the most distant, the most peripheral, species, was so distinct that ornithologists had described it as a separate genus or at least had not recognized at all its true relationship. In genus after genus I found the most peripheral species to be the most distinct. It is on this strictly empirical, strictly observational basis that I proposed in 1954 my theory of peripatric speciation. Nothing has been discovered in the more than 25 years since my publication that would have weakened my case. On the contrary, it has been enormously strengthened by the brilliant researches of Hampton Carson on speciation of *Drosophila* in the Hawaiian Islands.

My conclusion was that any drastic reorganization of the gene pool is far more easily accomplished in a small founder population than in any other kind of population. Indeed, I was unable to find any evidence whatsoever of the occurrence of a drastic evolutionary acceleration and genetic reconstruction in widespread, populous species. In view of frequent recent misrepresentations I must emphasize also what I did *not* claim.

> I did not claim that every founder population speciates. The reason is that the majority of them soon become extinct, but a second reason is that in the vast majority of them only minor genetic reorganizations occur.

> I did not claim that every genetic change in a founder population is a genetic revolution. It evidently requires special constellations for the occurrence of more drastic genetic reorganizations.

> I did not claim that every founder population undergoes a drastic change. All I claimed was that when a drastic change occurs, it occurs in a relatively small and isolated population.

> I did not claim that speciation occurs only in founder populations.

Those who have misinterpreted me seem to be ignorant of the distinction between necessary and sufficient conditions. What is very evident is the fact that one can neither predict what happens during speciation nor deal with all situations under a single heading. Typological thinking, unfortunately, is still exceedingly widespread, particularly among geneticists, and as a consequence pluralistic explanations are ignored.

It has been stated repeatedly that I have tried to replace genetic drift by my theory of peripatric speciation. This is not at all true. I have always been aware of the fact that all stochastic processes in evolution, including the founding of new populations, result in genetic drift as defined by Sewall Wright. However, Wright, whose essential interest has always been in vertical evolution, concentrated on processes occurring in temporarily isolated demes and described how this would affect genetic change in a species as a whole. To the best of my knowledge he did not prior to 1954 make this the theme of a theory of multiplication of species. In my case, I considered the stochastic aspect of the establishment of the founder population only as a baseline, and, as I emphasized from the very beginning, I consider selection to be the real cause of the genetic changes occurring in a founder population. Consequently, I never considered genetic revolutions or whatever else may happen during peripatric speciation as an alternate to genetic drift, since the two processes are entirely independent of each other and always coexist.

The Genetics of Peripatric Speciation

In 1954 when I proposed my theory of peripatric speciation, virtually nothing was known about the genetics of speciation except that in plants a single gene or merely a few genes were sometimes sufficient to establish a sterility barrier. In animals the situation was less clear. Most studies on behavioral isolating mechanisms as well as on sterility indicated the involvement of numerous genes. If one adds to them the genetic reconstruction needed to make the noncompetitive coexistence with mother and sister species possible, one must conclude that speciation probably requires a considerable genetic reorganization. According to the classical allopatric model of speciation, it would require very long periods of geographical isolation in order to pile up slowly the large number of genes required for speciation. This did not fit at all the situations to which my peripatric model applied, and I had to speculate on the possibility of a genetic process quite different from the slow piling up of gene differences. Even so, my explanation was essentially based on the then-prevailing assumption that all genes are in principle of the same kind. I envisioned a genetic process that consisted of the following elements.

(1) The founders (in many cases a single fertilized female) carry only a fraction of the total genetic variability of the parental population.

(2) The extreme inbreeding of the ensuing generations not only leads to increased homozygosity but also exposes many if not most of the recessive alleles (now made homozygous) to selection.

(3) The elimination of many of the previously existing allelic and epistatic balances may result in a considerable loosening up of the cohesion of the genotype.

(4) Such genetically unbalanced populations may be ideally suited to shift into new niches such as will be available under the changed environmental conditions of the location of the founder population.

(5) The genetic reorganization might be sufficiently drastic to have weakened genetic homeostasis sufficiently to facilitate the acquisition of morphological innovations.

(6) The drastically different physical as well as biotic environment of the founder population will exert greatly increased selection pressures.

Since the early generations will be rather small, stochastic processes will play an important role in the genetic reorganization. I concluded that the combination of all these different factors might result in a genetic turnover that was by several orders of magnitude larger than that occurring in a normal deme that is part of a populous widespread species. I referred to such a drastic reorganization as a *genetic revolution*. My concept of the genetic revolution was based on the idea of the genetic milieu of each gene. Since my ideas have often been misunderstood or misrepresented let me quote exactly from my 1954 paper: "Isolating a few individuals from a variable population . . . will produce a sudden change of the genetic environment of most loci. This change, in fact, is the most drastic genetic change . . . that may occur in a natural population, since it may affect all loci at once. Indeed, it may have the character of a veritable 'genetic revolution.' Furthermore, this 'genetic revolution,' released by the isolation of the founder population, may well have the character of a chain reaction. Changes in any locus will in turn affect the selective values at many other loci, until finally the system has reached a new state of equilibrium."

My systematic studies of literally thousands of peripherally isolated populations during the preceding 25 years had shown me that such a drastic change occurs only very occasionally. I did not claim in the least that every founder population experiences a genetic revolution. Neither did I claim that any genes had mutated. All I claimed was that by changing their genetic milieu the phenotypic expression and hence the selective values of many genes would be affected.

It has been questioned, with some justification, whether the term "revolution" was not too strong. The student of history, however, knows that many revolutions hardly touched any other institution of a country except the form of its government. Furthermore, nothing ever occurs in other kinds of populations that even approaches the drastic genetic turnover of those founder populations that experience a genetic revolution.

What I said in 1954 was all based on inference, and it is time to ask to what extent my speculations have been confirmed or refuted in the ensuing 28 years.

Perhaps the most important insight acquired since then is that of the heterogeneity of the DNA. Much misunderstanding in the past was generated by the continued insistence of some geneticists that all evolution is caused by "Mendelian characters." That, admittedly, was a useful designation in the early decades of this century when there was still a widespread belief in blending inheritance. However, today the term suggests that the category "Mendelian characters" is a uniform class. But the more we learn about kinds of DNA, the more obvious it becomes that this is far from the truth. The term "Mendelian characters" simply means that inheritance is controlled by the chromosomes on which the respective genetic factors are located. It tells us nothing, on the other hand, of the chemical, physiological, or evolutionary nature of these genetic factors. We now know that the genotype, even though all of it is composed of DNA, consists of highly heterogeneous classes of DNA, each of which is likely to have a somewhat or altogether different function. Those of us who for a long time have been on the road toward the explanation of speciation and evolution and who thought that we were nearing the goal now feel suddenly like the player in a parlor game who is told to go back to position zero. Indeed, as far as our understanding of the genetics of speciation is concerned we are almost at position zero.

But now let us look at some of the details. Since there are different classes of DNA and since we already have some evidence that they play different roles in speciation, we must look at each class separately.

(1) Enzyme genes. The method of electrophoresis supplied the extremely important evidence that in speciation there seems to be no major involvement, if any at all, of the classical enzyme genes. There are multiple proofs for this conclusion, perhaps the most convincing one being that the rate of isozyme replacement is no more rapid in actively speciating phyletic lines than in conservative ones. Also there are groups of related species and genera that are identical in all the enzyme loci that have been studied. Those who base their interpretation of the genetics of speciation on the findings on enzyme genes are surely barking up the wrong tree.

(2) Chromosomal reconstruction. Since in eukaryotes virtually all the genetic material is located on the chromosomes, no one will question that the chromosomes are important in speciation, the only question being, in what way? The frequency with which closely related species of most groups of organisms differ in chromosome structure testifies to the frequency at which chromosomal reorganization seems to accompany speciation. On the other hand, the occurrence of homosequential speciation in many Hawaiian species groups of *Drosophila* and in other genera of dipterans (Carson 1975) demonstrates that speciation is possible without gross chromosomal changes. In view of the fact that heterozygotes between the chromosome sets of two different species are often of lowered viability or are more or less sterile, there is little doubt that the accumulation of such differences would speed up the acquisition of isolating mechanisms between species. M. J. D. White has shown convincingly the important role this plays in the speciation of certain groups of organisms, for instance the morabine grasshoppers. But Carson's Hawaiian *Drosophila* show that chromosomal reorganization is not a necessary condition for speciation.

At this point we leave the solid ground of factually established evidence. For 15 or more years various authors have speculated on the importance for speciation of regulatory genes (in analogy to the findings in prokaryotes), and there seems to be indeed a great deal of evidence for the role of such regulatory mechanisms. Even though some geneticists are still reluctant to admit more than a minimum of gene interaction, it would seem to me that there is far more evidence for a rather tightly knit cohesion of the genotype than these classical Mendelians admit. Our ignorance of this postulated regulatory system is, however, still almost total. Single coding genes may be involved, with various functions such as to build up isolating mechanisms, and some of the recently discovered "mystery genes," such as copia and other moveable elements, may play a role. Middle repetitive DNA seems quite important in regulation and even highly repetitive DNA may be involved. Yet it is also possible that the entire single factor approach is wrong.

Genetic Milieu

For the last 60 years—but one could also say all the way back to Darwin (1859:11, 146; 1868 II:319–335)—two traditions of viewing the genotype can be distinguished. According to the atomistic ("beanbag") view, each gene is independent not only in its actions but also in the effects of selection on it. Evolutionary stasis of the phenotype, for instance, is explained by the stabilizing selection acting on individual genes. According to the holistic (integrative) view, genes perform as teams, and large numbers of other genes form the "genetic milieu" (Chetverikov 1926) of any given gene. Gene exchange at any locus may have an impact on the selective value of genes at other loci. Even though the atomists (reductionists) are fully aware of pleiotropy, polygeny, and other processes that produce the phenotype (the target of selection) and that automatically result in a selectional interaction of genes, they ignore these processes in their evolutionary interpretations. Those others, however, who have stressed the genetic milieu (Chetverikov), genetic homeostasis (Lerner), internal balance (Mather), or the cohesion of the genotype (Mayr) have consistently looked at the impact of natural selection in a different way from those for whom the role of genes is essentially additive. The atomists, for instance, would treat developmental constraints and stabilizing selection as two separate problems, while for the "holists" the major cause of stabilizing selection is precisely the set of development constraints generated by the cohesion of the genotype. Schmalhausen was fully aware of this, but the majority of geneticists, particularly the mathematical geneticists, ignore it because it makes calculations "messy." Indeed, there is still no adequate methodology to analyze the controlling factors of cohesion.

The holists, thus, have introduced one major new factor into evolutionary theory, the internal structure of the genotype. They claim that much of macroevolution cannot be explained by atomistic gene replacements or by selection pressures on single genes but only by a more or less drastic reorganization of the genotype made possible by loosening up the tight genetic cohesion of the genotype that characterizes widespread populous species. (I should not say *new* because Darwin already defended such a viewpoint and it has been suggested again and again, almost invariably by evolutionists from the naturalists' camp, although those who came to evolution via embryology, like Waddington and Goldschmidt, have expressed similar ideas.)

No one has made a stronger case in favor of the theory that a loosening up of the cohesion of the genotype is an important and perhaps the decisive component in much of speciation than Carson (1975). His arguments powerfully support similar earlier arguments made by authors from Darwin to Rensch, Lerner, Waddington, Mayr, and others.

Microevolution and Macroevolution

The classical view of macroevolution held by Darwin and the majority of paleontologists up to the present day is that species, in the course of their gradual evolution in time, change to such a degree that they will become different genera, or taxa of still higher rank, and acquire in the process all the adaptations and specializations of the world of organic diversity. Viewing evolution strictly in the vertical dimension, as was done by most paleontologists including Simpson, hardly permitted any other interpretation of evolution. When one carefully studies the writings of those geneticists who also gave thought to macroevolution, one finds that they adopted essentially the same view. Yet some of them felt that this left an unexplained area. This led Goldschmidt to his theory of "hopeful monsters"; it led to other saltational theories, and it led to various claims that the rethinking of evolutionary processes in the last 10 or 15 years had made the evolutionary synthesis obsolete. These are inevitable conclusions for those who think in terms of closed gene pools and of exclusively vertical evolution. Indeed, there are no visible connections between the phenomena studied by mathematical population geneticists and those macroevolutionary processes that are studied by paleontologists and comparative anatomists.

Such a connection can be established, however, by making use of the findings of population systematics, and of the horizontal approach to evolution, as studied by the new systematics, because they provide a perfect bridge between micro- and macroevolution. Rudiments of this demonstration can be found in Rensch's writings and in my 1942 book. In my theory of peripatric speciation I stated that peripherally isolated populations have various attributes that "are of great interest . . . to those who study major evolutionary changes. It seems to me that many puzzling phenomena, particularly those that concern paleontologists, are elucidated by a consideration of these populations. These phenomena include unequal (and particularly very rapid) evolutionary rates, breaks in evolutionary sequences and apparent saltations, and finally the origin of new 'types' . . . The genetic reorganization of peripherally isolated populations permits evolutionary changes that are many times more rapid than the changes within populations that are part of a continuous system. Here then is a mechanism that would permit the rapid emergence of macroevolutionary novelties without any conflict with the observed facts of genetics" (1954:206–207).

Curiously, this suggestion of mine continued to be totally ignored by the paleontologists for almost 20 years. However, in 1972 Eldredge and Gould published their theory of *punctuated equilibria* in which some of the most important findings of the paleontologists are explained in terms of my theory of peripatric speciation. In Essay 26 I discuss various versions of the theory of punctuationism. In some of Gould's discussions he seemed to imply that Goldschmidt (1940) had anticipated punctuationism. However, in their more recent papers Gould and Eldredge clearly distance themselves from Goldschmidtian saltationism. When postulating saltations, says Gould (1980), "I do not refer to the saltational origin of entire new designs complete in all their complex and integrated features . . . instead, I envisage a potential saltational origin for the essential features of key adaptations."

With this statement the argument now becomes largely a semantic one, as to how big is big or what is gradual? Let me emphasize: for a Darwinian, any genetic reconstruction of a population, which in a sexually reproducing diploid species always starts with the change of a single chromosome, is gradual by definition. It always has to pass through a stage of polymorphism or heterozygosity, regardless of the size of the building block with which the change is constructed. A mutation occurring early in ontogeny might produce a drastic change in the structure of the adult, as stated by the neo-Goldschmidtian. But as Goldschmidt himself pointed out, most macromutations would produce such developmental disturbances that the result would be a hope*less* monster. Therefore, fortunately, Goldschmidt's hypothesis allows a prediction. If there were such a neo-Goldschmidtian process, as I pointed out earlier (1963:438), one should find in nature such unsuccessful trial balloons in great abundance. However, contrary to the beliefs of the neo-Goldschmidtians, one fails to find a frequent polymorphism of hopeless or even hopeful monsters in nature. To be sure, not all mutations leading to evolutionary changes need to be minute (see Turner 1977 for mimicry-initiating mutations), but mutations producing the characters of whole new higher taxa are improbable. Consequently, there is no need to postulate any non-Darwinian processes to account for the origin of evolutionary novelties. All that is needed, as emphasized by Chetverikov, Mather, Lerner, Carson, and myself, is a less atomistic, less reductionist concept of the genotype.

For a paleontologist thousands, and even ten thousands, of years are like a moment. Hence, peripatric speciation is a more or less instantaneous phenomenon, a veritable saltation. Under the much higher magnification of the fine-grained analysis of living populations, a change requiring hundreds or thousands of years might appear not only gradual but actually slow. To be sure, some components of the genetic restructuring of a founder population, such as the shift in a chromosomal mutation from one to another homozygous condition, might be completed in a few generations. Likewise, if there is the replacement of one system of genetic homeostasis by another one, the transition might occur in relatively few generations. Nevertheless, the completion of the production of a genuine evolutionary novelty might require hundreds if not thousands of generations, yet even such a period is still too short to be discoverable by a paleontologist. The punctuationists have made fun of the traditional paleontologists for invoking the deficiency of the fossil record so often in their explanations. It is ironical that the punctuational appearance of speeded up evolution during peripatric speciation is likewise an artifact of the incompleteness of the fossil record. If all founder populations, which are exceedingly small, local, and of short duration, were preserved in the fossil record, they would surely document the gradual nature of the changes.

The preceding analysis of the function of speciation in macroevolution has shown that our understanding of this role has considerably advanced in recent years. At the same time it has also shown to what an extent we are still unable to provide satisfactory answers to important questions. Let me list some of these questions in order to illuminate the degree of our ignorance.

(1) What happens to the genotype during speciation, particularly during genetic revolutions in founder populations?

(2) What molecular forces are responsible for the seeming cohesion of the genotype?

(3) What role do the different kinds of DNA, such as middle and highly repetitive DNA, transposons, etc., play in speciation and evolution?

(4) What kind of constraints prevent an unlimited success of natural selection (Essay 8)?

(5) How drastic can the phenotypic effect of a mutation (*sensu lato*) be without encountering so much counter selection that the mutation is speedily eliminated? (The study of the variation [polymorphism] of thousands if not millions of natural populations by systematists has shown that neither the reductionist claim of many geneticists that only minimal mutations survive is necessarily correct nor the counter claim of the saltationists that whole evolutionary novelties may originate by a single mutation.)

(6) What percentage of well-established species experience total or at least nearly complete stasis?

(7) How can one explain mosaic evolution, that is, a more or less drastic change of part of the phenotype (and the portion of the genotype controlling it) without visible changes in the remainder of the phenotype? (Sibling species prove that reproductive isolation and new profiles of enzyme genes can be acquired without noticeable morphological reconstruction). "Living fossils" have hardly changed morphologically for hundreds of millions of years but have presumably steadily evolved in that part of the genome that has to do with physiological and ecological adaptation. The karyotype may, or may not, change drastically in different evolutionary lines. These observations lead to the question:

(8) Why do not all components of the genotype evolve at approximately the same rate? Is this a matter of chance or are these differences connected with particular selective advantages?

I have posed this set of questions in order to indicate how far we are still from a full understanding of the evolutionary process. What is comforting, however, is the fact that no matter what answer will eventually be given to these various questions, it is not likely to be in any conflict whatsoever with the basic theory of evolution, as generated by the evolutionary synthesis. The current thinking about Darwinian evolution can perhaps be summarized in a number of conclusions:

(1) The reductionist definition that evolution is a change of gene frequencies is meaningless. Evolution is a matter of phenotypes, structures, developmental pathways, functions, populations, and interacting

ecosystems. The nonreductionist tradition in evolutionary biology is old, beginning with Darwin's repeated references to the "mysterious laws of correlation," continuing with Chetverikov's concept of the genetic milieu, with Lerner's of genetic homeostasis, and my own of the cohesion of the genotype.

(2) The phenomena of macroevolution cannot be understood unless they are traced back to populations that are incipient species, and to neospecies. Major macroevolutionary processes are initiated during peripatric speciation.

(3) With the refutation of soft inheritance (inheritance of acquired characters, etc.) and of all finalistic processes (orthogenesis, etc.) natural selection remains as the only substantiated direction-giving process in evolution. The efficacy of natural selection, however, is reduced by numerous constraints. Stochastic processes, therefore, strongly influence the response to selection pressures.

(4) It is not sufficient to study macroevolution only with the evidence provided by genetics and paleontology-comparative anatomy. Crucial evidence is supplied also by neontology. For instance, virtually all the new structures that distinguish different genera can be found in many superspecies as the characters of particularly deviant peripheral neospecies. Why speculate what kind of embryological research might be revealing if the answer is already available in the ongoing experiments of nature? Bock (1970) has shown for living species of Hawaiian honeycreepers how far-reaching the macroevolutionary conclusions are which one can derive from the study of living biota. The one million known species of animals and 200,000 species of plants present literally thousands of examples of macroevolution by way of geographic speciation. It is to be hoped that in the future neontologists as well as paleontologists and geneticists make better use of this abundant material in order to shed new light on the processes of macroevolution.

(5) I do not know of any recent findings that would require a material modification of the concept of evolution acquired during the evolutionary synthesis.

NOTE

This essay is an abridged version of one which first appeared in *Evolution* 36(1982):1119–1132.

REFERENCES

Bateson, W. 1984. *Materials for the Study of Variation.* London: Macmillan.

Bock, W. J. 1970. Microevolutionary sequences as a fundamental concept in macroevolutionary models. *Evolution* 24:704–722.

Carson, H. L. 1975. The genetics of speciation at the diploid level. *Amer. Natur.* 109:83–92.

———— 1981. [Macroevolutionary Conference.] *Science* 211:773.

Chetverikov, S. S. 1926. On certain aspects of the evolutionary process from the standpoint of modern genetics. *J. Explt. Biol.* (Russian) A2. (Eng. trans.) 1961. *Proc. Amer. Phil. Soc.* 105:167–195.

Darwin, Charles. 1859. *On the Origin of Species by Means of Natural Selection or the Preservation of Favored Races in the Struggle for Life.* London: Murray.

———— 1868. *The Variation of Animals and Plants under Domestication.* London: Murray.

Eldredge, N., and S. J. Gould. 1972. Punctuated equilibria: an alternative to phyletic gradualism. In T. J. M. Schopf and J. M. Thomas, eds., *Models in Paleobiology,* pp. 82–115. San Francisco: Freeman, Cooper.

Gingerich, P. D. 1976. Paleontology and phylogeny: patterns of evolution at the species level in early Tertiary mammals. *Amer. J. Sci.* 276:1–28.

Goldschmidt, R. 1940. *The Material Basis of Evolution.* New Haven: Yale University Press.

Gould, S. J. 1977. The return of hopeful monsters. *Natur. Hist.* 86:22–30.

———— 1980. The promise of paleobiology. *Paleobiology* 6:96–118.

Gruber, H. E. 1974. *Darwin on Man.* New York: Dutton.

Haldane, J. B. S. 1957. The cost of natural selection. *J. Genet.* 55:511–524.

Kellogg, D. E. 1975. The role of phyletic change in the evolution of *Pseudocubus vema* (Radiolaria). *Paleobiology* 1:359–370.

Lerner, I. M. 1954. *Genetic Homeostasis.* Edinburgh: Oliver and Boyd.

Mayr, E. 1942. *Systematics and the Origin of Species.* New York: Columbia University Press.

———— 1954. Change of genetic environment and evolution. In J. Huxley, A. C. Hardy, and E. B. Ford, eds, *Evolution as a Process.* London: Allen and Unwin.

———— 1963. *Animal Species and Evolution.* Cambridge, Mass.: Harvard University Press.

———— 1970. *Populations, Species, and Evolution.* Cambridge, Mass.: Harvard University Press.

———— 1976. *Evolution and the Diversity of Life.* Cambridge, Mass.: Harvard University Press.

Ospovat, D. A. 1982. *The Development of Darwin's Theory: Natural History, Natural Theology, and Natural Selection, 1838–1859.* Cambridge: Cambridge University Press.

Rensch, B. 1947. *Neuere Probleme der Abstammungslehre.* Stuttgart: Enke.

Schindewolf, O. H. 1950. *Grundfragen der Paläontologie.* Stuttgart: Schweizerbart.

Scott, W. B. 1894. On variations and mutations. *Amer. J. Sci.* 48:355–374.

Simpson, G. G. 1944. *Tempo and Mode in Evolution.* New York: Columbia University Press.

Stanley, S. M. 1979. *Macroevolution: Pattern and Process.* San Francisco: Freeman.

Turner, J. R. G. 1977. Butterfly mimicry: the genetical evolution of adaptation. *Evol. Biol.* 10:163–206.

de Vries, H. 1901–1903. *Die Mutationstheorie. Versuche und Beobachtungen über die Entstehung der Arten im Pflanzenreich.* Vol. 1, *Die Entstehung der Arten durch Mutation.* Vol. 2, *Elementare Bastardlehre.* Leipzig: Veit. Eng. trans.: J. B. Farmer and A. D. Darbishire. Chicago: Open Court, 1909–1910.

Waddington, C. H. 1957. *The Strategy of the Genes.* London: Allen and Unwin.

White, M. J. D. 1978. *Modes of Speciation.* San Francisco: Freeman.

Willis, J. C. 1940. *The Course of Evolution by Differentiation or Divergent Mutation rather than by Selection.* Cambridge: Cambridge University Press.

SPECIATIONAL EVOLUTION THROUGH PUNCTUATED EQUILIBRIA

ONLY recently have we understood how different are the concepts to which the term *evolution* has been attached. With the wisdom of hindsight, we can now (250 years after Buffon) distinguish three exceedingly different concepts of evolution.

(1) Saltational evolution. Theories postulating this type of evolution are a necessary consequence of essentialism. If one believes in constant types, only the *sudden* production of a new type can lead to evolutionary change. That such saltations can occur and indeed that their occurrence is a necessity is an old belief. Almost all of the theories of evolution described by Osborn in his *From the Greeks to Darwin* (1894) were salta-tional theories, that is, theories of the sudden origin of new kinds. The Darwinian revolution (1859) did not end this tradition, which continued to flourish in the writings of T. H. Huxley, Bateson, de Vries, Willis, Goldschmidt, and Schindewolf. Traces of this idea can even be found in the writings of some of the punctuationists.

(2) Transformational evolution. According to this concept, clearly ar-ticulated by Lamarck, evolution consists of the gradual transformation of a thing from one condition of existence to another. Almost invariably transformational theories assume a progression from "lower to higher" and reflect a belief in cosmic teleology, resulting in an inevitable steady movement toward an ultimate goal, to an ultimate perfection. In biology all so-called orthogenetic theories, from K. E. von Baer to Osborn, Berg, and de Chardin are in this tradition.

(3) Variational evolution. As Lewontin (1983) has pointed out, Darwin introduced an entirely new concept of evolution. New gene pools are generated in every generation, and evolution takes place because the successful individuals produced by these gene pools give rise to the next generation. Evolution, thus, is merely contingent on certain processes

articulated by Darwin: variation and selection. No longer is a fixed object transformed, as in transformational evolution, but an entirely new start is made so to speak in every generation. Evolution is no longer necessarily progressive, it no longer strives toward perfection or any other goal. It is opportunistic, hence unpredictable.

What Darwin did not fully realize is that variational evolution takes place at two hierarchical levels, the level of the deme (population) and the level of species. Variational evolution at the level of the deme is what the geneticist deals with. It is effected by individual selection and leads minimally to the maintenance of fitness of the population through stabilizing selection.

The second level of variational evolution is that of the species. Owing to continuing (mostly peripatric) speciation, there is a steady highly opportunistic production of new species. Most of them are doomed to rapid extinction, but a few may make evolutionary inventions, that is, physiological, ecological, or behavioral innovations that give these species improved competitive potential. In that case they may become the starting point of successful new phyletic lineages and adaptive radiations. Such success is nearly always accompanied by the extinction of some competitor (see below under "Species Selection").

The change from transformational to variational evolution required a conceptual shift that was only imperfectly carried through by most Darwinians. As a consequence, geneticists described evolution simply as a change in gene frequencies in populations, totally ignoring the fact that evolution consists of the two simultaneous but quite separate phenomena of adaptation and diversification. The latter is due to a process of multiplication of species, a process almost totally ignored in the writings of Fisher, Haldane, Wright, and other leading evolutionary geneticists.

Transformational thinking likewise continued to dominate paleontology, expressed in the concept of phyletic gradualism. Since most paleontologists were typologists (in an almost Platonian sense), they subconsciously assumed that a species was everywhere the same and, thus, at any given time essentially uniform. Speciation consisted of the gradual transformation of such species in geological time. Since the gradualness of such phyletic transformation could be documented in the geological record only in the rarest cases, it was postulated that the absence of intermediates was a consequence of the notorious incompleteness of the fossil record.

The so-called evolutionary species definition adopted by most paleontologists (Simpson 1961; Willmann 1985) reflects the same focus on the vertical (time) dimension. If adopted, it leaves only two options: speciation is explained either by gradual phyletic evolution, with the gaps between species being due to the deficiency of the fossil record, or by sympatric saltational speciation. Indeed, most paleontologists adopted both options. Acceptance of phyletic gradualism does not require the acceptance of a constant rate of evolutionary change. Actually the rate may accelerate or slow down; still, change leads inexorably to the steady transformation of a lineage.

Even Darwin, for reasons that relate to his struggle against creationism, stressed the transformational aspect of evolution. He was, however, fully aware of highly different rates of evolution, from complete stasis to rates of change so fast that intermediates could not be discovered in the fossil record (Gingerich 1984; Rhodes 1983; and others). There is a revealing statement on punctuationism in the later editions of the *Origin:* "the periods during which species have been undergoing modification . . . have probably been short in comparison with the periods during which these same species remained without undergoing any change." Owing to his adoption of sympatric speciation, however, Darwin never needed to consider the geographical component in his theorizing. When he said that a new species might originate as a local variety, he did not claim that it was an isolated population. It seems to me that for Darwin the pulsing of evolutionary rates was a strictly vertical phenomenon.

The geneticists, with the exception of a few saltationists like de Vries and Bateson, usually ignored the problem of speciation altogether. The only geneticists who showed an interest in the multiplication of species were those who had been educated as taxonomists, like Dobzhansky and Stebbins. As far as relating speciation to macroevolution is concerned, this problem occupied primarily three zoologists, J. Huxley (1942), Mayr (1942; 1954), and Rensch (1947), who were neither geneticists nor paleontologists. Since these three were among the architects of the evolutionary synthesis, one can state that the problem of the relation between speciation and macroevolution was not entirely ignored by the evolutionary synthesis.

The widespread neglect of the role of speciation in macroevolution continued until 1972 when Eldredge and Gould proposed their theory of punctuated equilibria. Whether one accepts this theory, rejects it, or greatly modifies it, there can be no doubt that it had a major impact on paleontology and evolutionary biology.

The gist of the theory was "that significant evolutionary change arises in coincidence with events of branching speciation, and not primarily through the in toto transformation of lineages" (Gould 1982a; 1983). The contrast with the previously dominant view of evolutionary change was as follows: Traditionally, evolution had been seen as a single phase phenomenon of gradual change, sometimes slow, sometimes rapid. Now evolution was seen as an alternation between speciation events during which the major evolutionary (particularly morphological) change occurred and periods of lengthy stasis.

It has since become evident through historical studies that it was more the term *punctuated equilibrium* that was novel than the concept. A role for peripheral populations in speciation was already postulated by L. von Buch (1825) and fully substantiated by Darwin for the Galapagos mockingbirds. Unfortunately, by the time Darwin published the *Origin* (1859), he had adopted sympatric speciation (Mayr 1982). When he said that a new species might originate as a local variety, it did not at all necessarily refer to an isolated population. Nor are the changes in the rate of evolution to which Darwin refers brought in relation to speciation.

Before going any further in the analysis of the literature, it is important to call attention to a prevailing confusion between two quite distinct evolutionary phenomena, gradualism and uniformity of evolutionary rate. Darwin emphasized gradualism (Rhodes 1983), but as I shall show below, even that term is ambiguous, allowing for two very different interpretations. What Darwin did not insist upon was a uniformity of rates (Huxley 1982; Penny 1983; Rhodes 1983). The existence of so-called living fossils had been known to paleontologists early in the 19th century, and the occurrence of different rates of evolution in different phyletic lines was paleontological dogma in Darwin's lifetime. Simpson (1953) analyzed this phenomenon in great detail and even introduced a special terminology to characterize lineages with average, very rapid, and extremely slow evolutionary rates. Eldredge and Gould never claimed to have discovered this difference in rates, and the part of the ensuing polemic stressing these differences is therefore irrelevant for the evaluation of the punctuation theory.

It is not always easy to interpret Darwin's statements (Rhodes 1983) because isolation (at least during the process of speciation) had become unimportant for him owing to his adoption of sympatric speciation. I have been unable to discover in Darwin's writings any connection between allopatric speciation and change of evolutionary rate. Gould and Eldredge state correctly that Simpson likewise failed to make such a connection. His quantum evolution was a vertical (temporal) phenomenon, as it had to be considering his evolutionary species definition (Simpson 1944:207–217).

The fact that new species may originate in a very circumscribed area, and turn up in the fossil record only after having spread more widely, was well known to paleontologists and documented by Eldredge and Gould through a quotation from the writings of the French paleontologist Bernard (1895). This insight was made use of, however, only in stratigraphic researches and not in studies of macroevolution. On the contrary, the importance of peripatric speciation was minimized after Fisher (1930) and Wright (1931; 1932) had asserted (although for different reasons) that evolution was most rapid in populous, widespread species, a conclusion adopted also by Dobzhansky (1937; 1951) and by most evolutionists before the 1970s.

I believe I was the first author to develop a detailed model of the connection between speciation, evolutionary rates, and macroevolution (Mayr 1954). Although long ignored, the impact of my new theory of the importance of peripatric speciation in macroevolution is now widely recognized. "Mayr's hypothesis of peripheral isolates and genetic revolution must of necessity be a centerpiece of the punctuated equilibria theory" (Levinton 1983:113). This is also acknowledged by Eldredge and Gould. I once more presented my theory in great detail in 1963 (pp. 527–555). Under these circumstances it is most curious that the theory was completely ignored by the paleontologists until brought to light by Eldredge and Gould (1972).

The major novelty of my theory was its claim that the most rapid evolutionary change does not occur in widespread, populous species, as claimed by most geneticists, but in small founder populations. This conclusion was based on empirical observations gathered during my studies of the speciation of island birds in the New Guinea region and the Pacific. I had found again and again that the most aberrant population of a species—often having reached species rank, and occasionally classified even as a separate genus—occurred at a peripheral location, indeed usually at the most isolated peripheral location. Living in an entirely different physical as well as biotic environment, such a population would have unique opportunities to enter new niches and to select novel adaptive pathways.

My conclusion was that a drastic reorganization of the gene pool is far more easily accomplished in a small founder population than in any other kind of population (see Essay 25). Indeed, I was unable to find any evidence whatsoever for the occurrence of a drastic evolutionary acceleration and genetic reconstruction in widespread, populous, species.

I have attempted in Essay 25 to refute the numerous recent misrepresentations of my 1954 theory. I would add to the points made there that nowhere have I claimed that I chose the name peripatric because the founders came from the periphery of the parental range. I chose that name because the founder populations were peripherally isolated. My interpretation throughout was pluralistic and was naturally misunderstood in an age when singular, deterministic solutions were strongly preferred.

In 1954 I was already fully aware of the macroevolutionary consequences of my theory, saying that "rapidly evolving peripherally isolated populations may be the place of origin of many evolutionary novelties. Their isolation and comparatively small size may explain phenomena of rapid evolution and lack of documentation in the fossil record, hitherto puzzling to the paleontologist."

In 1982 supplemented my theory by pointing out (Essay 21) that peripatric speciation may occur not only in founder populations but also in any population going through a severe bottleneck such as refuge populations during Pleistocene glaciations (Haffer 1974).

The first to pick up my theory was Eldredge (1971), who found in his study of Paleozoic trilobites that the "majority of species . . . showed no change in species-specific characters throughout the interval of their stratigraphic occurrence" while new species appear quite suddenly in the strata. He therefore proposed "that the allopatric model . . . be substituted in the minds of paleontologists for phyletic transformation as the dominant mechanism of the origin of new species in the fossil record." This was followed in 1972 by the Eldredge and Gould paper, in which the term *punctuated equilibrium* was proposed. The Eldredge-Gould proposal was based on my 1954 theory, but placed a far stronger emphasis on stasis. They claimed, "The history of evolution is not one of stately unfolding, but a story of homeostatic equilibria, disturbed only 'rarely' (i.e., rather often in the fullness of time) by rapid and episodic events of speciation" (1972:84).

Objections

A modest theory of punctuationism is so strongly supported by facts and it fits, on the whole, so well into the conceptual framework of Darwinism that one is rather surprised at the hostility with which it was attacked. The controversy over punctuationism is, by now, fifteen years old, and it is possible to distinguish different classes of objections.

There are questions that deal with the core ideas of the theory: What is stasis? How can one account for it? Do all species experience stasis? Is all evolutionary change restricted to bouts of speciation? If so, why? What are the genetic aspects of speciation? These and other questions will be analyzed in the second part of this essay.

But not all the objections raised against punctuationism deal with these core ideas. Others were raised against rather specific claims made by Eldredge and/or Gould or against the way they treated their evidence. It will be helpful to deal with these objections first. They relate largely to claims that are not at all part of a punctuationist theory of evolution. To deal with them separately and to test them for their validity will clear the field for a subsequent testing of the core ideas of punctuationism.

Four aspects of the treatment of punctuationism by Gould and Eldredge were objected to most frequently.

(*1*) *The monolithic nature of the claims.*   Evolutionary change, they said, is never due to the gradual change of phyletic lineages ("phyletic gradualism") but is always due to speciation events occurring in peripherally isolated populations. They also claimed that once a new species had become successful it would invariably enter a period of stasis in which it would remain until it became extinct. In the early papers virtually no allowance was made for pluralism, such as occasional cases of phyletic speciation or the occurrence of evolutionary pulsations superimposed on gradually evolving lineages.

(*2*) *The claim of novelty.*   Nothing incensed some evolutionists more than the claims made by Gould and associates that they had been the first to have discovered, or at least to have for the first time properly emphasized, various evolutionary phenomena already widely accepted in the evolutionary literature. Stebbins and Ayala (1981), V. Grant (1982; 1983), and Levinton (1983) were fully justified in rejecting these claims of novelty. In particular, they showed that an insistence on gradualism by Darwin and his followers was a denial of saltationism but not a denial of different and changing rates of evolution.

(*3*) *Refutation of Darwinism.*   Particularly vigorous objection was also raised to the further claim that punctuationism would require a revision of the evolutionary synthesis. The findings of molecular biology together with various unorthodox theories of speciation and the new hierarchical view of evolution engendered by punctuationism, they said, have contributed to "a slow unravelling of the synthetic theory as a universal description of evolution." "I have been reluctant to admit it, but if Mayr's (1963:586) characterization of the synthetic theory is accurate, then that theory as a general proposition is effectively dead" (Gould 1980:120). The gist of my statement to which Gould refers was that, contrary to Goldschmidt and Schindewolf, nothing happens in macroevolution that does not happen in populations. What Gould actually attacks, and quite rightly so, is the completely reductionist characterization of evolution by the mathematical population geneticists. To equate these reductionist views with the theories of the evolutionary synthesis is, however, unjustified, as I pointed out in a critical review of similar statements published by Ho and Saunders (Mayr 1984). A rejection of the axiom of most population geneticists, "Evolution is a change of gene frequencies," is not a rejection of the evolutionary synthesis. The theory of the synthesis is much broader and constitutes in many respects a return to a more genuine Darwinism. The events which take place during peripatric speciation, no matter how rapid they may be, are completely consistent with Darwinism.

Curiously, some authors also mistakenly assumed that the occurrence of stasis would refute Darwinism. It is only teleological thinking that requires continuous evolutionary change, and Darwin both rejected teleology (see Essay 14) and accepted stasis (Rhodes 1983). An evolutionary lineage may continue to vary genetically without undergoing any major reconstruction. Alternatively, a stable lineage may continue to send out founder populations, some of which, through peripatric speciation, could become more or less distinct daughter species.

(4) *Revival of hopeful monsters.* The fourth reason why punctuationism faced so much opposition is that at one stage Gould pleaded for a revival of Goldschmidt's ideas and implied that they were akin to punctuationism. This claim indicated rather clearly that there was considerable conceptual confusion as to what punctuated equilibria really means. Before it can be discussed constructively whether or not Goldschmidt was a forerunner of punctuationism, it is necessary to discriminate among four possible interpretations of punctuationism.

(1) An evolutionary novelty originates by a systemic mutation: the individual produced by such a mutation is the representative of a new species or higher taxon.

(2) Evolutionary change is populational, but all substantial evolutionary changes take place during bouts of speciation. As soon as the process of speciation is completed, the new species stagnates ("stasis") and is unable to change in any significant way. Early statements by Eldredge and Gould (1972) and Gould and Eldredge (1977) gave the impression that this was their interpretation.

(3) Phyletic lineages ("evolutionary species") can evolve slowly and gradually into different species and even genera, but the more pronounced evolutionary changes and adaptive shifts take place during

speciational bouts in isolated populations. This has been all along my own interpretation (Mayr 1954; Essay 25) and is presumably that of many evolutionists familiar with geographic speciation.

(4) A multiplication of species (the branching of lineages) occurs but is of no greater evolutionary importance than changes within lineages. In fact, phyletic gradualism is responsible for most of evolutionary change. It was this view, held by the majority of paleontologists, which induced Eldredge and Gould (1972) to propose their theory of punctuated equilibria.

Only the first one of these four theories conflicts with Darwinism. It was Goldschmidt's theory, and considering how often Goldschmidt has been cited in connection with punctuationism, it is necessary to discuss his ideas in more detail.

To strengthen the punctuationist case, Gould cited with approval Goldschmidt's views on macroevolution, predicting that "during this decade Goldschmidt will be largely vindicated in the world of evolutionary biology" (Gould 1980:186). What Goldschmidt had claimed was that the differences among subspecies, and more broadly all geographic variation, was caused by minimal genetic changes, mutations of alleles, mostly being selected merely for climatic adaptation. Such changes would not permit any transgression of the ancestral type. Any genuine evolutionary novelty was due to the origin of a "hopeful monster," caused by a systemic mutation. This thesis followed from Goldschmidt's rather eccentric conception of the nature of chromosomes and of the genotype. A systemic mutation is "a complete change of the primary pattern or reaction system into a new one" and has the capacity to produce a strikingly different new individual that could serve as the founding ancestor of a new type of organism. As Maynard Smith (1983:276) pointed out, hopeful monster (see below) and systemic mutation are two different things. A complete restructuring of chromosomes in a single instant, as demanded by the theory of systemic mutation, is utterly impossible, at least as the origin of a new viable or even improved individual. Hopeful monsters, by contrast, are drastically altered phenotypes. They are at least in theory possible, and it should be possible to discover empirically how often they occur and how often (if ever) they are selectively superior.

Gould entirely misrepresents Goldschmidt's theory in claiming that Goldschmidt "argued that speciation is a rapid event produced by large genetic changes (systemic mutations) in small populations" (Gould and Calloway 1980:394). The whole concept of populations was alien to

Goldschmidt's thinking. According to him, a new type is produced by a single systemic mutation producing a unique individual. Gould (1982a) is also wrong in claiming that Goldschmidt never had the view "that new species arise all at once, fully formed, by a fortunate macromutation." Actually, this is what Goldschmidt repeatedly claimed. For instance, he cited with approval Schindewolf's suggestion that the first bird hatched out of a reptilian egg, and was even clearer on this point in a later paper (1952:91–92) than in his 1940 book.

In refutation of Goldschmidt's claims I demonstrated (Mayr 1942) that geographic variation in isolated populations could indeed account for evolutionary innovations. Such populations are of a very different evolutionary potential from contiguously distributed, clinally varying populations in a continental species. As I stated in 1954, and have reiterated in 1963 and 1982 (see Essay 25), one can defend a moderate form of punctuationism, based on strictly empirical evidence, without having to adopt Goldschmidt's theory of systemic mutations.

Some Basic Questions about Punctuationism

The theory of punctuationism, to repeat, consists of two basic claims: (1) that most or all evolutionary change occurs during speciation events, and (2) that species enter a phase of stasis after the end of the speciation process. The two claims are to some extent two separate theories.

The controversy that followed the proposal of this theory revealed that there are considerable conceptual and evidential difficulties in either substantiating or refuting this theory. First of all, the nature of the fossil record makes it exceedingly difficult, if not impossible, to obtain irrefutable evidence either for stasis or for a very short time span speciation. Second, throughout the controversy one encounters considerable terminological vagueness and equivocation, as for instance concerning the meaning of such words as gradual, stasis, speciation, and species selection. A careful analysis of the terms most frequently used in the punctuationism controversy is therefore indispensable.

WHAT IS GRADUAL?

The question whether or not evolution is gradual became the focus of a heated controversy in the punctuationism argument. Darwin, as everyone knew, had frequently emphasized the gradual nature of evolutionary change (1859:71, 189, 480), an emphasis largely due to his opposition to two ideologies dominant in his time, creationism and essentialism. After these ideologies lost their power, it was no longer necessary to be so singlemindedly opposed to the occurrence of discontinuities. Yet the recent controversy concerning the saltational versus gradual origin of evolutionary novelty revealed an equivocation.

Most modern authors failed to distinguish between two very different phenomena: (1) the production of a new taxon, and (2) the production of a new phenotype. If the production of a new taxon is gradual, it is taxic gradualness; if it is instantaneous, it is taxic saltation. Likewise, one can distinguish phenotypic gradualness and phenotypic saltation. What Darwin mostly argued against was the thesis that evolutionary novelties could originate through taxic saltation, that is, through the production of a single individual representing a new type, a new taxon. Instead, he proposed that all evolutionary innovation is effected through the gradual transformation of populations.

This distinction became important after Goldschmidt revived the essentialistic idea that a new higher taxon could be established as the product of a single systemic mutation. Even though the success of such a taxic saltation is too improbable to be endorsed by any contemporary evolutionist, it still leaves the possibility of the occurrence of phenotypic saltations. If a mutation with a drastic phenotypic change could be incorporated in a population and become part of a viable phenotypic polymorphism, it could lead to a seemingly saltational evolutionary change. Gould (1980:127) indeed envisages a "potential saltational origin for the essential features of key adaptations. Why may we not imagine that gill arch bones of an ancestral agnathan moved forward in one step to surround the mouth and form proto-jaws?" As Maynard Smith (1983:276) points out, the occurrence of "genetic mutations of large phenotypic effect is not incompatible with Darwinism." Stanley (1982) has argued quite persuasively that gastropod torsion might have originated through a single mutation. It would have had to pass through a stage of polymorphism until the new gene had reached fixation. Evidently such a process is feasible but its importance in evolution is contradicted by the fact that among the millions of existing populations and species, mutations with large phenotypic effects would have to be exceedingly frequent to permit the survival of the occasional hopeful monster among the thousands of hopeless ones. But this is not found. Furthermore, enough mechanisms for the gradual acquisition of evolutionary novelties are known (Mayr 1960) to make the occurrence of drastic mutations dispensable, at least as a normal evolutionary process.

The argument, thus, is not whether phenotypic saltations are possible, but rather whether evolution advances through the production of individuals representing new types or through the rapid transformation of populations. No matter how rapid, such a populational "saltation" is nevertheless Darwinian gradualism.

WHAT ACCOUNTS FOR STASIS?

Of all the claims made in the punctuationist theory of Eldredge and Gould, the one that encountered the greatest opposition was that of a complete stasis of all species, after having completed the phase of origination. Yet, it was this very claim which the authors designated as their most important contribution (Gould 1982a:86).

The extraordinary longevity of the so-called living fossils had, of course, been known since the early days of paleontology (Eldredge and Stanley 1984; de Ricqles 1983). But is such stasis the fate of all species? Evidence supporting this claim can be found in Stanley's book (1979), some review papers (Levinton 1983; Gould 1983) and in recent volumes of *Paleobiology, Systematic Zoology,* and other journals. On the other hand, the literature also reports numerous cases of seeming speciation by phyletic gradualism (Van Valen 1982:99–112). Perhaps most convincing are the cases of significant evolutionary transformation in continuous phyletic lineages reported by Rose and Bown (1984) for Eocene primates, and by Chaline and Laurin (1986) for Pliocene rodents. Such phyletic speciation seems to be more frequent in terrestrial than in marine organisms.

Two objections have been raised against the seeming cases of phyletic speciation. First, hiatuses and depositional breaks seem to occur even in the most complete sequences; second, the so-called species of these sequences may not be valid species since they usually differ only in rather minor characters of size and proportions. Be that as it may, Gould has recently seemed to concede that speciation by phyletic gradualism does occur.

I agree with Gould that the frequency of stasis in fossil species revealed by the recent analysis was unexpected to most evolutionary biologists. To quote only one example, 131 modern benthic foraminifera species with an adequate fossil record have an average duration of 20 million years (Buzas and Culver 1984). Admittedly stasis is measured in terms of morphological difference, and the possibility cannot be excluded that biological sibling species evolved without this being reflected in the morphotype. Let us tentatively assume that some species enter complete stasis while others evolve by phyletic gradualism.

The question "What percentage of new species adopts one or the other of these two options?" cannot be resolved either by genetic theory or through the study of living species. It can be decided only through an analysis of the paleontological evidence. And this poses great method-ological difficulties (Levinton and Simon 1980; Schopf 1982). For instance, in the analysis of the benthic foraminifera, the calculated average age of 20 million years was based on only 15 percent of the recent species. For all the others, the fossil record was too spotty to permit any determina-tions. In other words, the proof of stasis was based on a highly biased sample, consisting of common widespread species, which one could expect to have high longevity and which comprised a small minority of the entire fauna. It is conceivable that a considerable fraction of the remaining 85 percent of species underwent rapid phyletic speciation and thus became unavailable for analysis. The indications are that the vast majority of the so-called rare species are short-lived, probably not for reasons of rapid phyletic change but rather owing to extinction. The best one can do under the circumstances is to adopt an intermediate position by admitting the occurrence of some gradual phyletic speciation but pointing also to the unexpectedly large number of cases where fossil species showed no mor-phological change over very many millions of years.

The concept of stasis has been seriously compromised by recent discov-eries in molecular biology. The stasis found in morphological characters in such old genera as *Rana, Bufo, Plethodon,* or even *Drosophila* is not at all due to the retention of an entirely unchanged genotype. Through the electrophoresis method, countless changes in quasi-neutral enzyme genes have been discovered, but numerous other nonmorphological changes have also taken place in these genera, such as the acquisition of new isolating mechanisms, as well as of numerous adaptations to a changing environ-ment. The only thing that has remained stable is the morphotype, the basic *Bauplan.* Some of the lineages can be inferred to have separated 30–60 million years ago and yet the species belonging to them are morpho-logically still almost indistinguishable except in size, coloration, and minor differences in skeletal dimensions. This is just one more piece of evidence for the interesting phenomenon that organisms do not evolve as harmonious types but that different characters may evolve at highly dif-ferent rates. *Archaeopteryx* with its incongruous mixture of reptilian and avian characters illustrates this well, as does the peculiar mixture of similarities and differences revealed by a comparison of humans and chim-panzees.

The discovery of highly unequal rates of evolution of different components of the genotype does not, however, eliminate the problem. Why do certain components of the genotype and phenotype remain so stable for millions or tens of millions of years?

The fact that there is an almost inexhaustible source of variation makes this a particularly puzzling question. Not only is there a steady rate of mutation at all gene loci, but various phenomena have been discovered by molecular biology in recent years that would seem to lead almost inevitably to a frequent revamping of the genotype. For instance, merely isolating stocks of *Drosophila* in laboratory populations may lead to all sorts of mutual incompatibilities. Transposons and other genetic elements may change the mutability of adjacent loci, and a host of other molecular phenomena would seem to encourage genetic changes of evolutionary significance. Furthermore, selection pressures in an incessantly changing physical and biotical environment would seem inducive to continuing evolutionary change. That we encounter the stability found in the living fossils, and to a lesser extent in the majority of species and genera, is very puzzling indeed.

On the whole, three explanations have been advanced to explain stasis. Reductionist geneticists attribute stasis entirely to stabilizing (normalizing) selection. All mutants or recombinants that deviate from the norm are eliminated by natural selection. This is, of course, no explanation, because abundant normalizing selection also takes place in rapidly evolving lines. Obviously all zygotes with lowered viability are apt to be eliminated in any population either before birth or at least prior to reproduction. Such selection under the name of "elimination of degenerations of the type" was acknowledged by essentialists long before the establishment of evolutionism (Mayr 1982:488). The large mortality of zygotes both in static and in rapidly evolving species is evidence for such elimination. The "internal selection" of certain authors is largely such elimination, as pointed out by J. Remane (1983).

This, however, does not explain why, in spite of the universality of such normalizing selection, certain lineages evolve rapidly while others remain in total stasis. Nor does invoking "normalizing selection" make a distinction between the elimination of new deleterious mutations and that of deleterious recombinants, particularly those interfering with developmental constraints. It is misleading to say that stasis is caused "either because deviants were weeded out or because the developmental system prevented them from arising." Even in the latter case it is normalizing selection through which the developmental constraints operate.

Another explanation is that species in stasis had reached optimal adaptation and were no longer answering any directional selection. This is improbable for two reasons: The first is that selection virtually never succeeds in completely optimizing a genotype. Hence, a change in the degree of optimality should be recognizable during millions of years. Futhermore, the environment is known to have changed considerably in periods during which certain species displayed complete stasis. Why is this environmental change not reflected in an evolutionary response of these species?

Considering that both of these explanations are unsatisfactory, one must ask whether there is any other possible explanation. Yes, provided one gives up the atomistic view that each gene is independent both in its actions and in the effects of selection on it. If one adopts a more holistic (integrative) view of the genotype and assumes that genes perform as teams and that large numbers of other genes form the "genetic milieu" (Chetverikov 1926) of any given gene, one can suggest an explanation. It is that epistatic interactions form a powerful constraint on the response of the genotype to selection. Such epistatic interactions were already dimly appreciated by Darwin (1859:11,146; 1868II:319–335) and have since been stressed by Chetverikov (genetic milieu), Lerner (genetic homeostasis), Mather (internal balance), and Mayr (cohesion of the genotype; 1963; 1970; 1975; Essay 25).

Holists claim therefore that much of macroevolution cannot be explained by atomistic gene replacements or by selection pressures on single genes, but only by a more drastic reorganization, made possible by loosening the tight genetic cohesion of the genotype found throughout widespread populous species. Mayr, Carson (1975), Eldredge and Gould, and Stanley ascribe the stability of the phenotype, as observed in static species, to such an internal cohesion of the genotype, or parts of it. Significant evolutionary advance can take place only after a breaking up of previously existing epistatic balances.

There is, of course, no conflict whatsoever between this holistic view and Darwinism because the cohesive domains of the genotype must have come into existence through natural selection.

Unfortunately, current genetic techniques seem to be unable to analyze such cases of restructuring. Until such techniques become available—probably in the not too distant future—it is impossible to prove conclusively the existence of such genotypic domains and a general cohesion of the genotype, or at least of parts of it.

There are numerous aspects of geographic variation that make sense only if one accepts the notion of a cohesiveness of the genotype. For instance, how else can one explain the pattern of geographic variation of the superspecies *Tanysiptera galatea* (Mayr 1954)? *T. galatea* is distributed over all of New Guinea, where it is adapted in the northwest of the island to a purely tropical wet climate without any seasons, and in southeast New Guinea to a tradewind climate with a nine-month dry season. One would expect that two extremely different phenotypes of this species would have evolved at the two ends of New Guinea in response to two drastically different climatic selection pressures. Actually, there is only minimal geographic variation on the mainland, in contrast with a series of strikingly different species that have budded off *T. galatea* on nearby islands. What else could have been responsible for the unexpected stasis on the mainland, except some process akin to genetic homeostasis?

It has been claimed that this holistic view of the genotype was not "within the spirit of modern Darwinism." This is erroneous. The atomistic viewpoint was defended only by the mathematical geneticists from Fisher to Charlesworth and Lande. The more holistic viewpoint was promoted by numerous Darwinians from Chetverikov to Mather, Lerner, and Wallace, and was vigorously promoted by me from 1950 on. It was a strong tradition in evolutionary biology long before the introduction of the theory of punctuated equilibria.

Speciational Evolution

For the defenders of phyletic gradualism, speciation is the continuing change of a phyletic series until it becomes a different species (the total number of species remaining constant). For the punctuationist, speciation means a multiplication of species. If this is effected by the establishment of a founder population and its rapid genetic reconstruction, then such speciation can occur in a relatively short span of time. It is not sudden like saltational speciation but it may be "instantaneous" as far as the fossil record is concerned. Even if it takes hundreds or thousands of years, the paleontological analysis would record it as a sudden event.

In some of their early papers punctuationists referred to speciation as a sudden event, but later Gould provided the more conservative definition that "geologically instantaneous" is to be "defined as one per cent or less of later existence in stasis" (Gould 1982a:84). Hence 100,000 years would be instantaneous for a species experiencing a 10-million-year stasis. The semantic problem is evident when we consider that all populational evolution, which includes all evolution we are concerned with, is gradual. It is quite obvious from the recent controversy that the chronology of speciation events cannot be established by paleontological analysis. Rather, it will have to be inferred from an analysis of currently living speciating species, as I have repeatedly attempted to do (Mayr 1963). In freshwater fishes it may take less than 4,000 years (see Essay 22).

That peripatric speciation is by far the most common mode of speciation is indicated not only by the pattern of distribution of incipient recent species but also by the frequency by which new species, apparently having originated somewhere else, suddenly appear in the fossil record. Such cases are reported in almost every revision of a fossil genus. For instance, in a Tertiary genus of bryozoans in at least seven cases the ancestral species persisted after having given rise to a descendant species (Cheetham 1986).

THE GENETICS OF SPECIATIONAL EVOLUTION

The crucial question one must ask is how does peripatric speciation differ in its genetic impact from gradual phyletic speciation. Are the genetics of the two processes truly different? Does peripatric speciation speed up evolution? Is the evolutionary pulsing provided by peripatric speciation necessary for the origin of evolutionary innovations? Honesty demands that we admit a lack of concrete knowledge that would permit us to answer these questions. All that we can do at the present time is to hypothesize; and in that respect we have not made much progress since 1954.

The genetic interpretation I gave in 1954 (see also Essay 25) was based on the genetic views of that period. I was much impressed by the findings of Mather, Lerner, and Wallace, largely supported also by Dobzhansky, on the genetic homeostasis of the genotype and the constraints on evolutionary departures imposed by this internal balance. I postulated therefore that certain events in the founder population might help loosen this cohesion and liberate the founder population from the straitjacket imposed on it by the epistatic balances of its genotype. I designated such an event a genetic revolution. Curiously, several authors reporting on my papers have claimed that I had postulated macromutations or "thousands of mutations," when as a matter of fact I had invoked not even a single mutation. Rather, my interpretation was based on a developmental point of view that apparently was incompatible with then current thinking. Developmental considerations were at that time ignored by most evolutionary geneticists. To understand the mechanism I proposed, one must remember that natural selection is a two-step process. My theory dealt exclusively with the first step, the generation of variation. It suggested how additional variation might be made available in a founder population. The only possibility I could see on the basis of then available genetic theory was a change of epistatic balances. Thirty-five years later, such a change is still a conceivable process and is probably involved in some cases of peripatric speciation. On the other hand, many different kinds of DNA have been discovered which might control a drastically speeded up genetic reorganization in small populations, also effecting "genetic revolutions." However, these new discoveries do not weaken the basic message of my theory, the role of small founder populations in loosening up genetic cohesion and thus facilitating evolutionary change. A somewhat different version has been suggested by Carson and Templeton (1984), likewise based on the concept that recombinational rather than mutational events are the decisive factor in this restructuring, and likewise based on a change in epistatic balances.

Considering our ignorance of what happens to the genotype during peripatric speciation, it is only natural that models would be proposed by reductionist geneticists that can explain everything in terms of the simplest single-gene assumptions. These models cannot be refuted, but they leave far more natural phenomena unexplained than the theory of a genetic restructuring of founder populations. Most important, though, is that none of the recent attacks on punctuationism, particularly in its typological quasi-Goldschmidtian version, has affected the recognition of the great potential importance of founder populations. It reveals a complete misunderstanding if an author says of my 1954 theory that it was a "macromutation theory . . . an alternative to the selectionist program" (Turner 1984:351). Nothing was stressed as much in my 1954 theory as selection.

One more point must be emphasized. There is great pluralism in speciation events. Whenever sexual selection in an isolated population leads to the origin of new behavioral isolating mechanisms, new species may evolve that differ by only a few genes (Essay 22). Such species are usually almost indistinguishable morphologically. They did not experience a genetic revolution.

The punctuationists have rightly criticized the gradualists for using "the notorious imperfection of the fossil record" to support gradualism. It is ironic that the punctuationists use the same argument when claiming that the populations in which peripatric speciation takes place are too localized and ephemeral to leave a fossil record. The incompleteness of the fossil record is thus as much part of the argument of the punctuationists as it is of the gradualists.

Species Selection

As I described above, one of Darwin's brilliant insights was that evolution is variational rather than transformational, as was believed by

Lamarck and by many anti-Darwinians after 1859. In variational evolution it is the selection of certain favored individuals that leads to evolutionary change. This means that the emphasis is on the individual and the population. However, variational evolution also occurs at a higher hierarchical level, that of the species, a fact particularly stressed by punctuationism. Whenever important evolutionary innovations occur, they occur during speciation. Once a new species has carved out its niche and has become successful and stabilized, it will tend to change very little. The species therefore is considered the unit of evolution.

The recognition of the evolutionary importance of species long antedates punctuationism. I know of no better description of the role of species in evolution than the one I gave in my *Animal Species and Evolution* (1963): "It is the very process of creating [new] species which leads to evolutionary progress . . . Since each coadapted gene complex has different properties and since these properties are, so to speak, not predictable, it requires the creation of a large number of such gene complexes before one is achieved that will lead to real evolutionary advance. Seen in this light, it appears then that a prodigious multiplication of species is a prerequisite for evolutionary progress.

"Each species is a biological experiment. The probability is very high that the new niche into which it shifts is an evolutionary dead-end street. There is no way to predict, as far as the incipient species is concerned, whether the new niche it enters is a dead-end or the entrance into a large new adaptive zone."

"The evolutionary significance of species is now quite clear. Although the evolutionist may speak of broad phenomena, such as trends, adaptations, specializations, and regressions, they are really not separable from the progression of entities that display these trends, the species. The species are the real units of evolution as the temporary incarnation of harmonious, well-integrated gene complexes. And speciation, the production of new gene complexes capable of ecological shifts, is the method by which evolution advances . . . The species, then, is the keystone of evolution" (1963:621).

The continuous production of new species inevitably leads to competition among species and to a great deal of extinction. This process has recently been referred to as *species selection* (Stanley 1975; 1979). The term is new but the concept goes far back in the history of biology. Even in pre-Darwinian days Lyell postulated that the "introduction" of a new and better adapted species might lead to the extinction of an inferior species, or that the extinction of a species would be followed by the introduction of a better adapted species. Species extinction by competition was an important process also for Darwin. He illustrated it by the fate of the indigenous New Zealand fauna when it encountered species introduced from the British Isles (1859:201). Extinction of species caused by the appearance of better adapted species has been frequently discussed in the post-Darwinian evolutionary literature.

The long-standing recognition of species selection refutes Gould's (1982b:386) claims that such a "hierarchially based theory [of evolution] would not be Darwinism as traditionally conceived." This is a restriction of the term Darwinism to the most reductionist concept of the mathematical population geneticists. I agree with the nonreductionist Darwinians who believe in a hierarchical order of the evolutionary process and who have never seen species selection as in any conflict with Darwinism.

The recent arguments indicate that the answer to two important questions about species selection remain controversial.

(1) Is species selection in conflict with (and/or independent of) individual selection?

(2) Are there different kinds of species selection, with different authors defining species selection differently or even in a contradictory manner?

Several authors have rejected species selection, considering it a strict alternative to individual selection (Hoffman and Hecht 1986). Indeed, if one asks the uncompromising question, "Is the individual *or* the species the target of selection?" one is forced to reject species selection. One can express this even more unequivocally by asking whether there are properties of species that are not properties of the individuals of which the species is composed. Most authors have conceded that such situations indeed occur, even Maynard Smith (1983:280), who on the whole is hostile to the concept of species selection. By contrast, I must state that I do not know of a single species character that is not also part of the genotype of every individual. Furthermore, it had become a species character because it had been favored by individual selection.

In the classical situation one species is superior to another because its individuals are better adapted and better able to utilize the resources of the environment. Even though competition between two species is seemingly involved, analysis establishes that the superiority is due to the greater success of the individuals of the superior species, and that individual selection is involved. Whenever two species are competing with each other, the individuals of both species are so to speak merged into a single ecological population, and selection deals with the total of these individuals. The individuals of the "superior" species have a higher expectation of survival and successful reproduction than those of the "inferior" species, so that eventually only individuals of the superior species will survive. The traditional process of individual selection has thus led to the extinction of one species, that is, to species selection. In such a case there is no conflict between individual selection and species selection. This will be the case whenever the "characters in common" of the individuals of one species provide competitive superiority over the individuals of another species.

It has, however, also been argued that one species may be superior to another not because of any superiority of the composing individuals but because the species as a whole has characteristics that give superiority to its members. For instance, it has been argued that forms of reproduction or of dispersal and colonization that favor speciation, being species-specific, would give such species greater competitive superiority and thus constitute species selection. I am not persuaded by this argument. All the stated characteristics are also the characteristics of individuals and, when first appearing in evolution, entered populations in a polymorphic condition. It was the selective advantage of the individuals with the new characteristics (and of their offspring) which led to the spread and eventual universal incorporation of these new characteristics. In other words these species characters were established by individual selection. I have failed to find in the literature a single good example of a species characteristic that is not also a selectable characteristic of individuals.

It has become evident in recent years that the term species selection has been applied to diverse phenomena. I have distinguished three kinds of species selection (Essay 6), while Maynard Smith (1983) recognizes four kinds. It would seem rather irrelevant how many kinds one recognizes as long as one accepts the conclusion that species selection is not in opposition to individual selection but is an additional process at a higher hierarchical level.

An increased rate of speciation is believed by some to lead to increased species selection. However, there is considerable doubt that a high rate of speciation ("speciator selection") is in the long run necessarily a winning ticket in the evolutionary lottery. Many cases of very species-rich groups are known that ultimately became entirely extinct while contemporary species-poor groups, that is, slow speciators, lived on.

Gould and other strict punctuationists require species selection to explain evolutionary trends. Since they believe in a complete stasis of most species they must interpret evolutionary trends as a sequence of speciation events during which only those species are successful that had diverged in the direction of the favored trend (Gould and Eldredge 1986:146). "Species arise . . . with their differences established at the start, and do not change substantially thereafter. Trends must therefore [be due to a process of] sorting that operates via the differential birth and death of species considered as entities" (Gould 1982a:92). This is a conceivable scenario, but up to now it has been impossible to document it by an analysis of the fossil record. It may be an unnecessary postulate if more and more cases are found of a substantial change in phyletic lineages owing to gradual transformation.

We can conclude this discussion by stating that there is no conflict between individual and species selection. Individual selection is always involved, but individuals of certain species have "characters in common" that establish their unqualified competitive superiority over other species. I can see no reason why it should not be legitimate to refer to such cases as species selection. Such species, as individuals, are selectively superior to other species, as individuals.

The former confusion was aggravated by a failure to distinguish between survival selection and selection for reproductive success. Most of the characteristics usually listed as indicating or substantiating species selection are characteristics that contribute to reproductive success. Such characteristics may make themselves felt primarily in later generations.

PLURALISM AND SIMULTANEITY

My 1954 proposal of genetic revolutions during peripatric speciation was pluralistic. I emphasized that no major evolutionary innovations occurred in the origins of most new species. By contrast the first statements of the punctuationists were quite categorical and this led to polarization and resistance. Their more recent statements have been formulated in a more pluralistic manner: "Gradual phyletic transformation can and does occur" (Gould 1982a:84); and "The relative frequency of punctuated equilibrium differs across taxa and environments" (Gould 1986:439). This renunciation of absolute, all-or-nothing statements has greatly contributed to a reduction of opposition.

A second source of opposition to punctuationism is more difficult to overcome. It is the tendency of many authors to present two simultaneously occurring processes as a choice between one or the other. This is what Maynard Smith (1983) does when he asks whether the increased rate of evolutionary change in small populations is due to Mayr's genetic revolutions or due to natural selection. This is curious considering how strongly I stressed the role of natural selection in the reorganization of the genotype of founder populations. Maynard Smith apparently failed to realize that genetic revolutions deal with the first step of the selection process while selection proper is the second step. He does not accept the possibility that a "thawing out of the congealed part of the genotype," possible in founder populations, might make it easier for natural selection to achieve a stable new balance. Too often evolutionary factors or processes are presented as alternatives when in reality they occur simultaneously.

Speciation and Macroevolution

The great importance of the speciation event is that it links macro-evolution with microevolution. The fact that the individual is the target of selection and that the population and the species are the locale of evolutionary change automatically reduce all macroevolutionary processes to the microevolutionary level. The actors in this process, however, are not genes but genotypes and gene pools, entire cohesive systems of genes. The important insight is that whatever happens either in microevolution or in macroevolution, and whatever genetic phenomena are involved, proceeds through the selection of individuals (Mayr 1942:298). Admittedly, genetics has so far been unable to analyze that part of the genotype that does not ordinarily vary in a local population but is so tightly integrated that it gives the genus, the family, the order, the phylum, its particular character (Carson 1975). But even this part of the genotype, when it varies, varies in individuals and is subject to the recombination-selection cycles of ordinary allelic variation.

For the extreme reductionists among the geneticists, who look at evolution, even macroevolution, in terms of changing frequencies of genes, there is a complete continuity among all phenomena of evolution. But for those who think of evolution also as a change of species and higher taxa, and this includes Darwin, evolution has always been considered as hierarchical in structure. No other component of Darwin's thinking was as readily and widely adopted as his theory of common descent, a strictly hierarchical theory. And most of the paleontological literature, largely devoted to an elucidation of common descent, was hierarchical in approach. The term "hierarchical approach" introduces perhaps a new terminology but not a new concept. I agree with Verne Grant "that adherents of the synthetic theory have in fact [consistently] employed a hierarchical approach to problems of macroevolution" (1983:153).

Remaining Problems

The controversies about punctuationism have clearly revealed that there remain major gaps in our knowledge. Several of these have been mentioned above, but I want to single out two for more detailed discussion.

POPULATION SIZE AND RATE OF EVOLUTION

For the last 60 years biologists have debated whether evolution advances more rapidly in large populations (species) or in small ones. Beginning with R. A. Fisher and Sewall Wright, geneticists favored the large population as the locale of the most rapid evolutionary change. Under the assumption that the gene is the target of selection and that one can calculate the rate of evolution of a population or species by combining the rates of change of various independent genes, they concluded that evolution will be faster in larger populations. As recently as 1977 Wright has made this claim for widespread species consisting of many populations, a claim not based on observed facts but on theoretical considerations. This includes the acceptance of group selection among local demes, a type of selection questioned by nearly all those who have thoroughly studied the problem (Sober 1984).

Wright's scenario also assumed that the amount of gene flow among the postulated semi-isolated populations is very low, an assumption which has fared poorly in recent researches. To be sure, there are species with low gene flow, but these are species with spotty, isolated distributions. In all widespread populous species there seems to be far more unobserved gene flow than is usually believed (Slatkin 1985). Here the gene flow is often restricted to certain years in which the species has been particularly successful and the population has reached high densities. There is little evidence in most species for the semi-isolated demes postulated by Wright.

The evolutionary stability of large widespread species is supported by the observation that most of the species in the fossil record that display stasis are large widespread species with the samples taken from central populations. The claim of rapid evolutionary advance in large widespread populous species was questioned, on a theoretical basis, as far back as 1957 by Haldane, and more recently by some mathematical population geneticists (Newman et al. 1985; Lande 1985).

By contrast, Mayr (1942; 1954) and his followers were impressed by the striking difference between the parental population and many peripherally isolated founder populations and species. This evidence was enriched by the findings of Carson and associates on incipient species and neospecies of *Drosophila* in Hawaii. The relative rapidity with which new and often drastically different species of *Drosophila* originate through founder events suggests an enormously accelerated rate of evolution, as compared to the widespread and relatively uniform sibling species complexes of *Drosophila* with continental distributions (like *D. affinis, D. obscura*). The theory that evolution in small founder populations is more rapid than in large continental populations is based on solid observational evidence, in contrast with the theorizing of reductionist genetics. We have yet to find a situation where the isolated peripheral allospecies essentially retain their ancestral condition while the large central population has seemingly greatly diverged.

It is important to distinguish between the observational evidence and the genetic interpretation. I believe that the observational evidence for a greatly increased rate of evolution in small populations is far better established than is the evidence for a greatly increased rate in large populations. Whether this increased rate is due to the breaking up of previously existing balances and the formation of new ones, or simply due to stochastic processes (or to both) is still an open question. However, by merely invoking stochastic processes, Newman and Lande demonstrated that passage from one to another adaptive peak had a much better chance of success in a small population. By contrast, the chance of success can be vanishingly small in large populations. Hence, say Newman and his colleagues, their model "predicts stasis and punctuation for a small to moderate population but only stasis for a large population." This is consistent with the observations of the naturalists. The breaking up of epistatic balances in founder populations has now also been demonstrated experimentally (Goodnight 1987; Bryant et al. 1986).

There is one argument, seemingly based on observational evidence, which would seem to refute the thesis that large widespread species are uniform at a given time level, a uniformity continuing as stasis. It is the argument (Stenseth ms.) that geographic variation is prevalent in all widespread species, a notion which conflicts with the postulated temporal uniformity of such species. If this claim were substantiated it would be a valid objection to punctuationism. However, its factual support is quite limited. On the basis of my own analysis of the bird fauna of North America (Mayr and Short 1970) and of the New Guinea area (Mayr and Diamond 1988), three counterarguments can be made against this thesis.

(1) Most widespread species are remarkably uniform. This is particularly true for planktotrophic marine invertebrates.

(2) Most species displaying geographic variation vary only clinally in ecotypic characters, affecting primarily size, proportions, and coloration, but show no significant evolutionary departures. This is true for virtually all the illustrations of phyletic gradualism discovered by Gingerich (1977) in early tertiary mammals. Nor is the thesis of the evolutionary uniformity of widespread species refuted by the discovery of the localized selection of a few genes controlling mimicry in the genus *Heliconius* (Turner 1984). Such selection for local mimicry is the exact equivalent to substrate protective coloration in rodents and other animals selected for crypsis. Species with oligogenic geographic variation in crypsis may be exceedingly widespread, but the genes involved in this adaptation seem not to be a suitable basis for speciation. Interestingly, such genes are rarely combined into domains. The 14 or so variable characters of *Heliconius melpomene* that permit the mimicry of coexisting species are scattered over 9 chromosomes, and in the case of *Heliconius erato* some 17 or 18 variable characters are dispersed on 10 chromosomes (Turner 1984). Such independence of the gene loci greatly facilitates a shift to new patterns, but it does not encourage the evolution of a new stable evolutionary type.

(3) Finally, polytypic species with striking variation are invariably secondarily fused mosaics of former founder populations. Their variation cannot be used to argue against the evolutionary role of founder populations.

STRUCTURE OF THE GENOTYPE

Another unresolved argument concerns the structure of the genotype. As discussed above, mathematical geneticists base their calculations on the assumption that all genes are more or less independent of each other and that recombination following crossing over can produce a virtually unlimited assortment of genotypes. Other evolutionists, including such geneticists as Mather, Dobzhansky, Lerner, Wallace, and Carson, believe that there are cohesive domains in the genotype maintained by epistatic balances and that many evolutionary phenomena are best explained by such an assumption. Reductionist geneticists believe that they can explain all phenomena of seeming cohesion (domains of the genotype) in terms of single genes. Their opponents are yet unconvinced. Considering the rapid progress of molecular genetics and the ongoing discovery of new molecular structures and interactions of the genotype, there is every reason to believe that this argument will soon be resolved.

The Contributions of Punctuationism to Evolutionary Theory

Even some of its opponents admit that punctuationism has had an enormously stimulating impact on evolutionary biology (Rhodes 1984; Maynard Smith 1984a; Gould 1986). It has brought to light numerous equivocations and has helped to clarify distinctions between alternatives, such as between phyletic and allopatric speciation, between phenotypic and taxic saltations, between various types of group selection, between the evolutionary potential of small and large populations, between an uncompromisingly reductionist and a more holistic concept of the genotype, between various concepts of species selection, and more. To eliminate these equivocations it was not only necessary to clarify concepts but also to show that we needed a broader factual foundation. As Gould has emphasized, one of the most important contributions of punctuationism has been its stimulation of fruitful empirical research, much of it still ongoing.

To be sure, the extreme claims of some punctuationists, such as the universality of total stasis and the impossibility of evolutionary change without speciation, are clearly invalid. Furthermore, it has been shown that "speciational evolution" (perhaps a better term than punctuationism) is fully consistent with Darwinism; and finally, that seeming evolutionary saltations, as indicated by the fossil record, can be explained without invoking systemic mutations or other mechanisms in conflict with molecular genetics. In view of the pluralistic position of punctuationism, it is irrelevant for the theory of speciational evolution how relatively frequent evolutionary stasis is or how frequent the occasional occurrence of drastic genetic reorganization during peripatric speciation.

Most of all, punctuationism has shown how myopic has been the focusing of paleontologists and population geneticists on the one-dimensional, transformational, upward movement of evolution. It finally brought general recognition to the insight of the taxonomists (Poulton, Rensch, Mayr) that the lavish production of diversity is perhaps the most important component of evolution.

What had not been realized before is how truly Darwinian speciational evolution is. It was generally recognized that regular variational evolution in the Darwinian sense takes place at the level of individual and population. However, that a similar variational evolution occurs at the level of species was generally ignored. Transformational evolution of species (phyletic gradualism) is not nearly as important in evolution as the production of a rich diversity of species and the establishment of evolutionary advance by selection among these species. In other words, speciational evolution is Darwinian evolution at a higher hierarchical level. The importance of this insight can hardly be exaggerated.

NOTE

This essay first appeared in the *Journal of the Biology of Social Structure* 1(1988) under the title "Speciational evolution or punctuated equilibria."

REFERENCES

Ayala, F. J. 1983. Beyond Darwinism? The challenge of macroevolution to the synthetic theory of evolution. *PSA 1982* 2:275–291 (Philosophy of Science Association).

Bryant, E. H., S. A. McCommas, et al. 1986. The effect of an experimental bottleneck upon quantitative genetic variation in the housefly. *Genetics* 114:1191.

Buch, L. v. 1825. *Physicalische Beschreibung der Canarischen Inseln,* pp. 132–133. Berlin: Kgl. Akad. Wiss.

Buzas, M. A., and S. J. Culver. 1984. Species duration and evolution: benthic Foraminifera on the Atlantic continental margin. *Science* 225:829–830.

Carson, H. L. 1975. The genetics of speciation at the diploid level. *Amer. Nat.* 109:83–92.

Carson, H. L., and A. R. Templeton. 1984. Genetic revolutions in relation to speciation phenomena: the founding of new populations. *Ann. Rev. Ecol. Syst.* 15:97–131.

Chaline, J., and B. Laurin. 1986. Phyletic gradualism in a European Plio-Pleistocene *Mimomyx* lineage (Arvicolidae, Rodentia). *Paleobiology* 12:203–216.

Cheetham, A. H. 1986. Tempo of evolution in a Neogene bryozoan: rates of morphological change within and across species boundaries. *Paleobiology* 12:190–202.

Chetverikov, S. S. 1926. On certain aspects of the evolutionary process from the standpoint of modern genetics. *J. Exper. Biol.* A2:3–54 (Russian).

Darwin, C. 1859. *On the Origin of Species by Means of Natural Selection or the Preservation of Favored Races in the Struggle for Life.* London: Murray.

———— 1868. *The Variation of Animals and Plants under Domestication.* 2 vols. London: Murray.

de Ricqles, A., ed. 1983. Formes panchroniques et 'fossiles vivants'. *Bull. Soc. Zool. France* 108 (4):529–673.

Dobzhansky, Th. 1937. *Genetics and the Origin of Species.* 1st ed. New York: Columbia University Press.

———— 1951. *Genetics and the Origin of Species.* 3rd ed. New York: Columbia University Press.

Eldredge, N. 1971. The allopatric model and phylogeny in Paleozoic invertebrates. *Evolution* 25:156–167.

Eldredge, N., and S. J. Gould. 1972. Punctuated equilibria: an alternative to phyletic gradualism. In T. J. M. Schopf, ed., *Models in Paleobiology,* pp. 305–332. San Francisco: Freeman, Cooper.

Eldredge, N., and S. M. Stanley, eds. 1984. *Living Fossils.* New York: Springer.

Fisher, R. A. 1930. *The Genetical Theory of Natural Selection.* Oxford: Clarendon Press.

Gilinsky, N. L. 1986. Species selection as a causal process. *Evol. Biol.* 20:249–281.

Gingerich, P. D. 1977. Patterns of evolution in the mammalian fossil record. In A. Hallam, ed., *Patterns of Evolution,* pp. 469–500. Amsterdam: Elsevier.

———— 1984. Punctuated equilibria—where is the evidence? *Syst. Zool.* 33:335–338.

Goldschmidt, R. 1940. *The Material Basis of Evolution.* New Haven: Yale University Press.

———— 1952. Evolution, as viewed by one geneticist. *Amer. Sci.* 40:84–98.

Goodnight, C. J. 1987. On the effect of founder events on epistatic genetic variance. *Evolution* 41:80–91.

Gould, S. J. 1977. The return of hopeful monsters. *Nat. Hist.* 86:22–30.

———— 1980. Is a new and general theory of evolution emerging? *Paleobiology* 6:119–130.

———— 1982a. The meaning of punctuated equilibrium and its role in validating a hierarchical approach to macroevolution. In R. Milkman, ed., *Perspectives in Evolution,* pp. 83–104. Sunderland, Mass.: Sinauer.

———— 1982b. Darwinism and the expansion of evolutionary theory. *Science* 216:380–387.

———— 1983. [Answer to Schopf and Hoffman]. *Science* 219:439–440.

———— 1986. Punctuated equilibrium: empirical response. *Science* 232:439.

Gould, S. J., and C. B. Calloway. 1980. Clams versus brachiopods. *Paleobiology* 6:394.

Gould, S. J., and N. Eldredge. 1977. Punctuated equilibria: the tempo and mode of evolution reconsidered. *Paleobiology* 3:115–151.

———— 1986. Punctuated equilibrium at the third stage. *Syst. Zool.* 35:143–148.

Grant, Verne. 1982. Punctuated equilibria: a critique. *Biol. Zbl.* 101:175–184.

————— 1983. The synthetic theory strikes back. *Biol. Zbl.* 102:149–158.

Haffer, J. 1974. *Avian Speciation in Tropical South America.* Cambridge, Mass.: Nuttal Ornithological Club Publ. no. 14.

Haldane, J. B. S. 1957. The cost of natural selection. *J. Genet.* 55:511–524.

Hoffman, A. 1982. Punctuated versus gradual mode of evolution: a reconsideration. *Evol. Biol.* 15:411–436.

Hoffman, A., and M. K. Hecht. 1986. Species selection as a causal process: a reply. *Evol. Biol.* 20:275–281.

Huxley, A. 1982. Address of the President. *Proc. R. Soc. London* A379:IX–XVII.

Huxley, J. 1942. *Evolution: The Modern Synthesis.* London: Allen and Unwin.

Lande, R. 1985. Expected time for random genetic drift of a population between stable phenotype states. *Proc. Nat. Acad. Sci.* 82:7641.

————— 1986. The dynamics of peak shifts and the pattern of morphological evolution. *Paleobiology* 12:343–354.

Levinton, J. S. 1983. Stasis in progress: the empirical basis of macroevolution. *Ann. Rev. Ecol. Syst.* 14:103–137.

Levinton, J. S., and C. M. Simon. 1980. A critique of the punctuated equilibria model and implications for the detection of speciation in the fossil record. *Syst. Zool.* 29:130–142.

Lewontin, R. 1983. Kinds of evolution. *Scientia* 118:65–82.

Malmgren, B. A., W. A. Berggren, and G. P. Lohmann. 1984. Species formation through punctuated gradualism in planctonic Foraminifera. *Science* 225:317–319.

Mather, K. 1953. The genetical structure of populations. *Symp. Soc. Experim. Biol.* VII (Evolution):66–95.

Maynard Smith, J. 1983. Current controversies in evolutionary biology. In M. Grene, ed., *Dimensions of Darwinism,* pp. 273–286. Cambridge: Cambridge University Press.

————— 1984a. Palaeontology at the high table. *Nature* 309:401–402.

————— 1984b. The genetics of stasis and punctuation. *Ann. Rev. Genet.* 17:11–25.

Mayr, E. 1942. *Systematics and the Origin of Species.* New York: Columbia University Press.

————— 1954. Change of genetic environment and evolution. In J. Huxley, ed., *Evolution as a Process,* pp. 157–180. London: Allen and Unwin.

————— 1960. The emergence of evolutionary novelties. In S. Tax, ed., *Evolution after Darwin.* Vol. 1, pp. 349–380. Chicago: University of Chicago Press.

————— 1963. *Animal Species and Evolution.* Cambridge, Mass.: Harvard University Press.

————— 1970. *Populations, Species, and Evolution.* Cambridge, Mass.: Harvard University Press.

————— 1982. *The Growth of Biological Thought.* Cambridge, Mass.: Harvard University Press.

———— 1984. The triumph of evolutionary synthesis. *Times Literary Supplement* 257 (4):1261–1262.

Mayr, E., and J. Diamond. 1988. *Birds of Northern Melanesia.* (ms.).

Mayr, E., and L. Short. 1970. *Species Taxa of North American Birds: A Contribution to Comparative Systematics.* Cambridge, Mass.: Nuttall Orn. Club Publ. No. 9.

Newman, C. M., J. E. Cohen, and C. Kipnis. 1985. Neo-Darwinian evolution implies punctuated equilibria. *Nature* 315:400–401.

Osborn, H. F. 1894. *From the Greeks to Darwin.* New York: Columbia University Press.

Penny, D. 1983. Charles Darwin, gradualism and punctuated equilibria. *Syst. Zool.* 32:72–74.

Remane, J. 1983. Selektion und Evolutionstheorie. *Paläont. Zeitschr.* 57:205–212.

Rensch, B. 1947. *Neuere Probleme der Abstammungslehre.* Stuttgart: Enke.

Rhodes, F. H. T. 1983. Gradualism, punctuated equilibrium and the *Origin of Species. Nature* 305:269–272.

———— 1984. [Reply to Gingerich]. *Nature* 315:401–402.

Rose, K. D., and T. M. Bown. 1984. Gradual phyletic evolution at the generic level in early Eocene Omomyid primates. *Nature* 309:250–252.

Schopf, T. J. M. 1982. A critical assessment of punctuated equilibria. I. Duration of taxa. *Evolution* 36:1144 1157.

Schopf, T. J. M., and A. Hoffman. 1983. Punctuated equilibrium and the fossil record. *Science* 219:438–439.

Simpson, G. G. 1944. *Tempo and Mode.* New York: Columbia University Press.

———— 1953. *The Major Features of Evolution.* New York: Columbia University Press.

———— 1961. *Principles of Animal Taxonomy.* New York: Columbia University Press.

Slatkin, M. 1985. Gene flow in natural populations. *Ann. Rev. Ecol. Syst.* 16:393–430.

Sober, E. 1984. *The Nature of Selection.* Cambridge, Mass.: MIT Press.

Stanley, S. M. 1975. A theory of evolution above the species level. *Proc. Nat. Acad. Sci.* 72:646–650.

———— 1979. *Macroevolution: Pattern and Process.* San Francisco: Freeman.

———— 1982. Gastropod torsion: predation and the opercular imperative. *Neu. Jahrb. Geologie und Paläontologie. Abh.* 164:95–107.

Stebbins, L. G., and F. J. Ayala. 1981. Is a new evolutionary synthesis necessary? *Science* 213:967–971.

Turner, J. R. G. 1983. Mimetic butterflies and punctuated equilibria: some old light on a new paradigm. *Biol. J. Linn. Soc.* 20:277–300.

———— 1984. Darwin's coffin and Doctor Pangloss—do adaptationist models explain mimicry? in B. Shorrocks, ed., *Evolutionary Ecology,* pp. 313–361. Oxford: Blackwell Scientific Publications.

Van Valen, L. M. 1982. Integration of species: stasis and biogeography. *Evol. Theory* 6:99–112.

Wallace, B. 1985. Reflections on the still—"hopeful monster." *Quart. Rev. Biol.* 60:31–42.

Willmann, R. 1985. *Die Art in Raum und Zeit.* Berlin: Parey.

Wright, S. 1931. Evolution in Mendelian populations. *Genetics* 6:97–159.

———— 1932. The roles of mutation, inbreeding, crossbreeding, and selection in evolution. *Proc. Sixth. Int. Congr. Genet.* 1:356–366.

———— 1977. *Evolution and the Genetics of Populations.* Vol. 3. Chicago: University of Chicago Press.

part nine ❧

HISTORICAL PERSPECTIVE

Introduction

An analysis of almost any scientific problem leads automatically to a study of its history. As I hope the preceding essays in this volume have made plain, the many unresolved issues in evolutionary biology are no exception to this rule. To understand their history, one must deal not only with the state of factual knowledge at the given period but also with the *Zeitgeist* of the time. The interpretation any investigator gives to the results of his observations or experiments depends to a very large extent on this conceptual framework. For many years a major objective of my historical studies has been to discover the concepts—or sometimes, even more broadly, the ideologies—on which the theorizing of certain historical figures was based.

Particulate inheritance, for instance, was staring at Kölreuter in most of his crossing experiments, yet he failed to make the observation because this interpretation was completely incompatible with his essentialism and epigenetic thinking. Similarly, an ideological constraint was responsible for the failure of August Weismann to solve the problem of the source of genetic variation. Adhering to a rather strongly deterministic interpretation of causation, Weismann was unable to envisage a process of spontaneous mutation, although all of his theorizing actually pointed in the right direction. Yet in his analysis of other problems, Weismann's ability to speculate creatively and to alter his basic assumptions in the light of new evidence led him to insights that has earned him a rank among nineteenth-century evolutionists second only to Darwin himself. The career and accomplishments of this remarkable evolutionist are the subject of Essay 27.

In scientific controversies, there is rarely any argument about facts. It is rather their interpretation that is controversial. A careful study of such controversies will almost always reveal a difference in conceptual framework among the parties in dispute—and usually an inadequacy on one or both sides. For example, as discussed in Essay 28, modern criticisms of the evolutionary synthesis very often come from people who fail to make a distinction between proximate and evolutionary causation, or between the four very different processes that have been called teleological, or who fail to realize the independence of the first and second steps in natural selection. The last essay in this volume attempts to provide an overview of the conceptual framework and the achievements of the evolutionary synthesis and to show how unjustified most of the recent attacks have been. The synthesis was a major step in the development of evolutionary theory, but no one can deny that there are still many unknowns. However, the gaps in our knowledge can be filled only by those who have a sound understanding of what has already been achieved.

ON WEISMANN'S GROWTH
AS AN EVOLUTIONIST

AUGUST WEISMANN is one of the towering figures in the history of evolutionary biology. If we ask who in the nineteenth century after Darwin had the greatest impact on evolutionary theory, the unequivocal answer must be Weismann. It was he who was responsible for what Romanes later called neo-Darwinism. Curiously, all recent historical studies of Weismann deal with his impact on genetics, whereas a modern analysis of his evolutionary thought does not exist.

What makes Weismann's concern with the problems of evolution so fascinating is the gradual maturation of his thinking. There was an unfortunate tendency in the older histories of science to depict the thought of a great scientist as a monolithic structure, something poured out of one mold, impressively unchanging. More recent researches have shown how misleading such a presentation is, and this is true for almost any scientist one may mention. Lamarck, for instance, at the ripe age of fifty-six years, abandoned the concept of the fixity of species and of a constant world and became an evolutionist. When Weismann rejected the theory of an inheritance of acquired characters, surely a dramatic conversion, he was already over forty-seven years of age. This change required a fundamental rethinking of all of his previous assumptions on inheritance and evolution. It may be helpful to recognize three periods in Weismann's thinking: 1868–1881 or 1882, during which he accepted the inheritance of acquired characters; 1882–1895, when he searched for a source of genetic variation; and 1896–1910, when he recognized germinal selection as the aid of natural selection. Toward the very end of his life Weismann published a grand summary of his evolutionary thinking, a magisterial work in two volumes totaling 672 text pages, *Vorträge über Deszendenztheorie*. The popularity of this work is documented by the fact that it required three editions within a few years (1902, 1904, 1910). What is said about Weismann as evolutionist is usually based on this last summary of his thought. However, evolutionary discussions and speculations dominate the first and third of his research periods. When Weismann concluded in 1882 that the theory of an inheritance of acquired characters was untenable, he had to search for a new source of genetic variation and therefore devoted the second research period mostly to genetic studies. Indeed, he stated quite explicitly that his theory of inheritance "was so to speak only a means toward a higher purpose, [to establish] a foundation for the understanding of the transformation of organisms in the course of time" (1904:iii).

In 1859, when Darwin published his revolutionary *On the Origin of Species,* Weismann was twenty-five years old. As he mentioned in an autobiographical account, the subject of evolution was never raised during his student days at several German universities. Nor did Weismann's early zoological work have anything to do with evolution; it dealt instead with such topics as muscle histology and insect embryology. And yet, in 1868, Weismann chose Darwinism as the subject of his inaugural lecture at Freiburg. In no other country, not even in England, did Darwin's evolutionary theory have the impact it had in Germany, as is evident from the writings of contemporary zoologists, botanists, and anatomists.[1] The high importance Weismann attributed to this theory is documented by the fact that he compared the transmutation theory to the Copernican heliocentric theory and implied that no advance in human understanding since the acceptance of that theory had had as great an impact as Darwin's theory (1868:30).

Weismann understood from the beginning that one must distinguish two theories of evolution; evolution as such, called by Weismann the transmutation theory; and Darwin's explanatory theory, the theory of natural selection. The modern evolutionist makes even finer distinctions among Darwin's evolutionary theories, and it helps our understanding of Weismann's intellectual development to see how his thought relates to the five components of Darwinism as outlined in Essay 12 (above).

(1)Evolution as such. The alternative to evolution, said Weismann, is the creation hypothesis. Like Darwin in the *Origin* (Gillespie 1979), Weismann presented in his 1868 paper a long series of biological facts that are consistent with an evolutionary interpretation, but simply make no sense if the living world were the product of a single act of creation. Weismann's attitude toward evolution was close to that of modern evolutionist, for whom evolution is not a theory but an accepted fact. The various conclusions arrived at by evolutionary biology, said Weismann, "may be main-

tained with the same degree of certainty as that with which astronomy asserts that the earth moves around the sun; for a conclusion may be arrived at as safely by other methods as by mathematical calculation" (1886:255). The fact of evolution was so irrefutably established for Weismann that in his subsequent writings he virtually never bothered to list facts in support of evolution, but instead concentrated on the causal aspects of the evolutionary process.

(2) *The theory of common descent.* No other part of Darwin's theory was accepted by zoologists as enthusiastically as the theory of common descent. To establish the common descent of higher taxa was the overriding ambition of comparative anatomists, as well as of many zoologists and embryologists during the 1860s to 1880s. I shall not discuss the splendid successes, embarrassing failures, and violent criticisms of these endeavors, but say only that with one exception Weismann did not take part in this activity. The exception was that Weismann fully accepted the theory of recapitulation and based, for instance, his analysis of the ontogenetic stages of sphingid caterpillars entirely on this principle.[2] As he said, "The considerations previously set forth are entirely based on Fritz Müller's and Haeckel's view that the development of the individual presents the ancestral history in nuce, the ontogeny being a condensed recapitulation of the phylogeny" (1882:270). Each of the five larval stages of these caterpillars has its own markings, and Weismann was convinced that this developmental sequence reflected a phylogenetic sequence, with the youngest larvae representing the earliest ancestral condition. "This development of the markings in individuals very well reveals their phyletic development, since there can be no doubt but that we have here preserved to us in the ontogeny . . . a very slightly altered picture of the phyletic development" (1882:163). Weismann even went so far as to accept a series of phylogenetic laws that summarize the appearance or loss of characters at the various ontogenetic stages (1882:276). He repeated "the ontogeny of larval markings is a more or less condensed and occasionally falsified [= strongly modified] recapitulation of the phylogeny" (1882:273). For a further discussion of Weismann's views on recapitulation see Gould (1977:102–109).

Some modern critics of the theory of recapitulation might call Weismann's interpretation a "just-so story." Indeed, in the absence of any fossil record it is impossible to prove that the ontogenetic sequence recapitulates the phylogenetic one. Perhaps if one made a cladogram of the adult characters (the moths) and of the various molecular characters, one could test the probability of Weismann's reconstruction. There is a hidden conflict in Weismann's presentation, for he simultaneously interpreted the entire development of markings as the result of a series of selection pressures (1882:380–387). In his description of the sequence of ontogenetic stages Weismann often spoke as if he were dealing with an evolutionary sequence of species. Nowhere did he call attention to the fact that all of these stages are produced by the same genotype. This would mean, in terms of modern thinking, that different portions of the genotype are silent at each stage. If only this thought had occurred to Weismann, he would not have had so much difficulty in imagining that during mitotic cell division different portions of the germ plasm could be either activated or silenced. This is one more illustration of the fact that in science one sees only that which one is prepared to see. That is the reason why in the advance of science the development of new concepts is often more important than the making of new discoveries.

Weismann took another look at ontogeny in a later part of his 1882 paper, entitled "Incongruences in Other Orders of Insects" (1882:481–501). He pointed to the relative uniformity in the morphology of the adult hymenopterans and dipterans, and interprets the numerous specializations found in the larval stages as secondary adaptations. Evidently the evolution of imagines and larval stages is connected with differences in niche occupation; in other words, it is the result of different selection forces. Such developments, of course, completely negate the "law of recapitulation." What Weismann actually dealt with in these very interesting discussions is what today is usually referred to as mosaic evolution. It would be worthwhile to reanalyze Weismann's findings against a modern classification of the groups of insects with which he dealt.

(3) *The multiplication of species.* It is now understood that evolution consists of two major processes, the changes (usually adaptational) of populations in time, and the multiplication of species in space, that is, the origin of new organic diversity. The latter process, more often called speciation, has been clouded with confusion ever since 1859. Darwin in his early unpublished writings (1837 to 1844) had come to the conclusion that geographic isolation was a necessary prerequisite for speciation and that therefore allopatric speciation was the prevailing, if not the only, form of speciation (Kottler 1978; Sulloway 1979). However, by 1859 when he published the *Origin,* Darwin had concluded that sympatric speciation, the splitting of a single population without geographic isolation, was at least equally common.

This conclusion was vigorously attacked by the naturalist Moritz Wagner (1868, 1889), who insisted on the importance of geographic isolation.

In more than thirty years of traveling and exploring Wagner had found one case after another where two incipient species or new species were separated from each other by some geographic barrier—a river, a mountain range, an arm of the sea, or a vegetational discontinuity. No fault can be found with Wagner's empirical evidence. Unfortunately, he also claimed that no evolutionary change of any kind was possible without such isolation, and that natural selection could operate only in an isolated population. Wagner apparently believed that blending inheritance and gene flow would always wipe out any new genetic departure unless the population in which it occurred was strictly isolated.

The interchange between Wagner and Darwin has been well covered in the literature (Sulloway 1979). Weismann entered the fray and wrote an entire essay on the subject of the influence of isolation on species formation (1872). Like Wagner, he failed to see that two problems are involved in speciation, the localization of the incipient new species and the genetic process by which the origin of the new species is effected. More than a hundred years later M. J. D. White (1978) made the same mistake when he asked whether speciation occurred geographically or chromosomally (see Essay 21).

It is evident that Weismann, through his studies of butterflies of the Mediterranean Basin, must have been familiar with vicariant species, and indeed he described the double colonization of the islands of Corsica and Sardinia by *Papilio machaon*. Nevertheless, Weismann insisted that Wagner had attributed to the factor of isolation much too high a significance for species formation (1872:iv). Wagner could be clearly refuted, said Weismann, "if one could show that somewhere sometime a species had changed into another one in the middle of the area of distribution or had produced a new species at such a place" (1872:6–7). On the whole, Weismann did not argue so much against the importance of isolation in speciation as against Wagner's second claim, that even evolutionary change as such was impossible without isolation.

In the absence of all knowledge of genetics and owing to the rather universal acceptance of completely erroneous ideas such as blending inheritance, neither Wagner nor Weismann could develop any decisive arguments concerning sympatric speciation. Throughout their controversy both phrased their questions rather poorly. Wagner had had the right intuition because he saw (no matter how badly he stated it) that ordinarily isolating mechanisms, except in the case of polyploidy, cannot develop and be perfected except in geographic isolation. Because neither Darwin nor Weismann nor Wagner made a distinction between individual variants and geographic varieties, Wagner's claims could be read that no evolutionary change is possible even within a population; and this is indeed how Darwin and Weismann read it. Perhaps Wagner actually believed this.

To be sure, Weismann was familiar with vicariant species of butterflies that were separated by geographic barriers. But he argued that these same barriers had not succeeded in breaking other species of butterflies into vicariant species. He concluded therefore "that isolation must not necessarily lead to the formation of varieties" (1872:23). What Weismann overlooked was that Wagner had claimed only that if incipient species form, it is on either side of a geographic barrier.

In some of Weismann's comments on Wagner it is rather surprising that Weismann argued so vehemently against Wagner when his own findings to a large extent supported Wagner's claims. For instance, Weismann freely admitted that no less than nine species of butterflies have endemic varieties or species on Corsica and Sardinia, yet he rejected this as evidence in favor of Wagner because there are other species on these islands that do not differ from related populations on the mainland. Weismann used the same argument in comparing alpine species with their nearest arctic relatives in the north of Europe. He simply did not realize that these findings are no argument against Wagner's basic thesis. The failure of a species to develop endemism on Sardinia could result from either a more recent immigration of these species on the island or greater genetic or phenotypic stability. After all, sibling species also speciate without developing morphological differences.

Weismann did not limit himself to an attempted refutation of Wagner's arguments but devoted an entire chapter (1872:77–105) to the role of isolation in connection with a change of the environment. This section, as much of his earlier discussion, is badly flawed because Weismann made no distinction between general variation and the acquisition of isolating mechanisms. However, he admitted that in most insular areas the biota is highly unbalanced, that this may set up strong new selection pressures, and that these would lead to the change of most immigrants into new species. There is no place where natural selection has a better field for its activity than isolated areas (1872:81). Weismann did not seem to notice that this was one of Wagner's major arguments. Nevertheless, Weismann did not think that isolation really accelerates species formation because what happens during isolation is simply "the continuous elimination of the less well adapted individuals" (1872:82). There are indications that Weismann's genetic ideas of the 1870s were among the reasons he minimized the importance of isolation. He apparently could not envisage any mechanism for the origin of better-adapted germ plasms in the isolated population (see also 1872:102). In his final summary Weismann stated that he "must energetically deny that the process of selection is enhanced in any essential way by the prevention of gene flow through isolation" (1872:104). However, he provided no evidence whatsoever for this claim, nor did he show why the influence of gene flow into an insufficiently isolated population should be ignored.

Throughout this discussion Weismann stated repeatedly that he believes in the occurrence—perhaps the frequent occurrence—of sympatric speciation. The only evidence he advanced was Hilgendorf's (1867) study of temporal changes in the famous Steinheim planorbid snails.[3] Not clearly differentiating between species and varieties (following Darwin's example), Weismann said that it would be simplest to consider all of the nineteen varieties recognized by Hilgendorf as species because they differ in their morphology and are not connected by intermediates. Not making a distinction between individual variants and geographic races, Weismann was able to use even sexual dimorphism as an argument against Wagner. This phenomenon "proves irrefutably, that a species can split into two forms in one and the same area" without any isolation (1872:17). How little Weismann understood the populational nature of species is further documented by his statement "that the species is nothing absolute, and that the differences among different species are entirely of the same nature as the just mentioned differences among the sexes of one and the same species" (1872:19). And he used color polymorphism among caterpillars, pupae, and imagos of many species of butterflies as additional arguments. On the basis of the Steinheim snails and cases of polymorphism Weismann concluded that there are numerous cases "in which one species in one and the same area of occurrence has changed into another or several species" (1872:105).

Thirty years later Weismann returned to the problem of speciation. He devoted no less than four *Vorträge* to this subject (1904 2:235–295). In the meantime, important contributions to the subject had been published by the two leading lepidopteran biologists of the period, Karl Jordan and Edward B. Poulton.[4] They had not only greatly advanced the biological species concept but had also demonstrated the improbability of sympatric speciation. Weismann did not refer to their writings and still argued against Wagner's by now quite obsolete claim that "not selection but only isolation makes the . . . . splitting of a species into several forms possible" (1904 2:238). Even though Weismann claimed that "on this question I still adhere to the same views, as I did nearly 30 years ago" (1904 2:239), his discussion reveals that he now made major concessions to Wagner. He admitted how frequently and how rapidly speciation may occur on islands or in other isolated areas. More surprisingly, he agreed with Wagner on why isolation is so important. If there is any incipient variation, "such a variational trend can continue uninhibited for a long time, because it can not so easily be suppressed by admixture of strongly divergent germ plasm" (1904 2:241, 244). This had been Wagner's principal argument right along. However, Weismann still insisted on the occurrence of sympatric speciation: with the reasoning, "The polymorphism of the social insects alone supplies the proof that a species can split into several forms in one and the same area simply by natural selection" (1904 2:245).

Weismann's confused treatment of the problem of speciation, and the equally confused treatments by de Vries, Bateson, and other contemporaries, provides an excellent example in support of the thesis that no scientific problem can be solved unambiguously unless a thorough conceptual analysis is undertaken. Weismann failed to understand the difference between "forms" and species (in this he was a thorough typologist), between geographic isolation and reproductive isolation, between sympatric and allopatric speciation, or between mere morphological difference and the acquisition of isolating mechanisms. Nor did he understand that isolation is not a deterministic process nor the difference between genetic processes and their populational substrate. Although most components of this conceptual matrix were clarified individually or jointly by authors like Jordan, Poulton, Stresemann, and Rensch, it would seem that a full understanding of the problem of speciation was not achieved until the evolutionary synthesis (Mayr 1942). Philosophers of science often ask about the nature of scientific progress. There are not many problems for which the gradual acquisition of understanding can be studied more revealingly than that of speciation.

(4) *Evolutionary gradualism vs. a saltational origin of new taxa.* See below for discussion.

(5) *The causation of evolutionary change.* By far the most important of Darwin's evolutionary theories was natural selection. It attempted to explain with the help of material causes what had been explained previously by a supernatural cause, by design. This theory was so novel, so daring, that at first it was adopted by very few biologists. Weismann was one of the few and, as we shall see, eventually he went even further than Darwin in asserting the *Allmacht* of natural selection. Most evolutionists in the post-Darwinian period supported alternate theories. The three most

widely accepted of these opposing theories (Bowler 1983) were the following:

(a) A belief in an intrinsic driving force, or "phyletic force," resulting in evolution by "orthogenesis";

(b) Saltational evolution;

(c) Lamarckian factors (inheritance of acquired characters; see below).

(a) Orthogenesis. In the 1860s and 1870s the idea of some teleological, finalistic, phyletic force was perhaps the most popular of the three options, at least in Germany. In view of the fact that it had been adopted by some of Germany's most admired biologists, such as von Baer, Naegeli, and Kölliker; by the orthogenisists Haacke and Eimer; and by leading philosophers, particularly von Hartmann, Weismann found it necessary to emphasize again and again his resistance to the acceptance of such a metaphysical force, and he attempted to refute it by ever-new arguments. In view of the prestige of his opponents, he chose not to ridicule this school of thought but rather to show that it was inconsistent, in his opinion, with the known facts. "How could such intrinsic Triebkräfte produce such sophisticated patterns as found in leaf imitating butterflies, like Kallima?" he asked. Furthermore, if there were such an evolutionary force, he said, it should be possible to establish it by empirical research. But no one had succeeded in doing so; instead, he said, it was possible to trace all evolutionary changes "to known transforming factors" (1882:161).

(b) Saltationism. The second of the great antiselectionist theories of the post-Darwinian period was the origin of new types and organs by major saltations. This contention was diametrically opposed to evolutionary gradualism, one of Darwin's five evolutionary theories. Yet saltationism was favored by T. H. Huxley, Kölliker, and other Darwin contemporaries. Weismann was unalterably opposed to the possibility of saltational evolutionary change (1882:697). It is interesting that in 1882 and 1886 Weismann already anticipated the claims made a few years later by Bateson (1894) and de Vries (1901) and attempted to refute them. "An abrupt transformation of a species is inconceivable, because it would render the species incapable of existence" (1886:264), he wrote, believing that the existence of numerous coadaptations would make such an instantaneous total restructuring of the organism an impossibility. After all, when studying organisms one is amazed at the precision and ubiquity of adaptations. "It is evident that one can not possibly explain these innumerable

adaptations by the occurrence of a rare, accidentally at one time occurring, variation. The necessary variations with the help of which selection pieces together evolutionary change must always be offered again and again in many individuals" (1892:568). At the height of the popularity of de Vries's mutation theory, Weismann once more demonstrated the utter improbability of saltational evolution (1909:22–24).

Weismann was proud to be able to assert, "more definitely than Darwin has done," that all changes must have occurred "very gradually and by the smallest steps" (1886:264). He reiterated this assertion in many of his writings. For instance, "the long continued accumulation of imperceptibly small variations proves to be the magic means by which the forms of the organic world are so powerfully molded" (1882:87). Or, "The true cause of [the similarity of species and genera] is I believe to be found in the circumstance that all changes take place only by the smallest steps, so that greater differences can only arise in the course of longer periods of time" (1882:468). As empirical evidence for this claim, Weismann cited the Steinheim snails in which the "characters of a new species do not appear at once in their complete expression but increase gradually from generation to generation" (1872:41). It is obvious why Weismann had to insist on *gradual* evolutionary change when one remembers that at first he attributed much of it to use and disuse, and later considered evolutionary change as more or less due to quantitative changes in large numbers of determinants. Even after de Vries had published his mutation theory, Weismann postulated that the saltation of the phenotype was simply the making visible of a long series of preceding small genetic changes (1904 2:119).

The Darwin-Weismann tradition of gradualism and the Bateson-de Vries-Goldschmidt tradition of saltationism continued to battle each other for the first third of the twentieth century, and the argument has flared up again in recent years. Although this is not the place to examine the current controversy (see Essay 26), the importance of Weismann's contribution to strengthening the gradualistic tradition should be emphasized.

Natural Selection

Rejecting all other possible causations for evolutionary change, Weismann firmly adopted selectionism. He showed himself a convinced selectionist in his very first evolutionary publication (1868). At that time he followed Darwin by also believing in a minor contribution of use and disuse (see below); but from 1883 on, he was an uncompromising selectionist.

He supported his selectionism by various sets of arguments. In view of the evident invalidity of all teleological and saltational theories of evolution, there is really no other option than to accept natural selection, said Weismann. Natural selection "is the only conceivable natural explanation for the adaptation of organisms to the conditions of their environment" (1893:42). "When a character can with certainty be ascribed to adaptation, we can explain its origin in no other way than by the action of natural selection (1882:304; see also 1904 2:223). There is not a detail in the structure or physiology of an organism that has not been shaped by natural selection. Without a doubt, Weismann was the most consistent selectionist in the nineteenth century. It was not until far into the twentieth century that other authors committed themselves to natural selection to an equally unreserved degree.

Natural selection was the main topic of the series of papers Weismann published in the 1870s and 1880s dealing with the theory of descent. In his detailed studies of the markings of sphingid caterpillars, he unequivocally adopted the adaptationist program. He asked: "Have the markings of the caterpillars any biological value, or are they in a measure only sports of nature? Can they be considered as partially or entirely the result of natural selection? Or has this agency had no share in their production?" (1882:308). To be able to answer these questions, Weismann developed a rigorous methodology. When one wants to determine whether such markings have biological significance or not, one must examine "whether species with similar markings have any conditions of life in common which would permit of any possible influence as to the significance of the markings" (1882:309). And he tested his assumptions with carefully executed experiments in which caterpillars that had been placed on a variety of backgrounds were exposed to birds or lizards. Furthermore, Weismann always sought to determine whether or not there was a correlation between structural or color characters and behavioral characteristics.

The result of these studies was that he found abundant evidence in favor of selection. "It has been possible to show that each of the three chief elements in the markings of the Sphingidae have a biological significance, and their origin by means of natural selection has thus been made to appear probable" (1882:380). Caterpillars living under more variable circumstances than adults of these species are more variable in their adaptations. The phenotype of each larval stage (instar) is separately selected. In cases of small differences Weismann was able to apply Dohrn's principle of change of function (1882:365).[5]

Wherever Weismann looked, he found evidence for selection—not only in the animal kingdom but also among plants. He examined the form and color of flowers, so beautifully demonstrated by Darwin for orchids and by numerous other students of flower biology beginning with Sprengel; the venation of leaves; and the numerous protective devices of plants against herbivores. Weismann presented a superb description of the aquatic adaptations of marine mammals, demonstrating that all the differences between them and terrestrial mammals are clearly adaptations (1886:261).

Natural selection is active not only in the acquisition of new adaptations, but in the maintenance of existing ones. As soon as the selection pressure is relaxed, as in the case of the eyes of cave animals, individuals with imperfect structures will no longer be eliminated. "In my opinion every organ is kept at the peak of its conformation only by continuous selection. And it slides down from this height ceaselessly even though very slowly, as soon as it no longer is of value for the survival of the species" (1893:51). Such a loss of organs can result from either a relaxation of the maintenance selection or from an actual counter-selection exercised by competition for tissue substrate. Weismann presented a particularly impressive documentation of natural selection in his *Vorträge* (1904 1;36–170).

After presenting all this evidence, Weismann professed without hesitation that he was a panselectionist. "There is no part of the body of an individual or of any of its ancestors, not even the minutest and most insignificant part, which has arisen in any other way than under the influence of the conditions of life" (1886:260). He admitted, however, that "these are indeed only convictions, not real proofs." In the 1880s direct experimental proof for the force of selection was scanty. The analogy with artificial selection was still perhaps the most convincing evidence for selection; for, as Weismann said, the breeders had achieved almost any objective of their selection (1893:60). Otherwise the principle of exclusion is the best support for natural selection, a principle which states that a theory can be adopted for the time being if all competing theories have been refuted.

Weismann's panselectionism apparently developed only gradually. In 1872, in an analysis of the characteristics of a butterfly wing, he said that he begins "with the attempt to separate the purely morphological [neutral] characteristics in pattern and coloration from those that are of importance for the survival of the species" (1872:55). Since butterflies normally rest with closed wings, the underside is almost always cryptically colored. By contrast, the upper side is usually quite bright because there is no particular need of protective coloration (it is invisible in the closed wing), nor is there need for any special adaptation in the pattern (1872:57). Later, after Weismann had become a far more ardent selectionist, he realized (as had Darwin before him) that selection does not necessarily lead to perfection. "No device in nature is absolutely perfect, not even that beautifully constructed eye of man. Everything is only as perfect as necessary, at least as perfect as it must be in order to accomplish what it is supposed to accomplish. It will be the same with the transformation mechanisms of species which will be just as perfect as they must be in order to accomplish the transformation" (1892:571–572). This echoes similar statements made by Darwin (1859:201, 206).

In recent years several authors have proudly announced the discovery of constraints on the action of natural selection. Weismann was aware of the role of developmental constraints more than a hundred years ago, a sign of his deep understanding of the evolutionary process. As early as 1868 he stated "that the organism has the capacity to change only to a very limited degree and that it can vary only in such directions as are compatible with its chemical and physical constitution. It can therefore not produce all conceivable variations but only certain ones even though they are numerous" (1868:27). At that time Weismann evidently considered evolution a compromise between natural selection and the variation that a particular organism is able to produce (1868:28–29). The physical constitution of each species limits the action of natural selection because it "restrains the course of development, however wide the latter may be. [This is] why, under the most favorable external circumstances, a bird can never become transformed into a mammal—or, to express myself generally, why from a given starting point, the development of a particular species can not now attain, even under the most favorable external conditions, any desirable goal" (1882:113–114). These developmental constraints are the reason "why organic evolution has frequently proceeded for longer or shorter periods along certain developmental lines" (1886:258).

The question of what the target of selection is occupied Weismann all his life. At first he followed Darwin, recognizing only a single level of selection, and evidently considered the individual-as-a-whole as the target. He considered the genotype as a holistic system, the individual components of which could not be randomly interchanged or replaced. "What I have above called the physical constitution of a species is based upon these facts, and upon them depend the tout-ensemble of inherited characters, which are adapted to one another and woven together into a harmonious whole" (1886:288). Although seemingly selection deals with the phenotype ("the qualities of the finished organism"), it actually "deals with the Anlagen of these properties which are concealed in the germ cell" (1883:56). It will require a detailed ad hoc analysis to determine to what extent Weismann gave up this holistic view after he had developed his complex theories of the structure of the genotype, and his postulate of three levels of selection.

Sexual Selection

Darwin's theory of sexual selection had a mixed reception among his contemporaries. Alfred Russel Wallace, as is well known, denied it altogether. Weismann at first was quite enthusiastic, and in 1872 he attributed many differences between the sexes, even in insects, to sexual selection (1872:60); indeed he felt the theory should be applied even more extensively. The principle operates in the same manner as natural selection because the beneficiaries of it "gain an advantage in the struggle for existence by their new characteristics" (1872:61). Weismann saw clearly that sexual selection gave an advantage to certain individuals, and he remarked that the driving force of sexual selection is not the external environment but rather the taste of the individuals in their selection of a mate. Characters acquired by sexual selection do not offer any advantage in the daily struggle for existence (1872:62).

Ten years later Weismann had become rather confused about the causation of sexual dimorphism. He said that "Darwin ascribes too much power to sexual selection when he attributes the formation of secondary sexual characters to the sole action of this agency," and he thought that "the sexual dimorphism of butterflies is due in great part to the differences of physical constitution between the sexes" (1882:62). When, a few pages later, he again emphasized the unimportance of sexual selection (1882:101–102), the translator (Meldola) emphatically disagreed with him and listed in a footnote numerous instances that indicate the vast importance of sexual selection. Twenty years later Weismann returned completely to Darwin's position. He devoted an entire chapter of his *Vorträge* (1904 chap. 11) to sexual selection, definitely accepting the principle of female choice and attempting to refute all of Wallace's objections.

The subsequent history of the evaluation of sexual selection has been one of ups and downs. When the mathematical population geneticists declared the individual gene to be the unit of selection and defined fitness as the contribution of such genes to the gene pool of the next generation, there was little room left for sexual selection. This attitude characterized the period from the 1920s to the 1960s. In the past two decades the importance of sexual selection has again been acknowledged; it has been recognized that the individual as a whole is the principal target of selection and that such an individual might have a reproductive advantage owing to characteristics that do not contribute to general fitness. This topic has become one of the major concerns of sociobiology.

Weismann did not discuss in detail the role of behavior in evolution. However, he made it quite clear that in many instances he considered behavior to be the pacemaker of evolution. He illustrated this by different instars of the caterpillars of sphingid moths, which have different habitat preferences or daily cycles and which are protectively colored as required by the respective habitat or time of day. In the case of caterpillars that in the early instars are green and remain on the foliage but in later instars are brown and conceal themselves in the daytime, he wrote, "The species must have first acquired the habit of concealing itself by day underground and among dead herbage, before the original green color could have been changed into brown by natural selection" (1882:297).

Source of Genetic Variation

Weismann understood from the very beginning that variation was an indispensable prerequisite for the operation of natural selection. "Variability produces the material for the activity of natural selection, it produces small deviations in large numbers among which natural selection so to speak searches out those that are most useful" (1868:23). Variation was no problem for him, because he believed in "the origination of transformations by the direct action of external conditions of life" (1882:682). Consequently, "by considering each variation as the reaction of the organism to an external action, as a diversion of the inherited developmental direction, it follows that without a change in the environment no advance in the development of organic forms can take place" (1882:687). That variation was an indispensable component of the process of natural selection remained Weismann's firm conviction even after he had changed his mind completely about the causation of such variation. Even in 1893, he still referred to variation "as one of the main factors of natural selection" (1893:54).

There is probably no other evolutionist whom historians traditionally have considered as extreme a selectionist as Weismann. So it will come as a great surprise when they learn that, at least until 1881, Weismann like Darwin believed in an inheritance of acquired characters.[6] Not only did he believe in the heritable effect of use and disuse, he stated as late as November 1881 (1882:xvii): "Nor can the transforming influence of direct action, as upheld by Lamarck, be called in question, although its extent can not as yet be estimated with any certainty." He illustrated this with a specific example. In the butterfly *Araschnias levana* "the summer warmth, acting regularly on the second and third generations of the year, has, in the course of a lengthy period, stamped these two generations with a new form without the first generation being thereby changed" (1882:87). However, he added two pages later that the more drastic forms of alternation of generations must "have originated . . . indirectly through natural selection or adaptation" (1882:89). Since at that time Weismann did not draw a line between phenotype and genotype, he had no problem with transferring phenotypic changes gradually into the germ plasm. He summarized his conclusions in the statement: "The most general, and in so far chief result of these investigations, appears to lie in the conclusion, which may be thus formulated:—a species is only caused to change though the influence of changing external conditions of life, this change being in a fixed direction which entirely depends on the physical nature of the varying organisms, and is different in different species, or even in the two sexes of the same species" (1882:111–112). A stronger influence of the genetic potential is admitted in this statement than in others made either before or after 1882. On the whole, Weismann minimized innate genetic propensity, always seeming to be afraid of being accused of being a vitalist. One senses in many of his statements that he considered genetic propensity and natural selection as two opposing forces. For instance: "Although natural selection is the factor which has called into existence and perfected the three chief forms [of markings of caterpillars] . . . we can recognize a second factor which must be entirely innate in the organism, and which governs the uniformity of the bodily structure in such a manner that no part can become changed without exerting a certain action on the other parts—an innate law of growth (Darwin's 'correlation')" (1882:388). The importance of Darwin's correlation is also referred to in other Weismann writings.

At any rate, the inheritance of acquired characters was for Weismann the major source of variation in all of his evolutionary writings prior to 1882. He later admitted, "Twenty-five years ago I was still of the opinion that in addition to [other factors] also the inherited effect of use and disuse played a not unimportant role in evolution" (1893:43). Actually, in Weismann's pre-1883 writings one can find numerous suggestions about how the environment might influence the variation and inheritance of organisms. (Churchill 1968; Blacher 1982). What Weismann envisioned was that species and populations went through occasional periods of greatly increased variability, and that if segments of a population were isolated during such a period, then a process he called "amixia" would lead to differences. I have not analyzed in detail his concept of amixia, but some of Weismann's descriptions sound like that what later was called "genetic drift." His hypothesis of amixia permitted him to come to the following conclusion: "I believe that I have further shown that numerous local forms can be conceived to have arisen through the process of preventive crossing while they can not be explained by the action of climatic influences" (1882:110). Most of these early ideas Weismann dropped altogether when he abandoned his belief in an inheritance of acquired characters, but he retained the theory of an origin of varieties through random fixation (1904 2:241–242).

Because this is often misunderstood, let me emphasize that there was no real conflict between a belief in selection and a belief in an inheritance of acquired characters—for either Weismann or Darwin. Both believed from the beginning in the overwhelming importance of natural selection as the mechanism responsible for the production of adaptation, but they desperately needed to find some mechanism that would produce the variation needed for the operation of natural selection. It was the task of the inheritance of acquired characters to supply at least part of this variation. That both authors made use of this factor was only natural, for a belief in an inheritance of acquired characters was virtually universal until the beginning of the 1880s. This was as true for folklore as for science. The few dissenters (Galton, perhaps His) spoke with muted voices and were not heard.

Rejection of Inheritance of Acquired Characters

In view of Weismann's repeated Lamarckian statements in 1881, it is rather surprising how sweeping his repudiation of Lamarckism was in his 1883 lecture, *Über die Vererbung*. His refutation of Lamarckian claims is so broad, and his Darwinian interpretation of the numerous cases that had previously been cited as proofs for an inheritance of acquired characters is so well thought out, that one is almost forced to the conclusion that

Weismann must have been thinking about this problem for many years. Despite its title, Weismann's 1883 paper did not develop a theory of inheritance, but was devoted almost exclusively to the refutation of an inheritance of acquired characters. His strategy was remarkably similar to Darwin's, when he refuted creationism: Weismann took up one case after another that simply could not be explained by "use and disuse" and other Lamarckian mechanisms. How can the numerous special adaptations of the worker and soldier castes of ants be inherited by use, when these castes do not reproduce? How can habits become instincts through use, when a particular instinct is practiced only once in the whole life of the individual, as is so often the case for reproductive instincts among insects? How can the external structure of insects be modified by use and disuse, when the chitinous skeleton is laid down during the pupal stage and never changes afterward? During his long controversy with Herbert Spencer, Weismann (1893) produced numerous additional arguments against the plausibility of an inheritance of acquired characters. To a modern, fully convinced of the impossibility of an inheritance of acquired characters, Weismann's arguments seem most persuasive. But in Weismann's time the belief in the Lamarckian principle was so deeply ingrained that only a minority were converted. Use and disuse seemed far more convincing when considering the loss of extremities by snakes or of eyes by cave animals. It was not until the evolutionary synthesis of the 1940s that unreserved selectionism was more or less universally adopted by biologists; but the conclusive refutation of the principle of the inheritance of acquired characters was not achieved until the 1950s, through the so-called central dogma of molecular biology.

Weismann's strategy was to show not only that an inheritance of acquired characters encounters formidable difficulties, but also that cases cited in its favor could be explained quite well through the theory of natural selection. A structure that is used a great deal in an individual's lifetime is of course also exposed to strong selection forces. If the organ is inferior in a particular individual, its owner will be handicapped in the struggle for existence. Therefore "the improvement of an organ in the course of generations is not the result of a summation of the result of practice of individual lives, but of the summation of favorable genetic factors" (1883:26).

Weismann stated repeatedly that he was inclined to apply the principle of selection far more consistently than Darwin himself. The fact that in ducks and other domestic fowl the wings have somewhat degenerated while the legs have become stronger was explained by Darwin as the result of use and disuse. Weismann quite rightly pointed out that this can be explained even better by assuming that natural selection was the cause of this change in proportions.

It is rather ironical that natural selection is attacked in the current evolutionary literature not so much for its inability to explain certain adaptations or other evolutionary developments as for being a principle so successful that anything could be explained by natural selection–and therefore, to use Karl Popper's language, that it would be impossible ever to refute any evolutionary explanation based on the principle of natural selection. This was not the situation in Weismann's day, when people had not yet become accustomed to thinking in terms of natural selection and when they were far more comfortable explaining evolutionary developments in terms of an inheritance of acquired characters.

Weismann was in his mid-forties when he shifted from Lamarckism to an uncompromising selectionism. It would be most interesting to know what induced Weismann to revise his ideas so drastically, but this is not the place for such an analysis. An important role in his conversion was undoubtedly played by his own observation—and that of various cytologists and embryologists—that the future germ cells in various types of invertebrates are, so to speak, set aside after the first mitotic divisions of the developing embryo and no longer have any physiological connection with the body cells.

This observation led Weismann in 1885 to his theory of the "continuity of the germ plasm," which states that the "germ track" is separate from the body (soma) track from the very beginning, and thus nothing that happens to the soma can be communicated to the germ cells and their nuclei. We now know that Weismann's basic idea—a complete separation of the germ plasm from its expression in the phenotype of the body—was absolutely correct. His intuition to postulate such a separation was faultless. However, among two possible ways for effecting this, he selected the separation of the germ cells from the body cells, while we now know that the crucial separation is that between the DNA program of the nucleus and the proteins in the cytoplasms of each cell (Mayr 1982:700).

Weismann was correct in concluding that the germinal material is something entirely different from the body substance and "that the differentiation of the body cells is not acquired by them directly, but that it was prepared by changes in the molecular structure of the germ cell" (1883:14).

Weismann came close to the concept of a genetic program when he said that we will come to the right conclusions about development "if we consider all processes of differentiation that occur in the course of ontogeny as controlled by the chemical and physical molecular structure of the germ cell" (1883:18).

The Significance of Sex

The refutation of an inheritance of acquired characters seemingly left a serious void in evolutionary theory. Weismann fully understood that he had to find a new mechanism for the production of genetic variability. "We shall [therefore] be compelled to abandon the ideas as to the origin of individual variability which had been hitherto accepted, and shall be obliged to look for a new source of this phenomenon, upon which the processes of selection entirely depend" (1886:252–253). His solution was this: "I believe that such a source is to be looked for in . . . sexual reproduction . . . Two groups of hereditary tendencies are, as it were, combined. I regard this combination as the cause of hereditary individual characters, and I believe that the production of such characters is the true significance of [sexual] reproduction" (1886:272). Indeed, this process of genetic recombination through sexual reproduction was recognized by Weismann as being one of the most important processes in evolution, and to it he devoted an entire long essay (1886).

His conclusion was in total opposition to prevailing ideas. Since blending inheritance was widely accepted at that time, even by such authors as Darwin (Mayr 1982:779–781), who simultaneously believed in particulate inheritance, sexual reproduction was credited with assuring the uniformity of species. That Weismann could put forth such revolutionary claims was made possible by the discovery of van Beneden and other cytologists that maternal and paternal chromosomes do not fuse during fertilization but merely reestablish the diploidy of the zygote. What sexual reproduction thus achieves is not a homogenization of the parental characters, but their recombination. Genetic recombination together with natural selection can thus bring together previously separate and independent characteristics that greatly improve the selection value of their bearers. Concerning the origin of sexuality, Weismann's statements are vague, if not teleological.

To a remarkable extent Weismann was aware also of the drawbacks of sexuality. He describes the distinct advantage by which the temporary abandonment of sexual reproduction allows certain animals as aphids and cladocerans "a much more rapid increase in the number of individuals . . . in a given time" (1886:289). However, this is only a temporary advantage—a conclusion confirmed, according to Weismann, by the fact that "whole groups of purely parthenogenetic species or genera are never met with" (1886:290). Although this statement is no longer literally true, we know of only a single higher taxon of animals, the bdelloid rotifers, in which all species are asexual. All other asexually reproducing groups of animals seem to become extinct sooner or later.

In his later years Weismann emphasized that sexual reproduction in higher organisms produces "an inexhaustible source of ever new combinations of individual variation, such as are indispensable for the selection process" (1892:541). Curiously, this overwhelming importance of genetic recombination was almost totally ignored by most Mendelians, owing to their strongly reductionist position. The exceptions were a few schools, such as the Bussey Institute (Castle, East, and their students). The broad recognition of recombination in the evolutionary process had to await the evolutionary synthesis. Those mathematical population geneticists who emphasized the gene as the target of selection were particularly slow to appreciate fully the role of recombination. It is today quite evident that what is of particular importance in evolution is not so much allelic interactions as interactions among different loci and different chromosomes, which are constantly changed by genetic recombination. Weismann's (1891) championship of amphimixis, as he called it, is one of his most important contributions to evolutionary biology.

The Source of New Genetic Variation

The mixing of the genetic factors of both parents can supply an almost unlimited supply of genetically new individuals in every generation, but this process consists only in the intermingling of already existing variations. The origin of entirely new genetic factors remains unexplained. The origin of true genetic novelty became a crucial problem for anyone rejecting an inheritance of acquired characters, that is, a transfer of new somatic characters to the germ plasm. Weismann fully realized the seriousness of this problem and struggled with it from 1882 until far into the 1890s. He repeatedly stated that organisms normally give rise "only to exact copies of themselves" (1882:679, 682). Furthermore, natural selection is bound eventually to exhaust the supply of genetic variants available through recombination. Thus Weismann was forced to come up with a new solution; and for this, as well shall see, he was ill equipped.

Here we must make a short excursion into Weismann's biological philosophy. Weismann had a thoroughly mechanistic world view, perhaps acquired in the years when he was a chemist. He acknowledged only that the existence of physical forces and "vital phenomena are nothing but the reactions of the organism to the influences of the environment" (1882:688). When he was arguing against the philosopher von Hartmann, he explained that variation is caused "essentially by internal influences, i.e. by the underlying physical nature of the organism . . . to make use of a metaphor, the forces acting within the body are in equilibrium; if one organ becomes changed this causes a disturbance in the forces, and the equilibrium must be restored by changes in other parts, and these again entail other modifications, and so forth" (1882:653). This, Weismann explains, corresponds to Darwin's law of correlation.

As valid as Weismann's arguments were against unknown vital forces, they tended to prevent his adopting certain interpretations that would have facilitated the explanation of genetic variation. To be sure, Weismann realized that variation is the result of the propensity of an organism to vary and of the factors of the environment that elicit this variation. But he was so afraid of giving in to the orthogenisists that he consistently denied intrinsic potentialities. Thus when he saw oblique markings in the caterpillars of different genera of sphingids, he insisted that the tendency for such markings has "not been inherited from a remote period, but [they] have been independently acquired by this [species] or by some recent ancestral species. They have nothing to do genetically with the oblique stripes which occur [in a number of more or less unrelated genera]. They depend simply on analogous adaptation, that is on adaptation to an analogous environment" (1882:376).

Weismann's Concept of Causation

Throughout his life Weismann seems to have espoused the prevailing contemporary idea of causation. It was very much that of classical physics. To distinguish, as does the modern biologist, between proximate causations (controlled by a genetic program) and ultimate causations (caused by evolutionary forces) would have made no sense to Weismann. This is particularly obvious from comments he made on sexual selection. He was of the opinion that "Darwin ascribes too much power to sexual selection when he attributes the formation of secondary sexual characters to the sole action of this agency." Rather, he believed that "the sexual dimorphism of butterflies is due in great part to the differences of physical constitution between the sexes" (1882:62), and he contrasted these color characters with the stridulating organs of male orthopterans, which he felt truly were acquired through sexual selection. His inability to distinguish between proximate and ultimate causations is also evident in his discussion of the metamorphosis of the axolotl (1882:555–663) and of parthenogenesis (1885:83–112).

Weismann's uncertainties about causation were particularly evident in his question, "Are the markings of caterpillars purely morphological characters, produced entirely by internal causes? or, are they simply the response of the organism, to external influences?" (1882:287). Obviously the terms "purely morphological" and "response to external influences" had very different meanings for him than they have for us today. And equally different from the modern interpretation was his concept of the pathway between the germ plasm and the soma. I merely call attention to these difficulties and ambiguities; their full analysis would require a complete study of his entire work and of the gradual progression of his ideas.

The inability to distinguish between proximate and evolutionary causations affected Weismann's explanatory framework relative to a number of other biological problems. For instance, in agreement with other members of Germany's ruling mechanistic school, Weismann argued against all those—including Aristotle—who had claimed that more is needed for development than mere raw material (1882:691). Furthermore, he argued that if one accepts an ontogenetic vital force (that is, a genetic program), one must accept also a teleological program for all evolutionary change. At that time it was not understood that the teleonomy of the genetic program is something fundamentally different from an evolutionary teleology, that is, from the acceptance of a phyletic vital force.

A mechanistic conception of causation completely dominated Weismann's thinking about the source of genetic variation. For him a process such as spontaneous intramolecular mutation, not caused by any visible external force, was unthinkable. This caused him no difficulties, as long as he believed in an inheritance of acquired characters or in direct environmental induction. "A change arising from purely internal causes seems to me above all quite untenable . . . all changes from the least to the greatest appear to me to depend ultimately only on external influences; they are the response of the organism to external inciting causes" (1882:115). "By considering each variation as the reaction of the organism to an external action, as a diversion of the inherited developmental direction, it follows that without a change in the environment no advance in the development of organic forms can take place" (1882:687).

Somehow Weismann thought, certainly prior to 1883, that accepting spontaneous mutation, as we would now call it, would support "internal causations" and that this would open the door to what he called "metaphysical principles." He wanted to explain everything "mechanically," and his concept of mechanical was rather classical, implying the direct action of visible forces such as climate, nutrition, and the like. "Each individual variation must depend upon the power of the organism to react upon external influences, i.e. to respond by change of form and of function, and consequently to modify its original (inherited) developmental direction" (1882:678).

When, prior to 1883, Weismann gave up his belief in an inheritance of acquired characters, he had to find an entirely new explanation for the origin of genetic variation. Being involved at that time in a struggle with Naegeli, Kölliker, Eimer, and others about the existence of internal forces of evolution, Weismann was unable to adopt a process corresponding to what we now call "spontaneous mutation." Would not such a process have to be caused by internal forces? Instead, he postulated a number of other processes in the germ line, none of which turned out to be correct. Ultimately all these explanations were based on the idea of multiple replicas of determinants and biophores in the germ plasm and on quantitative shifts in the relative number of these various elements. I have discovered only a single reference in Weismann's writings that might be interpreted as hinting at spontaneous mutation, but it could equally well be interpreted as documenting induced variation: "Variations in the molecular structure of the germ cells will always occur in every species and they must be enhanced and fixed by selection if the results . . . are useful" (1883:18).

Weismann eventually decided that "the origin of hereditary individual variability can indeed not be found in the higher organisms–the Metazoa and Metaphyta; but it is to be sought for in the lowest—the unicellular organisms" (1886:277–278). In these organisms there is no separation into body cell and germ cell, and if such a single cell acquires a new character it is conveyed to the descendants by simple cell division. Through a rather farfetched train of thought Weismann was "thus driven to the conclusion that the ultimate origin of hereditary individual differences lies in the direct action of external influences upon organisms but only in the lowest unicellular organisms" (1886:279).

It is evident from his subsequent writings that this hypothesis by no means fully satisfied Weismann. Also, the vigorous claims of Herbert Spencer and other neo-Lamarckians that random variation was insufficient to supply the needed material for the exercise of natural selection left their mark on Weismann's thinking. He eventually agreed with the Lamarckians that a simple Darwinian selection of individuals is not sufficient to explain all the phenomena of evolution (1896:59), for instance, the continuing reduction of vestigial organs (1896:24). Finally, Weismann admitted that chance alone could not produce the right variation in the right species at the right time. When perfect adaptation is achieved, "chance is out of the question. The variations which are supplied to the natural selection of individuals must have been produced [within the germ] by the principle of the survival of the fittest" (1896:46). And this led Weismann to propose his hypothesis of germinal selection. This hypothesis cannot be understood within the framework of the Mendelian theory of inheritance, in which there is a one-to-one relationship between a feature of the phenotype and a single genetic determinant. Weismann's germ plasm theory (1892) by contrast not only allowed three levels of genetic factors—biophores, determinants, and ids—but postulated that the same nucleus could contain numerous homologous replicas of the same biophore or determinant.

This is not the place to analyze the rationale of Weismann's theory of inheritance, but one of its aspects must be mentioned because it vitally affected Weismann's evolutionary thinking. As part of the dominating influence of physicalism in nineteenth-century science, quality was considered an unscientific concept. Seemingly qualitative differences occurring in evolution had to be converted to quantitative differences. And this is precisely what Weismann believed his theory of inheritance could do. It permitted him to show that "all variation is in the last analysis quantitative, consisting of an increase of decrease of the living particles or of their constitutents, the molecules" (1904 2:128).

These changes do not occur spontaneously but are always caused by external factors, either by the differential nutrition of various determinants within the germ plasm or by the environment. As a result, the determinants can become larger or smaller, and thus more or less powerful. The detrminants change not only as wholes "but some of the kinds of biophore, of which they are composed replicate under changed conditions more rapidly than others, and the determinant itself is thus qualitatively changed . . . and consequently also the quality of the organs determined by the determinants" (1904 2:129). In this manner, according to Weismann, purely quantitative differences, that is, differences in the frequency and size of biophores and determinants, can become qualitative.

After 1895 Weismann postulated three levels of selection: (1) *Germinalselektion*, a struggle among determinants for the nutritive material available within the germ plasm; (2) struggles between cells, tissues, and organs with one another, as postulated in Roux's theory of the *Kampf der Teile,* and (3) Darwinian selection among individuals. The first two infra-individual levels would provide the flexibility that Weismann thought Darwinian selection lacked (1896:60).

But this was not enough. Weismann was particularly impressed by his observation that since "the needed variation which makes selection possible, is always available, there *must be a deep connection between the usefulness of a variation and its actual occurrence,* or in other words, *the direction of the variation of a part must be determined by its usefulness*" (1896:26; *italic* in original). But how could direction of variation be determined by selection? According to Weismann, the size of a determinant is controlled by its nutrition. Differences between determinants are largely a matter of chance, but when an increased or decreased size favors the survival (or reproductive success) of its bearer, this becomes a positive feedback and leads to a directional trend (1904 2:100). This was the core idea of Weismann's germinal selection. And germinal selection became for Weismann from 1896 on the *deus ex machina* of all puzzling evolutionary phenomena. In his *Vorträge* two entire chapters (1904 2:96–133) were devoted to germinal selection.

There is no need to explain the theory in much further detail. It is full of internal contradictions, and in 1900 it was refuted by the Mendelian theory. Only one other aspect need be mentioned. In addition to variations in the size of determinants caused by accidental variations in nutrition, Weismann recognized systematic influences of the environment "by which all homologous determinants will be affected in a similar manner, insofar they are at all sensitive to the respective change in nutrition" (1904 2:115). Weismann designated this process as "induced germinal selection," even though here no selection at all is involved. Indeed, he even mentioned that natural selection (of whole individuals) cannot influence evolutionary trends caused by induced germinal selection, because all determinants are changed in an identical manner (1904 2:116). Weismann managed even to explain de Vriesian mutation by cryptic, preceding germinal selection (1904 2:119).

Weismann apparently never realized that, notwithstanding the terminology, his was no longer a selection theory. Indeed, it approached Geoffroyism[7] rather dangerously. Even though Weismann continued to deny emphatically any ability of the soma (phenotype) to affect the germ plasm, in his theory of induced germinal selection he admitted a direct effect of the environment on the germ plasm.

The most fundamental component of any theory of inheritance is insistence on the basic constancy of the genetic material. Yet the theory of germinal selection abandoned a constancy of biophores and determinants. It seems that Weismann was forced into this change of mind by the claims of various Lamarckians that exposing the pupae of certain Lepidoptera to heat or cold shocks had not only changed the coloration of the emerging butterfly but also that of some of its untreated descendants (1904 2:230–231). These results are now known to have rested on faulty experimentation, and it is an irony of fate that Weismann accepted induced germinal selection and thus needlessly undermined the consistency of his theory of inheritance. It is worth recalling that Darwin proposed his ill-fated theory of pangenesis also quite needlessly, because it was expressly proposed to explain the effects of use and disuse, which one year after Darwin's death were shown to be illusory (Weismann 1883).

Weismann's theory of germinal selection was almost unanimously rejected by his contemporaries, although he pleaded for it once more as late as 1909 (pp. 36–37), as well as in the third edition of his *Vorträge* (1910). All possible support for it had by that time been swept away by the acceptance of Mendelian inheritance. This included major conceptual shifts, among them the acceptance of spontaneous mutations and of qualitative genetic changes.

Weismann retained an unshaken faith in Darwinian natural selection all his life, in spite of the vicious attacks of Driesch and other experimental zoologists. When the fiftieth anniversary of the publication of the *Origin of Species* was celebrated in 1909, Weismann published a strong manifesto on the power of natural selection. Such a declaration required considerable faith and courage, since natural selection at that time was at the lowest point of its scientific acceptance, owing to the attacks of de Vries, Bateson, Johannsen, and other Mendelians.

The Heritage of Weismann's Ideas

There can be little doubt that, after Darwin, no biologist in the nineteenth century had as great an impact on evolutionary thinking as Weismann. A combination of characteristics enabled Weismann to exercise this role. He had, on the one hand, a powerful analytic ability; he could build a logical argument step by step. On the other hand, like Darwin, he had an extraordinary facility for constructing hypotheses. This was by no means appreciated in the nineteenth century and, like Darwin, Weismann was ridiculed for his speculations. In view of the paucity of available facts on inheritance, some of his speculations would be considered rather daring even by a modern biologist who is constantly encouraged to try his hand at model building. When one reads the attacks by Oscar Hertwig, Wolff, Driesch, and others, one realizes how outraged his contemporaries were by Weismann's speculations.

What Weismann did, he did for good reason. He was aware of the intellectual poverty of the inductionism then dominant in Germany. "The time in which men believed that science could be advanced by the mere collection of facts has long passed away" (1886:295). Surely, it had passed away for Weismann, but most of his contemporaries had not yet perceived the message. Until Darwin, Haeckel, Weismann, and a few other theorizers had begun to develop a conceptual framework of biology, "the investigation of mere details had led to a state of intellectual shortsightedness, interest being shown only for that which was immediately in view. Immense numbers of detailed facts were thus accumulated, but . . . the intellectual bond which should have bound them together was wanting" (1882:xv). Weismann again and again emphasized that his hypotheses were not the ultimate truth but were proposed as heuristic devices. Of one of his theories he said, "Even if it should be later necessary to abandon this theory, nevertheless it seems to me to be a necessary stepping-stone in the development of our understanding, it was absolutely necessary that it had to be proposed, and it must be carefully analyzed, regardless of whether in the future it will be found to be correct or false" (1885:17). This is a clear statement of his hypotheticodeductive scientific philosophy.

His strong selectionist stance seemed to cause Weismann considerable anxieties. He was therefore eager to assert that natural selection does not lead to materialism. Nor does it by necessity require "the denial of a teleological Universal cause" (1882:716). We know nothing about this ultimate universal cause but we must insist, he claimed, that the forces found in nature, organic and inorganic, are all purely mechanical (1882:718).

One would do Weismann an injustice by treating him merely as a theoretician. He made major empirical contributions to histology, embryology, and cytology and was also a dedicated experimenter—in fact, one of the first to carry out selection experiments. Furthermore, he had been a passionate field naturalist from his earliest youth on. He had an excellent knowledge of butterflies and their food plants, not only in Germany but also in various areas of the Mediterranean Basin. In his theoretical controversies Weismann was able again and again to counter the arguments of his opponents by citing evidence based on his own broad experience as a naturalist. Finally, he was a skilled limnologist, who studied not only the diurnal cycle of planktonic crustaceans but also their life cycle throughout the year.

As brilliant as Weismann's thinking was, he was often far ahead of the factual information available in the 1880s and 1890s. Also, in some respects he had not been able to emancipate himself from some of the universal, but later rejected, dogmas of his time. He had a number of blind spots that are ultimately responsible for the invalidity of some of his theories. First and foremost, he had such a gross mechanical concept of causation that he could not conceive of spontaneous molecular changes of the genotype; for him it was always external causes that were responsible for such changes. A second blind spot in his thinking related to the problem of speciation. Expert naturalist that he was, he somehow failed to apply populational thinking to speciation. Also, in line with the conventional thinking of his time, Weismann was a strong determinist, unable to understand the probabilistic nature of allopatric speciation as well as that of adaptation.

These failures, however, were far outweighed by his successes. Let me try to list these contributions.

(1) Defense of natural selection.   There was a period, from about 1890 to 1910, when Darwin's theory was threatened to such an extent by various opposing theories that it was in the danger of going under. Weismann's undeviating support of natural selection at this time was a major contribution to the emergence of a strengthened Darwinism. Weismann helped to develop a methodology for the analysis of natural selection, showing that it would permit the making of predictions that would be confirmed if natural selection did indeed operate. He showed that the reduction or loss of structures, which greatly disturbed some of his contemporaries, could be explained as being caused by a relaxation of selection pressure.

Weismann was one of the first to test selection and environmental influences by experiment. For instance, he exposed caterpillars of different colorations to potential predators on differently colored substrates. In other experiments he tested the effect on the color of butterflies of the temperature at which the pupa was kept.

Perhaps even more effective than his evidence in favor of selection were his arguments against orthogenesis, saltationism, and Lamarckism, the three theories competing with selection (Bowler 1983). He was particularly convincing in his argument in favor of the gradual nature of evolutionary change.

(2) *Refutation of the theory of the inheritance of acquired characters.* Through his sweeping rejection of an inheritance of acquired characters Weismann established a new version of Darwinism. The inheritance of acquired characters never regained full credibility after Weismann's attack in 1883. We have seen that Weismann supported his case by three lines of evidence: (a) there is no cytological mechanism that could effect such a transfer from soma to germ plasm, (b) there are many adaptations that could not have been acquired by such an inheritance (for example, the soldiers of ants and termites), and (c) all reputed cases of inheritance of acquired characters can be explained by selection. Even though Weismann was occasionally wrong in detail, he was right in principle, and the basic Weismannian thought is now articulated in the so-called central dogma of molecular biology.

(3) *Firm establishment of particulate inheritance.* If there were blending inheritance (at the gene level), very different laws of genetics (Galton) would be valid than if inheritance were particulate. Furthermore, if each act of fertilization (zygote formation) consists in the combining (rather than fusion) of the paternal and maternal genomes, then this must be compensated by a reduction division, Weismann hypothesized (Churchill 1968; 1979; Farley 1982). These postulates laid the foundation for Mendelian genetics (as clearly stated by Correns), and Mendelian genetics in its turn validated Weismann's theories.

(4) *Recognition of the importance of sexual reproduction as a source of genetic variation.* The importance of genetic recombination depends on particulate inheritance. It was Weismann who first realized the vast importance of sexual reproduction as a mechanism for the production of almost unlimited genetic recombination (Churchill 1979). Although this factor was rather neglected in the mutationist heyday of Mendelism, it was revived from the 1930s on. Weismann, like the modern evolutionists, was a true follower of Darwin, for whom also the individual was the target of selection. Natural selection would be helpless if there were not an inexhaustible supply of genetic variation in the form of uniquely different individuals.

(5) *Constraints on natural selection.* As discussed above, Weismann strongly emphasized that there are severe developmental and other constraints on the power of selection.

(6) *Mosaic evolution.*[8] Weismann emphasized again and again that not only different components of the phenotype, but also different stages in the life cycle, vary in their rate of evolution. For instance, in butterflies the caterpillars often evolve faster and along entirely different lines than the imagos (1882:432). As a result, a classification based on larval char-

acters is not at all the same as one based on imagos. Weismann lists instances of mosaic evolution for many groups of insects (1882:481–501). In most cases it results from special adaptations of the larval stages. Indeed, Weismann offers a long list of evolutionary "incongruences" discovered when he compared either larval and adult stages or the corresponding stages of more closely or more distantly related higher taxa (1882:502–519).

(7) Cohesion of the genotype. Weismann stated repeatedly that, mosaic evolution notwithstanding, there are limits to the independence of different components of the genotype. Organisms must evolve more or less harmoniously, and a selection pressure on one organ very often results in a selection pressure on some other structure. Or a change in behavior results in the necessity for modification of a structure. In other words, it is the genotype as a whole that responds to the forces of natural selection, a consideration often ignored during the reductionist period of mathematical genetics.

More broadly, in an age of inductionism it was very important that there was someone who had the courage to speculate. Weismann pointed out significant problems and unanswered questions, even in cases where he himself was unable to find the right solution.

There was probably no one in the latter part of the nineteenth century who comprehended the basic thesis of Darwinism better than Weismann. He was the only one who understood the overwhelming role of natural selection. And he realized that the source of genetic variation was the great unknown in the process of selection and that a detailed theory of inheritance was the need of the hour. Even though he himself failed to meet that need, the intellectual preparation he gave to the area enabled the Mendelian theory to prosper. Even more than in his own lifetime, Weismann is today considered one of the very few truly outstanding evolutionary biologists.

NOTES

This essay is an expanded version of a lecture presented May 29, 1984, at a symposium held in Freiburg, Germany, to celebrate the 150th anniversary of the birth of August Weismann (1834–1914). The essay was first published in the *Journal of the History of Biology* 18 (1985):295–329, under the title "Weismann and evolution" (© 1985 by D. Reidel Publishing Company), and (in German) in *Freibourger Univerzitätsblätter*, heft 87–88, pp. 61–82.

1. See, for instance, J. W. Spengel, *Die Darwinsche Theorie* (Berlin: Wiegant und Hempel, 1872), and G. Seidlitz, *Literatur zur Deszendenztheorie,* an appendix to *Die Darwinsche Theorie,* 2nd ed. (Leipzig: Breitkopf und Härtel, 1875).

2. One of the conclusions derived from the theory of recapitulation was that the more similar two taxa are in their ontogenetic stages, the more closely related they are. Ontogeny thus provides clues to recency of common descent.

3. The taxonomic status of the varieties or species of Steinheim planorbid snails is as uncertain today as it was a hundred years ago. The two most recent analyses are by Mensink (1984) and Reif (1985).

4. For the classic papers on this subject see Jordan (1905) and Poulton (1908); see also Mayr (1955) and Mayr (1982: 411–414, 562–565).

5. A change of function can satisfactorily explain many seemingly saltational origins of new structures. For references to the relevant literature see Mayr (1982:610–611).

6. Weismann's conversion must have occurred between 1881 and 1883. He still supported an inheritance of acquired characters in at least three statements in the preface (dated November 1881) to the English edition of his *Studien zur Descendenztheorie* (1876), published in 1882. In this edition Weismann added a number of footnotes on various subjects, but made no disclaimers to some of the strongly Lamarckian statements of the original German text.

7. The proponents of Geoffroyism also believed in a direct induction of genetic changes by the environment but assumed that such changes were adaptive (Mayr 1982: 362, 687). For Weismann they were apparently random with respect to needs and merely constituted raw material for the operation of natural selection. However, a careful analysis of this point in Weismann's writing is still needed.

8. For a history of the concept of mosaic evolution see Mayr (1982:579, 613).

REFERENCES

Bateson, William. 1894. *Materials for the Study of Variation.* London: Macmillan.

Blacher, L. I. 1982. *The Problem of the Inheritance of Acquired Characters: A History of a priori and Empirical Methods Used to Find a Solution.* New Delhi: Amerind Publishing Company. Eng. trans. ed. F. B. Churchill. Russian original, Moscow: Nauka, 1971.

Bowler, Peter, J. 1983. *The Eclipse of Darwinism: Anti-Darwinian Evolution Theories in the Decades Around 1900.* Baltimore: Johns Hopkins University Press.

Churchill, F. B. 1968. "August Weismann and a break from tradition." *J. Hist. Biol.* 1:91–112.

———— 1979. "Sex and the single organism: biological theories of sexuality in mid-nineteenth century." *Stud. Hist. Biol.* 3:139–178.

Darwin, Charles. 1859. *On the Origin of Species by Means of Natural Selection; or, The Preservation of Favoured Races in the Struggle for Life.* London: Murray.

de Vries, H. 1901. *Die Mutationstheorie. Versuche und Beobachtungen über die Entstehung der Arten im Pflanzenreich.* Vol. 1, *Die Entstehung der Arten durch Mutation.* Leipzig: Veit. Eng. trans. J. B. Farmer and A. D. Darbishire. Chicago: Open Court Publishing Company, 1909–10.

Farley, John. 1982. *Gametes and Spores: Ideas about Sexual Reproduction, 1750–1914.* Baltimore: Johns Hopkins University Press.

Gillespie, N. C. 1979. *Charles Darwin and the Problem of Creation.* Chicago: University of Chicago Press.

Gould, Stephen, J. 1977. *Ontogeny and Phylogeny.* Cambridge, Mass.: Harvard University Press.

Hilgendorf, F. 1867. "Über *Planorbis multiformis* im Steinheimer Süsswasserkalk." *Monatsber. Preuss. Akad. Wiss. Berlin 1866,* 474–504.

Jordan, Karl. 1905. "Der Gegensatz zwischen geographischer und nichtgeographischer Variation." *Z. wiss. Zool.* 83:151–210.

Kottler, Malcolm, J. 1978. "Charles Darwin's biological species concept and theory of geographic speciation: the transmutation notebooks." *Ann. Sci.* 35:275–297.

Mayr, Ernst. 1942. *Systematics and the Origin of Species.* New York: Columbia University Press.

——— 1955. "Karl Jordan's contribution to current concepts in systematics and evolution." *Trans. Roy. Entomol. Soc. London* 107:45–66.

——— 1982. *The Growth of Biological Thought.* Cambridge, Mass.: Harvard University Press.

Mensink, H. 1984. "Die Entwicklung der Gastropoden im liozänen See des Steinheimer Beckens." *Palaeontographica,* Abt. A, 183:1–63.

Poulton, E. B. 1908. *Essays on Evolution.* Oxford: Clarendon Press.

Reif, W.- E. 1985. "Endemic evolution of Gyraulus kleini in the Steinheim Basin." In Bayer and A. Seilacher, eds., *Evolution in Marginal Basins.* Berlin: Springer.

Sulloway, Frank, J. 1979. "Geographic isolation in Darwin's thinking: the vicissitudes of a crucial idea." *Stud. Hist. Biol.* 3:23–65.

Wagner, M. 1868. *Die Darwin'sche Theorie und das Migrationsgesetz der Organismen.* Leipzig: Duncker und Humblot. *Eng. trans., The Darwinian Theory and the Law of Migration.* London: Stanford, 1873.

——— 1889. *Die Entstehung der Arten durch räumliche Sonderung.* Basel: Benno Schwalbe.

Weismann, August. 1868. *Über die Berechtigung der Darwinschen Theorie.* Leipzig: Engelmann.

——— 1872. *Über den Einfluss der Isolirung auf die Artbildung.* Leipzig: Engelmann.

——— 1875, 1876. *Studien zur Descendenztheorie* I, II. Leipzig: Engelmann.

——— 1882. *Studies in the Theory of Descent.* R. Mendola, trans. and ed. London: Sampson, Low, et al. (Eng. trans. of Weismann 1875, 1876.)

———— 1883. *Über die Vererbung*. Jena: G. Fischer.

———— 1885. *Die Continuität des Keimplasmas als Grundlage einer Theorie der Vererbung*. Jena: G. Fischer.

———— 1886. *Die Bedeutung der sexuellen Fortpflanzung für die Selektionstheorie*. Jena: G. Fischer.

———— 1891. *Amphimixis; oder, Die Vermischung der Individuen*. Jena: G. Fischer.

———— 1892. *Das Keimplasma: Eine Theorie der Vererbung*. Jena: G. Fischer. Eng. ed., 1893.

———— 1893. *Die Allmacht der Naturzüchtung. Eine Erwiderung an Herbert Spencer*. Jena: G. Fischer.

———— 1896. *Über Germinal-Selection. Eine Quelle bestimmt gerichteter Variation*. Jena: G. Fischer.

———— 1904. *Vorträge über Deszendenztheorie*. 2 vols. 2nd ed., Jena: G. Fischer. First published in 1902; third edition, 1910.

———— 1909. *Die Selektionstheorie. Eine Untersuchung*. Jena: G. Fischer ("The Selection Theory," in A. C. Seward, ed., *Darwin and Modern Science,* pp. 18–65. Cambridge: Cambridge University Press, 1909.)

White, M. J. D. 1978. *Modes of Speciation*. San Francisco: Freeman.

ON THE EVOLUTIONARY
SYNTHESIS AND AFTER

ADVANCE in science is rarely steady and regular. Nor can it necessarily be described in Thomas Kuhn's terms as a series of revolutions separated by long periods of steadily advancing normal science. Rather, when we study particular scientific disciplines we observe great irregularities: theories become fashionable, others fall into eclipse; some fields enjoy considerable consensus among their active workers, other fields are split into several camps of specialists furiously feuding with each other. This latter description applies well to evolutionary biology between 1859 and about 1935. In 1930 there seemed to be no hope of any consensus; yet a consensus was achieved to a large extent within a dozen years (1936–1947). The evolutionary theory that emerged has been referred to as the synthetic theory, and the process by which it was reached as "the evolutionary synthesis" (Huxley 1942).

What occurred during those years was not a revolution; rather it was a unification of a previously badly split field. I have little doubt that similar unification events must have happened in other branches of science, but at this moment I cannot recall one. The evolutionary synthesis is important because it has taught us how such a unification may take place: not so much by any revolutionary new concepts but rather by a process of housecleaning, by the final rejection of various erroneous theories and beliefs that had been responsible for the previous dissension. Among the constructive achievements of the synthesis was the finding of a common language among the participating fields and a clarification of many aspects of evolution and its underlying concepts.[1]

What the Synthesis Achieved

The period of the synthesis was not one of great innovations but rather of mutual education. Naturalists who had not known it before learned from the geneticists that inheritance is always "hard," never soft. There can be no heritable influence of the environment, no inheritance of acquired characters. Weismann's thesis was finally adopted universally more than 50 years after it had first been proposed. Another finding of genetics, its Mendelian (particulate) character, was also finally universally adopted. Many naturalists up to that time had divided characters into Mendelian ones, which they considered evolutionarily unimportant, and gradual or blending ones, which, following Darwin, they considered to be the true material of evolution.

Acceptance of these two findings of genetics helped in the rejection of the three major evolutionary theories that had competed with natural selection since the publication of the *Origin* (Bowler 1983). These theories were (1) neo-Lamarckism (the inheritance of acquired characters) and other forms of soft inheritance; (2) autogenetic theories based on the belief in a built-in drive toward evolutionary progress (orthogenesis, nomogenesis, aristogenesis, omega principle); (3) and saltational theories of evolution, which posited the sudden emergence of drastically new life forms (de Vriesian mutations). Perhaps no author contributed more to the refutation of these three theories than G. G. Simpson, whose *Tempo and Mode* (1944) consists in large part of evidence disproving them.

But the synthesis was not merely the general acceptance of the principles of theoretical population genetics. To understand the achievements of the synthesis, one must appreciate how typological and saltationist the evolutionary views of the original Mendelians were (Provine 1971). It is a Whiggish misrepresentation of history to equate Mendelism with genetics. To be sure, the two agreed in their rejection of soft and blending inheritance. On the other hand, the evolutionary views of the Mendelians Bateson, de Vries, and Johannsen were rejected during the synthesis (Mayr and Provine 1980). (Indeed, neo-Lamarckism seemed to explain evolution better than did the evolutionary theories of the Mendelians.) The claim, frequently made, that the evolutionary synthesis was nothing but the application of Mendelian inheritance to evolutionary biology overlooks how much the geneticists had to learn from the naturalists about the importance of population thinking, of the geographical dimension, and of the individual as the unit of selection (see below).

MUTATIONS

A major achievement of the synthesis was to develop a unified view on the nature of mutations. Darwin, accepting universal opinion on this subject, thought there were two kinds of variation: drastic ones, often referred to as sports, and small ones, represented by gradual or quantitative variation. For Darwin, particularly after the Fleeming Jenkin review, it was gradual variation that was important in evolution. By contrast, the Mendelians insisted that new species originated through drastic mutations. Genetic research during the first third of the twentieth century showed clearly that drastic and minutely differing variants were only extremes of a continuously varying spectrum, and that the same genetic mechanism was involved in mutations of all degrees of difference.

This finding had a number of important consequences. It permitted a reconciliation between the Mendelians and those who studied quantitative inheritance (Wright 1967). It also permitted the building of a bridge between micro- and macroevolution. Most important, it refuted the credo of essentialism. There is no uniform species essence, but rather each individual has a highly heterogeneous genotype (which varies from individual to individual). The belief in two kinds of variation was still widespread until the 1930s (Mayr and Provine 1980; Turner 1984), but the new way of interpreting genetic variation became completely victorious during the synthesis.

The realization that gradual variation could be explained in terms of Mendelian (particulate) inheritance also led to the end of any belief in so-called blending inheritance. This was helped by the clear recognition of a difference between genotype and phenotype, as was shown by Nilsson-Ehle, East, and others; a complete "blending" of characters of the phenotype was possible in spite of the discrete particulateness of the underlying genetic factors (Mayr 1982a: 790–792).

THE TRIUMPH OF NATURAL SELECTION

The synthesis was a reaffirmation of the Darwinian formulation that all adaptive evolutionary change is due to the directing force of natural selection on abundantly available variation. At the present time we are so completely used to the Darwinian formulation that we are apt to forget how different it is from the evolutionary explanations of Darwin's opponents.

The naturalists, in the tradition of Darwin, had been the staunchest defenders of natural selection from the very beginning, as documented in the writings of Wallace, Bates, F. Müller, Hooker, Poulton, K. Jordan, and others. Like Darwin, however, almost all of them tended to believe simultaneously in a certain amount of soft inheritance. Now that this form of inheritance was decisively refuted, the naturalists became the strongest supporters of natural selection.

There is, as yet, no good history of the acceptance of natural selection by the geneticists. Chetverikov, Timofeeff-Ressovsky, and Dobzhansky got it from a strong Russian tradition (Adams 1980). In England there was a strong selectionist tradition at Oxford (Lankester, Poulton, and so on; see Mayr and Provine 1980). But the classical Mendelians had no use for selection, and Morgan's thinking was very much in that tradition. A second tradition in the United States, however, centered in Harvard's Bussey Institution (Castle, East, Wright), apparently adopted selection without reservation. The ultimate source of the thinking of that group has not yet been investigated.

Gould (1983b) has developed the thesis that after the 1930s there was a "hardening of the synthesis," meaning a dogmatic insistence on the exclusive role of natural selection in bringing about evolutionary change. His primary evidence is that Wright in 1931 and 1932 had greatly stressed the nonadaptive nature of the characters of races and species (Provine 1986:289–291) and that this had apparently been understood by Dobzhansky (1937) and Simpson (1944) as the proposal that nonadaptive phases in evolution might alternate with adaptive ones. The existence of nonadaptive phases was stressed particularly by Simpson. After Wright in his later papers had considerably reduced the evolutionary importance of genetic drift, Dobzhansky and Simpson did likewise in their later publications. However, this does not mean a hardening of the synthesis, because neither Rensch nor I, nor any other evolutionist to my knowledge, had attributed to genetic drift much evolutionary significance, even though we adopted the concept.

To be sure, it took some time before the former neo-Lamarckians applied natural selection consistently. And in one respect I myself became a stronger selectionist after 1942: following a reexamination of cases that I had classified in 1942 as evidence for neutral polymorphism, I realized within a year or two that neutrality of the alleles could not explain cases of polymorphism that had remained stable for 50, 80, or more years, or that had stayed the same over large geographic distances. If such morphs were truly neutral, accidents of sampling should produce large fluctuations, while their absence would seem to indicate the presence of balancing selective forces (Mayr 1963).

It is understandable that in the early stages of the synthesis the universal presence of natural selection should have been emphasized strongly, since a considerable number of Lamarckians still existed among the older evolutionists. However, as soon as this stage had been overcome, one could observe a trend that was exactly the opposite of the one claimed by Gould.

More and more authors pointed out the existence of stochastic processes and all sorts of constraints that forever prevent the achievement of "perfection." Even though natural selection is indeed an optimization process, the existence of numerous opposing influences makes optimality quite unachievable (see Essay 14).

CONTRIBUTIONS BY THE NATURALISTS

Much of the advance described in the preceding section was due to the work of geneticists. It is therefore sometimes asserted that the synthesis was merely the application of Mendelian inheritance to evolutionary biology. This formulation ignores two aspects. One is that the geneticists, particularly the experimental and the mathematical ones, had to acquire just as much from the naturalists as the naturalists had to acquire from the geneticists. And second, the conceptual framework of evolutionary biology was greatly enriched by concepts and facts from natural history that were conspicuously absent in the writings of the geneticists (see Mayr and Provine 1980:32–38). For example, evolution, as defined by the geneticists, is "a change of gene frequencies in populations." It was visualized essentially as a strictly temporal phenomenon. This is reflected in Muller's statement: "Speciation represents no absolute stage in evolution, but is gradually arrived at, and intergrades imperceptively into racial differentiation beneath it and generic differentiation above" (Muller 1940:258). Throughout their writings the geneticists concentrated almost completely on the temporal component in evolution. The paleontologists, by necessity thinking in terms of vertical sequences, likewise confined themselves to the study of vertical evolution until they merged their thinking with the horizontal tradition of the naturalists (Eldredge and Gould 1972). One of the most important contributions of the naturalists, the heirs of Darwin, was to bring geographical thinking into the synthesis. The problem of the multiplication of species, the existence of polytypic species, the biological species concept, the role of species and speciation in macroevolution, and many other evolutionary problems can be dealt with only by invoking geographical evolution (Essay 23).

The incorporation of the geographical dimension was of particular importance for the explanation of macroevolution. Paleontologists had long been aware of a seeming contradiction between Darwin's postulate of gradualism, confirmed by the work of population genetics, and the actual findings of paleontology. Following phyletic lines through time seemed to reveal only minimal gradual changes but no clear evidence for any change of a species into a different genus or for the gradual origin of an evolutionary novelty. Anything truly novel always seemed to appear quite abruptly in the fossil record. This is not surprising, since new evolutionary departures seem to take place almost invariably in localized isolated populations that are not apt to leave a fossil record (Mayr 1954; 1963; Essay 25). Therefore, a purely vertical approach is unable to resolve the seeming contradiction.

An equally important contribution made by the naturalists was the introduction of population thinking into genetics. Mendelism was strongly typological—the mutation versus the wild type. And even later, in the thinking of many geneticists, not only Morgan and Goldschmidt but even Muller in his search for the perfect genotype, and R. A. Fisher, one can detect a strong typological component. Population thinking was brought into genetics by Chetverikov and his students (including Timofeeff-Ressovsky), by Dobzhansky, and by Baur.

With few exceptions (for example, polyploidy), every evolutionary phenomenon is simultaneously a genetic phenomenon and a populational one. It is curious how often this is overlooked almost up to the present, no matter how emphatically it has been pointed out by Dobzhansky and myself. This oversight was the reason for M. J. D. White's opposition to allopatric speciation (1978). It is meaningless to ask, as he did, whether speciation is chromosomal or geographic. In almost all cases where speciation is accompanied by chromosomal restructuring, this takes place in populations. Since such chromosomal restructuring can occur most easily and most rapidly in isolated founder populations, chromosomal speciation is simultaneously par excellence geographic speciation.

Finally, the naturalists, or at least some of them, attempted to replace the strictly reductionist formulation of most geneticists by a more holistic approach. Evolution, they said, is not merely a change in the frequency of alleles in populations, as the reductionists asserted, but is at the same time a process relating to organs, behaviors, and the interactions of individuals and populations. In this holistic attitude the naturalists agreed with the developmental biologists. (The importance of this emphasis on whole individuals, as against single genes, is discussed in Essay 6.)[2]

Post-Synthesis Clarification

The unification of evolutionary biology achieved by the synthesis painted its picture in bold strokes: Gradual evolution is due to the ordering of genetic variation by natural selection, and all evolutionary phenomena can be explained in terms of the known genetic mechanisms. This was an extreme simplification, considering that processes in organismic biology are usually highly complex, often involving several hierarchical levels and pluralistic solutions. The task of evolutionary biology after the synthesis of the 1940s was to convert the coarse-grained theory of evolution into a fine-grained, more realistic one. No longer constrained to a defense of Darwinism, the followers of the evolutionary synthesis, in the course of the more detailed analysis, began to tackle differences that still existed, not only between the reductionist tendencies of the geneticists and the organismic viewpoint of systematists and paleontologists but also concerning other aspects of evolutionary theory.

To take up first a more general aspect, in the post-synthesis period the deterministic tendencies brought into evolutionary biology either by the autogenetic camp (with their teleological thinking) or by authors with a physicalist background were gradually overcome. More attention was paid to stochastic processes and to pluralistic solutions (Mayr 1963). Hardly a single evolutionary phenomenon or outcome of an evolutionary challenge exists for which there are not different strategies found in different kinds of organisms (see below). Those who insist on accepting only a single answer for any evolutionary challenge are sorely tried by such pluralism.

A pluralism of modes of speciation had been accepted as far back as Darwin, but through the synthesis it became evident that (except for polyploidy) geographic speciation is by far the most frequent mode. In the post-synthesis period two rather different modes of geographic or allopatric speciation were described, the traditional one of the interruption of the continuous species range by a new barrier, and a newly proposed process, peripatric speciation, the establishment of a geographically isolated founder population (Essay 21).

In the decades following the synthesis it was discovered that the genotype is a far more complex system than implied by the early "beads on a string" metaphor. The more the individual as a whole came to be recognized as the principal target of natural selection, the better it was understood that recombination rather than mutation directly supplies the material for natural selection. This of course also meant an increasing attention to any structures or processes that affect recombination (particularly the nature of the karyotype).

A whole new dimension in the understanding of genetic variation was opened up, after 1953, by molecular biology. This included the discovery that, with minor exceptions, there is only a single genetic code for all organisms, from the simplest prokaryotes up. The discovery of the one-way transfer of information from the nucleic acids to the proteins provided the explanation for why an inheritance of acquired characters is impossible. Molecular genetics also demonstrated the neutrality or near-neutrality of much change in the molecular structure of genes (see below and Essay 6).

It might be useful to tabulate some of the post-synthesis revisions of the Darwinian theory, most of which are by nature a more precise formulation rather than a substantive change. They include the recognition that:

(1) The individual organism is the principal target of selection.

(2) Genetic variation is largely a chance phenomenon, indeed stochastic processes play a large role in evolution.

(3) The genotypic variation that is exposed to selection is primarily a product of recombination, and only ultimately of mutation.

(3) All speciation is simultaneously a genetic and a populational phenomenon.

(4) "Gradual" evolution means primarily populational evolution, but it may encompass major phenotypic discontinuities reflected in populational polymorphisms.

(5) Not every change of the phenotype is the necessary consequence of ad hoc natural selection.

(6) Evolution means changes in adaptation and diversity, not merely a change in gene frequencies.

(7) Selection is probabilistic, not deterministic.

(8) Every stage in the life cycle of an individual is exposed to natural selection, including all larval or embryonic stages.

(9) Therefore, development is of interest not only for the student of proximate causations but also for the evolutionist.

(10) Many genetic changes, primarily at the molecular level, are probably neutral or near neutral.

(11) Selection can operate in kin groups through inclusive fitness.

(12) In a consideration of group selection one must make a distinction between different kinds of groups (Essay 7).

(13) The history of life has been drastically affected by major extinctions that may have fallen largely at random.

Thus we see that the period from the evolutionary synthesis to the 1970s witnessed a steady improvement in our understanding of the nature of the genetic material, the constraints on natural selection, the processes of macroevolution, and in fact almost all aspects of evolutionary change. These developments were, as Thomas Kuhn would say, typical for normal science. In spite of differences in emphasis, the leading volumes on evolutionary biology published during this period presented a remarkable consensus regarding the mechanisms and the course of evolution (Dobzhansky 1951; 1970; Dobzhansky et al. 1977; Futuyma 1979; Grant 1977; Mayr 1963; 1970; Maynard Smith 1972; Stebbins 1966).

However, while agreeing in their firm adherence to Darwinism, these authors differed significantly in detail. The gap between the two camps that had accomplished the synthesis—the experimental geneticists and the naturalists—was only partially bridged. Both camps tended to hold on to certain ideas that were incompatible with those of the other camp. The geneticists tended to be atomistic reductionists, focusing on the gene; the paleontologists tended to disregard the consequences of horizontal evolution; the naturalists tended to concentrate on the individual as the target of selection and on the population and species as the units of evolution, and to think almost entirely in terms of phenotypes. These differences were eventually seized upon by some critics to claim that the synthesis was incomplete (which is true up to a point) or unsuccessful (which is not true). Other critics have maintained that the conclusions of the synthesis are now so utterly obsolete that they must be replaced by a totally new synthetic theory of evolution. These objections to the evolutionary synthesis are the subject of the next section.

Attacks on the Evolutionary Synthesis

Just as in the decade after the rediscovery of Mendel's rules, since about 1970 the claim has been made increasingly often that "Darwinism is dead." I shall not deal at all with the attacks by creationists, based on ideological commitments, since their ill-founded arguments have been decisively refuted by Futuyma (1983), Kitcher (1982), Montagu (1983), Newell (1982), Ruse (1982), Young (1985), and several other authors. However, claims that Darwinism is obsolete have also been made in numerous articles and books by several other nonbiologists, for instance Bethel, Macbeth, Hoyle, and Saunders. Their arguments are based on such ignorance of evolutionary biology that it is not worthwhile to provide references to their writings. More disturbing are the similar, although somewhat more muted, claims that have been made by some knowledgeable biologists—even evolutionary biologists. These include Gould (1977; 1980a), Eldredge (1983), White (1981), and Gutmann and Bonik (1981).

Different critics have singled out different aspects of the synthesis as particularly vulnerable. It has been claimed, for example, that:

(1) The findings of molecular biology are incompatible with Darwinism (see below).

(2) The new research on speciation shows that other modes of speciation are more widespread and more important than allopatric speciation, which the neo-Darwinians claim is the prevailing mode (see Essay 21).

(3) Newly proposed evolutionary theories, like punctuationism, are incompatible with the synthetic theory (see Essay 26).

(4) The synthetic theory, owing to its reductionist viewpoint, is unable to explain the role of development in evolution (see below).

(5) Even if one rejects the reductionist claim of the gene as the target of selection, Darwinism, by considering the individual the target of selection, is unable to explain phenomena at hierarchical levels above the individual, that is, it is unable to explain macroevolution (see Essay 23).

(6) By adopting the "adaptationist program," and by neglecting stochastic processes and constraints on selection, particularly those posed by development, the evolutionary synthesis paints a misleading picture of evolutionary change (see Part Three).

These highly diverse claims range from the extreme view that Darwinism as a whole has been refuted to milder versions such as that the synthesis is too narrowly adaptationist or that the concept of speciation has to be thoroughly revised. One by one these various criticisms have been refuted by Ayala (1981; 1983), Stebbins and Ayala (1981), Grant (1983), Maynard Smith (1982; 1983), Mayr (1984), A. Huxley (1982), Levinton and Simon (1981), and others.

The aspect of almost all of these objections that I wish to address here is an evident misunderstanding of the theory that emerged from the evolutionary synthesis. Often this misunderstanding arises from the assumption that the most extreme reductionist version of the synthesis, as represented by some of the mathematical population geneticists, is the basic dogma of the synthesis. When critics propose that the conclusions of the synthesis be replaced by a more modern view of evolution, one finds practically without exception that these supposedly novel views have been maintained all along by Rensch, myself, and several other representatives of the synthesis. Almost all the critics of the synthesis, I am sorry to say, quite conspicuously misrepresent the views of its leading spokesmen.

To take one example, the critics repeat forever the claim that "Darwinian evolution is due to the natural selection of random mutations." This completely ignores the fact that from Darwin on to the 1970s the individual as a whole was considered the target of selection for the organismic biologists, and therefore recombination and the structure of the genotype as a whole were viewed as being far more important for evolution than mutational events at individual loci. Furthermore, the critics completely misinterpret the word random (see Mayr 1963:176 and Essay 6).

Opponents of the synthesis consistently confound three schools of Darwinism: (1) neo-Darwinism, a term coined by Romanes in 1896 to designate Weismann's evolutionism or, as Romanes defined it, "Darwinism without an inheritance of acquired characters"; (2) early population genetics, a strongly reductionist school which defined evolution as the modification of gene frequencies by natural selection; and (3) the holistic branch of the synthesis which continued the traditions of the naturalists while accepting the findings of genetics.

The thinking of the reductionists was strongly influenced by R. A. Fisher, and this school has therefore sometimes been designated as Fisherian Darwinism. It is quite clearly the primary target of the opponents of the synthesis, but these critics very much confuse matters when they designate the reductionist school as neo-Darwinism or imply that it includes people like Huxley, Dobzhansky, Wright, Rensch, or myself, all of whom distinctly rejected the reductionist conclusions of the Fisherian school.

Darwinism is not a simple theory that is either true or false but is rather a highly complex research program that is being continuously modified and improved. This was true before the synthesis, and it continues to be true after the synthesis. Table 28.1 lists many of the significant stages in the modification of Darwinism that one might recognize. Yet recognizing such seemingly discontinuous periods is in many respects an artificial enterprise. To take but a few examples, the prevalence of particulate inheritance was obvious after 1886 but was not adopted until after 1900; the emphasis on the role of diversity in evolution was stressed by naturalists from Darwin on, but was almost entirely ignored by the Fisherians; the naturalists, for their part, rejected the beanbag genetics of the reductionists and during the post-synthesis period continued their

Table 28.1.   Significant stages in the modification of Darwinism.

Dates	Stage	Modification
1883	Weismann's influence	End of soft inheritance
1886		Diploidy and genetic recombination recognized
1900	Mendelism	Genetic constancy accepted and blending inheritance rejected
1918–1933	Fisherism	Evolution considered to be a matter of gene frequencies and the force of even small selection pressures
1936–1947	Evolutionary synthesis	Population thinking emphasized; interest in the evolution of diversity, allopatric speciation, variable evolutionary rates
1947–1970	Post-synthesis	Individual increasingly seen as target of selection; a more holistic approach; increased recognition of chance and constraints
1954–1972	Punctuated equilibria	Importance of speciational evolution
1970s–1980s	Rediscovery of sexual selection	Importance of reproductive success for selection

holistic tradition of emphasizing the individual as the target of selection. In short, each of these periods was heterogeneous to some extent, owing to the diversity in the thinking of different evolutionists. Most critics who have attempted to refute the evolutionary synthesis have failed to recognize this diversity of views and thus have succeeded in refuting only the reductionist fringe of the Darwinism camp. Their failure to appreciate the complexity of the evolutionary synthesis has led them to paint a picture of that period which is at best a caricature.

Another error made by most opponents of the synthesis is a failure to differentiate between proximate and evolutionary causations. For Darwin and all of his holistic followers, selection starts at fertilization and continues through all embryonic and larval stages. A Darwinian is truly puzzled when he reads in a critique by an embryologist that "development . . . comes to the fore as a problem unintelligible within neo-Darwinism." What aspect of development is this author talking about? If he is speaking of the translation of the genetic program into molecular chains of events during ontogeny, he is talking about proximate causations. Their study, indeed, has never been the job of the evolutionary biologist. But many other aspects of development do raise questions concerning evolutionary causations that have been of interest to evolutionists from Darwin on. They are of concern to the evolutionist, first, because each stage of development is a target of selection, and particularly so when the developmental stages (larvae) are free-living. Second, because embryonic stages may themselves serve as "somatic programs" in development (see below), and such stages tend to become highly conserved in evolution (for instance, the gill arch stage of the tetrapod embryo). Such highly conserved stages are often very helpful in the reconstruction of the phylogeny (recapitulation). No Darwinian will ever question the importance of development in evolution, but evolutionary interpretation is constrained by the extent to which the proximate causations of development have been elucidated by the embryologists. To undertake a study of such proximate causations is not the task of the evolutionist.

Curiously, the objection is sometimes raised that the evolutionary synthesis can shed light neither on the level of the gene nor on trans-specific levels because it is concerned with individuals (as targets of selection) and with populations (as incipient species). What is needed, it is claimed, is a hierarchical approach such as is not found in either neo-Darwinism or the evolutionary synthesis. That this objection is without basis has been shown by Grant (1983:153), Stebbins and Ayala (1981), and other defenders of the synthesis. Even Simpson's interpretation of evolution was strongly hierarchical, as correctly pointed out by Laporte (1983). None of these opponents of the synthesis has shown as clearly as Mayr (1963:621) that the species is the unit of action in macroevolution. (For more on this problem see Essay 23.)

And finally, for White (1981), the synthesis was premature, "because there was no molecular biology at the time and both the chemical nature of the hereditary material and the architecture of eukaryote chromosomes were yet unknown." This is like saying that the Darwinian revolution was premature because genetics had not been founded. Any scientific revolution or synthesis has to accept all sorts of black boxes, for if one would have to wait until all black boxes are opened, one would never have any conceptual advances.

MOLECULAR BIOLOGY

Early in the history of molecular biology there was a widespread feeling that its new discoveries might necessitate a complete rewriting of evolutionary theory. Considering how quickly molecular biology developed into a large and powerful field of its own, a call for an integration of this new field with classical evolutionary biology was to be expected. As White (1981) said, "We once again urgently need a new synthesis of two traditions—those of evolutionary and molecular biology." Other molecular biologists went even further, and stated that the findings of molecular biology had already refuted much of accepted Darwinism. Thus it was claimed, "Many of the observations [of molecular biology] (inducible mutation systems, rapid genomic changes involving mobile genetic elements, programmed changes in chromosome number) challenge the most fundamental assumptions which evolutionary theories make about the mechanisms of hereditary variations and the fixation of genetic differences" (Shapiro 1983).

Discoveries that seem to be in conflict with the picture of classical genetics are made daily in molecular biology. Perhaps none was as startling as the discovery that genes are highly complex systems consisting of exons, introns, and flanking sequences, and that there are numerous kinds of genes, some having seemingly no function at all, while several functional classes can be distinguished among the active genes. But have any of these discoveries required a revision of Darwinism? I do not think so.

There is no question whatsoever that molecular biology has given us numerous great new insights into the working of evolutionary causes, particularly the production of genetic variation. Gratifyingly, in most cases they were a confirmation or elaboration of existing views. Let me merely mention some of the most important molecular discoveries.

(1) The genetic program does not by itself supply the building material of new organisms, but only the blueprint for the making of the phenotype.

(2) The pathway from nucleic acids to proteins is a one-way street. Proteins and information that they may have acquired are not translated back into nucleic acids.

(3) Not only the genetic code but in fact most basic molecular mechanisms are the same in all organisms, from the most primitive prokaryotes up.

(4) Many mutations (changes in the base pairs) seem to be neutral or near-neutral, that is, without noticeable effect on the selective value of the genotype, but this varies from gene to gene.

(5) A critical comparative analysis of molecular changes during evolution provides a very large number of pieces of information suitable for the reconstruction of the phylogeny. This is particularly useful if the morphological evidence is indecisive. However, molecular characters are also vulnerable to homoplasy.

Interestingly, up to the present time, none of these findings have required an essential revision of the Darwinian paradigm. Moreover, the relation between molecular and evolutionary biology has not been entirely one-sided. Darwinian thinking has made a major contribution to the evolution of biochemistry into molecular biology. The study of the phylogeny of molecules and the search for the selective significance of molecular structure have greatly enriched molecular biology. No longer does one encounter in the literature of molecular biology any thorough study of a particular molecule or group of molecules that is not at the same time concerned with the evolutionary explanation of the molecular structures encountered during that analysis.

The Synthesis as Unfinished Business

The healthy turmoil that currently characterizes evolutionary biology should not be viewed as a death struggle but rather as the sort of lively activity one will find in any healthy and advancing branch of science. There is no justification whatsoever for such claims as "The modern synthesis has broken down" or "We need a new evolutionary theory."

Yet it must be admitted that, even though the refutation of the major anti-Darwinian theories drastically narrowed down the variation of evolutionary theory, some well-defined differences among the Darwinians still existed into the post-synthesis period, and many of these differences are still with us fifty years later.

For the geneticists—or at least all those influenced by Fisher and Haldane—the gene continued to be the primary target of selection, and most genes were believed to have constant fitness values. The whole problem of the origin of organic diversity (that is, the multiplication of species) was minimized by this school, if not ignored. Most of those who called themselves population geneticists worked with single, closed gene pools (Mayr 1963:177). Even Wright did not come to grips with the problem of the multiplication of species in his shifting balance theory, nor with the macroevolutionary problems generated by speciation.

By contrast, those who had come to evolutionary biology from systematics or some other area of natural history considered evolution to be a populational problem, and the whole (potentially reproducing) individual to be the target of selection. For them, the multiplication of species was the pathway to the solution of problems of macroevolution. Despite their tendency to think in terms of phenotypes, they eventually came to view the genotype as a system of gene interaction (cohesion of the genotype) and tended from the very beginning to deal with evolution hierarchically (Simpson, Rensch, Huxley, Mayr). For them, evolution was not a change in gene frequencies but the twin problem of adaptive change and the origin of diversity.

By no means are all current intra-Darwinian controversies remnants of the old geneticist-versus-naturalist feud. There are also differences among the geneticists concerning the relative frequency of neutral mutations and the amount of variation due to balance. Among the paleontologists there are disputes as to whether or not simple phyletic gradualism can give rise to higher taxa. And among other evolutionists there are disagreements on the validity of group selection, the extent of adaptation, the role of competition, the frequency of sympatric speciation, how continuous or punctuated evolution is, to what extent all components of the phenotype reflect ad hoc selection, what proportion of speciation is allopatric, what the target of selection is, how much variation is stored in populations, and to what extent the new findings of molecular biology require a revision of current theory. However, regardless of ultimate outcome, none of these disagreements affects any of the basic principles of Darwinism.

Although rarely completely wrong, the conclusions supported by the followers of the evolutionary synthesis were often incomplete and rather simplistic. Two kinds of processes, in particular, were often inadequately considered: (1) multiple simultaneous causations, and (2) pluralistic solutions.

Let me give some examples of simultaneous causations. In all selection phenomena—and selection is of course an anti-chance process—chance phenomena also occur simultaneously. Or, to give another example, speciation is never merely a matter of genes or chromosomes but also of the nature and geography of the populations in which the genetic changes occur. Geography and the genetic changes of populations affect the speciation process simultaneously.

Pluralism. Far more serious was a frequent neglect of the pluralism of evolutionary phenomena. The more one studies the processes of evolution, the more one is impressed by their diversity. Many controversies in evolutionary biology arise because of the inability of certain authors to appreciate this diversity. There seem to be multiple solutions to almost any evolutionary challenge. During speciation, pre-mating isolating mechanisms originate first in one group of organisms, in another post-mating mechanisms. Sometimes geographic races are phenotypically as distinct as good species yet not at all reproductively isolated. On the other hand, phenotypically indistinguishable sibling species may be fully isolated reproductively. Some species are extraordinarily young, having originated only 2,000 to 10,000 years ago, while others have not changed visibly in 10 to 50 million years. Polyploidy or asexual reproduction are important in some groups, and totally absent in others. Chromosomal restructuring seems to be an important component of speciation in some groups, such as the morabine grasshoppers of Australia, but seems not to occur in others. Interspecific hybridization is frequent in some groups but is rare or absent in others. Some groups speciate profusely, in others speciation seems to be a rare event.

Just because there is little gene flow in certain species, one cannot conclude that gene flow is irrelevant in all species. In fact, the amount of gene flow seems to differ greatly even in closely related species. The absence of major genetic reconstruction in some founder populations does not prove that genetic revolutions can never take place. Parapatry is frequent in groups in which post-zygotic isolating mechanisms evolve first while pre-zygotic ones are still very incomplete. However, this does not justify explaining all cases of parapatry by the same mechanisms. One phyletic lineage may evolve very rapidly while others, even closely related ones, may experience complete stasis for many millions of years. In short, there are several possible solutions for many evolutionary challenges, but all of them are compatible with the Darwinian paradigm.

Most evolutionists, particularly those who work on a single group of organisms, tend to neglect the extraordinary pluralism of evolution. As Francois Jacob (1977) has so rightly said, evolution is a tinkerer, and in a given situation makes use of that which is most readily available. One can take almost any evolutionary phenomenon and show how greatly it differs among the different groups of animals and plants. The lesson one learns from this is that sweeping claims are rarely correct in evolutionary biology. Even when something occurs "usually," this does not mean that it must occur always. One must remember the forever-present pluralism of evolutionary processes.

Almost every scientific theory is continuously in need of revision and supplementation, yet these changes do not necessarily touch the true core of the theory. The Mendelian theory of inheritance is an apt illustration of this. When it was rediscovered in 1900, it was expressed in the form of three laws. Within just a few years two of these laws, those of dominance and independent assortment, were found not to be universally valid. Still, no one claimed that Mendel's theory had therefore been refuted. The continuing minor revisions and supplementations of the Darwinian theory, including the version given to it during the synthesis, do not qualify as refutations either.

The architects of the evolutionary synthesis have been accused by some critics of claiming that they had solved all the remaining problems of evolution. This accusation is quite absurd; I do not know of a single evolutionist who would make such a claim. All that was claimed by the supporters of the synthesis was that they had arrived at an elaboration of the Darwinian paradigm which seemed to be sufficiently robust not to be endangered by the remaining puzzles. No one denied that there were many open questions, but there was the feeling that no matter what answer would emerge in response to these questions, it would be consistent with the Darwinian paradigm. Up to now, it seems to me, this confidence has not been disappointed.

New Frontiers in Evolutionary Biology

If someone were to ask me which frontiers of evolutionary biology are likely to see the greatest advances in the next 10 or 20 years, I would say the structure of the genotype and the role of development. I refrain from listing molecular evolution as a third topic because it is precisely the molecular structure of the genotype as well as the role of molecules in development that are of such crucial importance for genotype and development.

The simplistic reductionist view of the relations among genotype, development, and evolution is that each gene is translated into a corresponding component of the phenotype and that the contributions of these components to the fitness of the resulting organism determine the selective value of the genes. There is undoubtedly some truth in this view, but it greatly oversimplifies the actual connections. This had been realized all the way back to Darwin's day. Darwin himself spoke of the mysterious laws of correlation, and Haeckel converted the Meckel-Serres Law into the evolutionary principle of recapitulation. Again, this principle was a vast oversimplification, but it has a correct nucleus. The fact that the land-living vertebrates go through a gill arch stage in their development is a powerful clue for their descent from aquatic ancestors, to give only one example. Both of these generalizations call attention to holistic interactions. By contrast, the belief of some reductionists that the role of genes is exhaustively described by the largely independent contribution of each gene to some particular aspect of the phenotype neglects the fact that the genotype is a complexly interacting system.

Each year one or two books or symposia volumes are published on the relation between genotype, development, and evolution, and it is quite impossible in this short essay even to outline the basic problems. To make matters worse, even though the questions are increasingly better understood, no satisfying answers have yet been found for most of them. When answers have been suggested, they often contradict one another. Instead of attempting to present a detailed account of the problems and the opposing attempts to solve them, I will satisfy myself by merely indicating some of the general directions of future research.

DOMAINS OF THE GENOTYPE

We know now that there are different classes of genes, and they play not only different roles in ontogeny but also in evolution. Furthermore, certain genes seem to be tied together into functional units and seem to control development as units. They seem to represent well-circumscribed "domains"; others have spoken of a "hierarchical structure of the genotype," both terms attempting to express a similar concept of the heterogeneity of the genotype. The existence of such domains should not be in conflict with Mendelian segregation. How the developmental and evolutionary conservation of such domains is effected is not yet understood. However, the discovery of the widespread occurrence (from yeasts to mammals) of homeoboxes (Robertson 1985) and the study of complete sets of immune genes reveal possibilities.

At the present time the existence of such domains is primarily indicated by rather indirect evidence. There are a number of categories of such evidence, but I would like to discuss in some detail one particular phenomenon, so-called mosaic evolution.

Mosaic evolution. For a physicist one of the most important parameters of any process is its rate. In the endeavor to make evolutionary biology as similar to the physical sciences as possible, evolutionists have attempted from 1859 on to determine evolutionary rates (Simpson 1944; 1953). Alas, this has been a rather frustrating experience. As I have shown elsewhere (Mayr 1964), rates of evolution, speciation, and extinction may differ by several orders of magnitude in different organisms and under different circumstances. Perhaps most disturbing was the discovery that organisms do not evolve as harmonious types but that different characters (components of the phenotype) and domains of the genotype may evolve at highly different rates. This was already known to Lamarck and Darwin, and has since been confirmed by numerous evolutionists, particularly the paleontologists (Mayr 1963:596–598). *Archaeopteryx* is an apt illustration. In some of its characteristics (feathers and wings) it is already a typical bird, in others (teeth and tail) it is still a reptile, and in still other characters it is more or less intermediate between the two classes. A. C. Wilson (1974) has pointed out that frogs are morphologically highly conservative, and this is also true for their chromosomes, but that different phyletic lines of frogs seem to diverge in their enzyme genes at about the same rate as mammals, which evolve very rapidly in their morphological and chromosomal characters. What this suggests is that different domains of the genotype seem to evolve independently to a considerable degree. This is well illustrated, for instance, in a comparison of man with his anthropoid relatives. Over all, man is most similar to the chimpanzee, but in a few characters man's genotype is more similar to that of the gorilla or even the orang.

The term *mosaic evolution* has been used for the independent evolution of different domains of the genotype. At the molecular level it was found not only that different molecules have different evolutionary rates but also that the same molecule may change its rate at different stages in the evolution of a phyletic line. Whenever there is a period of major adaptive radiation, it is accompanied by a highly accelerated rate of molecular evolution, later followed by a deceleration to slow change. The so-called molecular clock may keep very different times for different molecules and may change its rate within a phyletic line. As Goodman (1982:377) has correctly pointed out, none of these findings on changes in molecular timing invalidates "the Darwinian thesis that natural selection is the principal force directing the evolution of molecules and organisms." Mosaic evolution, however, must be carefully considered in the construction of phylogenetic trees.

The cohesion of the genotype.   The opposite of the independent evolution of domains of the genotype is the integration of these domains and the seeming cohesion of the genes that belong to one of these domains. Just exactly what controls such a cohesion is still largely unknown, but the existence of such cohesion is abundantly documented (Mayr 1970; 1975). The almost total integrity of the gene complex that controls the number of extremities in tetrapods, insects, and arachnids is one of literally hundreds of examples. The capacity of the pre-Cambrian eukaryote genotype to form 70 or more distinct morphological types (phyla), at a time when this cohesion was still very loose, is another documentation. Throughout evolution there has been a tendency for a progressive "congealing" of the genotype, so that it has become more and more difficult to deviate from a long-established morphological type. This is one of the well-known constraints in evolutionary change, and one of the reasons why natural selection has such limited leeway. When we speak of the congealing of the genotype or, if we prefer, of the progressive integration of the genotype, we are simply using words to cover up our ignorance. Nevertheless, the evidence indicates that the older a taxon is, the more difficult it is to escape from the straitjacket of a highly integrated (congealed) genotype. This is why not a single new morphological type (phylum) has originated since the Cambrian.

At first sight this conclusion would seem to be contradicted by the adaptive radiations that have occurred in so many phyletic lines. After a stasis of 100 million years or longer a higher taxon may suddenly experience a burst into numerous new taxa, such as happened to the mammals in the Paleocene and Eocene. However, most of these were ad hoc adaptations and did not really deviate very far from the ancestral *Bauplan.* So far as I can judge, the same is true for most of the other well-known radiations. The songbirds, for instance, have produced over 5,000 living species since their origin, but except for their plumage and bills, they exhibit hardly any deviation from the standard *Bauplan.*

The power of these constraints is evident at every hierarchical level, sometimes even at that of the species, as documented by the frequency of sibling species. Indeed, most geographic variation deals merely with minor quantitative changes in characters, without affecting the genotype appreciably. Any restructuring, when it occurs, apparently takes place almost always in peripherally isolated founder populations (Essay 25).

It must be remembered, however, that evidence for increasing or decreasing cohesion of the genotype can also be found in the study of macroevolution. Evolutionary paleontologists have pointed out that evolutionary acceleration or deceleration may be displayed by all or at least most members of a new higher taxon. A well-known textbook case is evolution among the lungfishes, as worked out by Stanley Westoll (1949). Almost all the anatomical reconstruction in this new class of fishes took place in the earliest stages (during some 25 million years), while almost no change has taken place in the subsequent 200 million years. Such a drastic difference between the rates of evolutionary change in young and mature higher taxa is virtually the rule. Bats originated from insectivores within a few million years, but their morphology has hardly changed in the last 40 million years. Even though the fossil record is inadequate, it would seem that the same is true for the anatomical reconstruction during the origin of birds and whales. Exceptions are known, like the origin of the viperids, but they are very much in the minority.

THE ROLE OF SOMATIC PROGRAMS IN EVOLUTION

The traditional formula according to which development is programmed by the genotype leaves almost all questions unanswered. How independent are the genes in this program? How directly does a given gene control a component of the phenotype? And more importantly, how do evolutionary changes in individual components of the genotype affect the remainder of the genetic program?

The main trouble with the simplified formula is that it implies by far too direct a pathway from the gene to the endpoint of its action in the phenotype. Embryologists have long known (Kleinenberg 1886) that genes might effect the production of an embryonic structure that subsequently serves as part of the program for the later stages of development. That not all programs are purely genetic, and that there are also somatic programs, is best made clear by the example of behavior programs. When, for example, a male bird performs a certain courtship display, this action is not programmed directly by the genotype but rather by a secondary program laid down in the central nervous system during ontogeny. And it is this secondary—somatic—program that actually controls the behavior.

That there are such somatic programs is presumably not of major consequence for the evolution of behavior. The existence of somatic programs, however, is presumably of high significance in morphological evolution. It may help to explain many puzzling phenomena, of both ontogeny and evolution. For instance, it may explain most instances of so-called recapitulation. When an embryonic structure of the ancestors is maintained even though it no longer seems to be of functional value (as for instance the gill arches of the mammalian embryos), such an embryonic structure may serve as the somatic program for the subsequent ontogenetic stages. The existence of somatic programs imposes important constraints on evolution. Much development in the higher taxa seems to be constrained by such somatic programs, which seem to be highly resistant to evolutionary change. This statement is, of course, at the present time only words. However, recent research in molecular ontogeny has made promising beginnings. Progress is bound to be slow, because development involves highly complex interactions between different domains of the genotype and of somatic programs.

This subject area is of particular interest to the student of macroevolution. It is here that the connection will be made between the genetics of individuals and populations on the one hand and the major macroevolutionary processes and events on the other. It will have to pursue simultaneously reductionist approaches, that is, the study of the action of individual genes, and also holistic ones, that is, the study of domains of the genotype and of whole somatic programs.

NOTES

This essay is previously unpublished.

1. In order to determine exactly what had happened during (and immediately preceding) the evolutionary synthesis, two workshops were held in 1974 (Mayr and Provine 1980). Although the workshops took place 30 years after the synthesis, all the evolutionists who are traditionally mentioned as the major architects of the synthesis were still alive. The term "architect" has been applied in different ways. More widely conceived, it includes a number of pre-1936 geneticists and naturalists—Chetverikov, Fisher, Haldane, Wright, Timofeeff-Ressovsky, Sumner, Stresemann, and others; more narrowly, it includes only those who published "during the synthesis," that is, after 1936. In the present discussion I have adopted the latter delimitation. One hope of the organizers of the two workshops was to bring all of the architects together to make them available for searching questions raised by the attending historians and philosophers of science. Unfortunately, three of the architects, Julian Huxley, George Gaylord Simpson, and Bernhard Rensch, were unable to attend because of illness or other causes.

Altogether there were 33 participants, of whom 15 were historians and philosophers and 18 were scientists. Among the scientists were 10 geneticists, 3 paleontologists, 3 systematists (1 as observer only), 1 embryologist, and 1 anthropologist. Thus, the different branches of evolutionary biology were rather unevenly represented. Even more unbalanced was the representation of the countries in which the synthesis had taken place. All of the workshop participants were from the United States, except for two representing Great Britain (C. D. Darlington and E. B. Ford) and two representing France (E. Boesiger and C. Limoges).

It is thus evident that our attempt in 1980 to describe how the synthesis had occurred in various countries was incomplete. In fact, this was even true for Great Britain and the United States, because much-needed original analysis was not yet available. What was perhaps least successfully achieved for these two countries was a presentation of the still remaining differences among the Darwinians. There was never a uniform consensus in any country, and a complete account must always include an adequate report on minority views.

As far as the Soviet Union is concerned, the historian Mark Adams was able to present a remarkably complete picture from the Russian literature. But the account of the "failed" synthesis in France (Boesiger 1980; Limoges 1980) has been criticized subsequently, and new analyses of the situation in France are now in the process of preparation. In the case of Germany, the presentation in Mayr and Provine (1980) is particularly deficient, even though it contains a detailed statement of the development of Rensch's ideas and of his conversion to neo-Darwinism.

Rensch himself, after he had abandoned neo-Lamarckism around 1933–34, developed remarkably progressive ideas, showing in two important papers (1939; 1943, preceding Simpson's 1944 *Tempo and Mode*) that the phenomena of macroevolution could be interpreted as consistent with the known genetic mechanisms of microevolution. Unfortunately, Rensch's account was deficient in his discussion of two important questions, the first being to what extent taxonomic differences between races, species, and genera are either adaptive or simply incidental byproducts of evolutionary divergence. Rensch himself, Stresemann, and most German neo-Lamarckians apparently favored an adaptational explanation of geographic variation, as did J. A. Allen, Joseph Grinnell, and J. P. Chapin in the United States. And by extrapolation this would lead to a similar evolutionary explanation of species differences. But there were many other German taxonomists who, like Robson and Richards in England, considered species differences as nonadaptive accidents of variation. It will require a careful study of the *Archiv f. Naturgesch., Biol. Zentralblatt*, and various taxonomic journals to determine the relative frequency of those supporting the adaptive and those supporting the nonadaptive viewpoints.

Actually, this is a more complex problem than meets the eye. From Darwin on, taxonomists had always been advised to choose as the diagnostic characters of new species and genera those that were not related to niche occupation (Darwin 1859:414). Characters that were "adaptive" tended to originate convergently in unrelated taxa. Organisms thus displayed a mosaic of adaptive and nonadaptive characters. To deny an adaptive significance to certain diagnostic characters in taxonomic analysis did not signify a denial of the importance of selection and adaptation.

A second area in which Rensch's account was deficient was his failure to bring out sufficiently how advanced the thinking of some German evolutionary geneticists was, even prior to the synthesis. Outstanding among them was Erwin Baur, whose exemplary analyses of local populations of Spanish *Antirrhinum* are among the classics of the evolutionary literature. He especially stressed the large number of virtually invisible small mutations seemingly with a purely physiological effect. Throughout his work Baur had the populational aspects of evolution in mind. Furthermore, more than most of his contemporaries, he stressed genetic recombination, which, as he said (following Weismann), makes the importance of sexuality evident (Baur 1924; 1932).

Equally or even more critical for the development of the synthesis in Germany was the Russian geneticist N. W. Timofeeff-Ressovsky, who from 1925 had lived in Germany. In the years 1932 to 1940 he published no fewer than 28 papers dealing with population genetics and its relevance to evolution. As a student and early collaborator of S. S. Chetverikov, he brought population thinking into German genetics (independently of Baur) and decisively influenced Rensch and other German evolutionists. Timofeeff apparently played the same role in Germany that Dobzhansky played in the United States—indeed, not only in Germany but also in other countries that had sent young geneticists to work with Timofeeff (such as Buzzati-Traverso from Italy). Owing to Timofeeff's influence, an evolutionary synthesis took place in the 1930s in Germany, largely independent of the synthesis in the English-speaking countries.

The visible manifestation of this German synthesis was a volume, *Die Evolution der Organismen*, edited by G. Heberer (1943). What is remarkable in this volume is how completely it reflects the thinking of neo-Darwinism. Rensch (1980) and Reif (1983) make it perfectly clear, however, that it would be a mistake to conclude from this volume that all biologists in Germany had become neo-Darwinians. Actually, except for the camp of the geneticists, the neo-Darwinians were probably still very much in the minority at that time. And yet, with leading zoologists like Rensch and Ludwig and leading botanists like Zimmermann and Schwanitz having achieved consensus with the geneticists, the ultimate triumph of neo-Darwinism had now simply become a matter of time. This is not to deny that there was still strong resistance in certain biological disciplines, particularly in embryology (Hamburger 1980) and paleontology (Reif 1983).

The Heberer volume also documents the somewhat singular aspect of evolutionary biology in Germany. Ever since Haeckel, the major emphasis in evolutionary studies in Germany has been on phylogeny, and some of the longest contributions to the Heberer volume, 400 of 735 text pages, are devoted to phylogeny and its methodology. Unfortunately, those contributors of the volume who dealt with the causation of evolutionary change (genetics) with one exception were all on the staff of research institutes and failed to establish a school. As a consequence, right up to the present day research and teaching in the mechanisms of evolutionary change is still rather neglected at German universities.

2. An unexpected achievement of the synthesis was its effect on the prestige of evolutionary biology. The 1920s and 30s experienced an absolute low in the esteem of evolutionary biology within biology. This is not the place to analyze in detail the reasons for this condition, but it is well documented, not only by the comments made by geneticists, physiologists, and embryologists about evolutionary biology but also by the absence, or low number, of college courses devoted to this field. After the synthesis there was a remarkable upswing of interest. To be sure, it was most prominent in the United States, where the *Society for the Study of Evolution* (1946) as well as the journal *Evolution* (1947) were founded. But it was also evident in Great Britain and the Scandinavian countries. The number of journals, journal articles, general books, and textbooks dealing with evolutionary biology have steadily increased ever since.

3. Evolution seems to be a subject on which everybody thinks he is qualified to express an expert opinion. In fact, the Darwinian theory is a highly complex research program. Even though the basics of the theory of natural selection (the core of Darwinism) are now well understood, most individual components are still controversial among specialists. When lay people discuss aspects of evolution, they very often simply display their ignorance and misunderstanding. Most of the worst of the uninformed writings, authored by journalists, jurists, or other dilettantes, would never have been published if the press had higher standards. Alas, even some formerly reputable magazines now seem to be more interested in sensation than in the truth.

A respectable scientific journal such as *Nature* would never ask an embryologist or ethologist to discuss the theory of elementary particles. However, when *Nature* published an editorial entitled "How true is the theory of evolution?" it asked a person with remarkably little expertise to state the views of their magazine. How little that person knew about the subject is revealed by some of his statements (17 March 1981, Vol. 290, pp. 75–76): "Dobzhansky's classic work on the course of speciation among birds in Central America [D. never worked on the birds of Central America, or any other birds!] . . . it is also true that too little is known of the molecular structure of the chromosomes (chromatin) for the mechanism of speciation to be fully understood [actually, hardly a single one of the controversies on this subject will be solved by such knowledge] . . . the two components of Darwinism (evolution and natural selection) [a misleadingly simplistic characterization]." One would think that a journal like *Nature* should be able to find a better qualified person to discuss such a highly technical subject as evolution.

REFERENCES

Adams, M. 1980. Sergei Chetverikov, the Kolt'sov Institute and the evolutionary synthesis. In Mayr and Provine 1980:242–278.

Anon. 1981. How true is the theory of evolution? *Nature* 290:75–76.

Arthur, W. 1984. *Mechanisms of Morphological Evolution: A Combined Genetic, Developmental and Ecological Approach*. New York: Wiley. [See also J. S. Jones, *Nature* 312:386.]

Ayala, F. J. 1982. Beyond Darwinism? The challenge of macroevolution to the synthetic theory of evolution. *PSA 1982* 2:275–291.

———— 1983. Microevolution and macroevolution. In Bendall 1983:387–402.

Barigozzi, C., ed. 1982. *Mechanisms of Speciation*. New York: Alan R. Liss.

Baur, E. 1924. Untersuchungen über das Wesen, die Entstehung und die Vererbung von Rassenunterschieden bei Antirrhinum majus. *Bibliographia Genetica* 4:1–170.

———— 1932. Artumgrenzung und Artbildung in der Gattung Antirrhinum. *Zeitschr. Indukt. Abst. Vererbungslehre* 63:256–302.

Beatty, J. 1986. The synthesis and the synthetic theory. In Bechtel, ed., *Integrating Scientific Disciplines*, pp. 125–135. Dordrecht: Martinus Nijhoff.

Bendall, D. S., ed. 1983. *Evolution from Molecules to Men*. Cambridge, England: Cambridge University Press.

Boesiger, E. 1980. France. In Mayr and Provine 1980:309–321.

Bowler, P. J. 1983. *The Eclipse of Darwinism: Anti-Darwinian Evolution Theories in the Decades around 1900*. Baltimore: Johns Hopkins University Press.

Brace, C. L. 1985. [Review of Ho and Saunders 1984.] *Amer. J. Phys. Anthropology* 68:446–448.

Dobzhansky, Th. 1937. *Genetics and the Origin of Species*. 1st ed. New York: Columbia University Press.

———— 1951. *Genetics and the Origin of Species*. 3rd. ed. New York: Columbia University Press.

———— 1970. *Genetics of the Evolutionary Process*. New York: Columbia University Press.

Dobzhansky, Th., F. J. Ayala, G. L. Stebbins, and J. W. Valentine. 1977. *Evolution*. San Francisco: Freeman.

Eldredge, N. 1979. Alternative approaches to evolutionary theory. *Bull. Carnegie Mus. Nat. Hist.* 13:7–19.

———— 1985. *The Integration of Evolutionary Theory*. New York: Oxford University Press.

———— 1986. *Unfinished Synthesis: Biological Hierarchies and Modern Evolutionary Thought*. London: Oxford University Press.

Eldredge, N., and S. J. Gould. 1972. Punctuated equilibria: an alternative to phyletic gradualism. In T. J. M. Schopf 1972:82–115.

Eldredge, N., and J. Cracraft. 1980. *Phylogenetic Patterns and the Evolutionary Process*. New York: Columbia University Press.

Futuyma, D. 1979. *Evolutionary Biology*. Sunderland, Mass.: Sinauer. 2nd ed., 1986.

———— 1983. *Science on Trial: The Case for Evolution*. New York: Pantheon Books.

Gould, S. J. 1980a. Is a new and general theory of evolution emerging? *Paleobiology* 6:119–130.

———— 1980b. G. G. Simpson, paleontology, and the modern synthesis. In Mayr and Provine 1980.

———— 1981. Introduction in reprint of Th. Dobzhansky, *Genetics and the Origin of Species*, 1st ed. (1937). New York: Columbia University Press, xvii–xli.

———— 1982. Darwinism and the expansion of evolutionary theory. *Science* 216:380–387.

———— 1983a. The changing role of paleontology in Darwin's three centennials and a modest proposal for macroevolution. In Bendall 1983:347–366.

———— 1983b. The hardening of the modern synthesis. In Grene 1983:71–93.

Gould, S. J., and N. Eldredge. 1977. Punctuated equilibria: The tempo and mode of evolution reconsidered. *Paleobiology* 3:115–151.

Grant, V. 1977. *Organismic Evolution*. San Francisco: Freeman.

———— 1983. The synthetic theory strikes back. *Biol. Zentralbl.* 102:149–158.

Grene, M. 1983. *Dimensions of Darwinism*. Cambridge: Cambridge University Press.

Gutmann, W. F., and K. Bonik. 1981. *Kritische Evolutionstheorie. Ein Beitrag zur Überwindung altdarwinistischer Dogmen*. Hildesheim: Gerstenberg.

Hamburger, V. 1980. Embryology. In Mayr and Provine 1980:97–112.

Heberer, G., ed. 1943. *Die Evolution der Organismen*. Jena: G. Fischer.

Ho, M.-W., and P. T. Saunders. 1984. *Beyond Neo-Darwinism: An Introduction to the New Evolutionary Paradigm*. New York: Academic. [For critical reviews see Maynard Smith 1985, Mayr 1984, and Brace 1985.]

Huxley, Andrew. 1982. [Address of the President.] *Proc. R. Soc. London A:* 379:ix–xvii.

Huxley, J. 1942. *Evolution: The Modern Synthesis*. London: Allen and Unwin.

Jacob, F. 1977. Evolution and tinkering. *Science* 196:293–314.

Kitcher, P. 1982. *Abusing Science: The Case Against Creationism*. Cambridge, Mass.: M.I.T. Press.

Kleinenberg, N. 1886. Über die Entwicklung durch Substitution von Organen. In Kölliker, Albert v., und Ernst Ehlers, eds., *Zeitschrift für Wissenschaftliche Zoologie*, pp. 212–224. Leipzig: Wilhelm Engelmann.

Kuhn, T. 1962. *The Structure of Scientific Revolutions*. Foundations of the Unity of Science vol. 2, no. 2. Chicago: University of Chicago Press.

Laporte, L. F. 1983. Simpson's *Tempo and Mode in Evolution* revisited. *Proc. Amer. Phil. Soc.* 127:365–417.

Levinton, J. S. 1984. [Reviews.] *Paleobiology* 10:377–385.

Levinton, J. S., and C. Simon. 1980. A critique of the punctuated equilibria model and implications for the detection of speciation in the fossil record. *Syst. Zool.* 29:130–142.

Limoges, C. 1980. A second glance at evolutionary biology in France. In Mayr and Provine 1980:322–328.

Maynard Smith, J. 1972. *On Evolution*. Edinburgh: University of Edinburgh Press.

———, ed. 1982. *Evolution Now: A Century after Darwin*. London: *Nature*, with Macmillan Press.

——— 1983. Current controversies in evolutionary biology. In Grene 1983:273–286.

——— 1985. [Review of Ho and Saunders 1984.] *New Scientist* 14:38–39.

Mayr, E. 1942. *Systematics and the Origin of Species*. New York: Columbia University Press.

——— 1954. Change of genetic environment and evolution. In J. Huxley, ed., *Evolution as a Process*, pp. 157–180. London: Allen and Unwin.

——— 1963. *Animal Species and Evolution*. Cambridge, Mass.: Harvard University Press.

——— 1967. Evolutionary challenges to the mathematical interpretation of evolution. *Wistar Symposium Monograph* no. 5, pp. 47–58.

——— 1970. *Populations, Species, and Evolution*. Cambridge, Mass.: Harvard University Press.

——— 1982a. *The Growth of Biological Thought*. Cambridge, Mass.: Harvard University Press.

——— 1982b. Darwinistische Missverständnisse. In K. Bayertz, B. Heidtmann, and Hans-Jörg Rheinberger, eds., *Dialektik 5: Darwin und die Evolutionstheorie*, pp. 44–57. Köln: Pahl-Rugenstein.

——— 1984. The triumph of evolutionary synthesis. *Times Literary Supplement* 257 (4):1261–1262.

——— 1986. [Review of R. G. B. Reid. 1985. *Evolutionary Theory: The Unfinished Synthesis*. Ithaca, N.Y.: Cornell University Press.] *Isis* 77:358–359.

Mayr, E., and W. B. Provine, eds. 1980. *The Evolutionary Synthesis*. Cambridge, Mass.: Harvard University Press.

Milkman, R., ed. 1982. *Perspectives in Evolution*. Sunderland, Mass.: Sinauer.

Montagu, A., ed. 1983. *Science and Creationism*. Oxford: Oxford University Press.

Muller, H. J. 1940. Bearings of the 'Drosophila' work on systematics. In J. Huxley, ed., *The New Systematics*, pp. 185–268. Oxford: Clarendon Press.

Nei, M., and R. K. Koehn. 1983. *Evolution of Genes and Proteins*. Sunderland, Mass.: Sinauer.

Newell, N. D. 1982. *Creation and Evolution: Myth or Reality?* New York: Columbia University Press.

Provine, W. B. 1971. *The Origins of Theoretical Population Genetics*. Chicago: University of Chicago Press.

———— 1986. *Sewall Wright and Evolutionary Biology*. Chicago: Chicago University Press.

Reif, W. E. 1983. Evolutionary theory in German paleontology. In Grene 1983:173–203.

Rensch, B. 1939. Typen der Artbildung. *Biol. Rev.* 14:180–222.

———— 1943. Die paläontologischen Evolutionsregeln in zoologischer Betrachtung. *Biol. Generalis* 17:1–55.

———— 1947. *Neuere Probleme der Abstammungslehre*. Stuttgart: Enke.

———— 1980. Historical development of the present synthetic neo-Darwinism in Germany. In Mayr and Provine 1980:284–303.

Robertson, M. 1985. Mice, mating types, and molecular mechanisms of morphogenesis. *Nature* 318:12–13.

Ruse, M. 1982. *Darwinism Defended: A Guide to the Evolution Controversies*. Reading, Mass.: Addison-Wesley.

Schopf, T. J. M., ed. 1972. *Models in Paleobiology*. San Francisco: Freeman, Cooper.

Simpson, G. G. 1944. *Tempo and Mode of Evolution*. New York: Columbia University Press.

———— 1949. *The Meaning of Evolution*. New Haven: Yale University Press.

Stebbins, G. L. 1966. *Processes of Organic Evolution*. Englewood Cliffs, N.J.: Prentice-Hall.

Stebbins, G. L., and F. J. Ayala. 1981. Is a new evolutionary synthesis necessary? *Science* 213:967–971.

Turner, J. R. G. 1984. Darwin's coffin and Doctor Pangloss: do adaptationist modes explain mimicry? In B. Shorrocks, ed., *Evolutionary Ecology*. Palo Alto: Blackwell Scientific.

Westoll, S. 1949. On the evolution of the Dipnoi. In G. Jepsen, E. Mayr, and G. G. Simpson, eds., *Genetics, Paleontology, and Evolution*, pp. 121–184. Princeton, N.J.: Princeton University Press.

White, M. J. D. 1978. *Modes of Speciation*. San Francisco: Freeman.

———— 1981. Tales of long ago. *Paleobiology* 7:287–291.

———— 1982. Rectangularity, speciation, and chromosome architecture. In Barigozzi 1982:75–103.

Wilson, A. C., V. M. Sarich, and L. R. Maxson. 1974. The importance of gene rearrangement in evolution: evidence from studies on rates of chromosomal, protein, and anatomical evolution. *Proc. Nat. Acad. Sci.* 71:3028–3030.

Wright, S. 1967. The foundations of population genetics. In R. A. Brink, ed., *Heritage from Mendel*, pp. 245–263. Madison, WI: University of Wisconsin Press.

Young, Willard. 1985. *Fallacies of Creationism*. Calgary, Alberta: Detselig Enterprises Ltd.

INDEX

Achatinella, 140
Activities, goal-directed, 61
Adams, M., 528
Adaptation, 52, 96, 105, 118, *127–159;*
    literature on, 157*n;* perfect, 239; pheno-
    typic, 106; and selection, *133–147*
Adaptationist program, 54, 128, 130, *148–
    159,* 502
Adaptedness, 128
Adaptive zone, 135, 409
Affinity, 175
Agamospecies, 354
Agassiz, L., 169, 185, 200, 290
Alatalo, R. V., 131
Albatrosses, 156
Alexander, R. D., 79, 81
Allard, R. W., 429
Allospecies, 300, 325
Allozymes, 384
Altruism, 119; reciprocal, 78
Ancestor, common, 175
Animal breeding, 225
Anthropomorphism, 40
Anti-Darwinian beliefs, 186
Archaebacteria, 251
Archaeopteryx, 544
Archetypes, 406
Aristogenesis, 247
Aristotle, 38, 54–57, 60, 249, 261; *eidos*
    of, 56–57, 340; and teleology, 55
Asexuality, advantage of, 510
Ashlock, P. D., 286*n*
Atomism, 155, 449
Autapomorphies, 280
Ayala, F. J., 8, 10, 42, 48, 55, 77, 402,
    431

Bacon, F., 38
Baer, K. E. v., 59, 210, 242, 341

Balance: epistatic, 473; internal, 449; of na-
    ture, 221
Balme, D. M., 60
Barriers, 370
Barzun, J., 197
Bates, H. W., 137, 151
Bateson, W., 107, 204, 408
Bauplan, 109, 114; stability of, 405
Baur, E., 530, 549
Beatty, J., 93, 112*n,* 343
Beckner, M., 11, 38, 39, 55
Behavior, 45, 292, 408, 505; goal-directed,
    50; learned, 83
Behavioral program, openness of, 84
Belyaev, D. K., 113
Beneden, E. v., 510
Beniger, J. R., 4, 61
Bergson, H., 40, 248
Bertalanffy, L. v., 14
Biogeography, 181
Biological laboratories, 289
Biological species, *see* Species
Biology: autonomy of, 8–23; evolutionary,
    18, 25; fields of, 25; functional, 18, 25
Bioluminescence, 410
Biopopulation, 15, 351
Birds, classification of, *295–311.*
Blacher, L. I., 507
Blumenbach, J. F., 170
Boag, P. T., 153
Bock, W., 19, 129, 130, 134, 150, 155,
    302–305, 404, 410, 454
Böker, H., 130
Bowler, P. J., 237, 244, 499, 526
Braithwaite, R. B., 39
Branching, 174, 200
Brandon, R., 93, 112*n,* 119, 129
Bryant, E. H., 481
Bryozoans, 473